# Introduction to Embedded Systems

Manuel Jiménez · Rogelio Palomera
Isidoro Couvertier

# Introduction to Embedded Systems

## Using Microcontrollers and the MSP430

 Springer

Manuel Jiménez
Rogelio Palomera
Isidoro Couvertier
University of Puerto Rico at Mayagüez
Mayagüez, PR
USA

ISBN 978-1-4614-3142-8       ISBN 978-1-4614-3143-5   (eBook)
DOI 10.1007/978-1-4614-3143-5
Springer New York Heidelberg Dordrecht London

Library of Congress Control Number: 2013939821

Printed on acid-free paper

Springer is part of Springer Science+Business Media (www.springer.com)

*To my family, for all your support and all we missed while I was working on this project*

Manuel Jiménez

*To my family*

Rogelio Palomera

*To my Father, in Whom I move and live and exist, and to my family*

*Microcontrollers*
*What is it that You seek which such tenderness to Thee?*
*Microcontrollers I seek, microcontrollers belong to Me*
*Why do you so earnestly seek microcontrollers to Thee?*
*Microcontrollers I seek, microcontrollers to freed*

*Microcontrollers you did set free when for them you died*
*Microcontrollers I did set free, microcontrollers are Mine*
*Why do they still need to be free if they are already Thine?*
*Microcontrollers are alienated still, alienated in their minds*

Isidoro Couvertier

# Preface

The first years of the decade of 1970 witnessed the development of the first microprocessor designs in the history of computing. Two remarkable events were the development of the 4004, the first commercial, single-chip microprocessor created by Intel Corporation and the TMS1000, the first single-chip microcontroller by Texas Instruments. Based on 4-bit architectures, these early designs opened a whole new era in terms of technological advances. Several companies such as Zilog, Motorola, Rockwell, and others soon realized the potential of microprocessors and joined the market with their own designs. Today, we can find microprocessors from dozens of manufacturers in almost any imaginable device in our daily lives: from simple and inexpensive toys to sophisticated space ships, passing through communication devices, life supporting and medical equipment, defense applications, and home appliances.

Being such a ubiquitous device, it is not surprising that more and more people today need to understand the basics of how microprocessors work in order to harness and exploit their capacity. Among these people we find enthusiastic hobbyists, engineers without previous experience, electronic technology students, and engineering students.

Throughout the years, we have encountered many good books on microprocessors and embedded systems. However, most of these books share the same basic type of problems: they were written either for one specific microprocessor and their usefulness was limited in scope to the device they presented, or were developed without a specific target processor in mind lacking the practical side of teaching embedded systems. Aware of these realities, we have developed an introductory-level book that falls in the middle of these two extremes.

This book has been written to introduce the reader to the subjects of microprocessors and embedded systems, covering architectural issues, programming fundamentals, and the basics of interfacing. As part of the architectural aspects, it covers topics in processor organization, emphasizing the typical structure of today's microcontrollers, processor models, and programming styles. It also covers fundamentals on computer data representations and operations, as a prelude to subjects in embedded software development. The presented material is rounded off with discussions that cover from the basics of input/output systems to using all sorts of embedded peripherals and interfacing external loads for diverse applications.

Most practical examples use Texas Instruments MSP430 devices, an intuitive and affordable platform. But the book is not limited to the scope of this micro-controller. Each chapter has been filled with concepts and practices that set a solid design foundation in embedded systems which is independent from the specific device being used. This material is then followed by a discussion of how these concepts apply to particular MSP430 devices. This allows our readers to first build a solid foundation in the underlying functional and design concepts and then to put them in practice with a simple and yet powerful family of devices.

## Book Organization

The book contents are distributed across ten chapters as described below:

Chapter 1: *Introduction* brings to the reader the concept of an embedded system. The chapter provides a historical overview of the development of embedded systems, their structure, classification, and complete life cycle. The last section discusses typical constraints such as functionality, cost, power, and time to market, among others. More than just a mere introduction, the chapter provides a system-level treatment of the global issues affecting the design of embedded systems, bringing awareness to designers about the implications of their design decisions.

Chapter 2: *Number Systems and Data Formats* reviews basic concepts on number formats and representations for computing systems. The discussion includes subjects in number systems, base conversions, arithmetic operations, and numeric and nonnumeric representations. The purpose of dedicating a whole chapter to this subject is to reinforce in a comprehensive way the student background in the treatment given by computers to numbers and data.

Chapter 3: *Microcomputer Organization* covers the topics of architecture and organization of classical microcomputer models. It goes from the very basic concepts in organization and architecture, to models of processors and micro-controllers. The chapter also introduces the Texas Instruments MSP430 family of microcontrollers as the practical target for applications in the book. This chapter forms the ground knowledge for absolute newcomers to the field of micropro-cessor-based systems.

Chapter 4: *Assembly Language Programming* sets a solid base in micropro-cessor's programming at the most fundamental level: Assembly Language. The anatomy of an assembly program is discussed, covering programming techniques and tips, and illustrating the application cases where programming in assembly language becomes prevalent. The chapter is crowned with the assembly pro-grammer's model of the MSP430 as a way to bring-in the practical side of the assembly process. We use the IAR assembler in this chapter. Yet, the discussion is complemented by Appendix C with a tutorial on how to use Code Composer Studio[1] (CCS) for programming the MSP430 in assembly language.

Chapter 5: *C Language Programming* treats the subject of programming embedded systems using a high-level language. The chapter reviews fundamental programming concepts to then move to a focussed discussion on how to program the MSP430 in C language. Like in Chap. 4, the discussion makes use of IAR and other tools as programming and debugging environments.

Chapter 6: *Fundamentals of Interfacing* is structured to guide first time embedded systems designers through the process of making a microprocessor or microcontroller chip work from scratch. The chapter identifies the elements in the basic interface of a microprocessor-based system, defining the criteria to implement each of them. The discussion is exemplified with a treatment of the MSP430 as target device, with an in-depth treatment of its embedded modules and how they facilitate the basic interface in a wide range of applications.

Chapter 7: *Embedded Peripherals* immerses readers into the array of peripherals typically found in a microcontroller, while also discussing the concepts that allow for understanding how to use them in any MCU- or microprocessor-based system. The chapter begins by discussing how to use interrupts and timers in microcontrollers as support peripherals for other devices. The subjects of using embedded FLASH memory and direct memory access are also discussed in this chapter, with special attention to their use as a peripheral device supporting low-power operation in embedded systems. The MSP430 is used as the testbed to provide practical insight into the usage of these resources.

Chapter 8: *External World Interface* discusses one of the most valuable resources in embedded microcontrollers: general purpose I/O lines. Beginning with an analysis of the structure, operation, and configuration of GPIO ports the chapter expands into developing user interfaces using via GPIOs. Specific MSP430 GPIO features and limitations are discussed to create the basis for safe design practices. The discussion is next directed at how to design hardware and software modules for interfacing large DC and AC loads, and motors through GPIO lines.

Chapter 9: *Principles of Serial Communication* offers an in-depth discussion of how serial interfaces work for supporting asynchronous and synchronous communication modalities. Protocols and hardware for using UARTs, SPI, I²C, USB, and other protocols are studied in detail. Specific examples using the different serial modules sported in MSP430 devices provide a practical coverage of the subject.

Chapter 10: *The Analog Signal Chain* bridges the digital world of microcontrollers to the domain of analog signals. A thorough discussion of the fundamentals of sensor interfacing, signal conditioning, anti-aliasing filtering, analog-to-digital

---

[1] CCS is a freely available integrated development environment provided by Texas Instruments at that allows for programming and debugging all members of the MSP430 family in both assembly and C language.

and digital-to-analog conversion, and smoothing and driving techniques are included in this chapter. Concrete hardware and software examples are provided, including both, Nyquist and oversampled converters embedded in MSP430 MCUs are included, making this chapter an essential unit for mixed-signal interfaces in embedded applications.

Each chapter features examples and problems on the treated subjects that range form conceptual to practical. A list of selected bibliography is provided at the end of the book for those interested in reading further about the topics discussed in each chapter.

Appendix materials provide information complementing the book chapters in diverse ways. Appendix A provides a brief guide to the usage of flowcharting to plan the structure of embedded programs. Appendix B includes a detailed MSP430 instruction set with binary instruction encoding and cycle requirements.

A detailed tutorial on how to use Code Composer Essentials, the programming and debugging environment for MSP430 devices, is included in Appendix C. The extended MSP430X architecture and its advantages with respect to the standard CPU version of the controller is the subject of Appendix D.

## Suggested Book Usage

This book has been designed for use in different scenarios that include under-graduate and graduate treatment of the subject of microprocessors and embedded systems, and as a reference for industry practitioners.

In the academic environment, teaching alternatives include EE, CE, and CS study programs with either a single comprehensive course in microprocessors and embedded systems, or a sequence of two courses in the subject. Graduate programs could use the material for a first-year graduate course in embedded systems design.

For a semester-long introductory undergraduate course in microprocessors, a suggested sequence could use Chaps. 1– 5 and selected topics from Chaps. 6 and 8. A two-course sequence for quarter-based programs could use topics in Chaps. 1 through 4 for an architecture and assembly programming-focused first course. The second course would be more oriented towards high-level language programming, interfacing, and applications, discussing select topics from Chaps. 1 and 5–10.

Besides these suggested sequences, other combinations are possible depending on the emphasis envisioned by the instructor and the program focus.

Many EE and CE programs offer a structured laboratory in microprocessors and embedded systems, typically aligned with the first course in the subject. For this scenario, the book offers a complementary laboratory manual with over a dozen experiments using the MSP430. Activities are developed around MSP430 launchpad development kits. Instructions, schematics, and layouts are provided for building a custom designed I/O board, the eZ-EXP. Experiments in the lab manual are structured in a progressive manner that can be easily synchronized with the subjects taught in an introductory microprocessors course. All experiments can be completed using exclusively the inexpensive launchpad or TI eZ430 USB boards, IAR and the freely available IAR or CCS environments, and optionally the eZ-EXP attachment.

For industry practitioners, this book could be used as a refresher on the concepts on microprocessors and embedded systems, or as a documented reference to introduce the use of the MSP430 microcontroller and tools.

## Supplemental Materials

Instructor's supplemental materials available through the book web site include solutions to selected problems and exercises and power point slides for lectures. The site also includes materials for students that include links to application examples and to sites elsewhere in the Web with application notes, downloadable tools, and part suppliers.

We hope this book could result as a useful tool for your particular learning and/or teaching needs in embedded systems and microprocessors.

Enjoy the rest of the book!

# Acknowledgments

We would like to thank Texas Instruments, Inc. for giving us access to the MSP430 technology that brings up the practical side of this book. In particular, our gratitude goes to the people in the MSP430 Marketing and Applications Group for their help and support in providing application examples and documentation during the completion of this project.

Completing this work would have not been possible without the help of many students and colleagues in the Electrical and Computer Engineering Department of the University of Puerto Rico at Mayagüez who helped us verifying examples, revising material, and giving us feedback on how to improve the book contents. Special thanks to Jose Navarro, Edwin Delgado, Jose A. Rodriguez, Jose J. Rodriguez, Abdiel Avilés, Dalimar Vélez, Angie Córdoba, Javier Cardona, Roberto Arias, and all those who helped us, but our inability to remember their names at the time of this writing just evidences of how old we have grown while writing this book. You all know we will always be grateful for your valuable help.

Manuel
Rogelio
Isidoro

# Contents

# Chapter 1
# Introduction

## 1.1 Embedded Systems: History and Overview

An *embedded system* can be broadly defined as a device that contains tightly coupled hardware and software components to perform a single function, forms part of a larger system, is not intended to be independently programmable by the user, and is expected to work with minimal or no human interaction. Two additional characteristics are very common in embedded systems: reactive operation and heavily constrained.

Most embedded system interact directly with processes or the environment, making decisions on the fly, based on their inputs. This makes necessary that the system must be reactive, responding in real-time to process inputs to ensure proper operation. Besides, these systems operate in constrained environments where memory, computing power, and power supply are limited. Moreover, production requirements, in most cases due to volume, place high cost constraints on designs.

This is a broad definition that highlights the large range of systems that fall into it. In the next sections we provide a historic perspective in the development of embedded systems to bring meaning to the definition above.

### 1.1.1 Early Forms of Embedded Systems

The concept of an embedded system is as old as the concept of a an electronic computer, and in a certain way, it can be said to precede the concept of a general purpose computer. If we look a the earliest forms of computing devices, they adhere better to the definition of an embedded system than to that of a general purpose computer. Take as an example early electronic computing devices such as the Colossus Mark I and II computers, partially seen in Fig. 1.1. These electro-mechanical behemoths, designed by the British to break encrypted teleprinter German messages during World War II, were in a certain way similar to what we define as an embedded system.

M. Jiménez et al., *Introduction to Embedded Systems*,
DOI: 10.1007/978-1-4614-3143-5_1,
© Springer Science+Business Media New York 2014

**Fig. 1.1** Control panel and
paper tape transport view of
a Colossus Mark II computer
(public image by the British
Public Record Office, London)

Although not based on the concept of a stored-program computer, these machines
were able to perform a single function, reprogrammability was very awkward, and
once fed with the appropriate inputs, they required minimal human intervention
to complete their job. Despite their conceptual similarity, these early marvels of
computing can hardly be considered as integrative parts of larger system, being
therefore a long shot to the forms known today as embedded systems.

One of the earliest electronic computing devices credited with the term "embedded
system" and closer to our present conception of such was the Apollo Guidance
Computer (AGC). Developed at the MIT Instrumentation Laboratory by a group of
designers led by Charles Stark Draper in the early 1960s, the AGC was part of the
guidance and navigation system used by NASA in the Apollo program for various
spaceships. In its early days it was considered one of the riskiest items in the Apollo
program due to the usage of the then newly developed monolithic integrated circuits.

The AGC incorporated a user interface module based on keys, lamps, and seven-
segment numeric displays (see Fig. 1.2); a hardwired control unit based on 4,100
single three-input RTL NOR gates, 4 KB of magnetic core RAM, and 32 KB of core
rope ROM. The unit CPU was run by a 2.048 MHz primary clock, had four 16-bit
central registers and executed eleven instructions. It supported five vectored interrupt
sources, including a 20-register timer-counter, a real-time clock module, and even
allowed for a low-power standby mode that reduced in over 85 % the module's power
consumption, while keeping alive all critical components.

The system software of the Apollo Guidance Computer was written in AGC
assembly language and supported a non-preemptive real-time operating system that
could simultaneously run up to eight prioritized jobs. The AGC was indeed an
advanced system for its time. As we enter into the study of contemporary applications, we will find that most of these features are found in many of today's embedded
systems.

Despite the AGC being developed in a low scale integration technology and assembled in wire-wrap and epoxy, it proved to be a very reliable and dependable design.
However, it was an expensive, bulky system that remained used only for highly

**Fig. 1.2** AGC user interface module (public photo EC96-43408-1 by NASA)

specialized applications. For this reason, among others, the flourishing of embedded systems in commercial applications had to wait until another remarkable event in electronics: the advent of the microprocessor.

### 1.1.2 Birth and Evolution of Modern Embedded Systems

The beginning of the decade of 1970 witnessed the development of the first microprocessor designs. By the end of 1971, almost simultaneously and independently, design teams working for Texas Instruments, Intel, and the US Navy had developed implementations of the first microprocessors.

Gary Boone from Texas Instruments was awarded in 1973 the patent of the *first single-chip microprocessor architecture* for its 1971 design of the TMS1000 (Fig. 1.3). This chip was a 4-bit CPU that incorporated in the same die 1K of ROM and 256 bits of RAM to offer a complete computer functionality in a single-chip, making it the first microcomputer-on-a-chip (a.k.a microcontroller). The TMS1000 was launched in September 1971 as a calculator chip with part number TMS1802NC.

The i4004, (Fig. 1.4) recognized as the *first commercial, stand-alone single chip microprocessor,* was launched by Intel in November 1971. The chip was developed by a design team led by Federico Faggin at Intel. This design was also a 4-bit CPU intended for use in electronic calculators. The 4004 was able to address 4K of memory, operating at a maximum clock frequency of 740 KHz. Integrating a minimum system around the 4004 required at least three additional chips: a 4001 ROM, a 4002 RAM, and a 4003 I/O interface.

The third pioneering microprocessor design of that age was a less known project for the US Navy named the Central Air Data Computer (CADC). This system implemented a chipset CPU for the F-14 Tomcat fighter named the MP944. The system

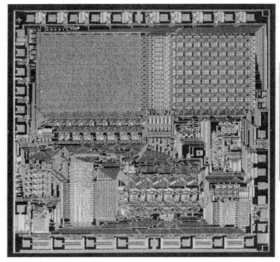

**Fig. 1.3** Die microphotograph (*left*) packaged part for the TMS1000 (*Courtesy of Texas Instruments, Inc.*)

**Fig. 1.4** Die microphotograph (*left*) and packaged part for the Intel 4004 (*Courtesy of Intel Corporation*)

supported 20-bit operands in a pipelined, parallel multiprocessor architecture designed around 28 chips. Due to the classified nature of this design, public disclosure of its existence was delayed until 1998, although the disclosed documentation indicates it was completed by 1970.

After these developments, it did not take long for designers to realize the potential of microprocessors and its advantages for implementing embedded applications. Microprocessor designs soon evolved from 4-bit to 8-bit CPUs. By the end of the 1970s, the design arena was dominated by 8-bit CPUs and the market for microprocessors-based embedded applications had grown to hundreds of millions of dollars. The list of initial players grew to more than a dozen of chip manufacturers that, besides Texas Instruments and Intel, included Motorola, Zilog, Intersil, National Instruments, MOS Technology, and Signetics, to mention just a few of the most renowned. Remarkable parts include the Intel 8080 that eventually evolved into the famous 80 × 86/Pentium series, the Zilog Z-80, Motorola 6800 and MOS 6502. The evolution in CPU sizes continued through the 1980s and 1990s to 16-, 32-, and 64-bit designs, and now-a-days even some specialized CPUs crunching data at 128-bit widths. In terms of manufacturers and availability of processors, the list has grown to the point that it is possible to find over several dozens of different choices for processor sizes 32-bit and above, and hundreds of 16- and 8-bit processors. Examples of manufacturers available today include Texas Instruments, Intel, Microchip, Freescale (formerly Motorola), Zilog, Advanced Micro Devices, MIPS Technologies, ARM Limited, and the list goes on and on.

Despite this flourishing in CPU sizes and manufacturers, the applications for embedded systems have been dominated for decades by 8- and 16-bit microprocessors. Figure 1.5 shows a representative estimate of the global market for processors including microprocessors, microcontrollers, DSPs, and peripheral programmable in recent years. Global yearly sales approach $50 billion for a volume of shipped units of nearly 6 billion processor chips. From these, around three billion units are 8-bit processors. It is estimated that only around 2 % of all produced chips (mainly in the category of 32- and 64-bit CPUS) end-up as the CPUs of personal computers (PCs). The rest are used as embedded processors. In terms of sales volume, the story is different. Wide width CPUs are the most expensive computer chips, taking up nearly two thirds of the pie.

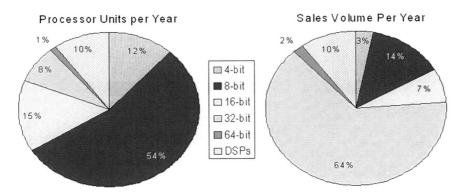

**Fig. 1.5** Estimates of processor market distribution (*Source* Embedded systems design—www.embedded.com)

### 1.1.3 Contemporary Embedded Systems

Nowadays microprocessor applications have grown in complexity; requiring applications to be broken into several interacting embedded systems. To better illustrate the case, consider the application illustrated in Fig. 1.6, corresponding to a generic multimedia player. The system provides audio input/output capabilities, a digital camera, a video processing system, a hard-drive, a user interface (keys, a touch screen, and graphic display), power management and digital communication components. Each of these features are typically supported by individual embedded systems integrated in the application. Thus, the audio subsystem, the user interface, the storage system, the digital camera front-end, and the media processor and its peripherals are among the systems embedded in this application. Although each of these subsystems may have their own processors, programs, and peripherals, each one has a specific, unique function. None of them is user programmable, all of them are embedded within the application, and their operation require minimal or no human interaction.

The above illustrated the concept of an embedded system with a very specific application. Yet, such type of systems can be found in virtually every aspect of our daily lives: electronic toys; cellular phones; MP3 players, PDAs; digital cameras; household devices such as microwaves, dishwasher machines, TVs, and toasters; transportation vehicles such as cars, boats, trains, and airplanes; life support and medical systems such as pace makers, ventilators, and X-ray machines; safety-critical

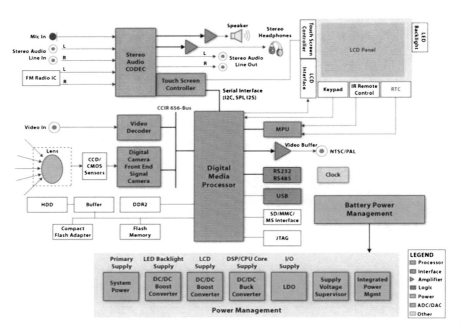

**Fig. 1.6** Generic multi-function media player (*Courtesy of Texas Instruments, Inc.*)

**Fig. 1.7** General view of an embedded system

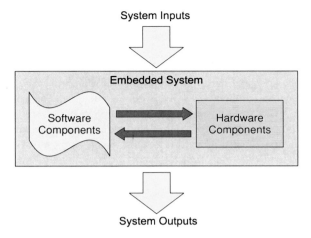

systems such as anti-lock brakes, airbag deployment systems, and electronic surveillance; and defense systems such as missile guidance computers, radars, and global positioning systems are only a few examples of the long list of applications that depend on embedded systems. Despite being omnipresent in virtually every aspect of our modern lives, embedded systems are ubiquitous devices almost invisible to the user, working in a pervasive way to make possible the "intelligent" operation of machines and appliances around us.

## 1.2 Structure of an Embedded System

Regardless of the function performed by an embedded system, the broadest view of its structure reveals two major, tightly coupled sets of components: a set of hardware components that include a central processing unit, typically in the form of a microcontroller; and a series of software programs, typically included as firmware,[1] that give functionality to the hardware. Figure 1.7 depicts this general view, denoting these two major components and their interrelation. Typical inputs in an embedded system are process variables and parameters that arrive via sensors and input/output (I/O) ports. The outputs are in the form of control actions on system actuators or processed information for users or other subsystems within the application. In some instances, the exchange of input-output information occurs with users via some sort of user interface that might include keys and buttons, sensors, light emitting diodes (LEDs), liquid crystal displays (LCDs), and other types of display devices, depending on the application.

---

[1] Firmware is a computer program typically stored in a non-volatile memory embedded in a hardware device. It is tightly coupled to the hardware where it resides and although it can be upgradeable in some applications, it is not intended to be changed by users.

Consider for example an embedded system application in the form of a microwave oven. The hardware elements of this application include the magnetron (microwave generator), the power electronics module that controls the magnetron power level, the motor spinning the plate, the keypad with numbers and meal settings, the system display showing timing and status information, some sort of buzzer for audible signals, and at the heart of the oven the embedded electronic controller that coordinates the whole system operation. System inputs include the meal selections, cooking time, and power levels from a human operator through a keypad; magnetron status information, meal temperature, and internal system status signals from several sensors and switches. Outputs take the form of remaining cooking time and oven status through a display, power intensity levels to the magnetron control system, commands to turn on and off the rotary plate, and signals to the buzzer to generate the different audio signals given by the oven.

The software is the most abstract part of the system and as essential as the hardware itself. It includes the programs that dictate the sequence in which the hardware components operate. When someone decides to prepare a pre-programmed meal in a microwave oven, software picks the keystrokes in the oven control panel, identifies the user selection, decides the power level and cooking time, initiates and terminates the microwave irradiation on the chamber, the plate rotation, and the audible signal letting the user know that the meal is ready. While the meal is cooking, software monitors the meal temperature and adjusts power and cooking time, while also verifying the correct operation of the internal oven components. In the case of detecting a system malfunction the program aborts the oven operation to prevent catastrophic consequences. Despite our choice of describing this example from a system-level perspective, the tight relation between application, hardware, and software becomes evident. In the sections below we take a closer view into the hardware and software components that integrate an embedded system.

## 1.2.1  Hardware Components

When viewed from a general perspective, the hardware components of an embedded system include all the electronics necessary for the system to perform the function it was designed for. Therefore, the specific structure of a particular system could substantially differ from another, based on the application itself. Despite these dissimilarities, three core hardware components are essential in an embedded system: The Central Processing Unit (CPU), the system memory, and a set of input-output ports. The CPU executes software instructions to process the system inputs and to make the decisions that guide the system operation. Memory stores programs and data necessary for system operation. Most systems differentiate between program and data memories. Program memory stores the software programs executed by the CPU. Data memory stores the data processed by the system. The I/O ports allows conveying signals between the CPU and the world external to it. Beyond this point, a number of other supporting and I/O devices needed for system functionality might be present, depending on the application. These include:

- Communication ports for serial and/or parallel information exchanges with other devices or systems. USB ports, printer ports, wireless RF and infrared ports, are some representative examples of I/O communication devices.
- User interfaces to interact with humans. Keypads, switches, buzzers and audio, lights, numeric, alphanumeric, and graphic displays, are examples of I/O user interfaces.
- Sensors and electromechanical actuators to interact with the environment external to the system. Sensors provide inputs related to physical parameters such as temperature, pressure, displacement, acceleration, rotation, etc. Motor speed controllers, stepper motor controllers, relays, and power drivers are some examples of actuators to receive outputs from the system I/O ports. These are just a few of the many devices that allow interaction with processes and the environment.
- Data converters (Analog-to-digital (ADC) and/or Digital-to-Analog (DAC)) to allow interaction with analog sensors and actuators. When the signal coming out from a sensor interface is analog, an ADC converts it to the digital format understood by the CPU. Similarly, when the CPU needs to command an analog actuator, a DAC is required to change the signal format.
- Diagnostics and redundant components to verify and provide for robust, reliable system operation.
- System support components to provide essential services that allow the system to operate. Essential support devices include power supply and management components, and clock frequency generators. Other optional support components include timers, interrupt management logic, DMA controllers, etc.
- Other sub-systems to enable functionality, that might include Application Specific Integrated Circuits (ASIC), Field Programmable Gate Arrays (FPGA), and other dedicated units, according to the complexity of the application.

Figure 1.8 illustrates how these hardware components are integrated to provide the desired system functionality.

## 1.2.2 Software Components

The software components of an embedded system include all the programs necessary to give functionality to the system hardware. These programs, frequently referred to as the system *firmware*, are stored in some sort of non volatile memory. Firmware is not meant to be modifiable by users, although some systems could provide means of performing upgrades. System programs are organized around some form of operating system and application routines. The operating systems can be simple and informal in small applications, but as the application complexity grows, the operating system requires more structure and formality. In some of these cases, designs are developed

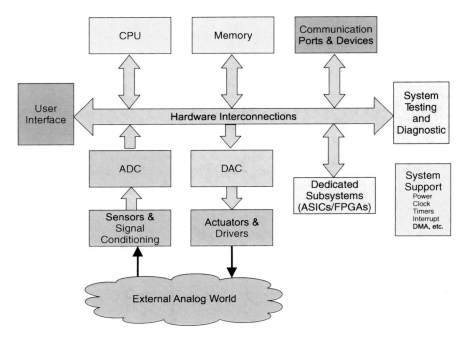

**Fig. 1.8**  Hardware elements in an embedded system

around *Real-Time Operating Systems* (RTOS). Figure 1.9 illustrates the structure on an embedded system software.

The major components identified in a system software include:

**System Tasks**. The application software in embedded systems is divided into a set of smaller programs called *Tasks*. Each task handles a distinct action in the system and requires the use of specific *System Resources*. Tasks submit service requests to the *kernel* in order to perform their designated actions. In our microwave oven example the system operation can be decomposed into a set of tasks that include reading the keypad to determine user selections, presenting information on the oven display, turning on the magnetron at a certain power level for a certain amount of time, just to mention a few. Service requests can be placed via registers or interrupts.

**System Kernel**. The software component that handles the system resources in an embedded application is called the *Kernel*. System resources are all those components needed to serve tasks. These include memory, I/O devices, the CPU itself, and other hardware components. The kernel receives service requests from tasks, and schedules them according to the priorities dictated by the *task manager*. When multiple tasks contend for a common resource, a portion of the kernel establishes the resource management policy of the system. It is not uncommon finding tasks that need to exchange information among them. The kernel provides a framework

**Fig. 1.9** Software structure
in an embedded system

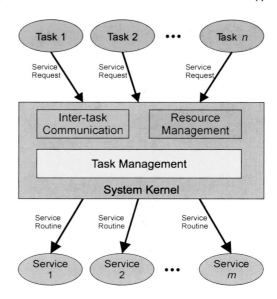

that enables a reliable inter-task communication to exchange information and to coordinate collaborative operation.

**Services**. Tasks are served through *Service Routines*. A service routine is a piece of code that gives functionality to a system resource. In some systems, they are referred to as device drivers. Services can be activated by polling or as interrupt service routines (ISR), depending on the system architecture.

## 1.3 Classification of Embedded Systems

The three pioneering microprocessor developments at the beginning of the 1970s, besides initiating the modern era of embedded systems, inadvertently created three defining categories that we can use to classify embedded systems in general: *Small, Distributed, and High-performance*. Figure 1.10 graphically illustrates the relationships among these classes.

### Small Embedded Systems

Texas Instruments, with the TMS1000 created the microcontroller, which has become the cornerstone component of this type of embedded systems, which is by far, the most common type. This class is typically centered around a single microcontroller chip that commands the whole application. These systems are highly integrated,

**Fig. 1.10** Classification of
embedded systems

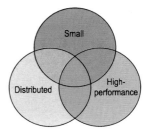

adding only a few analog components, sensors, actuators, and user-interface, as needed. These systems operate with minimal or no maintenance, are very low cost, and produced in mass quantities. Software in these systems is typically single-tasked, and rarely requires an RTOS. Examples of these systems include tire pressure monitoring systems, microwave oven controllers, toaster controllers, and electronic toy controllers, to mention just a few.

### Distributed Embedded Systems

The style created by Intel with the 4004 is representative of this type of embedded systems. Note that in this class we are not referring to what is traditionally known as a distributed computing system. Instead, we refer to the class of embedded systems where, due to the nature of the data managed by these systems and the operations to be performed on them, the CPU resides in a separate chip while the rest of components like memories, I/O, co-processors, and other special functions are spread across one or several chips in what is typically called the processor chipset. These are typically board-level systems where, although robustness is not a critical issue, maintenance and updates are required, and include some means of systems diagnosis. They typically manage multiple tasks, so the use of RTOS for system development is not uncommon. Production volume is relatively high, and costs are mainly driven by the expected level of performance. Applications might require high-performance operations. Applications like video processors, video game controllers, data loggers, and network processors are examples of this category.

### High-Performance Embedded Systems

The case of the CADC represents the class of highly specialized embedded systems requiring fast computations, robustness, fault tolerance, and high maintainability. These systems usually require dedicated ASICS, are typically distributed, might include DSPs and FPGAs as part of the basic hardware. In many cases the complexity of their software makes mandatory the use of RTOS' to manage the multiplicity of tasks. They are produced in small quantities and their cost is very high. These are the type of embedded systems used in military and aerospace applications, such as flight controllers, missile guidance systems, and space craft navigation systems.

As Fig. 1.10 illustrates, the categories in this classification are not mutually exclusive. Among them we can find "gray zones" where the characteristics of two or the three of them overlap and applications might become difficult to associate to a single specific class. However, if we look at the broad range of embedded applications, in most cases it becomes generally easy to identify the class to which a particular application belong.

## 1.4 The Life Cycle of Embedded Designs

Embedded systems have a finite lifespan throughout which they undergo different stages, going from system conception or birth to disposal and in many cases, rebirth. We call this the *Life Cycle of an Embedded System*. Five stages can be identified in the cycle: Conception or Birth, Design, Growth, Maturity, and Decline. Figure 1.11 illustrates the sequence of these stages, and the critical phases that compose each stage.

An embedded design is first conceived by the identification of a need to solve a particular problem, or the opportunity to apply embedded technology to an old problem. Many embedded systems are conceived by the need of changing the way old problems are solved or the need of providing cost-effective solutions to them. Other

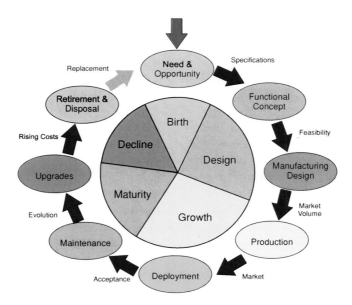

**Fig. 1.11** Life cycle of an embedded system

systems are conceived with the problems themselves. Opportunity arises with the realization of efficient solutions enabled by the use of embedded systems. In any case, the application where the design needs to be embedded dictates the specifications that lead to the design stage.

In the design stage, the system goes first through a phase of functional design, where proof-of-concept prototypes are developed. This phase allows to tune the system functionality to fit the application for which it was designed. A feasibility analysis that considers variables such as product cost, market window, and component obsolescence determines whether or not a manufacturing and product design proceeds. This second design phase establishes the way a prototype becomes a product and the logistics for its manufacturing. Design is by far the most costly stage in the life cycle of an embedded system, since most of the non-recurrent engineering costs arise here. The expected market volume guides the entrance into the growth stage.

The growth stage initiates with the production of the system to supply an estimated market demand. Production is scheduled to deploy the system at the beginning of its market window. The deployment phase involves distribution, installation, and set-up of the system. The fitness of the system to its originating needs defines its acceptance, driving the embedded system into its mature stage.

In the maturity stage, the product is kept functional and up to date. This involves providing technical support, periodic maintenance, and servicing. As the application evolves, the system might require fitness to the changing application needs through software and/or hardware upgrades. The maturity stage is expected to allow running the product with minimal costs. As the system begins to age, maintenance costs begin to increase, signaling its transition into the decline stage.

In the decline stage, system components become obsolete and difficult to obtain, driving the cost of maintenance, service, and upgrades to levels that approach or exceed those of replacing the whole system at once. This is the point where the system needs to be retired from service and disposed. Product designs need to foresee this stage to devise ways in which recycling and reuse might become feasible, reducing the costs and impact of its disposal. In many cases, the system replacement creates new needs and opportunities to be satisfied by a new design, re-initiating the life cycle of a new system.

## 1.5  Design Constraints

A vast majority of embedded systems applications end up in the heart of mass produced electronic applications. Home appliances such as microwave ovens, toys, and dishwasher machines, automobile systems such as anti-lock brakes and airbag deployment mechanisms, and personal devices such as cellular phones and media players are only a few representative examples. These are systems with a high cost sensitivity to the resources included in a design due to the high volumes in which they are produced. Moreover, designs need to be completed, manufactured and launched in time to hit a market window to maximize product revenues. These constraints shape the design of embedded applications from beginning to end in their life cycle. Therefore, the list of constraints faced by designers at the moment of conceiving an embedded solution to a problem come from the different perspectives. The most salient constraints in the list include:

**Functionality**: The system ability to perform the function it was designed for.
**Cost**: The amount of resources needed to conceive, design, produce, maintain, and discard an embedded system.
**Performance**: The system ability to perform its function on time.
**Size**: Physical space taken by a system solution.
**Power and Energy**: Energy required by a system to perform its function.
**Time to Market**: The time it takes from system conception to deployment.
**Maintainability**: System ability to be kept functional for the longest of its mature life.

Aside from functionality, the relevance of the rest of the constraints in the list changes from one system to another. In many cases, multiple constraints must be satisfied, even when these can be conflicting, leading to design tradeoffs. Below we provide a more detailed insight into these constraints.

### 1.5.1 Functionality

Every embedded system design is expected to have a functionality that solves the problem it was designed for. More than a constraint, this is a design requirement. Although this might seem a trivial requirement, it is not to be taken lightly. Functional verification is a very difficult problem that has been estimated to consume up to 70 % of the system development time and for which a general solution has not been found yet. The task of verifying the functionality of the hardware and software components of an embedded system falls in the category of NP-hard problems. Different methods can be applied towards this end, but none of them is guaranteed to provide an optimal solution. In most cases, the combination of multiple approaches is necessary to minimize the occurrence of unexpected system behavior or system bugs. The most commonly used methods include the following:

- Behavioral Simulation: The system is described by a behavioral model and suitable tools are used to simulate the model. This approach is more suited for embedded systems developed around IP cores where a behavioral description of the processor is available, along with that of the rest of the system, allowing verification to be carried early in the design process.
- Logic- and circuit-level Simulation: These methods apply to dedicated hardware components of the system. When custom hardware forms part of an embedded application, their functionality must be verified before integration with the rest of the components. Analog and digital simulation tools become useful for these chores.

- Processor Simulation: This method uses programs written to simulate the functionality of software developed for a target processor or microcontroller on a personal computer or workstation. Simulators become useful for verifying the software functionality off the target hardware. The functionality that can be verified through this method is limited to some aspects of software performance, but they provide an inexpensive way to debug program operation. Software simulators simulate hardware signals via I/O registers and flags. In general, they allow to examine and change the contents of registers and memory locations; and to trace instructions, introduce breakpoints, and proceed by single steps through the program being debugged.

- JTAG Debuggers: JTAG is an acronym for Joint Test Action Group, the name given to the Standard Test Access Port and Boundary Scan Architecture for test access ports in digital circuits. Many present development tools provide ways of debugging the processor functionality directly on the target system through a JTAG port. This is achieved through the use of internal debug resources embedded in the processor core. The IEEE-1149.1 standard for JTAG is one of the most commonly used ways of accessing embedded debugging capabilities in many of today's microcontrollers. JTAG debuggers allow to perform the same actions listed for simulators, but directly on the embedded system through the actual processor.

- In-circuit Emulation (ICE): Hardware Emulation is the process of imitating the behavior of a hardware component with another one for debugging purposes. An ICE is a form of hardware emulation that replaces the processor in a target system with a test pod connected to a piece of hardware which emulates the target processor. This box connects to a personal computer or workstation that gives the designer a window of view into the embedded application being debugged. ICEs are one of the oldest, most powerful, and most expensive solutions to debug and verify the functionality of embedded systems. They provide the designer with the ability of tracing, single stepping, setting breakpoints, and manipulating I/O data, memory, and register contents when debugging the application. They also give the designer control and visibility into the processor buses, the ability to perform memory overlay, and to perform bus-triggered debugging, where breakpoints are activated by the contents of specific hardware signals in the system. These are features not available through other hardware verification methods.

- Other Verification Tools: Other tools used for embedded system verification include Background Mode Emulators (BMEs) and ROM Monitors. A BME is similar to a JTAG debugger in the sense that it relies on a dedicated chip port and on-chip debugging resources to exert control over the CPU. However BMEs are not as standardized as JTAG ports. BMEs change in format and capabilities from one chip to another. ROM monitors are another type of debuggers that utilize special code within the application to provide status information of the CPU via a serial port. This last type relies on processor capabilities to introduce hardware breakpoints.

## *1.5.2 Cost*

Developing an embedded system from scratch is a costly process. In a broad sense, the the total cost $C_T$ of producing a certain volume $V$ of units an embedded system can be obtained as,

$$C_T = NRE + (RP * V),  \tag{1.1}$$

where *NRE* represent the Non-recurring Engineering (NRE) costs for that product, (*RP*) represents the Variable or Recurrent Production costs , and the production volume *V* is the number of marketed units.

The NRE costs include the investment made to complete all the design aspects of an embedded solution. This component, independent of the production volume, results from the addition of the costs of all engineering aspects of the system development including research, planning, software/hardware design and integration (prototyping), verification, manufacturing planning, production design, and product documentation.

The traditional design process of most embedded system solutions is developed around *commercial, off-the-shelf* (COTS) parts. Although other methodologies exist, standard COTS methods are the default choice for embedded system designers having commercial components as the hardware resources to assemble an embedded solution. These methods allow for minimizing the hardware development costs through tight constraints on physical parts components, usually at the expense of larger software design and system verification cycles. Figure 1.12 shows a representative diagram of a typical COTS-based design process for an embedded solution. The diagram details the steps in the system conception, design, and prototyping stages. It also includes the estimated time duration of the different stages in the process. It can be observed that although hardware and software design could be carried in parallel, their functional verification can only happen on a functional prototype. The combination of these steps can take up to 50 % of the product development time, with up to 75 % of the NRE costs.

Recurrent costs include all the expenses involved in producing the units of a given production volume. Recurrent costs include the cost of parts, boards, enclosures and packaging, testing, and in some cases even the maintenance, upgrade, and system disposal costs. Recurrent costs are affected by the production volume as indicated in Eq. 1.1, and are significantly lower than NRE costs. Representative figures for these terms on a given embedded product can be in the order of millions of dollars for the NRE costs while the RP cost could be only a few dollars or even cents. As an example consider the development of the first iPod, by Apple and sub-contractor PortalPlayer. Coming up with the functional product idea, defining its specifications, the product architecture; designing the hardware for each supported feature, the software for the desired functionality, debugging the system, planning its packaging, form factor, manufacturing plan, production scheme, and customer support took a team of nearly 300 engineers working for about one year. A conservative estimate puts NRE costs anywhere between five and and ten million dollars. The recurrent

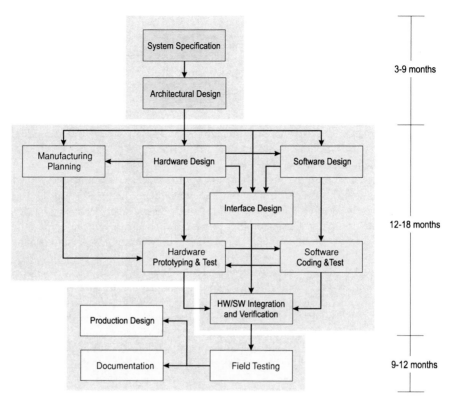

**Fig. 1.12** Embedded systems design flow model based on COTS parts [8]

costs for a first generation player can be estimated around $120, considering that its architecture was based on two ARM7 MCUs, one audio codec, a 5 GB hard drive, lithium batteries, and miscellaneous knobs, buttons, display, and enclosure. Apple reported selling 125,000 iPods in the first quarter 2002, so let's assume their first batch of units consisted of exactly that number of units—that is a least a lower bound. With these numbers, and NRE set at $10 M, we can make a rough estimate of a total cost of at least $25 M for producing that first batch of units. Note that this is a simplistic estimate since it does not include the cost of items such as production testing, marketing costs, distribution, etc.

So, if it is so expensive to produce that kind of embedded system, how come they can be sold for as cheap as $300? The explanation for that is their production in large quantities. A common measure of the cost of an embedded system is the per-unit cost, $U_C$, obtained by dividing the total cost by the volume. Applying this formula we can estimate the cost of producing each unit in a batch of a given volume $V$, for a break-even. The selling price is then set considering the expected company revenues, among other factors.

$$U_C = \frac{C_T}{V} = \frac{NRE}{V} + RP \tag{1.2}$$

Equation 1.2 reveals that NRE costs are diluted by the production volume, making the unit cost of electronic systems affordable for the average individual. This formula also justifies the approach refining a design to minimize the cost of hardware, as is done with COTS parts, since the extra NRE investment is brought down when the production is large enough. Care must be exercised when deciding the level of optimization introduced into the design, since reduced production volumes could make the product cost prohibitively high. For our iPod example, the estimated unit cost would drop to about $200. As the product matures, bringing new versions into the market becomes much cheaper since IP can be reused bringing down the NRE costs. This explains why the next generation of a gadget, even when bringing new and more powerful features can be launched to the market at the same or even lower price than the previous generation.

### 1.5.3 Performance

Performance in embedded systems usually refers to the system ability to *perform* its function on time. Therefore, a measure of the number of operations per unit time will somehow always be involved.

Sometimes, performance is associated with issues such as power consumption, memory usage, and even cost; however, in our discussion we will restrict it to the system ability to perform tasks on time. Now, this does not make performance measurement an easy task. A number of factors need to be considered to measure or to establish a performance constraint for a system.

Given our definition, one might naively think that the higher the system clock frequency, the better the performance that can be obtained. Although clock frequency does have an impact on the system performance, it could end up being a deceiving metric if we do not have a more general picture of the system. Performance in embedded systems depends on many factors, some due to hardware components, some other due to software components. Hardware related factors include:

- Clock Frequency, the speed at which the master system clock ticks to advance time in the system. Given two identical systems, executing the exact same software, changing the clock frequency will have a direct impact on the system speed to finish a given task. However, decisions in most cases are not as simple. Changing for example from processor A to processor B we might experience that the same instructions take more clock cycles in B than in A, and therefore processor A could have a higher performance than B, even when B might be running at a higher clock frequency.

- Processor Architecture has a deep impact in the system performance. The use of multiple execution units, pipelining, caches, style (CISC[2] versus RISC[3]) buses, among other architectural differences affect the time the system takes to complete a task.
- Component Delay, i.e. the time it takes a component to react at its output to an input stimulus, will affect system performance. In the operation of synchronous bus systems, the maximum attainable speed will be limited by that of the slowest component. Semi-synchronous or asynchronous bus systems would slowdown only when accessing slow devices. A typical example of component delay affecting system performance is the memory access time. The time taken by a memory device to respond to a read or write operation will determine how fast the processor can exchange data with it, therefore determining the memory throughput and ultimately the system performance.
- Handshaking Protocols, i.e. the way transactions begin and end with peripheral devices and how frequently they need to occur will take their toll in the time the system takes to complete a task. Consider for example a device for which each word transfer requires submitting a request, waiting for a request acknowledgment, making the transfer and then waiting again for a transfer acknowledgment signal to ensure that no errors occurred in the transfer would function significantly slower than another where a session could be established, transferring blocks of words per handshaking transaction.
- Low-power Modes in hardware components also affect performance. When a component is sent to a low-power or sleep mode, either by gating its clock, its power lines, or some other mechanism, waking the device up consumes clock cycles, ultimately affecting its response time. In the case of devices able to operate at different voltage levels to save power, the lower the voltage the longer the response time, therefore reducing the system performance.

One consideration that needs to be addressed when dealing with hardware factors that affect performance is that high speed is expensive. The higher the speed of the hardware, the higher the system cost and in many cases the higher also its power consumption. Therefore the designer must exercise caution when choosing the hardware components of a system to satisfy specific performance requirements in not overspending the design. Once our embedded design reaches a level of performance that satisfies the expected responsiveness of its functionality, a higher performance becomes waste. Consider for example an embedded system design to give functionality to keyboard. The system must be fast enough to not miss any keystroke by the typist. A fast typist could produce perhaps 120 words per minute. With an average of eight characters per word, each new event in the keyboard would happen every 200 ms. We do not need the fastest embedded processor in the market to fulfill the performance requirements of this application. A different story would result if our embedded design were to interface for example a high-speed internet channel to a video

---

[2] Complex Instruction Set Computer.
[3] Reduced Instruction Set Computer.

processing unit. With data streaming at 100 Mbps or faster, the application requirements would demand a fast and probably expensive, power hungry, hardware design.

Software factors also affect system performance. Unlike hardware, we always
want to make the software faster because that will make the system more efficient not
only in terms of performance, but also in terms of power consumption and hardware
requirements, therefore helping to reduce recurrent costs. Software factors affecting
system performance include:

- Algorithm Complexity. This is perhaps the single most relevant software factor impacting system performance. Although algorithm complexity traditionally
  refers to the way the number of steps (or resources) needed to complete a task scales
  as the algorithm input grows, it also provides a measure of how many steps it would
  take a program to complete a task. Each additional step translates into additional
  cycles of computation that affect not only time performance, but also power consumption and CPU loading. Therefore the lower the complexity, the faster, in the
  long run, is the system expected performance. Some particular situations may arise
  for small input. For these cases it might occur that a higher complexity algorithm
  completes faster than one with lower complexity. This implies that a documented
  decision needs to consider aspects such as typical input size, memory usage, and
  constant factors among others.
- Task Scheduling determines the order in which tasks are assigned service priorities
  in a multitasking environment. There exist different scheduling algorithms that can
  be used when a system resource needs to be shared among multiple tasks. The rule
  of thumb is that one resource can be used by only one task at a time. Therefore,
  the way in which services are scheduled will affect the time it takes for services to
  be provided thereby affecting the system performance. Although in a single-task
  embedded system this consideration becomes trivial, multitasking systems are far
  common and meeting performance requirements makes it necessary to give careful
  consideration to the way application tasks will be scheduled and their priorities.
- Intertask Communication deals with the mechanisms used to exchange information
  and share system resources between tasks. The design of this mechanism defines
  the amount of information that needs to be passed and therefore the time overhead
  needed for the communication to take place. Therefore, the selected mechanisms
  will impact the system performance.
- Level of Parallelism refers to the usage given by the software to system resources
  that accept simultaneous usage. For example, an embedded controller containing
  both a conventional ALU and a hardware multiplier can have the capacity of
  performing simultaneous arithmetic operations in both units. A dual-core system
  can process information simultaneously in both cores. It is up to the software how
  the level of parallelism that can be archived in the system is to be exploited to
  accelerate the completion of tasks, increasing the system performance.

In summary, meeting performance constraints in an embedded system requires
managing multiple factors both in hardware and software. The system designer needs
to plan ahead during the architectural design the best way in which software and

hardware elements will be combined in the system to allow for an efficient satisfaction of the specific performance requirements demanded by an application.

## 1.5.4 Power and Energy

Power in embedded systems has become a critical constraint, not only in portable, battery operated systems, but for every system design. The average power dissipation of an embedded design defines the rate at which the system consumes energy. In battery powered applications, this determines how long it takes to deplete the capacity of its batteries. But aside from battery life, power affects many other issues in embedded systems design. Some of the most representative issues include:

- System Reliability: In digital electronic systems, which operate at a relatively fixed voltage level, power translates into current intensities circulating through the different system traces and components. This current circulation induces heat dissipation, physical stress, and noise in the operation of system elements, all associated to system failures. Heat dissipation is a major concern in system reliability due to the vulnerability of semiconductors to temperature changes. At high temperatures, device characteristics deviate from their nominal design values leading to device malfunction and ultimately system failure. The physical stress induced by large current densities on traces, contacts, and vias of integrated circuits and printed circuit boards is the primary cause of electromigration (EM). EM describes the transport of mass in metal under the stress of high current densities, causing voids and hillocks, a leading source of open and short circuits in electronic systems. Noise is generated by the pulsating nature of currents in inductive and capacitive components of digital electronic systems. As the current intensities increase with the power levels, so does the noise energy irradiated by reactive components, reaching levels that could exceed the system noise margins, leading to system malfunction.
- Cooling Requirements: A large power consumption leads to large levels of heat dissipation. Heat needs to be removed from the embedded electronics to avoid the failures induced by high temperature operation. This creates the necessity of including mechanisms to cool down the hardware. Increasing power levels rapidly scales forced air ventilation in static heatsinks to active heat removal mechanisms that include fans, liquid pumps, electronic and conventional refrigeration devices, and even laser cooling techniques. The necessity of including such components in embedded systems increases the system recurrent costs, and imposes physical requirements to the mechanical design in the form of increased size and weight, and shape limitations.
- Power Supply Design: All embedded systems need a source of energy. Portable systems rely on batteries. Stationary systems need power supplies. The larger the power dissipation of a system, the larger the size of its batteries or power supply,

and the lesser the chances of using energy scavenging techniques or alternate sources of energy. This also impacts system costs and mechanical product design.

- System Size, Weight, and Form: The mechanical design characteristics are affected not only by the requirements of power supply and cooling mechanisms, but also by the minimum admissible separation between components to limit heat density and reduce the risk of heat induced failures.

- Environmental Impact: The number of embedded systems around us continues to grow at accelerated rates. Estimates by the Consumer Electronics Association (CEA) show that nowadays an average individual uses about sixty embedded systems per day. Twenty years ago this number was less than ten. This growing trend is expected to continue for decades to come as embedded systems become more pervasive in contemporary lifestyles. Although this trend is fueled by the improvements they bring to our quality life, it is also placing a real burden on the planet energy resources and leaving a mark in the environment. A study of energy consumption by consumer electronics in the US commissioned by the CEA found that in 2006 11 % of the energy produced in the US, this is 147 TWh, was destined to household consumer electronics. Figure 1.13 shows the consumption distribution of the types the CEA included in the study. It can be observed that excluding personal computers, 65 % of the energy was consumed by devices we have catalogued as embedded system applications. It deserves to mention that the numbers in this study do not include digital TV sets, the electronics in major appliances such as dishwashers, refrigerators, ovens, etc.; nor the consumption by other segments of the population such as corporations, schools, universities, etc. Despite these exclusions, the numbers portrait the environmental impact of embedded systems and highlight the importance of tighter constraints in their power consumption.

Meeting the power constraints of an embedded application can be approached from multiple fronts:

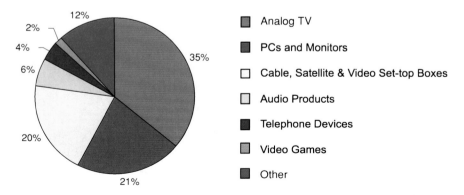

**Fig. 1.13** Distribution of U.S. residential consumer electronics (CE) energy consumption in 2006. The total CE energy consumed was 147 TWh (*Source* Consumer Electronics Association)

- Hardware Design: Utilizing processors and hardware components specifically designed to minimize power consumption. Low-voltage, low-power peripherals, processors with power efficient standby and sleep modes, and low-power memories are some of the most salient types of components available for reducing the power consumption by the system hardware.
- Software Design: The way in which software is designed has a significant impact on the system power consumption. A given software function can be programmed in different ways. If each program were profiled for the energy it consumes during execution, each implementation would have its own energy profile, leading to the notion that software can be tailored to reduce the energy consumed by an application. This concept can be applied during programming if the programmer is conscious of the energy level associated to a programming alternative or exploited through the use of power optimizing compilers. The bottom line consideration here is that every cycle executed by the processor consumes energy in the system, therefore, the least number of execution cycles by the processor, the lower the energy consumed by the system.
- Power Management Techniques: Power management combines software strategies for power reduction with the hardware capabilities of the system to reduce the amount of energy needed to complete a task. This is perhaps the most effective way of developing power efficient applications.

### 1.5.5  Time-to-Market

Time to market (TTM) is the time it takes from the conception of an embedded system until it is launched into the market. TTM becomes a critical constraint for systems having narrow market windows. The market window $W$ is defined as the maximum sales life of a product divided by two. Thus, the totals sales life of a product would be $2W$. The market window for embedded systems, as for any commercial product, is finite with profits usually following a Gaussian distribution with mean around $W$. Figure 1.14 shows a typical market window curve, illustrating an on-time product entry for a well managed time-to-market.

The beginning of the sales life imposes a strong constraint on the time-to-market since the total revenues, obtained as the area under the curve, are maximized when an on-time market entry is achieved. A delayed product deployment causes large losses of revenues. Figure 1.15 illustrates this situation with a simplified revenue model using linear market rise and fall behaviors.

When a product enters the market with delay $D$, its peak revenue $R'_{max}$ is scaled by a factor $(1 - D/W)$. This model presumes that the delayed peak revenue occurs at the same point in time as the on time peak, and a maximum delay not exceeding $W$. The loss of revenue $L$ can be calculated as the difference between the on-time total revenues and the delayed total revenues, obtained as indicated by Eq. 1.3.

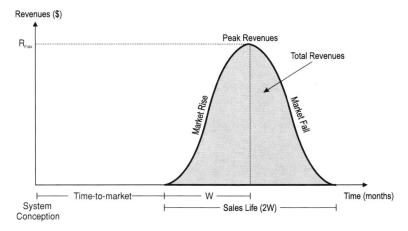

**Fig. 1.14** Typical revenue-time curve for embedded products, denoting the time-to-market and market window

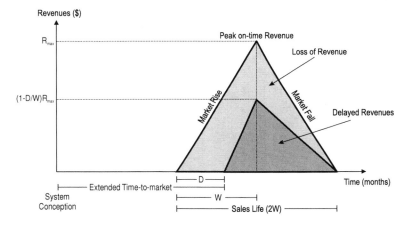

**Fig. 1.15** Linear revenue model with a delayed system deployment

$$L = \frac{R_{max} D}{2W}(3W - D) \tag{1.3}$$

It is common to express losses as a percentage of the maximum on-time revenues. This can be obtained by dividing Eq. 1.3 by $R_{max} W$, resulting the expression in Eq. 1.4.

$$L\% = \frac{D(3W - D)}{2W^2} * 100\% \tag{1.4}$$

To numerically illustrate the impact of a violation to the time-to-market constraint, consider a product with a sales life of 24 months. A TTM extended by 10 weeks

($D = 2.5$) would cause a total revenue loss of 29 %. A 4-month delay would reduce revenues by nearly 50 %.

## *1.5.6 Reliability and Maintainability*

Maintainability in embedded systems can be defined as a property that allows the system to be acted upon, to guarantee a reliable operation throughout the end of its useful life. This property can be regarded as a design constraint because, for maintainability to be enabled, it has to be planned from the system conception itself.

The maintainability constraint can have different levels of relevance depending on the type of embedded system being considered. Small, inexpensive, short lived systems, such as cell phones and small home appliances are usually replaced before they need any maintenance. In these type of applications, a reliable operation is essentially dependent on the level of verification done on them prior to their deployment. Large, expensive, complex embedded systems, such as those used in airplanes or large medical equipment, are safety critical and expected to remain in operation for decades. For this class of systems, a reliable operation is dependent on the system ability to be maintained.

From a broad perspective, maintainability needs to provide the system with means for executing four basic types of actions:

- *Corrective Actions*: that allow to fix faults discovered in the system.
- *Adaptive Actions*: which enable the introduction of functional modifications that keep the system operating in a changing environment.
- *Perfective Actions*: to allow for adding enhancements to the system functionally attending to new regulations or requirements.
- *Preventive Actions*: that anticipate and prevent conditions that could compromise the system functionality.

These actions need to be supported throughout the hardware and/or software components of the embedded system, but their implementation needs to deal with constraints inherent to each of these domains (Fig. 1.16).

### Issues in Hardware Maintainability

Hardware designers face several limiting factors when it comes to integrating maintainability in their systems. These include:

- The cost overhead in the system NRE related to the design and validation of maintenance and testing structures.
- The impact on the time-to-market caused by additional development time required for the inclusion of maintainability features. This is also a factor in the software design.

**Fig. 1.16** The four actions supporting system maintainability

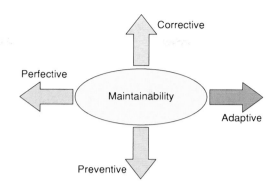

- The increase in recurrent costs caused by the additional hardware components needed in the system to support testing and maintenance.
- Other factors affecting hardware maintainability in deployed systems include problems of component obsolescence, accessibility to embedded hardware components, and availability of appropriate documentation to guide changes and repairs.

**Issues in Software Maintainability**

Supporting software maintainability in embedded systems represents an even greater challenge than hardware because of the way embedded programs are developed and deployed in such systems. In the application design flow, changes occurring either on the hardware or software components during their concurrent design, sharing software modules among product variants, the low-level requirements and tight interaction with the hardware, and the hardware constraint themselves are some of the characteristics that make embedded software maintainability particularly difficult. Results from representative case studies that include consumer electronics, communication equipment, aerospace applications, and industrial machinery have indicated that there are 10 major issues that need to be addressed in embedded software development to increase its maintainability.

1. Unstable requirements: Requirements change during the software development phase, which usually are not effectively communicated to all partners, creating problems in the hardware software integration, and therefore making maintenance more complex.
2. Technology changes: New technologies require new development tools which not necessarily preserve compatibility with early software versions, making difficult maintenance activities.
3. Location of errors: A common scenario in software maintenance is that the person assigned to fix a problem is not the one who reported the problem and also

different from the one who wrote the software, making the process of locating the error very time consuming. Moreover, the uncertainty of whether the error is caused by a software or a hardware malfunction adds to the difficulty.

4. Impact of Change: When a problem is identified and the remedial change is decided, identifying the ripple effect of those changes, such that it does not introduce additional problems becomes a challenge.

5. Need for trained personnel: The tight interaction between hardware and software components in embedded applications makes the code understanding and familiarization a time consuming activity. This calls for specialized training in an activity that results less glamorous and attractive to a new professional than designing and developing new code.

6. Inadequate documentation: All too common, the software development phase does not produce adequate, up-to-date documentation. Maintenance changes are also poorly documented, worsening the scenario.

7. Development versus application environment: Embedded software is typically developed on an environment other than that of the target application. A problem frequently caused by this situation is that real-time aspects of the system functionality might not be foreseen by the development environment, leading to hard to find maintenance problems.

8. Hardware constraints: Due to the direct impact of hardware in the system recurrent costs, stringent constraints are placed on hardware components. This requires software written in a constrained hardware infrastructure that leaves little room for supporting maintenance and updates.

9. Testing process: Due to the cost of testing and verification, it is common that only the basic functionality is verified, without a systematic method to ensure a complete case software test. This, albeit of the intrinsic complexity of the software verification process itself, accumulates to problems to appear later when the system is deployed.

10. Other problems: Many other issues can be enumerated that exacerbate the tasks of maintenance. These include software decay, complexity in upgrading processes, physical accessibility, and integration issues, among others.

## 1.6  Summary

This chapter presented a broad view of the field of embedded systems design. The discussion on the history and overview helped to identify the differences between traditional computing devices and embedded systems, guiding the reader through the evolution of embedded controllers from their early beginning to the current technologies and trends.

Early in the discussion, the tight interaction between embedded hardware and software and software was established, providing an insight into their broad structure and relationship. A broad classification of embedded systems was then introduced to identify typical application scopes for embedded systems.

The last part of the chapter was devoted to the discussion of typical design constraints faced by embedded system designers. In particular, issues related to system functionality, cost, performance, energy, time-to-market, and maintainability were addressed in attempt to create awareness in the mind of the designer of the implications carried by the design decisions, and the considerations to be made in the design planning phase of each application.

## 1.7 Problems

1.1 What is an Embedded System? What makes it different from a general purpose computer?

1.2 List five attributes of embedded systems.

1.3 What was the first electronic computing device designed with the modern structure of and embedded system?

1.4 What were the three pioneering designs that defined first microprocessors?

1.5 Which design can be credited as the first microcontroller design? Why is this design considered a milestone in embedded systems design?

1.6 What proportion of all designed microprocessors end up used in personal computers? From the rest, which CPU size has dominance in terms of units per year? Which generates the largest sale volume in dollars?

1.7 Write a list of at least ten tasks the *you* perform regularly that are enabled by embedded systems.

1.8 What are the fundamental components of an embedded system? Briefly describe how each is structured.

1.9 Think about the functionality of one simple embedded system enabled device you regularly interact with (for example, a video game console, cellular phone, digital camera, or similar) and try to come up with the following system-level items:

    a.  Functional hardware components needed.

    b.  Software tasks performed by the device.

    c.  Identifiable services that could be requested by each task and the hardware resource it uses.

1.10 List three embedded system applications for each of the classes identified in the discussion in Sect. 1.3. Provide at least one example of an application falling between two classes, for each possible combination of two classes. Can you identify one example with characteristics of the three classes? Briefly explain why you consider each example in the class where you assigned them.

1.11 Select one specific embedded system application you are familiar with and enumerate at least three important issues arising in each of its five life cycle stages. Rank them in order of relevance.

1.12 Perform an internet search to identify at least three commercially available verification tools for the Texas Instruments MSP430 family of microcontrollers.

Categorize them by the verification method they use. Comment on the cost-effectiveness of each.

1.13 What is the difference between non-recurrent (*NRE*) costs and recurrent production costs (*RP*). Use the diagram in Fig. 1.11 and identify which stages generate *NRE* costs and which stages generate recurrent *RP* costs.

1.14 A voice compression system for an audio application is being designed as an embedded system application. Three methods of implementation are under consideration: Under method A, the design would use a Digital Signal Processor (DSP); in method B, the design would be completed with an embedded microcontroller; while in method C, the design will be completed with an Application Specific Integrated Circuit (ASIC) where all functions will reside in hardware. Alternative A has $NRE_a$ of \$250,000 and a recurrent cost per unit $RP_a$ of \$75. Alternative B has $NRE_b = \$150,000$ and $RP_b = \$130$; while the ASIC solution has $NRE_c = \$2,000,000$ and $RP_c = \$20$.

   a. For each of the three alternatives, plot the per-unit cost break-even price of the system production volume $V$.
   b. Determine the production volume $V$ where the ASIC solution becomes the most cost effective solution.
   c. If the market for this product had an expected volume sales of 1,500 units, which implementation alternative would you recommend? Justify you recommendation.
   d. In part c, what should the market price of the system be if the company expects a profit margin of 25 %? How much would the expected revenues be if all marketed units were sold?

1.15 Explain the difference between clock frequency and system performance in an embedded design. Depict a scenario where increasing the clock speed of an application would not result in an increase in performance. Explain why.

1.16 Enumerate three approaches each at the hardware and software level design that could impact the amount of energy consumed by an embedded application. Briefly explain how each impacts the energy consumption.

1.17 Consider the voice compression system of problem 1.14. For a market window of two years, and expected profit margin of 27 % in the marketing of 50,000 units implemented with alternative A, determine the following:

   a. Determine the expected company revenues in dollars.
   b. What would the percent loss of revenue be if the time-to-market were extended, causing a deployment delay of 6 months? How much is the loss in dollars?
   c. Plot the cumulative loss of revenues of this product for delays in its time-to-market in a per-week basis.

1.18 Identify three issues each at the software and hardware design levels to consider when planning the maintainability of an embedded system application. Explain briefly how to integrate them in the design flow of the system.

# Chapter 2
# Number Systems and Data Formats

To understand how microcontrollers and microprocessors process data, we must adopt symbols and rules with proper models to mimic physical operations at a human friendly level. Digital systems operate with only two voltage levels to process information. As part of the modeling process, two symbols are associated to these levels, '0' and '1'. Any consecutive concatenation of these two symbols, in any order, is referred to as a word. Words are used as mathematical entities and abstract symbols to represent all sort of data. We refer to words as integers, thereby simplifying not only the description and manipulation of the strings, but also programming notation and hardware processes.

The meaning of a word is context dependent and may be completely arbitrary for user-defined data or assignments, set by the hardware configurations or else follow conventions and standards adopted in the scientific community. We limit our discussion here to numerical, text, and some particular cases of special interest.

This chapter contains more material than needed in a first introduction to the topic of embedded systems. A first time reader may skip Sects. 2.9 and 2.10. Arithmetic operations with integers, on the other hand, are needed to go beyond transfer operations.

## 2.1 Bits, Bytes, and Words

A *bit* is a variable which can only assume two values, 0 or 1. By extension, the name bit is also used for the constants 0 or 1. An ordered sequence of $n$ bits is an *$n$-bit word*, or simply *word*. Normally, $n$ is understood from the context; when it is not, it should be indicated. Particular cases of $n$-bit words with proper names are:

- *Nibble*: 4-bit word
- *Byte*: 8-bit word
- *Word*: 16-bit word
- *Double word*: 32-bit word.
- *Quad*: 64-bit word.

M. Jiménez et al., *Introduction to Embedded Systems*,
DOI: 10.1007/978-1-4614-3143-5_2,
© Springer Science+Business Media New York 2014

These lengths are associated to usual hardware. In the MSP430, for example, registers are 16 bits wide. Notice the specific use of the term "word" for 16 bits.

Individual bits in a word are named after their position, starting from the right with 0: bit 0 (b0), bit 1 (b1), and so on. Symbolically, an $n$-bit word is denoted as

$$b_{n-1}b_{n-2}\ldots b_1 b_0 \tag{2.1}$$

The rightmost bit, $b_0$, is the *least significant bit* (lsb), while the leftmost one, $b_{n-1}$, is the *most significant bit* (msb). Similarly, we talk of least or most significant nibble, byte, words and so on when making reference to larger blocks of bits.

Since each bit in (2.1) can assume either of two values, there are $2^n$ different $n$-bit words. For easy reference, Table 2.1 shows the powers of 2 from $2^1$ to $2^{30}$.

Certain powers of 2 have special names. Specifically:

- *Kilo* (K): $1\,K = 2^{10}$
- *Mega* (M): $1\,M = 2^{20}$
- *Giga* (G): $1\,G = 2^{30}$
- *Tera* (T): $1\,T = 2^{40}$

Hence, when "speaking digital", $4\,K$ means $4 \times 2^{10} = 2^{12} = 4,096$, $16\,M$ means $16 \times 2^{20} = 2^{24} = 16,077,216$ and so on.[1]

**Table 2.1**  Powers of 2

| N | $2^N$ | N | $2^N$ | N | $2^N$ |
|---|---|---|---|---|---|
| 1 | 2 | 11 | 2,048 | 21 | 2,097,152 |
| 2 | 4 | 12 | 4,096 | 22 | 4,194,304 |
| 3 | 8 | 13 | 8,192 | 23 | 8,388,608 |
| 4 | 16 | 14 | 16,384 | 24 | 16,777,216 |
| 5 | 32 | 15 | 32,768 | 25 | 33,554,432 |
| 6 | 64 | 16 | 65,536 | 26 | 67,108,864 |
| 7 | 128 | 17 | 131,072 | 27 | 134,217,728 |
| 8 | 256 | 18 | 262,144 | 28 | 268,435,456 |
| 9 | 512 | 19 | 524,288 | 29 | 536,870,912 |
| 10 | 1,024 | 20 | 1,048,576 | 30 | 1,073,741,824 |

## 2.2 Number Systems

Numbers can be represented in different ways using 0's and 1's. In this chapter, we mention the most common conventions, starting with the normal binary representation, which is a positional numerical system.

---

[1] The notation K, M, G and T was adopted because those powers are the closest to $10^3$, $10^6$, $10^9$ and $10^{12}$, respectively. However, manufacturers of hard drives use those symbols in the usual ten's power meaning. Thus, a hard drive of 120 MB means that it stores $120 \times 10^6$ bytes.

## 2.2.1 Positional Systems

Our decimal system is *positional*, which means that any number is expressed by a permutation of digits and can be expanded as a weighted sum of powers of ten, the base of the system. Each digit contributes to the sum according to its position in the string. Thus, for example,

$$32.23 = 3 \times 10^1 + 2 \times 10^0 + 2 \times 10^{-1} + 3 \times 10^{-2},$$
$$578 = 8 \times 10^0 + 7 \times 10^1 + 5 \times 10^2$$

This concept can be generalized to any base. A *fixed-radix*, or *fixed-point* positional system of *base r* has *r* ordered digits $0, 1, 2, \ldots$ "r−1". Number notations are composed of permutations of these *r* digits. When we run out of single-digit numbers we add another digit and form double-digit numbers, then when we run out of double-digit numbers we form three-digit numbers, and so on. There is no limit to the number of digits that can be used. Thus, we write

Base 2 : 0, 1, 10, 11, 100, 101, 110, $\ldots$;

Base 8 : 0, 1, 2, 3, 4, 5, 6, 7, 10, 11, $\ldots$, 17, 20, $\ldots$, 77, 100, 101, $\ldots$;

Base 12 : 0, 1, 2, 3, 4, 5, 6, 7, 8, 9, $A$, $B$, 10, 11, $\ldots$, $BB$, 100, $\ldots$, $BBB$, 1000, $\ldots$;

To avoid unnecessary complications, the same symbols are used for the digits $0, \ldots, 9$ in any base whenever are required, and letters $A, B, \ldots$ are digits with values of ten, eleven, etc. To distinguish between different bases, the radix is placed as a subscript to the string, as in $28_9, 1A_{16}$.

Numbers are written as a sequence of digits and a point, called *radix point*, which separates the *integer* from the *fractional part* of the number to the left and right side of the point, respectively. Consider the expression in equation (2.2):

$$a_{n-1}a_{n-2}\ldots a_1a_0.a_{-1}a_{-2}\ldots a_m \qquad\qquad (2.2)$$

Here, each subscript stands for the exponent of the weight associated to the digit in the sum. The leftmost digit is referred to as the *most significant digit* (msd) and the rightmost one is the *least significant digit* (lsd). If there were no fractional part in (2.2), the radix point would be omitted and the number would be called simply an integer. If it has no integer part, it is customary to include a "0" as the integer part.

The number denoted by (2.2) represents a power series in *r* of the form

$$\underbrace{a_{n-1}r^{n-1} + a_{n-2}r^{n-2} + \cdots + a_1r^1 + a_0r^0}_{\text{integer part}}$$

$$\underbrace{+ a_{-1}r^{-1} + \cdots + a_{-m}r^{-m}}_{\text{fractional part}} = \sum_{i=-m}^{n-1} a_i r^i \qquad\qquad (2.3)$$

An interesting and important fact follows from this expansion: *In a system of base r, $r^n$ is written as 1 followed by n zeros.* Thus, $100_2 = 2^2$, $1000_5 = 5^3$, etc.

The systems of interest for us are the *binary* (base 2), *octal* (base 8), *decimal* (base 10), and *hexadecimal* (base 16) —*hex* for short,—systems. Table 2.2 shows equivalent representations to decimals 0–15 in these systems, using four binary digits, or bits, for binary numbers.

To simplify keyboard writing, especially when using text processors, the following conventions are adopted:

- For binary numbers use suffix 'B' or 'b'
- For octal numbers use suffix 'Q' or 'q'
- For hex numbers use suffix 'H' or 'h', or else prefix 0x. Numbers may not begin with a letter
- Base ten numbers have no suffix.

Hence, we write 1011B or 1011b instead of $1011_2$, 25Q or 25q for $25_8$, 0A5H or 0A5h or 0xA5 for $A5_{16}$.

**Table 2.2** Basic numeric systems

| Decimal | Binary | Octal | Hex | Decimal | Binary | Octal | Hex |
|---------|--------|-------|-----|---------|--------|-------|-----|
| 0 | 0000 | 0 | 0 | 8 | 1000 | 10 | 8 |
| 1 | 0001 | 1 | 1 | 9 | 1001 | 11 | 9 |
| 2 | 0010 | 2 | 2 | 10 | 1010 | 12 | A |
| 3 | 0011 | 3 | 3 | 11 | 1011 | 13 | B |
| 4 | 0100 | 4 | 4 | 12 | 1100 | 14 | C |
| 5 | 0101 | 5 | 5 | 13 | 1101 | 15 | D |
| 6 | 0110 | 6 | 6 | 14 | 1110 | 16 | E |
| 7 | 0111 | 7 | 7 | 15 | 1111 | 17 | F |

## 2.3 Conversion Between Different Bases

The power expansion (2.3) of a number may be used to convert from one base to another, performing the right hand side operations in the target system. This requires, of course, that we know how to add and multiply in systems different from the common decimal one. This is why the decimal system is usually an intermediate step when a conversion is done by hand between non decimal systems. Yet, any algorithm is valid to work directly when knowledge or means are available.

### 2.3.1 Conversion from Base r to Decimal

The expansion (2.3) is the preferred method to convert from base $r$ to decimal. Let us illustrate with some numbers. No subscript or suffix is used for the decimal result.

**Example 2.1** *The following cases illustrate conversions to decimals:*

$$214.23_5 = 2 \times 5^2 + 1 \times 5^1 + 4 \times 5^0 + 2 \times 5^{-1} + 3 \times 5^{-2} = 59.52$$
$$10110.01B = 1 \times 2^4 + 0 \times 2^3 + 1 \times 2^2 + 1 \times 2^1 + 0 \times 2^0$$
$$+ 0 \times 2^{-1} + 1 \times 2^{-2} = 22.25$$
$$B65FH = 11 \times 16^3 + 6 \times 16^2 + 5 \times 16^1 + 15 \times 16^0 = 46687$$

*Notice that for hexadecimal conversion, all hex digits are interpreted in their decimal values for the sum.*

Since binary numbers are so common in embedded systems with digits being only 0 and 1, it is handy to have a quick reference for weights using positive and negative powers of 2, as illustrated in Fig. 2.1 for some cases.

**Example 2.2** *Let us convert 1001011.1101B to decimal using the powers as shown in Fig. 2.1, as shown in the following table, taking for the sum only those powers for bit 1:*

| Number    | 1  | 0  | 0  | 1 | 0 | 1 | 1. | 1    | 1    | 0     | 1      |
|-----------|----|----|----|---|---|---|----|------|------|-------|--------|
| Exponents | 6  | 5  | 4  | 3 | 2 | 1 | 0  | −1   | −2   | −3    | −4     |
| Power     | 64 | 32 | 16 | 8 | 4 | 2 | 1  | 0.5  | 0.25 | 0.125 | 0.0625 |

*Therefore,* $1001011.1101B = 64 + 8 + 2 + 1 + 0.5 + 0.25 + 0.0625 = 75.8125$

## 2.3.2 Conversion from Decimal to Base r

Conversion into base $r$ is easier if the integer and fractional parts are treated separately.

**Integer Conversion**

One popular procedure for converting decimal integers into base $r$ is the *repeated division* method. This method is based on the division algorithm, and consists in successively dividing the number and quotients by the target radix $r$ until the quotient

| Weights: | 128 | 64  | 32  | 16  | 8   | 4   | 2   | 1   |
|----------|-----|-----|-----|-----|-----|-----|-----|-----|
| Bits:    | b7  | b6  | b5  | b4  | b3  | b2  | b1  | b0  |

| Weights: | 0.5   | 0.25  | 0.125 | 0.0625 | 0.03125 | 0.015625 |
|----------|-------|-------|-------|--------|---------|----------|
| Bits:    | b(-1) | b(-2) | b(-3) | b(-4)  | b(-5)   | b(-6)    |

**Fig. 2.1** Weights for binary number $b_7 b_6 b_5 b_4 b_3 b_2 b_1 b_0 . b_{-1} b_{-2} b_{-3} b_{-4} b_{-5} b_{-6}$

is 0. The successive remainders of the divisions are the digits of the number in base $r$, starting from the least significant digit: divide the number by $r$ and take the remainder as $a_0$; divide the quotient by $r$, and the remainder as $a_1$, and so on.[2] Let us illustrate with a pair of examples.

**Example 2.3** *Convert decimal* 1993 *to expressions in bases 5 and 16.*

*To convert to base 5, divide by 5. The remainders are the digits of the number we are looking for, going from lsd to msd:*

| Division | Quotient | Remainder |
|---|---|---|
| 1993/5: | 398 | $a_0 = 3$ |
| 398/5: | 79 | $a_1 = 3$ |
| 79/5: | 15 | $a_2 = 4$ |
| 15/5: | 3 | $a_3 = 0$ |
| 3/5: | 0 | $a_4 = 3$ |

*Hence,   $1{,}993 = 30{,}433_5$*

*To convert to base 16, repeat the procedure using 16 as divisor. If the remainder is greater than or equal to 10, convert to hex equivalent digit.*

| Division | Quotient | Remainder |
|---|---|---|
| 1993/16: | 124 | $9 \; a_0 = 9$ |
| 124/16: | 7 | $12 \; (a_1 = C)$ |
| 7/16: | 0 | $a_2 = 7$ |

*The result is then* $1993 = 7C9_{16} = 7C9h.$

## Fractional Part

Conversion of decimal fractions can be done by the *repeated multiplication* method: Multiplying the number by $r$, the integer part of the first product becomes most significant digit $a_{-1}$. Discard then the integer part and multiply again the new

---

[2] Again, the algorithm is valid between any bases. Let $M_k$ be the integer in a base $k$ system, which we want to express in $r$ base system. The power sum expansion means $M_k = (a_{h-1}a_{h-2} \ldots a_1 a_0)_r = a_{h-1} \times r^{h-1} + \cdots a_1 \times r + a_0$, where the $a_i$'s are the digits in the $r$ base system. Dividing by $r$, the quotient is $a_h \times r^{h-1} + \cdots a_1$ and the remainder $a_0$. Repeating division, the quotient is $a_h \times r^{h-2} + \cdots a_2$ with a remainder $a_1$. Continuing this way, $a_2, a_3, \ldots a_h$ will be extracted, in that order.

fractional part by $r$ to get the next digit $a_{-2}$. The process continues until one of the following conditions is met[3]:

- A zero decimal fraction is obtained, yielding a finite representation in radix $r$; or
- A previous fractional part is again obtained, having then found the periodic representation in radix $r$; or
- the expression in base $r$ has the number of digits allocated for the process.

Since we are converting finite decimal fractions, the result must be either finite or periodic. The mathematical theory behind this assertion does not depend on the radix. Yet, periodicity might appear only after too many multiplications, so the third option above is just a matter of convenience. Let us look at some examples.

**Example 2.4** *Convert the following decimal fractions to binary, limited to 8 digits if no periodicity appears before: (a) 0.375, (b) 0.05, (c) 0.23.*

*(a) Let us begin with 0.375:*

$$2 \times 0.375 = 0.750 \quad \rightarrow \quad a_{-1} = 0$$
$$2 \times 0.750 = 1.50 \quad \rightarrow \quad a_{-2} = 1$$
$$2 \times 0.5 = 1.00 \quad \rightarrow \quad a_{-3} = 1 \quad \text{Zero fractional part. Stop}$$

*Therefore, since the decimal fractional part in the last product is 0.00, the equivalent expression in binary is finite, specifically, $0.375 = 0.011_2 = 0.011B$.*

*(b) Converting 0.05:*

$$2 \times 0.05 = 0.1 \quad \rightarrow \quad a_{-1} = 0$$
$$2 \times 0.10 = 0.2 \quad \rightarrow \quad a_{-2} = 0$$
$$2 \times 0.2 = 0.4 \quad \rightarrow \quad a_{-3} = 0$$
$$2 \times 0.4 = 0.8 \quad \rightarrow \quad a_{-4} = 0$$
$$2 \times 0.8 = 1.6 \quad \rightarrow \quad a_{-5} = 1$$
$$2 \times 0.6 = 1.2 \quad \rightarrow \quad a_{-6} = 1 \quad \text{Repeatingfractionalpart0.2.Stop}$$

*The decimal fraction 0.2 in the last line has appeared before (third line), so the pattern 0011 will be periodic. Therefore, $0.05 = 0.0\overline{0011}_2 = 0.0000110011\ldots B$.*

*(c) Converting 0.23:*

$$2 \times 0.23 = 0.46 \quad \rightarrow \quad a_{-1} = 0$$
$$2 \times 0.46 = 0.92 \quad \rightarrow \quad a_{-2} = 0$$
$$2 \times 0.92 = 1.84 \quad \rightarrow \quad a_{-3} = 1$$

---

[3] Let $M_{10} = (a_{-1}a_{-2}\cdots a_{-m})_r = a_{-1} \times r^{-1} + a_2 \times r^{-2} \cdots + a_{-m} \times r^{-m}$. Multiplying by $r$, the product is $a_{-1} + a_2 \times r^{-1} \cdots + a_{-m} \times r^{-m+1)}$. The integer part here is $a_{-1}$ and the fractional part follows the same pattern but starting with $a_{-2}$. Continuing with multiplication to the fractional part only, $a_{-2}, a_{-3}, \ldots$ will be extracted, in that order.

$$2 \times 0.84 = 1.68 \rightarrow a_{-4} = 1$$
$$2 \times 0.68 = 1.36 \rightarrow a_{-5} = 1$$
$$2 \times 0.36 = 0.72 \rightarrow a_{-6} = 0$$
$$2 \times 0.72 = 1.44 \rightarrow a_{-7} = 1$$
$$2 \times 0.44 = 0.88 \rightarrow a_{-8} = 0 \quad \text{Stop because of predefined limit.}$$

*We have reached 8 digits without finding the decimal fraction that will repeat.*[4] *Within this approximation,* $0.23 \approx 0.00111010B$

**Mixed Numbers with Integer and Fractional Parts**
In this case, the conversion is realized separately for each part. Let us consider an example.

**Example 2.5**   *Convert* $376.9375_{10}$ *to base 8.*
*First convert the integer part by successive divisions by 8:*

| Division | Quotient | Remainder |
|----------|----------|-----------|
| 376/8:   | 47       | $a_0 = 0$ |
| 47/8:    | 5        | $a_1 = 7$ |
| 5/8:     | 0        | $a_2 = 5$ |

*which yields* $376 = 570Q.$
*We now convert the fractional part by successive multiplications:*

| Multiplication | product | Int. Part. |
|----------------|---------|------------|
| $0.9375 \times 8 =$ | 7.5 | $a_{-1} = 7$ |
| $0.5 \times 8 =$    | 4.0 | $a_{-2} = 4$ |

*which yields* $3.9375 = 0.74Q.$ *Therefore the final result is* $376.9375 = 570.74Q.$

### 2.3.3  Binary, Octal and Hexadecimal Systems

The lower the base, the more digits required to represent a number. The binary numerical system, the 'natural' one for digital systems, is quite inconvenient for people. A simple three digit decimal number such as 389, becomes 110000101B in binary, a nine digit number. The octal and hexadecimal systems provide simpler alternatives for representing binary words. Since $8 = 2^3$ and $16 = 2^4$, conversion between binary and octal or binary and hex systems is straightforward:

**Octal and Binary**   Associate each octal digit to three binary digits, from right to left in the integer part and left to right in the fractional part.

---

[4]  The reader can continue and verify that the periodic pattern will eventually appear.

**Hex and Binary**    Associate each hex digit to four binary digits, from right to left in the integer part and from left to right in the fractional part.

Table 2.2 shows the equivalences between octal and hex digits with binary numbers, where zeros to the left can be conveniently added. Using this table, let us look at some examples.

**Example 2.6**  *Convert (a)* 0x4AD.16 *to binary and octal expressions; (b)* 37.25Q *to binary and hex expressions.*
*(a) Using Table 2.2 as reference,* 4AD.16H *is first converted into binary as follows:*

$$\underbrace{0100}_{4}\underbrace{1010}_{A}\underbrace{1101}_{D}.\underbrace{0001}_{1}\underbrace{0110}_{6} \Rightarrow 10010101101.0001011B$$

*Notice that the left zeros in the integer part and the right zeros of the fractional part have been suppressed in the final result. Now, to convert to octal, take the binary part and partition in groups of three bits as shown next, adding extra zeros as needed:*

$$\underbrace{010}_{2}\underbrace{010}_{2}\underbrace{101}_{5}\underbrace{101}_{5}.\underbrace{000}_{0}\underbrace{101}_{5}\underbrace{100}_{4} \Rightarrow 2255.054Q$$

*(b) For the second part, proceed as before, splitting the octal expression into binary, and then converting into hex expression.*

$$\underbrace{011}_{3}\underbrace{111}_{7}.\underbrace{010}_{2}\underbrace{101}_{5} \Rightarrow 11111.010101B$$

*Now to hex:*
$$\underbrace{0001}_{1}\underbrace{1111}_{F}.\underbrace{0101}_{5}\underbrace{0100}_{4} \Rightarrow 1F.54H$$

## 2.4  Hex Notation for *n*-bit Words

The previous example showed the convenience of hexadecimal system for humans: *there is a one to one correspondence between each hex digit and a group of four bits, a nibble.* For this reason, to simplify writing and reading, it is customary to represent an *n*-bit word by a hex equivalent just as it were a binary integer, *irrespectively of the actual meaning of the string.* Thus, we refer to 1010011 as 53h or 0x53, to 1101100110000010 as 0xD982, and so on.

This convention is universally adopted in embedded systems literature, and also in debuggers. Thus, memory addresses, register contents, and almost everything is expressed in terms of hexadecimal integers, without any implication of them being a number.

**Fig. 2.2** Simplified functional connection diagram of a 7-segment display to a microcontroller port

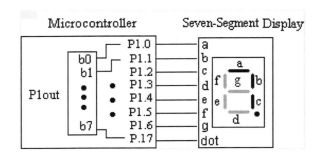

**Example 2.7** *The objective of this example is to illustrate how integers are used to represent hardware and data situations. Fig. 2.2 represents a functional connection, no hardware schematic shown, of a seven-segment display to Port 1 of a microcontroller. The port consists of eight pins, named* P1.0 *to* P1.7, *which are internally connected to a byte size register called* P1Out. *Let us call* b0, b1, ..., b7 *the outputs of the register, connected to* P1.0 *to* P1.7 *in the same order. The high and low voltage states of the register outputs are then applied to the seven-segment through the port pins.*

*The seven-segment display consists of eight light-emitting diodes (LED), in the form of seven segments called* a, b, c, d, e, f, *and* g *plus a* dot. *Each diode is connected to one port pin as shown in the figure. When the pin shows a high voltage state, the respective LED is turned on.*

*In the diagram, the seven-segment display shows* 7., *i. e., it has LED's* a, b,c *and the* dot *on, while the others are off. This means that pins* P1.7, P1.6, P1.5, P1.4, P1.3, P1.2, P1.1, P1.0, **in that order**, *show the respective states* HIGH, LOW, LOW, LOW, LOW, HIGH, HIGH, HIGH. *Using the symbol* 1 *for the* HIGH *state and* 0 *for the* LOW *state, and suppressing commas, the register output state can be represented as* 10000111.

*We use integers to represent this state, and say then that the contents of the register* P1Out *is* 10000111B, *or* 87h, *or* 207q, *or* 135. *The binary number gives a one to one description of each bit inside the register. The hex and octal representations give the same information in a simplified form, using the property of the one to one relationship with the binary digits. The decimal representation is just the conversion of any of the previous expressions.*

*If we want to display* 6, *the register contents should be* 7Dh. *Verify it.*

The decimal and octal systems can also be used to this end (see Problem 2.27). The octal system is not so popular nowadays because of hardware, which has made nibbles, bytes, 16-bit words and double words the normal environment.

## 2.5  Unsigned Binary Arithmetic Operations

The familiar techniques applied to arithmetic operations of non negative decimal numbers are also valid for binary numbers. We review next the basic operations for this system.

### 2.5.1 Addition

The addition rules are

$$0 + 0 = 0 \tag{2.4a}$$
$$0 + 1 = 1 + 0 = 1 \tag{2.4b}$$
$$1 + 1 = 10 \tag{2.4c}$$

As in the decimal system, we start adding the two least significant digits of the operands and carry any excess digit into the sum of the next more significant pair of digits. Therefore, a sum of numbers with more than one digit should consider effectively the addition of three digits to include the carry of the previous column. This addition of three bits yields the so called complete adder case illustrated by Table 2.3.

**Table 2.3**  Rules for complete binary addition

| $a_i$ | + | $b_i$ | + | $carry_i$ | = | $carry_{i+1}$ | $sum_i$ |
|---|---|---|---|---|---|---|---|
| 0 |   | 0 |   | 0 |   | 0 | 0 |
| 0 |   | 0 |   | 1 |   | 0 | 1 |
| 0 |   | 1 |   | 0 |   | 0 | 1 |
| 0 |   | 1 |   | 1 |   | 1 | 0 |
| 1 |   | 0 |   | 0 |   | 0 | 1 |
| 1 |   | 0 |   | 1 |   | 1 | 0 |
| 1 |   | 1 |   | 0 |   | 1 | 0 |
| 1 |   | 1 |   | 1 |   | 1 | 1 |

**Example 2.8**  *Add the binary equivalents of (a) 27 and 18; (b) 152.75 and 236.375.*

*(a)  The binary addition of 27 = 11011B and 18 = 10010B is illustrated next.*

| carries | | → | | 1 | 0 | 0 | 1 | 0 | | |
|---|---|---|---|---|---|---|---|---|---|---|
| 27 | + | → | | 1 | 1 | 0 | 1 | 1 | + |
| 18 | = | → | | 1 | 0 | 0 | 1 | 0 | = |
| 45 | | → | 1 | 0 | 1 | 1 | 0 | 1 | |

*(b) Since* 152.75 = 10011000.11B *and* 236.375 = 11101100.011B, *the addition is done as illustrated below. Here, the weight of each digit is shown for easy conversion.*

| Weights: | | 256 | 128 | 64 | 32 | 16 | 8 | 4 | 2 | 1 | 0.5 | 0.25 | 0.125 |
|---|---|---|---|---|---|---|---|---|---|---|---|---|---|
| Carries | | 1 | 1 | 1 | 1 | 1 | 0 | 0 | 0 | 1. | 1 | 0 | |
| 152.75 | + | | 1 | 0 | 0 | 1 | 1 | 0 | 0 | 0. | 1 | 1 | |
| 236.375 | = | | 1 | 1 | 1 | 0 | 1 | 1 | 0 | 0. | 0 | 1 | 1 |
| 389.125 | | 1 | 1 | 0 | 0 | 0 | 0 | 1 | 0 | 1. | 0 | 0 | 1 |

## 2.5.2 Subtraction

Subtraction follows the usual rules too. When the minuend digit is less than the subtrahend digit, a borrow should be taken. For subtraction, the rules are

$$0 - 0 = 1 - 1 = 0 \qquad\qquad (2.5a)$$
$$1 - 0 = 1 \qquad\qquad (2.5b)$$
$$0 - 1 = 1 \text{ with a } \mathbf{borrow}\,1 \qquad\qquad (2.5c)$$

When a borrow is needed, it is taken from the next more significant digit of the minuend, from which the borrow should be subtracted. Hence, actual subtraction can be considered to be carried out using three digits: the minuend, the subtrahend, and the borrowed digit. The result yields again a difference digit and a borrow that needs to be taken from the next column. This comment is better illustrated by the following Table 2.4 and example 2.9 below.

The table can be considered a formal expression of the usual procedure. Let us illustrate with an example.

**Table 2.4** Complete subtraction

| $a_i$ | $-$ | $b_i$ | $-$ | $borrow_i$ | $=$ | $borrow_{i+1}$ | $difference_i$ |
|---|---|---|---|---|---|---|---|
| 0 | | 0 | | 0 | | 0 | 0 |
| 0 | | 0 | | 1 | | 1 | 1 |
| 0 | | 1 | | 0 | | 1 | 1 |
| 0 | | 1 | | 1 | | 1 | 0 |
| 1 | | 0 | | 0 | | 0 | 1 |
| 1 | | 0 | | 1 | | 0 | 0 |
| 1 | | 1 | | 0 | | 0 | 0 |
| 1 | | 1 | | 1 | | 1 | 1 |

**Example 2.9** *Subtract* 137 *from* 216 *using binary subtraction.*

*Since* 137 = 10001001B *and* 216 = 11011000B, *we have the following operation, where the least significant borrow is 0 by default:*

| | | | | | |
|---|---|---|---|---|---|
| minuend | 216 | − | → | 1 1 0 1 1 0 0 0 | − |
| subtrahend | 137 | = | → | 1 0 0 0 1 0 0 1 | |
| Borrows | 110 | | | 0 0 0 1 1 1 1 0 | = |
| | 79 | | → | 0 1 0 0 1 1 1 1 | |

The binary system has been introduced up to now only for non negative numbers. Hence, to this point, the operation 137 − 216 = −79 has no meaning. However, it is illustrative to look at the process when Table 2.4 is used.

**Example 2.10** *Subtract* 216 *from* 137 *using binary subtraction according to Table 2.4 of the complete subtraction rule.*

*We arrange again the operands and borrows, with the least significant borrow 0 by default:*

| | | | | | |
|---|---|---|---|---|---|
| minuend | 137 | − | → | 1 0 0 0 1 0 0 1 | − |
| subtrahend | 216 | = | → | 1 1 0 1 1 0 0 0 | |
| Borrows | | | | 1 1 1 0 0 0 0 0 | = |
| | −79 | | →? | 1 1 0 1 1 0 0 0 1 | |

Two interesting points to remark: First, the result contains one bit more than the operands; this "one" comes from the borrow, since the subtrahend is greater than the minuend. This feature is used to identify inequalities. Second, a binary interpretation of the result using the numeric transformation previously studied is not valid, since it would yield 481. Even discarding the most significant digit, we are still without a good result, 225.

Of course, we can proceed just as we have done it in our usual arithmetic practice: actually subtract 216 − 137 and then append the negative sign in front of the result. However, we have no way of using a negative sign in embedded systems. Hence, we need a method to identify here −79. This method is discussed in Sect. 2.8, where the signed number representation is introduced. For the moment, let us introduce an associated concept, subtraction by complement's addition.

### 2.5.3 Subtraction by Complement's Addition

Let us consider two decimal numbers $A$ and $B$ with the same number of digits, say $N$. If necessary, left zeros can be appended to one of them. From elementary Algebra, it follows that

$$A - B = [A + (10^N - B)] - 10^N \tag{2.6}$$

The term $10^N - B$ in the addition is known as *ten's complement of B*. If $A > B$, then the addition $A + (10^N - B)$ results in $N + 1$ digits, i.e., with a carry of 1, which is eliminated by $10^N$. If $A < B$, the addition will contain exactly $N$ digits with no extra carry, and the result will be the negative of the tens complement of the addition. Let us illustrate with 127 and 31 (three digits required):

$$127 - 31 = (127 + 969) - 1000 = \underline{1}096 - 1000 = 96$$

and

$$31 - 127 = (31 + 873) - 1000 = 904 - 1000 = -(1000 - 904) = -96$$

**Two's Complement Concept and Application**
The method described above is applicable to any base system. This is so, because operations in (2.6) can be referred to any reference to base, and for any base $r$, $(r^N)_{10} = (10^N)_r$. When the base is 2, the complement is specifically called *two's complement*.

Notice that when considering two's complements, we need to specify the number of bits. For example, with four bits, the two's complement of 1010B is 10000B − 1010B = 0110B, but with 6 bits this becomes 1000000B − 1010B = 110110B. If you feel comfortable using decimal equivalents for powers of 2, these two examples can be thought as follows: 1010B is equivalent to decimal 10. For four bits, $2^4 = 16$ so the two's complement becomes $16 - 10 = 6 = 0110B$; with 6 bits, $2^6 = 64$ and the two's complement is $64 - 10 = 54 = 110110B$.

**Example 2.11**  *The operations (a) 216 − 137 and (b) 137 − 216 using binary numbers and the two's complement addition, and interpret accordingly, expressing your result in decimal numbers.*

*The binary equivalents for the data are* $216 = 11011000B$ *and* $137 = 10001001B$, *respectively. For the two's complement, since both numbers have 8 bits, consider* $2^8 = 256$ *as the reference, so the two's complement of 216 is* $256 - 216 = 40 = 00101000B$, *and that of 137 is* $256 - 137 = 119 = 01110111B$. *With this information, we have then:*
(A) *For 216 − 137:*

$$
\begin{array}{r}
11011000\ + \\
01110111\ = \\
\hline
101001111
\end{array}
$$

*Since the answer has a carry (9 bits), the result is positive. Dropping this carry, we have the solution to this subtraction* 01001111B = 79.
(B) *For 137 − 216:*

$$10001001 \; +$$
$$00101000 \; =$$

$$10110001$$

*Since the sum does not yield a carry (8 bits), the result is negative, with an absolute value equal to its two's complement.* $10110001B = 177$, *and the two's complement is* $256 - 177 = 79$. *Therefore, the solution is* $-79$, *as expected.*

*Notice that with only eight bits,* $10110001$ *is the same result obtained in example 2.10 when considering only eight least significant bits.*

**Important Remark** When the subtraction of $A - B$ is realized by complement addition, the presence of a carry in the addition means that the difference is non-negative, $A \geq B$. On the other hand, if the carry is 0, there is a need for a borrow, $A < B$, and the result is the negative of the complement of the sum.

### Calculating two's Complements

The operand's binary two's complement was calculated in the previous example through the interpretation of the algorithm itself. Two common methods to find the two's complement of a number expressed in binary form are mentioned next. The first one is practical for hardware realization and the second one is easy for fast hand conversion.

- *Invert-plus-sum*: Invert all bits (0–1 and viceversa) and add arithmetically 1.

  - To illustrate the proof, let us consider 4 bits, $b_3b_2b_1b_0$. The complement, in binary terms is

$$10000 - b_3b_2b_1b_0 = (1111 - b_3b_2b_1b_0) + 1$$

  According to the rules of subtraction (2.5a), the subtraction in parenthesis of the right hand side consists in inverting each bit.

- *Right-to-left-scan*: Invert only all the bits to the left of the rightmost '1'.

  - To illustrate with an example, assume that the six least significant bits of the original number are a 1 followed by five zeros, *****100000. After inverting the bits, all the bits to the right of the original rightmost 1 (now 0) will be 1's, and all those to the left will be inverted of the original bits, that is, after inversion we have xxxxxxx011111.... When we add 1, we will get xxxxxxx100000...., where the 'x' are the inverted bits of original number.

Both methods can be verified from the above examples, where N = 11011000B is the original number. For the two's complement take 00100111B + 1 = 00101000B.

## 2.5.4 *Multiplication and Division*

Multiplication and division also follow the normal procedures used in decimal system. We illustrate with two examples.

**Example 2.12** *Multiply 19 by 26, and divide 57 by 13.*
*For the multiplication, since 26 = 11010b and 19 = 10011b,*

$$
\begin{array}{r}
11010 \ \times \\
10011 = \\
\hline
11010 \\
11010 \\
11010 \qquad \\
\hline
111101110
\end{array}
$$

*The result is* $111101110b = 1EEh = 494$, *as it should be.*
*Notice that the result here has 9 bits. In general, the number of bits in the product is at most equals to the sum of the number of bits in the operands.*
*For the division, where 57 = 111001B and 13 = 1101B, the operation is shown next:*

$$
\begin{array}{r}
100 \\
1101 \quad | \overline{111001} \\
-1101 \\
\hline
101
\end{array}
$$

*This division yields a quotient 4 (100b) and residue 5 (101b), as it should be. The process followed the same rules as in decimal division.*

## 2.5.5 *Operations in Hexadecimal System*

It was pointed out the convenience of using the hexadecimal system as a method to simplify reading and writing binary words. This convenience extends to binary operations when dealing with numbers. While the rules can be set up for hexadecimal system, it is generally easier to "think" in decimal terms and write down in hexadecimal notation. Any result greater than 15 (F) yields a carry. Let us illustrate these remarks with examples 2.8 and 2.9 in hexadecimal notation.

**Example 2.13** *Using the hexadecimal system, add* (a) *27 and 18, and* (b) *152.75 and 236.375; and* (c) *subtract 137 from 216.*
(a) *The binary addition of 27 = 1Bh and 18 = 12h is illustrated next.*

$$
\begin{array}{rclcl}
27 & + & \rightarrow & 1\text{Bh} & + \\
18 & = & \rightarrow & 12\text{h} & = \\
\overline{45} & & \rightarrow & \overline{2\text{Dh}} &
\end{array}
$$

*For this addition, we can say 'B – eleven – plus 2 is thirteen – D –' and '1+1 = 2'.*
(b) *Since* 152.75 = 98CH *and* 236.375 = EC.6H, *the addition is illustrated next.*

$$
\begin{array}{lclcl}
\text{Carries} & & & 1\,1\,1.0 & \\
152.75 & + & \rightarrow & 98.\text{Ch} & + \\
236.375 & = & \rightarrow & \text{EC.6h} & = \\
\overline{389.125} & & \rightarrow & \overline{1\,8\,5.2\text{h}} &
\end{array}
$$

*Reading:* '12 (C) *plus* 6 *is* 18 (12h) *which makes* 2 *and carry* 1 *in hex.* 1 + 8 *is* 9, *plus* 12 (C), *is* 21 (15 h), *sum* 5 *and carry* 1, *and so on.*
(c) *Since* 137 = 89 h *and* 216 = D8h, *we have, with the least significant borrow* 0 *by default:*

$$
\begin{array}{lrclcl}
\text{minuend} & 216 & - & \rightarrow & \text{D 8h} & - \\
\text{subtrahend} & 137 & = & \rightarrow & 8\ 9\text{h} & \\
\text{Borrows} & 110 & & & 1\ 0\text{h} & = \\
& \overline{79} & & \rightarrow & \overline{4\ \text{Fh}} &
\end{array}
$$

*Again, let us read this subtraction:* 8 *minus* 9 *needs a borrow,* 8 + <u>*sixteen from borrow*</u>, 24, *minus* 9 *makes* 15 (F). *D is* 13, *minus* 8 *minus the borrow* 1 *yields* 4.

**Sixteen's Complements and Binary Two's Complements in Hex Notation**

The concept of $10^N - X$ as complement of $X$ when this one consists of N digits is independent of the base. The base is used as a reference to talk about the system. Thus, decimal if base ten, binary if base two, and so on. Hence, for hexadecimal systems we talk of *sixteen's complement* of $A$. Now, since there is a one to one correspondence to binary numbers, it can be shown that if $A$ is transformed into binary expressions, than the complement will be the hex equivalent of the two's complement. Hence, one way to work fast the two's complements of bytes and words, is to work with their hex equivalents directly. A simple procedure to get the complement of a hex expression is as follows:
**Sixteen's Complement:** find the difference for each hex digit with respect to F and then add 1 to the result.

**Example 2.14** *A set of data for bytes and words is given in hex notation as* $0\times7A, 0\times9F24, 91H$ *and* CFA20h *Find the two's complement of each one using hex notation and verify the response in binary.*
*For* $0\times7A$ *we find* 100 h - 7Ah *as follows:* (FFh − 7Ah = 85 h) *and* (85h + 1) = 86 h. *Hence, the hex expression for the two's complement is* 86 h. *To verify:*

$$7Ah = 01111010B \Rightarrow \text{Two's Complement} : 10000110B = 86\,h$$

*For* 9F24h, *take the difference for each digit with respect to* F *and add* 1 : 60 DBh+ 1 = 60 DCh.
*For* 91h : 6Eh + 1 = 6 Fh.
*For* 0xCFA20 : 305DFh + 1 = 305 E0h.

*The verification in binary form is left to the reader.*

**Important Remark:** Hex notation not always represents four bits as is the case of the most significant hex digit. For example, 1EFh may be the hex representation for a binary expression of 12, 11, 10 or 9 bits and the two's complement will be different in each case. The hex subtraction to consider is different, depending on the case: 1000, 800, 400 and 200 h, respectively, since these are the hex expressions for $2^{12}$, $2^{11}$, $2^{10}$ and $2^9$.

**Example 2.15** *The hex expression* 26ACh *is being used to represent a binary number. Find the hex expression of the binary two's complement if* (a) *we are working with a* 16-*bit,* (b) 15-*bit, and* (c) 14-*bit binary number.*

(a) *In the 16-bit number, all hex digits are true digits, so we use the same method as in example 2.14 above:* 26ACh $\Rightarrow$ D953h + 1 = D954h.
(b) *For the 15-bit number, the most significant hex digit represents actually three bits, so the complement is with respect to* 8000h = 7FFFh + 1. *Hence, for the digit* 2 *we take* 7 *as the initial minuend, getting,* 5953h + 1 $\Rightarrow$ 5954h *as result.*
(c) *In the 14-bit number, the most significant hex digit represents actually two bits, so the complement is with respect to* 4000h = 3FFFh + 1. *Hence, for the digit* 2 *we take* 3 *as the initial minuend, getting,* 1953h + 1 $\Rightarrow$ 1954h *as result.*

*The reader can verify the results by transforming to binary and taking two's complement.*

## 2.6  Representation of Numbers in Embedded Systems

Let us now see some methods to represent numbers in a digital environment. First for *integers* and then for *real* numbers; these are those containing a fractional component. Beginners may skip the representation of real numbers in the first experience, and come back to this chapter when the concepts may become necessary.

If only non-negative numbers are considered in the machine representation, we speak of *unsigned numbers*. If negative, zero and positive numbers are considered, then we talk of *signed numbers*. It is always important to define what type of representations we are considering, since this fact defines the interpretation of the results as well as the programming algorithms.

When trying to represent numbers with a digital system we have important constraints which can be summarized as follows:

1. Only 0's and 1's are allowed. No other symbols such as minus sign, point, radix, exponent, are available.
2. In a given environment or problem, all representations have a fixed length of $N$ bits, so only a finite set of $2^N$ numbers can be represented.
3. Arithmetic operations may yield meaningless results irrespectively of their theoretical mathematical validity.

The first constraint follows from the very nature of digital systems, since we are using the "0" and "1" associated to voltage states. The second constraint tells us that the number of bits cannot be extended indefinitely and all numbers have the same number of bits, $N$, including left zeros. This value may be dictated either by hardware constraints or by a careful planning in our application. In any case, there is finite list of representable numbers, no matter how large or small the number is and we can always point out the "next" number following or preceding a given one. Thus, some properties of real numbers such as density[5] are lost.

The third constraint is exemplified by *overflow*, which occurs when the result of an operation is beyond the finite set of allowable numbers representable by finite-length words. An example in daily life of this fact can be appreciated in the car odometers, with a limited number of digits. An odometer with 4 digits cannot represent more than 9999 miles. Hence, adding 30 miles to 9990 does not yield 10020, but only 0020. We will have the opportunity to take a closer look to this phenomena.

## 2.7 Unsigned or Nonnegative Integer Representation

There are several ways in which unsigned integers can be represented in digital systems. The most common ones are the normal binary method and the 8421 Binary Coded Digit representation. Others such as Gray code, excess codes, etc., are suitable for certain applications.

### 2.7.1 Normal Binary Representation

In this convention, $n$-bit words are interpreted as normal nonnegative binary numbers in a positional system as discussed in Sect. 2.2.1. The interval of integers represented goes from 0 to $2^n - 1$. In particular,

- For bytes, n = 8 : 0 (00h) to 255 (0xFF)
- For words, n = 16 : 0 (0000h) to 65,535 (0xFFFF)
- For double words, n = 32 : 0 (00000000h) to 4,294,967,295 (0xFFFFFFFF)

---

[5] Density of the real number system (or of rational numbers) means that between any two numbers there is an infinite quantity of real numbers.

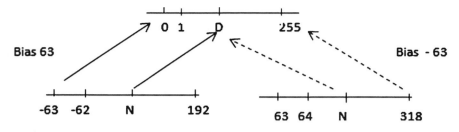

**Fig. 2.3** Biased representation for number $N : N + B = D$

One advantage of this representation is that the hardware for the arithmetic addition is usually designed with a positional system in mind. Therefore, when using other representations it is often necessary to convert into normal binary to do the arithmetic, or else devise a special algorithm. When limited to $n$ bits using unsigned representation, overflow in addition occurs when there is a carry.

### 2.7.2  Biased Binary Representation

To work "representable" number intervals that fall outside the normal binary interpretations of $n$-bit words, the convention of *biased* or *excess* notation is very useful. By "representable" intervals, we mean that the number of elements in the set is less than or equal to $2^n$, so it is possible to assign an $n$-bit word to each element. The interval represented with bias $B$ is $[-B, 2^n - B - 1]$. We illustrate the principle of bias representation with Fig. 2.3 for a byte.

If the number we want to represent is $N$, we add a bias $B$ to this number and map it to $D$, which is then coded by normal binary convention. In other words:

$$N + B = D \tag{2.7}$$

Hence, we are mapping the interval of interest so that the lower bound corresponds to 0. Thus, the word 0 does not represent a decimal 0, but a nonzero integer $-B$, where $B$ is the *bias*. The bias is also sometimes called *excess*, although the number can be negative.

**Example 2.16**  (a) *Represent* 35 *with bias* 63 *and bias* 127; (b) *What is the decimal equivalent of* 110011 *if it is biased by* 63?

(a)  *For bias* 63, *we must find the normal binary form for* $35 + 63 = 98$, *which is* 1100010. *For bias* 127 *we should work* $35 + 127 = 162$, *for which the normal binary equivalent is* 10100010.

(b)  *The decimal equivalent for a normal binary coding* 110011 *is* 51. *Since it is in excess* 63, *the actual equivalent is* $51 - 63 = -12$.

**Example 2.17** *What is the interval that can be coded with bytes using a bias of (a) 127 and (b) −63?*

(a) *The normal binary codification for bytes goes from 0 to 255. Therefore, with bias 127, the range is from −127 to 128.*
(b) *Proceeding as before, i.e., subtracting the bias from the bounds, the range is +63 to +318.*

Using biased representations, we can include negative numbers, as seen in the above examples, or any other interval without the need of increasing the number of bits. Notice however that when arithmetic operations are involved, we need to compensate for the bias effect. This is usually done by software.

## 2.7.3 Binary Coded Decimal Representation -BCD-

In the *Binary Coded Decimal* representation (BCD), each digit of a decimal number is represented separately by a 4-bit binary string. It is common to use its equivalent 4-bit binary number, as shown in Table 2.2. This format is also referred to as *"8421 BCD"* to emphasize the normal weights for the individual bits: 8, 4, 2, and 1. This also differentiates it from other encoding by digits using other weights for the bits such as the 8 4 −2 −1, or codes by excess. Notice that 8421–BCD representation is similar to hex notation, except that nibbles 1010 to 1111 are excluded. This will be appreciated in the example below.

BCD encoding is very important, and it is the preferred one for many applications, especially those in which the exact representation of decimal numbers becomes an absolute need.[6] Many electronic systems are especially designed in hardware to work in this system, although most small embedded ones are based on binary hardware.

It is sometimes convenient to use bytes even when coding BCD. When using a byte, if the least significant nibble is used for the code and the most significant nibble has all 0's or all 1's, the representation is a *zoned BCD* or *unpacked BCD*. If the two nibbles of the byte are used for coding, it is a *packed BCD.* or *compressed BCD*.

**Example 2.18** *Using bytes, represent the decimal 236 in (a) binary, (b) normal BCD and (c) packed BCD.*

(a) *By conversion, since 236 < 256, one byte suffices and 236 = 11101100b(0 × EC)*
(b) *In normal BCD, three bytes will be required. Using the most significant nibble all 0's we get the following representation*

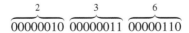

$$\underbrace{00000010}_{2} \ \underbrace{00000011}_{3} \ \underbrace{00000110}_{6}$$

*In hex notation:* 02 h, 03 h, *and* 06 h

---

[6] An introductory website for what is called *Decimal Arithmetic*, which makes all operations based on BCD encoding, is found at http://speleotrove.com/decimal/.

*(c)  In packed BCD, two bytes will be necessary, with the most significant nibble taken as 0:*

$$\overbrace{0000}^{0}\,\overbrace{0010}^{2}\,\,\overbrace{0011}^{3}\,\overbrace{0110}^{6}$$

*That is,* 02 h 36 h

Sometimes, performing arithmetic directly from BCD without going through BCD-to-binary and binary-to-BCD conversions may be advantageous. However, even though the digits are individually coded, actual operations are done in binary arithmetic. There is hence the need to understand how to manipulate the binary operations in this case to yield the BCD representation of the actual result. This is explained next. For easiness, we use hexadecimal notation for the nibbles, recalling that for digits 0-9 both BCD and hex notations are identical.

**BCD Arithmetic: Addition**
When two decimal digits -or two digits plus a carry—are added, the difference in the result with respect to the hex notation appears as soon as the addition is greater than 9. That is, whenever the decimal addition falls between 10 and 19, generating a digit and a carry, which corresponds to 0Ah to 13h in the hexadecimal addition. To obtain the digits again, adding 6 is required, since 0Ah + 6h = 10h, and 13h + 6h = 19h. We summarize with the following
**Rule for BCD addition**: When adding hex digits (nibbles) in BCD representation

1. If the sum is between 0 and 9, leave it as it is;
2. Else, if it is between Ah to 13h, add 6 to the result of that sum.
3. These rules are applied separately to each column –units, tens, etc –.

**Example 2.19**  *Illustrate the addition rules for BCD representation with the hex digit representation for 32 + 46, 56 + 35, and 85 + 54; 67 + 93; 99 + 88. Since 32 + 46 yields no result greater than 9 in either column, the addition here is direct. On the other hand, the units column for 56 + 35 is greater than 10, so 6 should be added to that column. We omit carries for simplicity:*

| Decimal | Hex   |  | Decimal | Hex   |
|---------|-------|--|---------|-------|
| 32 +    | 32h + |  | 56 +    | 56h + |
| 46 =    | 46h = |  | 35 =    | 35h = |
| 78      | 78h   |  | 91      | 8Bh+  |
|         |       |  |         | 06h = |
|         |       |  |         | 91h   |

Next, 85 + 54 *yields a result greater than 9 in the tens column, so we add* 60h *to the hex addition. On the other hand,* 67 + 93 *and* 99 + 88 *have both columns yielding results greater than 9 and we need then to add 66h to the hex addition in order to obtain the correct BCD representation as the final sum:*

| Decimal | Hex    |
|---------|--------|
| 85 +    | 85h +  |
| 54 =    | 54h =  |
| 139     | D9h    |
|         | 60h =  |
|         | 139h   |

| Decimal | Hex    |
|---------|--------|
| 67 +    | 67h +  |
| 93 =    | 93h =  |
| 160     | FAh +  |
|         | 66h =  |
|         | 160h   |

| Decimal | Hex    |
|---------|--------|
| 99 +    | 99h +  |
| 88 =    | 88h =  |
| 187     | 121h+  |
|         | 66h =  |
|         | 187h   |

**BCD Arithmetic: Subtraction**

When taking the difference of decimal digits, the difference in the result with respect to the hex notation appears as soon as there is a need of a borrow. Since a borrow in hexadecimal system means 16 decimal units, while in decimal system means 10 units, we subtract 6 units more. Summarizing

**Rule for BCD subtraction**: When subtracting hex digits in BCD code,

1. If the minuend is greater or equal than the subtrahend, including a borrow, leave the difference that results;
2. Else, if a borrow is needed, subtract another 6h to the difference.
3. These rules are applied to each column –units, tens, etc –.

**Example 2.20** *Illustrate the subtraction rules for BCD representation with the hex digit representation for* 59 – 27, 83 – 37, 129 – 65, *and* 141 – 48. *The subtractions are done below, without explicitly showing the borrows. The reader is invited to verify the operations*

| Decimal | Hex    |
|---------|--------|
| 59 –    | 59h –  |
| 27 =    | 27h =  |
| 32      | 32h    |

| Decimal | Hex    |
|---------|--------|
| 83 –    | 83h –  |
| 37 =    | 37h =  |
| 46      | 4Ch –  |
|         | 06h =  |
|         | 46h    |

| Decimal | Hex    |
|---------|--------|
| 129 –   | 129h – |
| 65 =    | 65h =  |
| 64      | C4h    |
|         | 60h =  |
|         | 64h    |

| Decimal | Hex    |
|---------|--------|
| 141 –   | 141h – |
| 48 =    | 48h =  |
| 93      | F9h –  |
|         | 66h =  |
|         | 93h    |

# 2.8 Two's Complement Signed Integers Representation:

To represent positive, negative, and zero numbers using strings of 0's and 1's, there are different methods, each one with its own advantages for particular applications. The most popular convention is the *two's complement* method, based on the two's complement operation introduced in Sect. 2.5.3. This method has the following important characteristics:

**Number of bits:** All strings must have the same number of bits and any interpretation of results must be limited to that length.

**Range** Of the total set of $2^n$ words, one half correspond to negative number representations. With $n$ bits, the interval of integers is between $-2^{n-1}$ and $2^{n-1} - 1$.

**Backward compatibility and two's complement:** To keep previous results with unsigned representations, we have the following constraints:

1. Representation of nonnegative integers are equivalent to their unsigned counterparts in normal binary system.
2. The binary representations for A and $-A$ are two's complements of each other.
3. The representation of $-2^{n-1}$, has no two's complement within the set, so this number is represented by 1 followed by $n-1$ 0's.

**Sign bit:** If the most significant bit is 0, the number is non-negative, if it is 1 the number is negative. The most significant bit is called *sign bit*.

**Backward compatibility 2:** Addition and subtraction follow the same rules as in the unsigned case.

Let us see how these principles generate the signed set of integers with four bits.

**Example 2.21** *Derive the set of signed integers representable with 4 bits using the two's complement conditions. The range of integers is from $-2^{4-1} = -8$ to $2^{4-1} - 1 = +7$. From the backward compatibility property, integers 0 to $+7$ have the representation 0000, 0001, ..., 0111. Integers $-1, -2, \ldots -7$ are the two's complements for those strings. These representations are summarized in the following table:*

| Decimal | $\rightarrow$ | Nibble | Decimal | $\rightarrow$ | Nibble |
|---------|---------------|--------|---------|---------------|--------|
| +1 | $\rightarrow$ | 0001 | −1 | $\rightarrow$ | 1111 |
| +2 | $\rightarrow$ | 0010 | −2 | $\rightarrow$ | 1110 |
| +3 | $\rightarrow$ | 0011 | −3 | $\rightarrow$ | 1101 |
| +4 | $\rightarrow$ | 0100 | −4 | $\rightarrow$ | 1100 |
| +5 | $\rightarrow$ | 0101 | −5 | $\rightarrow$ | 1011 |
| +6 | $\rightarrow$ | 0110 | −6 | $\rightarrow$ | 1010 |
| +7 | $\rightarrow$ | 0111 | −7 | $\rightarrow$ | 1001 |

*On the other hand, 1000 represents $-2^3 = -8$.*

*Notice that 0000 and 1000 are, respectively, two's complements of themselves. Thus, it is not possible to have representations for both $+8$ and $-8$ in this set.*

Because half of the integers are nonnegative, and must have the same representation as in the unsigned case, all of them are represented with a leading 0, covering exactly half of the set. Hence, all the rest, i.e., the words representing negative numbers, have the most significant bit equal to 1. This is why the most significant bit indicates whether the number is negative or nonnegative.

On the other hand, since the addition and subtraction rules are valid, we can decompose a string as follows:

$$b_{n-1}b_{n-2}\cdots b_1b_0 = \underbrace{b_{n-1}00\cdots 0}_{\text{msb followed by n−1 zeros}} + 0b_{n-2}\cdots b_1b_0$$

Now, since the second term in the right is a "normal" binary representation of a positive number, for which the power series (2.3) is applicable, and the first term is 0 if $b_{n-1} = 0$, and $-2^{n-1}$ if $b_{n-1} = 1$, we arrive to the following conclusion:

When the word $A = b_{n-1}b_{n-1}\cdots b_1b_0$ represents a signed number in two's complement convention, it can be expanded in the following weighted sum, similar to the positional expansion:

$$A = -2^{n-1}\cdot b_{n-1}+2^{n-2}\cdot b_{n-2}+\cdots+2^1\cdot b_1+b_0 = -2^{n-1}\cdot b_{n-1} + \sum_{i=0}^{n-2} 2^i\cdot b_i \quad (2.8)$$

For example, 10110111 represents $-128 + (32 + 16 + 4 + 2 + 1) = -73$.

For bytes, words and double words, the range of signed integers becomes, respectively:

- Byte: $-2^7 = -128$ to $2^7 - 1 = +127$;
- Word: $-2^{15} = -32,768$ to $2^{15} - 1 = +32,767$
- Double Word: $-2^{31} = -2,147,403,608$ to $2^{31} - 1 = +2,147,403,607$

**Sign Extension:** Signed numbers using $n$-bit words can be extended to $m$-bit words, with $m > n$ bit strings, by appending to the left the necessary number of bits, all equal to the original sign bit. This procedure is called *sign extension*.

Applying, $-5$ with four bits has the code 1011 and with six bits it becomes 111011, while $+5$ becomes 000101. The proof of this property is left as an exercise.

The weighted sum (2.8), together with sign extension when applicable, can be applied for reading and generating representations of negative numbers. If the sign bit is 0, the number is non-negative and can be read as always. If the sign bit is 1, reduce the length scanning from left to right until you find the first 0. Read the "positive" component formed by the $M$ remaining bits, including the 0, and subtract it from $2^M$. For example, to read 1111101, consider only 01, which stands for 1, and $2^2 = 4$, so the result is $-(4 - 1) = -3$.

Let us illustrate again reading numbers and generating opposites with examples. This type of exercises become useful when working with programs and validating them. Remember that the programmer needs to verify the correctness by working examples with data that will be used in the process.

**Example 2.22** *Find the signed decimal integer represented by the 12-bit word FD7h. Various methods are illustrated:*

(a) *Since FD7h represents 111111010111 and the sign bit is 1, the number is negative. By the sign extension principle, this number would be the same one represented with 7 bits,* 1010111. *Here,* 010111B *is the binary representation for* 23, *so the expansion yields* $-2^6 + 23 = -41$. *Hence,* FD7h *is the representation for* $-41$.

(b) *Logical complement plus one: The two's complement of* 111111010111 *is*

$$000000101000 + 1 = 101001$$

*which is the binary equivalent of* 41. *Hence,* FD7h *stands for* -41.

(c) *Working with hex complements: The hex complement for* FD7h *with* 12 *bits is* 029h, *with decimal equivalent* $2 \times 16 + 9 = 41$. *Hence,* FD7h *stands for* $-41$

While the previous example was concerned with reading representations, now let us work the conversion process from signed decimal expressions. Positive ones have been previously considered, so let's just take the negative numbers now.

**Example 2.23** *Find the two's complement representation for* $-104$ *with* 16 *bits and express it in hex notation.*

*We solve by three methods:*

(a) *By the two's complement of* 104: *With* 16 *bits,* $+104$ *is* 0000000001101000. *The two's complement becomes then* 1111 1111 1001 1000, *or* 0×FF98.

(b) *Hex complement directly:* $+104 = 0068h$ *and the hex complement becomes* 0×FF98h.

(c) *By the weighted sum, starting with the minimum number of bits required and then sign extending: Since* $-104 > -2^{8-1} = -128$, *eight bits are enough. Taking (2.8) as a reference,* $128 - 104 = 24 = 11000B$. *Therefore, inserting* 0's *as needed, with eight bits we have* $-128 + 24 = -104 \to 10011000$. *After sign extension with eight additional* 1's *to the left to complete the* 16 *bits we arrive at the hex notation* 0xFF98.

**Example 2.24** *A byte* 83h *is used to represent numbers, but we need to extend the representation to* 16-*bit word size. What is the new representation and meaning, if (a) we are working with unsigned numbers, and (b) we are working with signed numbers.*

(a) *Unsigned numbers are extended to any number of bits by simply adding leading zeros. Hence, the new representation would be* 0×0083. *The decimal equivalent in both cases is* $8 \times 16 + 3 = 131$.

(b) *The most significant bit is* 1, *so the number is negative,* 83h $\to$ $-128 + 3 = -125$. *Since we are working with signed numbers, the sign extension rule should be applied, and the new representation becomes* 0×FF83.

## 2.8.1 Arithmetic Operations with Signed Numbers and Overflow

When adding and subtracting signed number representations, a carry or borrow may appear. This one is discarded if we are to keep the fixed length representation. *Overflow* will occur, however, if in this fixed length the result is nonsense. In particular;

- Addition of two numbers of the same sign should yield a result with the same sign;
- a positive number minus a negative number should yield a positive result and
- a negative number minus a positive number should yield a negative result.

When the results do not comply with any of these conditions, there is an overflow. More explicitly:

**Overflow** occurs if (a) addition of two numbers of the same sign yields a number of opposite sign or (b) subtraction involving different signed numbers yields a difference with the sign of the subtrahend.

Let us illustrate operations and overflow with 4-bit numbers, which were generated in Sect. 2.8.

**Example 2.25**    (a) *Check the validity of the following operations for signed numbers using two's complement convention with four bits:* $3 + 2, 4 + (-4), (-6) + 7,$ $(-3)+(-5), 6-2, 4-4, (-2)-(-8), 3-(-4).(b)$ *Verify overflow in the following cases:* $3 + 5, (-5) + (-8), 4 - (-6), (-6) - (+3).$

(a) *All the following operations yield valid results discarding any carry or borrow when present. For example,* $3 - (-4)$ *in binary yields* $10111,$ *but only* $0111$ *is considered yielding* $+7,$ *as expected. Notice that* $(+4) + (-4)$ *and* $4-4$ *both yield* $0$ *when only four bits are taken, but the former yields a carry.*

| | | | | | |
|---|---|---|---|---|---|
| $3+$ | $\rightarrow$ | $0011+$ | $4+$ | $\rightarrow$ | $0100+$ |
| $2 =$ | $\rightarrow$ | $0010 =$ | $(-4) =$ | $\rightarrow$ | $1100 =$ |
| $5$ | $\rightarrow$ | $0101$ | $0$ | $\rightarrow$ | $1\,0000$ |
| $(-6)+$ | $\rightarrow$ | $1010+$ | $(-3)+$ | $\rightarrow$ | $1101+$ |
| $7 =$ | $\rightarrow$ | $0111 =$ | $(-5) =$ | $\rightarrow$ | $1011 =$ |
| $1$ | $\rightarrow$ | $1\,0001$ | $(-8)$ | $\rightarrow$ | $1\,1000$ |
| $6-$ | $\rightarrow$ | $0110-$ | $4-$ | $\rightarrow$ | $0100-$ |
| $2 =$ | $\rightarrow$ | $0010 =$ | $4 =$ | $\rightarrow$ | $0100 =$ |
| $4$ | $\rightarrow$ | $0100$ | $0$ | $\rightarrow$ | $0000$ |
| $(-2)-$ | $\rightarrow$ | $1110-$ | $3-$ | $\rightarrow$ | $0011-$ |
| $(-8) =$ | $\rightarrow$ | $1000 =$ | $(-4) =$ | $\rightarrow$ | $1100 =$ |
| $6$ | $\rightarrow$ | $0110$ | $7$ | $\rightarrow$ | $1\,0111$ |

(b) *Overflow now occurs when the result of operation falls outside the range covered by the set of strings and is mainly shown by a result with a sign bit different from what was expected. Since 4-bit words cover from* $-2^3 = -8$ *to* $2^3 - 1 = 7,$ *the operations* $3 + 5 = 8, (-5) + (-8) = (-13)$ *and* $4 - (-6) = 10$ *do not make sense in this set. Let us look at the results in binary form interpreted from the standpoint of the two's complement convention:*

| | | | | | | | | |
|---|---|---|---|---|---|---|---|---|
| 0011+ | → | +3 | 1011+ | → | −5 | 0100− | → | +4 |
| 0101 = | → | +5 | 1000 = | → | −8 | 1100 = | → | −6 |
| 1000 | → | −8 | 10011 | → | +3 | 11010 | → | −6 |

*We see in the first case an addition of two numbers with leading bit 0 (non negative) yielding a number with a sign bit 1. In the second case, two negative numbers add up to a positive result. In the third case, we subtract a negative number from a positive one, but the result is negative instead of positive. All these cases are deduced after analyzing the sign bits of the operands and results.*

### 2.8.2 Subtraction as Addition of Two's Complement

From elementary Algebra it is known that $A - B = A + (-B)$. On the other hand, for the two's complement signed integer representations in digital systems, $-B = \overline{B} + 1$. Therefore, the subtraction operation can be reduced to an addition in the form

$$A - B = A + \overline{B} + 1 \tag{2.9}$$

In other words, both in unsigned and signed conventions, using two's complement addition for a subtraction is valid. The main difference is that this principle is limited to the $n$ bits used in the signed representation and the result is directly interpreted for both positive and negative cases.

This principle allows us to use the same hardware for both representations. When performed in software, it is the programmer's responsibility to interpret results. One important fact to not forget is that in two's complement addition for a subtraction, carry and borrow are opposite. Look at cases $4 - 4$ and $4 + (-4)$ in the example above. This difference should be taken into account when programming.

**Remark:** In the MSP430 microcontrollers, as in most microcontrollers, subtraction is actually realized through the addition of the two's complement of the subtrahend. Hence, a carry $= 0$ corresponds to the presence of a borrow.

## 2.9 Real Numbers

Previous sections emphasized the representation of integers using $N$-bit words. Now we consider real numbers, that is, numbers containing a fractional part. Two methods are the most common: the fixed-point and the floating-point representations.

**Fig. 2.4** Format $Fm_1 \cdot m_2$ for an $n$-bit word, where $m_1 + m_2 = n$

| Integer part, $m_1$ bits | Fractional part, $m_2$ bits |
|---|---|
| $b_{N-1} \quad b_{N-2} \cdots b_{m_2}$ | $b_{m_2-1} \quad \cdots \quad b_0$ |

Fixed-point notation is preferred in situations where huge quantities of numbers are handled, and the values are neither very large nor very small. An advantage is that operations are the same as with integers, without changes in hardware, since the only difference is the presence of an implicit point. On the other hand, very large and very small numbers require a large number of bits and floating-point is preferred. The arithmetic with floating point number representation is more complicated than that of integer numbers and will be left out of the scope of this text.

### 2.9.1 Fixed-Point Representation

In a *fixed-point notation*, or a *fixed-point representation system*, numbers are represented with $n$-bit words assuming the radix point to always be in the same position, established by convention. The notation can be used to represent either unsigned or signed real numbers.

Let $A$ be the $n$-bit word $b_{n-1}b_{n-2}\ldots b_1 b_0$ and $n = m_1 + m_2$. The format $Fm_1 \cdot m_2$ for $A$ means that the $m_2$ least significant bits constitute the fractional part and the $m_1$ most significant bits the integer part, as illustrated in Fig. 2.4.

From this interpretation and the power expansion (2.3) as well as the expansion (2.8) for signed numbers, we have

$$
\begin{aligned}
A &= \left( b_{n-1}\, b_{n-2} \cdots b_{m_2} b_{m_2-1} \cdots b_1 b_0 \right)_{Fm_1 \cdot m_2} \\
&= \pm b_{n-1} 2^{m_1-1} + b_{n-2} 2^{m_1-2} + \cdots \\
&\quad + b_{m_2} + b_{m_2-1} 2^{-1} + b_{m_2-2} 2^{-2} + \cdots b_0 2^{-m_2}
\end{aligned}
\tag{2.10}
$$

Factoring $2^{-m_2}$, this can also be written as

$$
\begin{aligned}
A &= (b_{n-1} b_{Nn-2} \cdots b_1 b_0)_{Fm_1 \cdot m_2} \\
&= \frac{1}{2^{m_2}} \times \left( \pm b_{n-1} \times 2^{n-1} + b_{n-2} \times 2^{n-2} + \cdots + b_1 \times 2^1 + b_0 \right)
\end{aligned}
\tag{2.11}
$$

In these equations, the plus $(+)$ sign in the highest term applies to the unsigned case and the minus $(-)$ to the signed one.

Two points are worth mentioning here. First, numbers increase in *steps* of $1/2^{m_2}$. So density in the approximation depends largely on the number of bits in the fractional part. Second, the number being represented is equal to the integer equivalent of the word divided by $2^{m_2}$, both for signed and unsigned numbers. Based on this last property, the interval represented by $n$-bit word $b_{n-1}b_{n-2}\cdots b_1 b_0$ in format $Fm_1 \cdot m_2$ is

- 0 to $(2^n - 1)/2^{m_2}$ for unsigned numbers, and
- $-2^{-m_2} \cdot 2^{n-1}$ to $2^{-m_2} \cdot (2^{n-1} - 1)$ for signed numbers

Let us now work some examples.

**Example 2.26** *(a) Describe the unsigned numbers represented by 4-bit words in format F2.2 and format F0.4. (b) Repeat for signed numbers.*

(a) **Unsigned numbers**: *Format F2.2 covers from 0 up to $2^{-2} \times (2^4 - 1) = 3.75$ in step size of $2^{-2} = 0.25$. On the other hand, format F0.4 covers from 0 to $2^{-4} \times (2^4 - 1) = 0.9375$ with a step size of $2^{-4} = 0.0625$.*

(b) **Signed numbers**: *The F2.2 format range is from $-2^{-2} \times 2^3 = -2$ to $2^{-2} \times (2^3 - 1) = 1.75$ in steps of $2^{-2} = 0.25$. On the other hand, the F0.4 format range is from $-2^{-4} \times 2^3 = -0.5$ to $2^{-4} \times (2^3 - 1) = 0.4375$ in steps of $2^{-4} = 0.0625$. With this information, we can generate the representations for each case as described in the following tables:*

| Unsigned numbers | | | | Signed numbers | | | |
|---|---|---|---|---|---|---|---|
| Format | F2.2 | Format | F0.4 | Format | F2.2 | Format | F0.4 |
| 0000 | 0.00 | 0000 | 0.0000 | 1000 | −2.00 | 1000 | −0.5000 |
| 0001 | 0.25 | 0001 | 0.0625 | 1001 | −1.75 | 1001 | −0.4375 |
| 0010 | 0.50 | 0010 | 0.1250 | 1010 | −1.50 | 1010 | −0.3750 |
| 0011 | 0.75 | 0011 | 0.1875 | 1011 | −1.25 | 1011 | −0.3125 |
| 0100 | 1.00 | 0100 | 0.2500 | 1100 | −1.00 | 1100 | −0.2500 |
| 0101 | 1.25 | 0101 | 0.3125 | 1101 | −0.75 | 1101 | −0.1875 |
| 0110 | 1.50 | 0110 | 0.3750 | 1110 | −0.50 | 1110 | −0.1250 |
| 0111 | 1.75 | 0111 | 0.4375 | 1111 | −0.25 | 1111 | −0.0625 |
| 1000 | 2.00 | 1000 | 0.5000 | 0000 | 0.00 | 0000 | 0.0000 |
| 1001 | 2.25 | 1001 | 0.5625 | 0001 | 0.25 | 0001 | 0.0625 |
| 1010 | 2.50 | 1010 | 0.6250 | 0010 | 0.50 | 0010 | 0.1250 |
| 1011 | 2.75 | 1011 | 0.6875 | 0011 | 0.75 | 0011 | 0.1875 |
| 1100 | 3.00 | 1100 | 0.7500 | 0100 | 1.00 | 0100 | 0.2500 |
| 1101 | 3.25 | 1101 | 0.8125 | 0101 | 1.25 | 0101 | 0.3125 |
| 1110 | 3.50 | 1110 | 0.8750 | 0110 | 1.50 | 0110 | 0.3750 |
| 1111 | 3.75 | 1111 | 0.9375 | 0111 | 1.75 | 0111 | 0.4375 |

**Example 2.27** *What is the range of unsigned and signed numbers covered with a 16-bit word in formats F8.8 and F0.16? What is the step in each case? The steps for the F8.8 and F0.16 formats are, respectively, $2^{-8} = 0.00390625 \approx 3.91 \times 10^{-3}$ and $2^{-16} = 0.0000152587290625 \approx 1.53 \times 10^{-5}$. The intervals are shown in the following table.*

| Number type | Format | Lower limit | Upper limit |
|---|---|---|---|
| Unsigned | F8.8 | 0 | $\dfrac{2^{16} - 1}{2^8} = 255.9960938$ |
| Unsigned | F0.16 | 0 | $\dfrac{2^{16} - 1}{2^{16}} = 0.9999847412109375$ |
| Signed | F8.8 | $-\dfrac{2^{15}}{2^8} = -128$ | $\dfrac{2^{15} - 1}{2^8} = 127.99609375$ |
| Signed | F0.16 | $-\dfrac{2^{15}}{2^{16}} = -0.5$ | $\dfrac{2^{15} - 1}{2^{16}} = 0.4999847412109375$ |

These examples show how the number of bits in the fractional part defines the precision between steps. This precision gets better at the expense of the total range. We can also appreciate the effect of the step. In the example for unsigned numbers in format F0.4, for example, 0110 can be used to represent any number between 0.3750 up to, but not including, 0.4375. Hence, the step indicates the maximum error which can be introduced with this representation. Therefore, *to consistently keep an error lower than $\epsilon$, the number of bits $m_2$ in the fractional part must satisfy $2^{m_2} \geq \epsilon$.*

**Very small numbers:** For very small numbers without integer part *extra scaling* may be used. This consists in implicit additional zeros to the left after the point, including the step. Thus, with an extra scaling of three decimals, the range for signed numbers in the format F0.4 (see example 2.26) becomes [–0.0005, 0.0004375].

### Two's complement in fixed point representation

When working (unsigned) integers with $n$ bits, it was pointed out that the two's complement of $A$ is $2^n - A$. Now, for format Fm1.m2, following (2.11) it can be deduced that the difference is now to be considered with respect to $2^{m1}$. This can be seen in Example 2.26. In format F2.2, numbers adding $2^2 = 4$ are represented by two's complement strings. The same can be said for F0.4 and $2^0 = 1$.

**Example 2.28**  *What is the signed decimal equivalent for* 10110010 *if the string is in format F3.5?*

*(a) By power expansion,*

$$10110010 \rightarrow -2^2 + 1 + 0.5 + 0.0625 = -2.4375$$

*(b) By two's complement*

$$01001110 \rightarrow -(2 + 0.25 + 0.125 + 0.0625) = -2.4375$$

*(c) By two's complement in decimal representation:*

$$10110010 = 5.5625 \rightarrow -(8 - 5.5625) = -2.4375$$

*(d) By integer division:*

$$10110010 = -78 \rightarrow \frac{-78}{32} = -2.4375$$

## 2.9.2 Floating-Point Representation

Floating-point is preferred for very large or very small numbers, or to cover large intervals and better precision when two large numbers are divided. There are several conventions for floating point representation, but IEEE 754 Floating Point Standard is the most common format. We discuss this one here. Texas Instruments uses both this one and another one adjusted to its products.

Floating-point representation is similar to the scientific notation, where a number is written in the form $b.a_{-1}a_{-2}\ldots \times 10^m$, with $m$ an integer and $0 \le |b| \le 9$. For example, 193.4562 may be expressed as $1.934562 \times 10^2$, and $0.00132 = 1.32 \times 10^{-3}$. When $b \ne 0$, we say that the decimal point is in *normal position* and the notation is *normalized*. The number $b$ is the *base* and the fractional part $a_{-1}a_{-2}\ldots$ the *mantissa*.

A similar principle is used for binary numbers. For easiness in reading, we use mixed notation, with the base in binary and power of 2 in decimal, as in $110.101 = 1.10101 \times 2^2$. As an extension, the mantissa is *normalized* if the decimal point is normalized, and *unnormalized* or *denormalized* when the integer part is 0. Using denormalized mantissa, $111.101 = 0.110101 \times 2^3$.

IEEE standard convention provides two representations: *single-precision floats* using 32 bits , and *double-precision floats* with 64 bits. Only the sign, the exponent, and the mantissa need to be included, with the following rules:

**Sign:**     The most significant bit is the *sign bit*: 0 for positive, 1 for negative numbers
**Exponent:**     The exponent is biased:

- Single precision: 8 bits [bits 30-23], with bias 127
- Double precision: 11 bits [bits 62-52], with bias 1023

**Mantissa:** The mantissa is normalized, except for exponent 0

- Single precision: 23 bits [bits 22-0]
- Double precision: 52 bit [bits 51-0]

The distribution is illustrated in Fig. 2.5.

The exponent is biased to avoid the need of representing negative exponents. Thus, an exponent of 00h in single precision floats means $-127$, and 000h is $-1023$ in double-precision numbers. All one's exponents (FFh in single-precision and 7FFh in double-precision floats) have special meaning. Moreover, all 0's exponent assume denormalized mantissa. The conventions for all 0's and all 1's exponents are shown in Table 2.5.

**(a)**

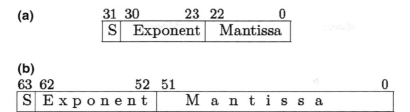

**(b)**

**Fig. 2.5** Component distribution in real numbers

**Table 2.5** Special convention for IEEE standard floating point

| Sign | Exponent | Mantissa | Meaning |
|------|----------|----------|---------|
| 0 | 0 | 0 | 0 |
| 1 | 0 | 0 | −0 |
| 0 | All 1's | 0 | Infinity |
| 1 | All 1's | 0 | −Infinity |
| X | All 1's | ≠ 0 | NaN (Not a number) |
| X | 00h | ≠ 0 | Mantissa denormalized |

Let us work some examples of floating point numbers to get the feeling for this convention.[7]

**Example 2.29** *Convert the following single-precision floating numbers to their decimal equivalents: (a)* 4A23100CH, *(b)* 002A2000H.

*(a)* 4A23100CH *stands for* 0100 1010 0010 0011 0001 0000 0000 1100. *Let us now separate the bits into the three components:*

$$\begin{array}{ccc} \text{Sign} & \text{Exponent} & \text{Mantissa} \\ 0 & 10010100 & 010\ 0011\ 0001\ 0000\ 0000\ 1100 \end{array}$$

*The sign is 0, which means a positive number.*
*The decimal equivalent to the binary in the exponent,* $10010100b = 148$ *is biased by 127, so the actual exponent is* $148 - 127 = 21$.
*Including the leading 1, and using the mixed notation mentioned before, we have* $1.01000110001000000001100 \times 2^{21} = 101000110001000000001100.00B = 2,671,619.$

*(b)* *The hex notation* 80300000H *means* 1000 0000 0011 0000 0000 0000 0000 0000.
*Proceeding as before, the sign bit is 1, so the number is negative. The exponent part is 0, and the mantissa is* 011, *disregarding the rightmost zeros. Since the*

---

[7] The reader may refer to the Prof. Vickery's website http://babbage.cs.qc.edu/courses/cs341/IEEE-754.html in Queen's College of CUNY for automatic conversions. The program was devised by the student Quan Fei Wen.

*exponent part is 0, the mantissa is denormalized and the actual exponent is −127. Therefore the decimal equivalent number is computed as* $− \left(0.011 \times 2^{-127}\right) = − \left(2^{-129} + 2^{-130}\right) \approx −2.20405 \times 10^{-39}$.

**Example 2.30** *Obtain the single-precision floating point representation for 532.379 and for* $−2.186 \times 10^{-5}$.

*To obtain floating point representation in single-precision, we follow four steps:*

1. *Express the binary equivalent of 532.379 with 24 bits:*

   - *532 = 1000010100B yields ten bits;*
   - *0.379 with 14 bits: 0.379 = .01100001000001B*
   - *Hence, 532.379 = 1000010100.01100001000001b*

2. *Express the 24-bit equivalent in binary 'scientific' form:*

   - $532.79 = 1.00001010001100001000001 \times 2^{9}$

3. *Identify each component of the floating-point representation, biasing exponent by 127:*

   - *Sign: 0 for positive number*
   - *Exponent:* $9 + 127 = 136 = 10001000b$
   - *Mantissa or significant: 00001010001100001000001*

4. *Combine the elements:*

   - *Result: 0100 0100 0000 0101 0001 1000 0100 0001*
   - *In hex-notation: 44051841h*

   *Now for* $−2.186 \times 10^{-5}$:

1. *Express the binary equivalent of 0.00002186 with 24 significant bits (that is, zeros before the first 1 do not count):*

   - *0.0000000000000000101101110101111111111110*

2. *Express the 24-bit equivalent in binary 'scientific' form:*

   - $2.186 \times 10^{-5} = 1.01101110101111111111110 \times 2^{-16}$

3. *Identify each component of the floating-point representation, biasing exponent by 127:*

   - *Sign: 1 for negative number*
   - *Exponent:* $−16 + 127 = 111 = 01101111b$
   - *Mantissa or significant: 01101110101111111111110*

4. *Combine the elements:*

   - *Result: 1011 0111 1011 0111 0101 1111 1111 1110*
   - *In hex-notation: B7B75FFEh*

**Example 2.31** *What are the smallest and largest positive decimal numbers representable in floating-point with (a) single-precision mode, and (b) double-precision mode?*

(a) *To find the smallest decimal number, the exponent in the floating-point should be 0, with the smallest non-zero mantissa. This number is therefore* 00000001H. *Interpreting this expression as before, the smallest number becomes*

$$0.00000000000000000000001 \times 2^{-127} = 2^{-23} \times 2^{-127} = 2^{-150} \approx 7.006 \times 10^{-46}$$

*Any attempt to represent a smaller number results in* underflow.
*For the largest number, the exponent and mantissa should be as large as possible. Since the exponent cannot be all 1's, the largest possible exponent is* 1111 1110, *equivalent to* $254 - 127 = +127$ *after unbiasing. Hence, the number we are looking for is* 7F7FFFFFH, *which corresponds to* $1.11111111111111111111111 \times 2^{127} = (2^{24} - 1) \times 2^{104} \approx 3.4 \times 10^{38}$.

(b) *For double-precision mode, similar considerations follow. For the smallest number,* 0000000000000001H, *using for convenience mixed hex and decimal notations,*

$$0.0000000000000001h \times 2^{-1023} \approx 5 \times 10^{-324}$$

*For the largest number the exponent is* 11111111110B, *equivalent to* 1023 *after unbiasing. Hence, the largest number is* 0x7FEFFFFFFFFFFFFF, *which is calculated as*

$$\left(2^{53} - 1\right) \times 2^{1023} \approx 1.8 \times 10^{308}$$

## 2.10 Continuous Interval Representations

Embedded systems are also applied in the measurement or monitoring of analog signals. This application requires encoding an analog signal with an $n$ bit-word. Electronically, this is done with an Analog to Digital Converter (ADC), as discussed in Chap. 10. We introduce here the elementary theoretical basis.

*Quantization* of an interval of real numbers $[a, b]$ is the process of representing it by a finite set of values $Q_0, Q_1, \ldots Q_m$ called Quantization Levels. The interval $[a, b]$ is partitioned in $m$ subintervals, one per quantization level. For digital encoding using $n$ bit words, $m = 2^n$ subintervals. Each quantization level is representative of any value $x$ in the associated sub interval.

The difference $x - Q_k$ between the actual value and its associated quantization level is the *Quantization Error*. The same $n$-bit word is used to encode both $x$ and $Q_k$.

The most common way to define the quantization levels is with a uniform distribution in the interval of interest. With this method, the difference between two consecutive quantization levels is

$$\Delta = \frac{b-a}{2^n} \tag{2.12}$$

$\Delta$ is called *resolution* or *precision* of the quantization. Most often, the resolution is referred to by the number of bits, thus talking of an $n$-bit resolution. Figure 2.6 illustrates this concept with 3 bits. Using this figure as reference, let us illustrate several general properties.

1. The upper limit of the quantized interval, $b$, is not a quantization level. This value will be "represented" by level $Q_{2^n-1}$, with an error of $1\,\Delta$.
2. Since any value $x$ will be interpreted by one level $Q_k$, any quantization error in the even distributed levels satisfies

$$Quantization\,error\ =\ x - Q_k \le \Delta \tag{2.13}$$

3. If we start at $a$ with $Q_0$ and enumerate successively the levels, then the quantization levels are defined by

$$Q_k = a + k\Delta \quad k = 0, 1, \ldots, 2^n - 1 \tag{2.14}$$

That is,

$$a, \quad a + \Delta, \quad a + 2\Delta, \ldots, \quad a + \left(2^n - 1\right)\Delta$$

The level $Q_k$ in (2.14) may be encoded with the unsigned binary equivalent $n$-bit word for $k$. In this case the encoding is said to be *straight binary*. Since we can go from one encoding to the next one by adding 1 to the LSB, the resolution $\Delta$ defined by (2.12) is also called 1*LSB resolution* . .

If a straight binary encoding is assigned with $Q_0 = a \neq 0$, as illustrated in Fig. 2.6, it is called *offset quantization*. If $a = 0$, then it is a *normal* quantization. It can be demonstrated that the midpoint of interval $[a, b]$ is always a quantization level $Q2^{n-1}$. In straight binary codes is encoded with $1000\ldots00$.

**Example 2.32** *An interval between $-10$ and $+10$ is quantized with using equally separated quantization levels. (a) Find the LSB resolution, and the quantization levels for a 4-bit resolution. (b) For an 8-bit resolution, find the LSB resolution and enumerate the lowest four and the highest four quantization levels.*

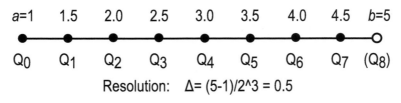

**Fig. 2.6** Uniform distribution of quantization levels for 3 bits

***Solution:*** *With 4 bits, the LSB resolution is found using (2.12):*

$$\Delta_4 = \frac{10 - (-10)}{2^4} = 1.25$$

*The sixteen quantization levels are then:*

| | | | |
|---|---|---|---|
| $Q_0 = -10.00$ | $Q_4 = -5.00$ | $Q_8 = 0.00$ | $Q_{12} = +5.00$ |
| $Q_1 = -8.75$ | $Q_5 = -3.75$ | $Q_9 = +1.25$ | $Q_{13} = +6.25$ |
| $Q_2 = -7.50$ | $Q_6 = -2.50$ | $Q_{10} = +2.50$ | $Q_{14} = +7.50$ |
| $Q_3 = -6.25$ | $Q_7 = -1.25$ | $Q_{11} = +3.75$ | $Q_{15} = +8.75$ |

*Notice that the greatest quantization value is 1 $\Delta$ less than + 10, as expected.*
*(b) With 8 bits,*

$$\Delta_8 = \frac{10 - (-10)}{2^8} = 0.078125$$

*The first four quantization levels are $Q_k = -10 + k\Delta$ for $k = 0, 1, 2,$ and 3:*

$$Q_0 = -10, \ Q_1 = -9.921875 \ Q_2 = -9.84375, \ and \ Q_3 = -9.765625$$

*The upper four apply the same formula with $k = 252, 253, 254, 255$:*

$$Q_{252} = 9.6875 \ Q_{253} = 9.765625 \ Q_{254} = 9.84375 \ Q_{255} = 9.921875$$

**Encoding with regular intervals**

It was pointed out that a subinterval is associated uniquely with a quantization level $Q_k$ and any value $x$ in the subinterval is encoded as $Q_k$. Now, encoding will depend in how we assign the subintervals.

As Fig. 2.6 illustrates, the quantization levels defined by Eq. (2.14) automatically determine a set of $2^n$ equal subintervals. Each of this has a quantization level as a lower bound. Let us associate then the subinterval to this bound. With this assignment, for an analog $x$ we can find the encoding as the decimal equivalent $k$ of the associated $n$-bit word by

$$k = \begin{cases} \mathrm{int}\left[2^n \frac{x-a}{b-a}\right] & \text{if } x < b \\ 2^n - 1 & \text{otherwise} \end{cases} \tag{2.15}$$

where "int[$\bullet$]" denotes the integer less than or equal to the argument value.

Let us illustrate the full process with an example.

**Example 2.33** *Continuing with example 2.32, find the encoding, the quantization level, and the quantization errors for $-2.20$ and $7.4$ for the 4 and 8-bit resolutions, when regular sub intervals are used in the quantization.*

*Solution:*  *Since we have already the list of quantization levels from the previous example, we could look at it and find the associated quantization level for −2.20 as* $Q_6 = -2.50$, *The encoding is therefore* 0110.

*However, for the sake of illustrating the process, let us use Eq. (2.15) as*

$$N_Q = int\left[2^4\frac{-2.20 - (-20)}{20}\right] = int[6.24] = 6$$

*with the quantization level,* $-10 + 6(1.25) = -2.50$. *The quantization error is* $-2.20 - (-2.5) = 0.3$.

*The table below summarizes the results for all requested cases. For later comparison, we have included the percentage of the error with respect to the 1 LSB resolution.*

| Parameter | 4-bits | | 8-bits | |
|---|---|---|---|---|
| | −2.20 | 7.40 | −2.20 | 7.40 |
| $k$ | 6 | 13 | 99 | 222 |
| $Q_k$ | −2.5 | 6.25 | −2.265625 | 7.34375 |
| Error | 0.30 | 1.15 | 0.065625 | 0.05625 |
| Error/$\Delta$ | 24 % | 92 % | 84 % | 72 % |

This previous example illustrates two facts. One is that the quantization error decreases as we increase the number of bits. This process has practical limitations such as hardware size, noise, and others. The second is that the maximum error is 1 LSB, which sometimes cannot be tolerated. Let us look at the following alternative which is most common.

### 2.10.1 Encoding with $\frac{1}{2}$LSB Offset Subintervals

The bound for a quantization error is minimum if the level is at the center of the subinterval. Since the distance between two levels is the 1 LSB resolution, then we can reduce the errors by shifting the subintervals half LSB. This is illustrated in Fig. 2.7.

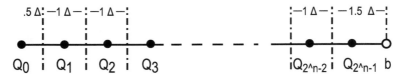

**Fig. 2.7**  Half LSB partition of interval $[a, b]$

In this case, with the exception of values in the upper subinterval, for any value $x$ the quantization error is within $\pm\frac{1}{2}$LSB, which is the best we can get. Only those values falling in the range $[Q_{2^n-1} + 0.5\Delta, b]$ may yield a greater error, but always less than 1 LSB.

The encoding for $x$ follows a similar computation, except that now we round to the nearest integer, and not to the lower bound integer. That is,

$$k = \begin{cases} \text{ROUND}\left(2^n \frac{x-a}{b-a}\right) & \text{if } x < a \text{ and } k < 2^n \\ 2^n - 1 & \text{Otherwise} \end{cases} \qquad (2.16)$$

where rounding is to the nearest integer.

**Example 2.34** *Repeat example 2.33 using half LSB partition.*

**Solution:** *We now use equation (2.16) instead of (2.15). The reader can verify the following table.*

| Parameter | 4-bits | | 8-bits | |
|-----------|--------|--------|--------|--------|
|           | −2.20  | 7.40   | −2.20  | 7.40   |
| $k$       | 6      | 14     | 100    | 223    |
| $Q_k$     | −2.5   | 7.5    | −2.1875 | 7.421875 |
| Error     | 0.30   | −0.1   | −0.0125 | −0.021875 |
| Error/$\Delta$ | 24 % | −8 %  | −16 %  | −28 %  |

*Notice the better approximation results. All errors are less than 50 % $\Delta$.*

## 2.10.2 Two's Complement Encoding of [−a, a]

In many applications, the interval to be encoded is symmetrical around the origin. In this case, it may be more convenient to encode using signed two's complement numbers, as illustrated in Fig. 2.8 for three bits. Notice that the quantization values remain the same.

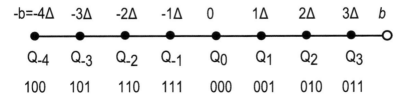

**Fig. 2.8** Half LSB partition of an analog interval

In the encoding process, we apply formulas similar to the previous ones, but applying a 0 offset. Namely, we now have the following formulas where $k$ may be a positive, negative, or zero integer, and its encoding is done with two's complement convention.

For quantization levels

$$Q_k = k\Delta \quad k = -2^{n-1}, \ldots - 1, 0, 1, \ldots, 2^{n-1} - 1 \qquad (2.17)$$

For regular subintervals, we have

$$k = \begin{cases} \text{int}\left[2^n \frac{x}{2b}\right] & \text{if } x < b \text{ and } k < 2^{n-1} \\ 2^{n-1} - 1 & \text{Otherwise} \end{cases} \qquad (2.18)$$

And for half LSB partition

$$k = \begin{cases} \text{ROUND}\left(2^n \frac{x}{2b}\right) & \text{if } x < b \text{ and } k < 2^{n-1} \\ 2^{n-1} - 1 & \text{Otherwise} \end{cases} \qquad (2.19)$$

## 2.11  Non Numerical Data

Besides numerical data, digital systems work non numerical information as well, represented with only 0's and 1's. In this section some of these codes are introduced.

### 2.11.1  ASCII Codes

There are many character codes, or *alphanumeric* codes, as they are also known. The most popular one for embedded systems is the *ASCII* code. ASCII is the acronym of *American Standard Code for Information Interchange*. Using seven bits, representations are given for the 26 upper and 26 lower case letters of the English alphabet, the ten numerals (0–9), miscellaneous symbols for punctuation such as, ":", ";", and other items.

ASCII was initially introduced for information exchange, it includes codes for control characters and operations used for rounding data and arranging printed text into a specific format. As such this set includes codes for layout control such as "backspace", "horizontal tabulation", and so on. Other control characters are used to separate data into divisions like paragraphs, pages, files; examples are "record separator", "file separator". Others are used for communications, such as "start of text", "end of text", etc.

**Table 2.6**  ASCII code chart

|           |   | Most Significant Digit |     |     |     |     |     |     |     |
|-----------|---|------|-----|-----|-----|-----|-----|-----|-----|
|           |   | **0** | **1** | **2** | **3** | **4** | **5** | **6** | **7** |
|           | **0** | NUL | DLE | SP | 0 | @ | P | ` | p |
|           | **1** | SOH | DC | ! | 1 | A | Q | a | q |
|           | **2** | STX | DC2 | " | 2 | B | R | b | r |
|           | **3** | ETX | DC3 | # | 3 | C | S | c | s |
|           | **4** | EOT | DC4 | $ | 4 | D | T | d | t |
|           | **5** | ENQ | NAK | % | 5 | E | U | e | u |
| Least     | **6** | ACK | SYN | & | 6 | F | V | f | v |
|           | **7** | BEL | ETB | ' | 7 | G | W | g | w |
| Significant | **8** | BS | CAN | ( | 8 | H | X | h | x |
|           | **9** | HT | EM | ) | 9 | I | Y | i | y |
| Digit     | **A** | LF | SUB | * | : | J | Z | j | z |
|           | **B** | VT | ESC | + | ; | K | [ | k | { |
|           | **C** | FF | FS | , | < | L | \ | l | \| |
|           | **D** | CR | GS | − | = | M | ] | m | } |
|           | **E** | SO | RS | . | > | N | ^ | n | ~ |
|           | **F** | SI | US | / | ? | O | _ | o | DEL |

Table 2.6 shows the ASCII code chart in hex notation, and Table 2.7 explains the acronyms for the different control functions. The seven bit code for a particular character or control function is found reading up the column for the most significant hex digit (three bits) and horizontally for the least significant one, or least significant four bits.

**Example 2.35**  *(a) Find the ASCII code for characters 5 and S; (b) Assuming an additional leading 0 to form bytes in an ASCII coding, decode* 01001101 01010011 01010000 00110100 00110011 00110000.

*(a) From table 2.6 "5" is in the intersection of column 3 and row 5. Hence, its ASCII code is 35h. Proceeding similarly, the ASCII code for S is 53h.*

*01001101 01010011 01010000 00110100 00110011 00110000 can be expressed in hex notation as 4Dh 53h 50h 34h 33h 30h. From Table 2.6 we can verify that this code stands for MSP430.*

Programmers usually don't have to memorize ASCII codes, since compilers take care of them with the *character constants* or *strings*, almost always enclosed by apostrophes. Thus, instead of writing 0x4D or 4Dh, the programmer writes `'M'`, both in C or asembly language. Similarly, it is easier to write `'The ball'` than `0x54 0x68 0x65 0x20 0x62 0x61 0x6C 0x6C`. Several of the ASCII control codes have made it into C, for example, some of the character escape codes, as illustrated in Table 2.8; others are simply normal ASCII's found on the keyboard, but included as escape characters because of the C syntax.

**Parity bit**. Notice that the ASCII code uses seven bits, so bit 7 in a byte becomes a don't care as far as the interpretation concerns. Thus, both 35h and B5h can be

**Table 2.7**  Meaning of control functions in ASCII code

| NUL: | Null | SOH: | Start of heading |
|------|------|------|------------------|
| STX: | Start of text | ETX: | End of text |
| EOT: | End of transmission | ENQ: | Enquiry |
| ACK: | Acknowledge | BEL: | Bell |
| BS: | Backspace | HT: | Horizontal tab |
| LF: | Line feed | VT: | Vertical tab |
| FF: | Form feed | CR: | Carriage return |
| SO: | Shift out | SI: | Shift in |
| SP: | Space | DLE: | Data link escape |
| DC1: | Device control 1 | DC2: | Device control 2 |
| DC3: | Device control | DC4: | Device control 4 |
| NAK: | Negative acknowledge | SYN: | Synchronous idle |
| ETB: | End of transmission block | CAN: | Cancel |
| EM: | End of medium | SUB: | Substitute |
| ESC: | Escape | FS: | File separator |
| GS: | Group separator | RS: | Record separator |
| US: | Unit separator | DEL: | Delete |

**Table 2.8**  ASCII values of character escapes in C

| Constant | Meaning | Hex | Constant | Meaning | Hex |
|----------|---------|-----|----------|---------|-----|
| '\a' | BEL (Bell, "alert") | 07 | '\r' | CR (carriage return) | 0D |
| '\b' | BS (backspace) | 08 | '\"' | double quote | 22 |
| '\t' | HT (horizontal tab) | 09 | '\'' | single quote | 27 |
| '\n' | LF (line Feed, "new line") | 0A | '\?' | question mark | 3F |
| '\v' | VT (vertical tab) | 0B | '\\' | backslash | 5C |
| '\f' | FF (form feed) | 0C | | | |

used for character 5. ASCII was devised initially as a transmission code. Since the information can be corrupted, there is the need to devise ways to verify that the correct signal has been sent. One of these is the use of a *parity bit*, which is added to the ASCII code to form a byte. A word has *even parity* if the number of 1's is even, and *odd parity* otherwise. Thus $41H \rightarrow 0010\ 0001$ is an even parity code for letter A, while $A1H \rightarrow 1010\ 0001$ is an odd parity code. for the same letter.

## 2.11.2 Unicode

While English is the most important technical language nowadays, and undoubtely enough for many embedded applications, it is not the only one used in communications. Hence, the need for more codes. *Unicode*[8] is a computing industry standard for the consistent encoding, representation, and handling of text expressed in most of the world's writing systems. The development of this standard is coordinated by the Unicode Consortium, whose official website is http://www.unicode.org. The origins of Unicode date back to 1987 and the development has continued since then without interruption. Nowadays, the Unicode Standard is the universal character encoding standard for written characters and text. It defines a consistent way of encoding multilingual text that enables the exchange of text data internationally and creates the foundation for global software.

Unicode characters are represented in one of three encoding forms: a 32-bit form (UTF- 32), a 16-bit form (UTF-16), and an 8-bit form (UTF-8). The 8-bit, byte-oriented form, UTF-8, has been designed for ease of use with existing ASCII-based systems.

The Unicode Standard contains 1,114,112 code points, most of which are available for encoding of characters. The majority of the common characters used in the major languages of the world are encoded in the first 65,536 code points, also known as the Basic Multilingual Plane (BMP). The overall capacity for more than 1 million characters is more than sufficient for all known character encoding requirements, including full coverage of all minority and historic scripts of the world.

The latest version, in 2011, is Unicode 6.0, and contains more than 109,384 characters and symbols, with ASCII as a subset code. These characters are more than sufficient not only for modern communication for the world languages, but also to represent the classical forms of many languages. The standard includes the European alphabetic scripts, Middle Eastern right-to-left scripts, and scripts of Asia and Africa. Many archaic and historic scripts are encoded. The Han script includes 74,616 ideographic characters defined by national, international, and industry standards of China, Japan, Korea, Taiwan, Vietnam, and Singapore. In addition, the Unicode Standard contains many important symbol sets, including currency symbols, punctuation marks, mathematical symbols, technical symbols, geometric shapes, dingbats, and emoji.

The reader is highly encouraged to visit the Unicode website to learn more about this standard.

## *2.11.3 Arbitrary Custom Codes*

Users can create their own coding for specific applications. Also, reconfigurable or programmable hardware imposes special encodings for applications based on bit interpretations. In most situations, bits are interpreted either individually or by groups.

We illustrate this last comment using the Supply Voltage Supervisor (SVS) control register found in the MSP430 microcontrollers with the exception of the first generation family '3xx. The bit distribution is shown in Fig. 2.9.

---

[8] This subsection is based on http://www.unicode.org/versions/Unicode6.0.0/ch01.pdf

| 7 6 5 4 | 3 | 2 | 1 | 0 |
|---------|-------|-------|-------|-------|
| VLDx | PORON | SVSON | SVSOP | SVSFG |

**Fig. 2.9**  SVS control register of the MSP430x1xx family (Courtesy of Texas Instruments Inc.)

The interpretation of bits is done working with the most significant nibble as a group, and the other bits individually as follows:

**Bits 7-4 VLDx, Voltage level detect**: They turn on the SVS and select the nominal SVS threshold voltage level:

| | | | | |
|---|---|---|---|---|
| 0000 : | $SVSis off$ | 1000 : | 2.8 V |
| 0001 : | 1.9 V | 1001 : | 2.9 V |
| 0010 : | 2.1 V | 1010 : | 3.05 V |
| 0011 : | 2.2 V | 1011 : | 3.2 V |
| 0100 : | 2.3 V | 1100 : | 3.35 V |
| 0101 : | 2.4 V | 1101 : | 3.5 V |
| 0110 : | 2.5 V | 1110 : | 3.7 V |
| 0111 : | 2.65 V | 1111 : | $Compares external input voltage SVSIN to 1.2 V.$ |

**Bit 3 PORON**: This bit enables the SVSFG flag to cause a POR device reset: 0 SVSFG does not cause a POR; 1 SVSFG causes a POR

**Bit 2 SVSON, SVS on**: This bit <u>reflects</u> the status of SVS operation, it DOES NOT turn on the SVS. (The SVS is turned on by setting VLDx $> 0$) This bit is 0 when the SVS is Off and 1 when the SVS is On

**Bit 1 SVSOP, SVS output**: This bit reflects the output value of the SVS comparator. Its value is 0 when SVS comparator output is low, and 1 when it is high.

**Bit 0 SVSFG, SVS flag**: This bit indicates a low voltage condition. SVSFG remains set (1) after a low voltage condition occurs until reset (0) by software or a brownout reset.

The above interpretation is defined from hardware design, and used to configure the peripheral. Hence, to select a 3.5 V threshold voltage and enable a power-on-reset we must force the word 11011xxx in the register. Notice that the three least significant bits are mostly read-only.

## 2.12 Summary

- A bit is a 0 or a 1. A sequence of bits is a word. A nibble is a 4-bit word, a byte is an 8-bit word, a double word is a 32-bit word, and a quad a 64-bit word. The term word is also used to denote a 16–bit word.

- Unsigned integers are most commonly represented using the normal positional binary system.
- The BCD system is a representation where each integer is represented separately by a four digit normal binary equivalent.
- With $n$ bits, the range of unsigned integers that can be covered is between 0 and $2^n - 1$.
- The hexadecimal system has the characteristic of a one to one relationship between each hex digit and four bits. This property is used to introduce the hex notation.
- Signed integers are normally represented with the 2's complement notation. With $n$ bits the covered range is from $-2^{n-1}$ to $2^{n-1} - 1$.
- Overflow occurs when the result of an operation falls outside the range covered by the representation.
- Real numbers are coded by either the fixed point format or the floating point format. The Fm.n format for fixed point format codifies the real number with the most significant m bits for the integer part, and the n least significant bits for the fractional part, in steps of $2^{-n}$.
- Floating point formats are single–precision, with 32 bits, and double–precision, with 64 bits. Floating point is preferred for very large and very small data.
- The ASCII code is the most popular for alphanumeric encoding. It does not include encoding for non-English letters. Unicode and other codes are used for most general cases.
- Continuous intervals are discretized by assigning words to intervals. The number of bits in the words determine the resolution of the codification.
- Non conventional codes are needed for specific applications or determined by the hardware of the system.

## 2.13 Problems

2.1  Express the following numbers as powers of 2:

    a. 256
    b. 64K
    c. 32M
    d. 512
    e. 4G

2.2  Express the following powers of 2 in terms of K, M G:

    a. $2^{14}$
    b. $2^{16}$
    c. $2^{24}$
    d. $2^{32}$
    e. $2^{20}$

2.3  Convert the following binary (base 2) integers to decimal (base 10) equivalents.

    a. 010000B
    b. 11111B
    c. 011101B
    d. 11111111B
    e. 10111111B
    f. 10000000B

2.4 Convert the following decimal (base 10) numbers into their equivalent binary (base 2) and hexadecimal (base 16) numbers.

    a. 10
    b. 32
    c. 40
    d. 64
    e. 156
    f. 244

2.5 Convert the following binary (base 2) numbers to hexadecimal (base 16) numbers.

    a. 10B
    b. 101B
    c. 1011B
    d. 1010111B
    e. 11101010B
    f. 10000000B
    g. 1011010110101110B

2.6 Convert the following unsigned hexadecimal (base 16) numbers into their equivalent decimal (base 10) and binary (base 2) numbers.

    a. 1Ch
    b. 7ABCDh
    c. 1234h
    d. 1FD53h
    e. 9D23Ah
    f. 0A1B2Ch

2.7 Construct a table for the equivalence of powers of 2, from $2^1$ to $2^{24}$ in hexadecimal numbers.

2.8 Starting with the fact that $r^n$ is represented in base $r$ as a 1 followed by $n$ zeros, demonstrate that the decimal equivalent for the binary number consisting of $n$ 1's is $2^n - 1$.

2.9 Make the following conversions:

    a. 10011.101B to decimal, hexadecimal, and octal
    b. 2A1F.Bh to binary, octal, and decimal
    c. 275.32Q to binary, hexadecimal, and decimal
    d. 1327.76 to binary (up to 6 binary fractional digits).

2.10 The following words using bits have no special meaning. Express the words in hexadecimal and octal notations.

    a. 101011011010
    b. 1011011001010
    c. 01101011011

2.11 Construct a table for example 2.7 on page 47 showing the different contents of register P1Out to display numbers 0 to 9, with and without dot.

2.12 Perform the following decimal additions and subtractions using the binary system.

    a. $381 + 214$
    b. $169 + 245$
    c. $279 + 37$
    d. $130 - 78$
    e. $897 - 358$

2.13 Repeat the previous exercise with the hexadecimal system.

2.14 Express the 2's complement of 1 in each of the following formats

    a. 4-bit binary (base 2) and hexadecimal (base 16) number
    b. 8-bit binary (base 2) and hexadecimal (base 16) number
    c. 16-bit binary (base 2) and hexadecimal (base 16) number

2.15 Find the two's complements of the following numbers with the specified number of bits.

    a. 79 with 8 and 10 bits
    b. 196 with 12 bits
    c. 658 with 16 bits
    d. 102 with 8, 10 and 16 bits.

2.16 Repeat the previous example using hex notation from the beginning. Notice that not all cases correspond to true hex representations.

2.17 Perform the following subtractions, both directly and with addition of complements, using binary and hex notation. For the case of negative results, take the complement of the result to interpret the subtraction. Notice in each case the difference between the borrow in the direct subtraction and the carry in the algorithm using two's complement addition.

    a. $198 - 87$, with eight bits.
    b. $56 - 115$, with eight bits.
    c. $38496 - 6175$, with sixteen bits.
    d. $1904 - 876$, with sixteen bits.
    e. $2659 - 14318$, with sixteen bits.

2.18  Encode the following decimal numbers in (a) BCD and in (b) 127 biased binary code.

  a.  10
  b.  28
  c.  40
  d.  64
  e.  156
  f.  244

2.19  What is the minimum number of bits required to represent all integers in the closed interval [180, 307]? What is the bias needed to represent those integers in the interval using the minimum number of bits?

2.20  Encode the following negative numbers using 2's complement representation in the binary and hexadecimal number systems using 8 and 16 bits.

  a.  $-12$
  b.  $-68$
  c.  $-128$

2.21  Find the corresponding decimal numbers for each of the following words encoded as 2's complement signed numbers.

  a.  1000b
  b.  1111b
  c.  11000110b
  d.  10111110b
  e.  01010101b
  f.  A2h
  g.  7Ch
  h.  43h
  i.  BEh
  j.  62AFh
  k.  CCCCh
  l.  3333h

2.22  Demonstrate the sign extension principle. Namely, show that if an $n$-bit word is converted into an $(n + h)$-bit word by appending the $h$ bits to the left equal to the original sign bit, the signed decimal equivalent remains the same. (*Hint: use expansion* (2.8) *with the extended word, and reduce it to the equivalent of the n-bit one*)

2.23  Code the following decimal signed numbers in the best approximation using fixed point formats F4.5 and F4.7. Find the absolute and relative errors in each case when the result is interpreted again as decimal equivalent.

  a.  $-5.125$
  b.  7.312
  c.  $-0.772$

  d. 6.219

  e. −3.856

  f. 2.687

2.24 One disadvantage of floating point representations is that the steps are not uniform, but depend on the range in which the number is. To see this, find the decimal equivalent of the two consecutive numbers in single precision, hex notation, 41A34000 and 41A34001. Do the same thing for 4DA34000 and 4DA34001. How do the steps in consecutive numbers compare?

2.25 Using the ASCII code find the binary streams represented by following character strings.

  a. abcd

  b. ABCD

  c. 123aB

  d. ASCII

  e. Microprocesors and Microcontrollers

  f. YoUr nAMe

2.26 Convert the following binary streams into their equivalent character strings using the ASCII code.

  a. "How Dy"

  b. "university"

  c. "012abCD"

  d. "blank space"

2.27 In order for a computer or host to be able to connect to the Internet it must have a "logical address" known as "IP address". IP addresses are expressed using the dotted decimal notation where the decimal equivalent of each byte is separated from the following using a dot (.). Thus, when using the IPv4 version of the Internet Protocol (IP) a 32-bit IP address is expressed as four (4) decimal numbers separated by dots. For example, IP address 127.0.0.1, known as the loopback or local host IP address, would be expressed in binary as 01111111.00000000.00000000.00000001. Express each of the following IPv4 IP addresses as four (4) binary numbers separated by dots and viceversa.

  a. 240.0.0.1

  b. 10.240.68.11

  c. 192.168.40.163

  d. 11111111.11111111.11110000.00000000

  e. 00001010.00000000.00000000.00000001

  f. 10101100.10110010.00001111.11111111

2.28 Convert the following single-precision numbers to decimal.

  a. AB1C3204

  b. 01798642

2.29 Obtain the single-precision floating point representation for the following numbers:

a.  641.13
b.  −1253.462

2.30 Find the dynamic range (smallest and largest positive and negative numbers) that can be expressed using 32-bits using with a floating point notation and with an integer notation.

2.31 A voltage takes values between 0 and 3 V. Conversion is done with 4-bit resolution. Specify the intervals and assignments.

# Chapter 3
# Microcomputer Organization

The minimal set of components required to establish a computing system is denominated a *Microcomputer*. The basic structural description of an embedded system introduced in Chap. 1 showed us the integration between hardware and software components. This chapter discusses the basic hardware and software elements required to establish a computer system and how they interact to provide the operation of a stored program computer.

## 3.1 Base Microcomputer Structure

The minimal hardware configuration of a microcomputer system is composed of three fundamental components: a Central Processing Unit (CPU), the system memory, and some form of Input/Output (I/O) Interface. These components are interconnected by multiple sets of lines grouped according to their functions, and globally denominated the *system buses*. An additional set of components provide the necessary power and timing synchronization for system operation. Figure 3.1 illustrates the integration of such a basic structure.

The components of a microcomputer can be implemented in diverse ways. They could be deployed with multiple chips on a board-level microcomputer or integrated into a single chip, in a structure named a microcomputer-on-a-chip, or simply a *Microcontroller*. Nowadays most embedded systems are developed around microcontrollers.

Regardless of the implementation style, each component of a microcomputer has the same specific function, as described below.

*Central Processing Unit* (CPU): The CPU forms the heart of the microcontroller system. It retrieves instructions from program memory, decodes them, and accordingly operates on data and/or on peripherals devices in the Input-Output subsystem to give functionality to the system.

M. Jiménez et al., *Introduction to Embedded Systems*,
DOI: 10.1007/978-1-4614-3143-5_3,
© Springer Science+Business Media New York 2014

**Fig. 3.1** General architecture
of a microcomputer system

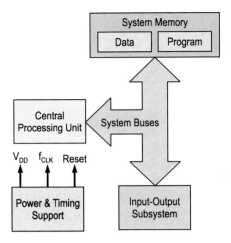

*System Memory*: The place where programs and data are stored to be accessed by
   the CPU is the system memory. Two types of memory elements are identified
   within the system: Program Memory and Data Memory. Program memory stores
   programs in the form of a sequence of instructions. Programs dictate the system's
   operation. Data Memory stores data to be operated on by programs.

*Input/Output subsystem* The I/O subsystem, also called *Peripheral Subsystem*
   includes all the components or peripherals that allow the CPU to exchange information with other devices, systems, or the external world. As was introduced in
   Chap. 1, the I/O Subsystem includes all the components that complement the CPU
   and memory to form a computer system.

*System Buses* The set of lines interconnecting CPU, Memory, and I/O Subsystem
   are denominated the system buses. Groups of lines in the system buses perform
   different functions. Based on their function the system bus lines are sub-divided
   into address bus, data bus, and control bus.

Looking back at the hardware structure of an embedded system as was introduced
in Sect. 1.2.1, we can notice that all the hardware components of Fig. 1.8, other
than CPU and Memory, were condensed, in Fig. 3.1 into the block representing the
Input/Output subsystem. This allows us to see that both diagrams are equivalent,
confirming our asseveration that embedded systems are indeed microcomputers.

## 3.2 Microcontrollers Versus Microprocessors

Before we delve any deeper into the structure of the different components of
a microcomputer system, let's first establish the fundamental difference between
microprocessors and microcontrollers.

### 3.2.1 Microprocessor Units

A Microprocessor Unit, commonly abbreviated MPU, fundamentally contains a general purpose CPU in its die. To develop a basic system using an MPU, all components depicted in Fig. 1.8 other than the CPU, i.e., the buses, memory, and I/O interfaces, are implemented externally.[1] Other characteristics of MPUs include an optimized architecture to move code and data from external memory into the chip such as queues and caches, and the inclusion of architectural elements to accelerate processing such as multiple functional units, ability to issue multiple instructions at once, and other features such as branch prediction units and numeric co-processors. The general discussion of these features fall beyond the scope of this book. Yet, whenever appropriate, we may introduce some related concepts.

The most common examples of systems designed around an MPUs are personal computers and mainframes. But these are not the only ones. There are many other systems developed around traditional MPUs. Manufacturers of MPU's include INTEL, Freescale, Zilog, Fujitsu, Siemens, and many others. Microprocessor design has advanced from the initial models in the early 1970s to present day technology. Intel's initial 4004 in 1971 was built using $10\,\mu$m technology, ran at $400$ kHz and contained $2,250$ transistors. Intel's Xeon E7 MPU, released in 2011, was built using $32$ nm technology, runs at $2$ GHz and contains $2.6 \times 10^9$ transistors. MPUs indeed, represent the most powerful type of processing components available to implement a microcomputer.

Most small embedded systems, however, do not need the large computational and processing power that characterize microprocessors, and hence the orientation to microcontrollers for these tasks.

### 3.2.2 Microcontroller Units

A microcontroller unit, abbreviated MCU, is developed using a microprocessor core or central processing unit (CPU), usually less complex than that of an MPU. This basic CPU is then surrounded with memory of both types (program and data) and several types of peripherals, all of them embedded into a single integrated circuit, or chip. This blending of CPU, memory, and I/O within a single chip is what we call a microcontroller.

The assortment of components embedded into a microcontroller allows for implementing complete applications requiring only a minimal number of external components or in many cases solely using the MCU chip. Peripheral timers, input/output (I/O) ports, interrupt handlers, and data converters are among those commonly found in most microcontrollers. The provision of such an assortment of resources inside the same chip is what has gained them the denomination of *computers-on-a-chip*. Figure 3.2 shows a typical microcontroller configuration.

---

[1] Hence, the name peripheral.

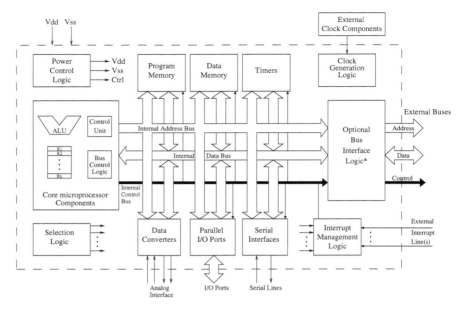

**Fig. 3.2**   Structure of a typical microcontroller

Microcontrollers share a number of characteristics with general purpose micro-processors. Yet, the core architectural components in a typical MCU are less complex and more application oriented than those in a general purpose microprocessor.

Microcontrollers are usually marketed as family members. Each family is developed around a basic architecture which defines the common characteristics of all members. These include, among others, the data and program path widths, architectural style, register structure, base instruction set, and addressing modes. Features differentiating family members include the amount of on-chip data and program memory and the assortment of on-chip peripherals.

There are literally hundreds, perhaps thousands, of microprocessors and micro-controllers on the market. Table 3.1 shows a very small sample of microcontroller family models of different sizes from six companies. We will concentrate on the MSP430 family for practical hands on introduction.

### 3.2.3 RISC Versus CISC Architectures

As we have already emphasized, microcomputer systems run with software which is supported by its hardware architecture. These systems are designed according to which of the two components, hardware or software, should be optimized. Under this point of view, we speak of CISC and RISC architectures.

**Table 3.1** A sample of MCU families/series

| Company | 4-bits | 8-bits | 16-bits | 32-bits |
|---------|--------|--------|---------|---------|
| EM Microelectronic | EM6807 | EM6819 | | |
| Samsung | S3P7xx | S3F9xxx | S3FCxx | S3FN23BXZZ |
| Freescale semiconductor | | 68HC11 | 68HC12 | |
| Toshiba | | TLCS-870 | TLCS-900/L1 | TLCS-900/H1 |
| Texas instruments | | | MSP430 | TMS320C28X |
| | | | TMS320C24X | Stellaris line |
| Microchip | | PIC1X | PIC2x | PIC32 |

CISC (*Complex Instruction Set Computing*) machines are characterized by variable length instruction words, i.e., with different number of bits, small code sizes, and multiple clocked-complex instructions at machine level. CISC architecture focuses in accomplishing as much as possible with each instruction, in order to generate simple programs. This focus helps the programmer's task while augmenting hardware complexity.

RISC (*Reduced Instruction Set Computing*) machines, on the other hand, are designed with focus on simple instructions, even if that results in longer programs. This orientation simplifies the hardware structure. The design expects that any single instruction execution is reduced—at most a single data memory cycle—when compared to the "complex instructions" of a CISC system.

It is usually accepted that RISC microcontrollers are faster, although this may not be necessarily true for all instructions.[2]

## 3.2.4 Programmer and Hardware Model

Most readers might already have programming experience with some high level language such as Java, C, or some other language. Most probably, the experience did not require knowledge of the hardware system supporting the execution of the program.

Embedded systems programmers need to go one step forward and consider both the hardware and software issues. Hence, they need to look at the system both from a hardware point of view, the *hardware model*, as well as a software point of view, the *programmer's model*.

In the hardware model, the user focuses on the hardware characteristics and subsystems that support the instructions and the interactions with the outer world. This knowledge is indispensable from the beginning especially because of the intimate relationship with the programming possibilities. In this model we speak of hardware

---

[2] Since speed has become an important feature to consider in applications, the term "RISC" has become almost a buzzword in the microcontroller market. The designer should check the truth of such labeling when selecting a microcontroller.

subsystems, characteristics of peripherals, interfacing with memory, peripherals and outer world, timing, and so on. The hardware supports the programmer's model.

In the programmer's model, we focus on the instruction set and syntaxis, addressing modes, the memory map, transfers and execution time, and so on. Very often, when a microcontroller is designed from scratch, the process starts with the desired instruction set.

This chapter focuses first on a general system view of the hardware architecture, and then the software characteristics.

## 3.3 Central Processing Unit

The Central Processing Unit (CPU) in a microcomputer system is typically a microprocessor unit (MPU) or core. The CPU is where instructions become signals and hardware actions that command the microcomputer operation. The minimal list of components that define the architecture of a CPU include the following:

- Hardware Components:

    - An Arithmetic Logic Unit (ALU)
    - A Control Unit (CU)
    - A Set of Registers
    - Bus Interface Logic (BIL)

- Software Components:

    - Instruction Set
    - Addressing Modes

The instructions and addressing modes will be defined by the specifics of the hardware ALU and CU units. In this section we concentrate on the hardware components. Instructions and addressing modes are considered in Sect. 3.7.

Figure 3.3 illustrates a simplified model view of the CPU with its internal hardware components. These components allow the CPU to access programs and data stored somewhere in memory or input/output subsystem, and to operate as a stored program computer. The sequence of instructions that make a program are chosen from the processor's instruction set. A memory stored program dictates the sequence of operations to be performed by the system. In the processing of data, each CPU component plays a necessary role that complements those of the others.

The collection of hardware components within the CPU performing data operations is called the processor's *datapath*. The CPU datapath includes the ALU, the internal data bus, and other functional components such as floating-point units, hardware multipliers, and so on. The hardware components performing system control operations are designated *Control Path*. The control unit is at the heart of the CPU control path. The bus control unit and all timing and synchronization hardware components are also considered part of the control path.

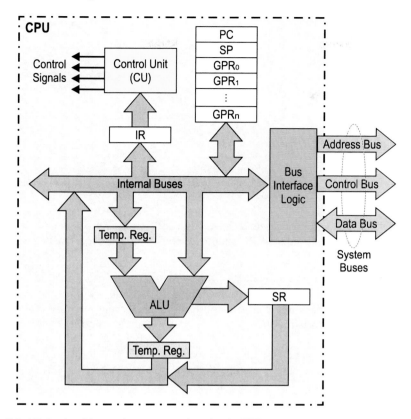

**Fig. 3.3** Minimal architectural components in a simple CPU

## 3.3.1 Control Unit

The control unit (CU) governs the CPU operation working like a finite state machine that cycles forever through three states: fetch, decode, and execute, as illustrated in Fig. 3.4. This fetch-decode-execute cycle is also known as *instruction cycle* or *CPU cycle*. The complete cycle will generally take several clock cycles, depending on the instruction and operands. It is usually assumed as a rule of the thumb that it takes at least four clock cycles.[3] Since the instruction may contain several words, or may require several intermediate steps, the actual termination of the execution process may require more than one instruction cycle.

---

[3] Often, CPU have internal clocks that run faster than system clocks, at least four times. Thus, in literature we may see that an instruction takes only one clock cycle. This is true with respect to the system cycle, but the process was driven by the internal CPU clock which actually took several cycles.

**Fig. 3.4**  States in control unit operation: fetch, decode, and execute

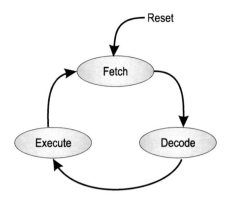

Several CPU blocks participate in the fetch-decode-execute process, among which we find special purpose registers PC and IR (see Fig. 3.3). The cycle can be described as follows:

1. **Fetch State**:  During the fetch state a new instruction is brought from memory into the CPU through the bus interface logic (BIL). The *program counter* (PC) provides the address of the instruction to be fetched from memory. The newly fetched instruction is read along the data bus and then stored in the *instruction register* (IR).
2. **Decoding State**:  After fetching the instruction, the CU goes into a decoding state, where the instruction meaning is deciphered. The decoded information is used to send signals to the appropriate CPU components to execute the actions specified by the instruction.
3. **Execution State**:  In the execution state, the CU commands the corresponding CPU functional units to perform the actions specified by the instruction. At the end of the execution phase, the PC has been incremented to point to the address of the next instruction in memory.

After the execution phase, the CU commands the BIL to use the information in the program counter to fetch the next instruction from memory, initiating the cycle again.

The cycle may require intermediate cycles similar to this one whenever the decoding phase requires reading (fetching) other values from memory. This is dictated by the addressing mode use in the instruction, as it will be explained later.

Being the CU a finite state machine, it needs a *Reset* signal to initiate the cycle for the first time. The program counter is hardwired to, upon reset, load the memory address of the first instruction to be fetched. That is how the first cycle begins operation. The address of this first instruction is called *reset vector*.

### 3.3.2 Arithmetic Logic Unit

The arithmetic logic unit (ALU) is the CPU component where all logic and arithmetic operations supported by the system are performed. Basic arithmetic operations such as addition, subtraction, and complement, are supported by most ALUs. Some may also include hardware for more complex operations such as multiplication and division, although in many cases these operations are supported via software or by other peripherals, such as a hardware multiplier.

Logic operations performed in the ALU may include bitwise logic operations AND, OR, NOT, and XOR, as well as register operations like SHIFT and ROTATE. Bitwise operations are an important tool for manipulating individual bits of a register without affecting the other ones. This is possible by exploiting the properties of logic Boolean operations.

The CU governs the ALU by specifying which particular operation is to be performed, the source operands, and the destination of the result. The width of the operands accepted by the ALU of a particular CPU (datapath width) is typically used as an indicator of the CPU computational capacity. When for example, a microprocessor is referred to as a 16-bit unit, its ALU has the capability of operating on 16-bit data. The ALU data width shapes the CPU datapath architecture establishing the width of data bus and data registers.

### 3.3.3  Bus Interface Logic

The Bus Interface Logic (BIL) refers to the CPU structures that coordinate the interaction between the internal buses and the system buses. The BIL defines how the external address, data, and control buses operate. In small embedded systems the BIL is totally contained within the CPU and transparent to the designer. In distributed and high performance systems the BIL may include dedicated peripherals devoted to establish the CPU interface to the system bus. Examples of such extensions are bus control peripherals, bridges, and bus arbitration hardware included in the chip set of contemporary microprocessor systems.

### 3.3.4  Registers

CPU registers provide temporary storage for data, memory addresses, and control information in a way that can be quickly accessed. They are the fastest form of information storage in a computer system, while at the same time they are the smallest in capacity. Register contents is *volatile*, meaning that it is lost when the CPU is de-energized. CPU registers can be broadly classified as general purpose and specialized.

*General purpose registers* (GPR) are those not tied to specific processor functions and may be used to hold data, variables, or address pointers as needed. Based on this usage, some authors classify them also as *data* or *address registers*. Depending on the processor architecture, a CPU can contain from as few as two to several dozen GPRs.

*Special purpose registers* perform specific functions that give functionality to the CPU. The most basic CPU structure includes the following four specialized registers:

- Instruction Register (IR)
- Program Counter (PC), also called Instruction Pointer (IP)
- Stack Pointer (SP)
- Status Register (SR)

**Instruction Register (IR)**

This register holds the instruction that is being currently decoded and executed in the CPU. The action of transferring an instruction from memory into the IR is called *instruction fetch*. In many small embedded systems, the IR holds one instruction at a time. CPUs used in distributed and high-performance systems usually have multiple instruction registers arranged in a queue, allowing for concurrently *issuing*[4] instructions to multiple functional units. In these architectures the IR is commonly called an *instruction queue*.

**Program Counter (PC)**

This register holds the address of the instruction to be fetched from memory by the CPU. It is sometimes also called the *instruction pointer (IP)*. Every time an instruction is fetched and decoded, the control unit increments the value of the PC to point to the next instruction in memory. This behavior may be altered, among others, by *jump instructions* which, when executed, replace the contents of the PC with a new address. Being the PC an address register, its width may also determine the size of the largest program memory space directly addressable by the CPU.

The PC is usually not meant to be directly manipulated by programs. This rule is enforced in many traditional architectures by making the PC not accessible as an operand to general instructions. Newer RISC architectures have relaxed this rule in an attempt to make programming more flexible. Nevertheless, this flexibility has to be used with caution to maintain a correct program flow.

**Stack Pointer (SP)**

The *stack* is a specialized memory segment used for temporarily storing data items in a particular sequence. The operations of storing and retrieving the items according to this sequence is managed by the CPU with the stack pointer register (SP). Few MCU's models have the stack hardwired defined. Most models, however, allow the

---

[4] Instruction issue is a term frequently used in high-performance computer architectures to denote the transfer of decoded instruction information to a functional unit like an ALU, for execution.

user to define the stack within the RAM section, or else it is automatically defined during the compiling process.

The SP contents is referred to as the Top of Stack (TOS). This one tells the CPU where a new data is stored (*push operation*) or read (*pull operation* or *pop operation*). A detailed explanation of how SP works with these operations is given in Sect. 3.7.4.

**Status Register (SR)**

The status register, also called the *Processor Status Word (PSW)*, or *Flag Register* contains a set of indicator bits called *flags*, as well as other bits pertaining to or controlling the CPU status. A flag is a single bit that indicates the occurrence of a particular condition.

The number of flags and conditions indicated by a status register depends on the MCU model. Most flags in the SR reflect the situation just after an operation is executed by the ALU, although in general they can also be manipulated by software. This dependence is emphasized in Fig. 3.3 by the position of the SR.

The status of the flags depends also on the size of the ALU operands. Generally, both operands will have the same size, $n$ bits, while the ALU operation may produce $n + 1$ bits. By "ALU result" we mean then the $n$ least significant bits while the most significant bit is the Carry Flag. This remark is clarified Example 3.1 below, but let us first identify the most common flags to be found in almost all MCUs. These are

- *Zero Flag (ZF)*: Also called the zero bit. It is set when the result of an ALU operation is zero, and cleared otherwise. It may also be tied to other instructions.
- *Carry Flag (CF)*: This flag is set when an arithmetic ALU operation produces a carry. Some non arithmetic operations may affect the carry without having a direct relation to it.
- *Negative* or *sign flag (NF)*: This flag is set if the result of an ALU operation is negative and cleared otherwise. This flag in fact reflects the most significant bit of the result.
- *Overflow Flag (VF)*: This flag signals overflow in addition or subtraction with signed numbers (see Sect. 2.8.1). The flag is set if the operation produces an overflow, and cleared otherwise.[5]
- *Interrupt Flag (IF)*: This flag, also called *General Interrupt Enable* (GIE), is not associated to the ALU. It indicates whether a program can be interrupted by an external event (interrupt) or not. Interrupts blocked by the IF are called *maskable*. Section 3.9 discusses the subject of interrupts in microprocessor based systems.

MCUs may have more flags than the one mentioned. The user must consult the specifications of the microcontroller or microprocessor being used to check available flags and other bits included in the SR. The following example shows how the ALU affects flags after an addition.

---

[5] There are other types of overflow, like when the result of an arithmetic operation exceeds the number of bits allocated for its result. This condition however is not signaled by this flag, although particular CPUs may have another flag for this purpose.

**Example 3.1** *The following operations are additions performed by the ALU using 8-bit data. For each one, determine the Carry, Zero, Negative, and Overflow flags.*

$$
\begin{array}{llll}
01001010\,+ & 10110100\,+ & 10011010\,+ & 11001010\,+ \\
01111001\,= & 01001100\,= & 10111001\,= & 00011011\,= \\
\hline
0\ 11000011 & 1\ 00000000 & 1\ 01010011 & 0\ 11100101 \\
\uparrow\ \uparrow & \uparrow\ \uparrow & \uparrow\ \uparrow & \uparrow\ \uparrow \\
\text{C N} & \text{C N} & \text{C N} & \text{C N}
\end{array}
$$

*Solution:* *The operands have eight bits, so this length is our reference for the flags when we look at the result. The most significant bit in this group is flag N. The bit to the left is C. In hex form, these additions are, respectively, 4Ah + 79h = C3h; B4h + 4Ch = 100h; 9Ah + B9h = 153h; and CAh + 1Bh = E5h. The zero flag is set if the result is 0, discarding the carry, and the overflow flag is set if the addition of numbers of the same sign (that is, with equal most significant bit) yield a result of different sign (signaled by N). With this information we have then:*

*Operation 4Ah + 79h = C3h :  C = 0, N = 1, Z = 0 and V = 1.*
*Operation B4h + 4Ch = 100h :  C = 1, N = 0, Z = 1 and V = 0.*
*Operation 9Ah + B9h = 153h :  C = 1, N = 0, Z = 0 and V = 1.*
*Operation CAh + 1Bh = E5h :  C = 0, N = 1, Z = 0 and V = 0.*

Notice that there is no overflow whenever both operands in an addition are of different sign[6]; that is, when their most significant bits are different. When both operands are negative and yield a positive result, as in the third case of the above example, we say that an *underflow* has happened. The term overflow is generic, though, and can be used in any situation.

Although the interpretation of what a flag means depends ultimately on the programmer, the meaning of the N, Z, and V flags is, we think, clear. The N and V flags are mainly related to the use of signed numbers in arithmetic operations of subtraction and addition. The Z flag tells us if the result is 0 or not, no matter the operation, arithmetic or logic or any of other type. The C flag needs however further remarks.

**The Carry Flag**

The Carry flag may be of interest in several operations besides addition and subtraction. Let us first discuss these two operations.

When an addition is performed, the C flag may or may not be considered part of the result. In the first case, it also provides a sign extension when working with signed numbers, as it can be appreciated in the previous example. Or it can also be used to work with numbers larger than the ALU capacity. When it is not part of the addition result, then its meaning will depend on the programmer's intention.

---

[6] One can visualize this with the full interval in a numerical line. Starting at the midpoint, two consecutive walks in opposite directions, neither greater than half the interval length, will keep you inside the interval.

**Table 3.2** Flags and number comparison with $A - B$

| Comparison | Unsigned Numbers | Signed numbers |
|---|---|---|
| $A = B$ | Z = 1 | Z = 1 |
| $A \neq B$ | Z = 0 | Z = 0 |
| $A \geq B$ | C = 1[1] | N = V |
| $A > B$ | C = 1 and Z = 0 | N = V and Z = 0 |
| $A < B$ | C = 0 | N $\neq$ V |
| $A \leq B$ | C = 0 or Z = 1 | N $\neq$V or Z = 1 |

Remember: C = 1 no borrow needed, C = 0 borrow is needed

On the other hand, in subtraction we are more interested in knowing if there has been a borrow or not. Some CPU's may have another flag to work with borrow. But it is also normal to use the Carry flag in a dual role to signal a carry in addition operations and a borrow in subtraction operations. Since subtraction is usually hardware realized with addition of two's complement, two standpoints have been adopted:

(1) The C is set (C = 1) after a subtraction if no borrow is generated. This mechanism is adopted in some families like PIC, Atmel, and INTEL microcontrollers.
(2) C is reset (C = 0) if the subtraction needs a borrow. This is usual in RISC microcontrollers, including the MSP430, since it needs less hardware realization.

*We adopt hereafter the second option, following our choice of MCU for practical introduction, the MSP430.*

The C flag is associated to other instructions, depending on the CPU. A common application is to use it as a mid step for bits in shifting and rotation operations, connecting different registers. Other instructions depend on the microcontroller selected by the user and should be consulted in the proper documentation. For the MSP430, these instructions will be introduced as we study them.

## SR Flags and Number Comparison

A very useful characteristic of the flags is the information they provide when comparing numbers. In particular, this feature is used by the CPU to decide actions to take in conditional jumps, thereby making it possible to write non sequential programs.

When comparing two numbers by subtraction, $A - B$, the status of the flags allows us—and the CPU—to decide the relationship between the two numbers, as illustrated in Table 3.2. This table assumes that subtraction is carried out by two's complement addition, which is the case in the majority of CPU's.

Let us explain briefly why the above interpretations. Remember that all subtractions are made with binary words, whose interpretation as signed or unsigned numbers pertains to the user, not to the machine.

The first two lines in the table are obvious, since A–B = 0 if and only if A = B. For the inequalities, let us consider the unsigned and signed cases separately.

For unsigned cases, the sign and overflow flags are meaningless. However, we know that the presence of a carry in a subtraction done with two's complement addition means that the result is not negative (See remark on Sect. 2.5.3). That is, $A - B \geq 0$ if and only if there is a carry, $(C = 1)$. Strict inequality requires the result to be non-zero, i.e., $Z = 0$. This proves the third and fourth lines, respectively. The other two lines are just the opposite of the previous cases—$A < B$ means not$(A \geq B)$ and $A \leq B$ means not$(A > B)$ –.

Now, consider the case for signed numbers. The sign flag N, which reflects the most significant bit of the result, has sense only when subtraction is correctly performed, meaning by this that $V = 0$. In this case: (a) $N = 0$ means that the difference A–B is not negative, so $A \geq B$ iff $N = V = 0$, and (b) $N = 1$ means that the difference is negative, so $A < B$ iff $N = 1 \neq V = 0$, where iff means *if and only if*.

On the other hand, overflow occurs in the subtraction when A and B have different sign, but A–B has the same sign as B, triggering $V = 1$. In this case consider the following: (a) If $B < 0$, we can say that $A > 0 > B$, and $V = 1 = N = 1$. (b) if $B > 0$ then $A < B$ and $N = 0 \neq V = 1$.

The two paragraphs above prove then the third and fifth lines for signed numbers. Strict $>$ inequality requires $Z = 0$ because the subtraction cannot yield 0. The last case is the opposite of this strict inequality.

**Example 3.2** *Consider the following bytes:* X = 3Ch, Y = 74h, W = A2h, *and* Z = 89h.

(a) *Write down the correct relationships (excluding $\neq$) substituting the question mark sign in the following expressions when the bytes represent unsigned numbers, and when they represent signed numbers. Do it intuitively.*

(b) *Verify that the flags provide the same information as in (a) when comparing the numbers by subtraction using two's complement addition.*

$$X?Y \quad Y?W \quad W?Z \quad Z?X$$

**Solution:** *(a) For easier analysis, let us translate the given bytes into their decimal equivalents.*

*Unsigned case: X = 60; Y = 116; W = 162; Z = 137. Hence,*

$$X < Y(also\ X \leq Y); \quad Y < W(also\ Y \leq W);$$
$$W > Z(also\ W \geq Z); \quad Z > X(also\ Z \geq X)$$

*Signed case: X = 60; Y = 116; W = −94; Z = −119. Therefore,*

$$X < Y(also\ X \leq Y); \quad Y > W(also\ Y \geq W);$$
$$W > Z(also\ W \geq Z); \quad Z < X(also\ Z \leq X)$$

(b) *To subtract, we use the two's complements of the given bytes:* $X' = C4h$; $Y' = 8Ch$; $W' = 5Eh$; $Z' = 97h$ *in the following table:*

| Subtraction | Operation | Unsigned numbers | | Signed numbers | |
|---|---|---|---|---|---|
| | | Flags | Relation | Flags | Relation |
| X-Y | 3Ch + 8Ch = C8h | C = 0 | X < Y | V = 0, N = 1 | X < Y |
| | | (Z = 0) | X $\leq$ Y | (Z = 0) | X $\leq$ Y |
| Y − W | 74h + 5Eh = D2h | C = 0 | Y < W | V = 1, N = 1 | Y $\geq$ W |
| | | (Z = 0) | Y $\leq$ W | & Z = 0 | Y > W |
| W − Z | A2h + 77h = 119h | C = 1 | W $\geq$ Z | V = 0, N = 0 | W $\geq$ z |
| | | & Z = 0 | W > Z | & & Z = 0 | W > Z |
| Z − X | 89h + C4h = 14Dh | C = 1 | Z $\geq$ X | V = 1, N = 0 | Z < X |
| | | & Z = 0 | Z > X | & (Z = 0) | Z $\leq$ X |

*In the case of* **or** *decisions, the second flag to consider was put in parenthesis. Thus (Z = 0) should be considered part of the process of verifying C = 0 or Z = 1, which in the shown cases was valid because of the C = 0 compliance.*

### 3.3.5 MSP430 CPU Basic Hardware Structure

The MSP430 family is based on a 16-bit CPU which was introduced in the early models of the series 3xx. Later on the architecture was extended to 20-bits, the CPUX, keeping full compatibility with the original 16-bit CPU. Thus, the instruction set of the CPU is fully supported by the CPUX, which works like a 16-bit CPU for these operations. There are special instructions proper to the CPUX targeting 20-bit data.

In this book, hands on programming is focused on the MSP430 16-bit CPU. Accordingly, this section provides a quick overview on its characteristics. An explanation for the MSP430X control processing unit CPUX is given in Appendix D.

**MSP430 registers:** There are sixteen 16-bit registers in the MSP430 CPU named R0, R1 ..., R15. Registers R4 to R15 are of the general purpose type. The specialized purpose registers are:

- Program Counter register, named R0 or PC.
- Stack Pointer Register, named R1 or SP.
- Status Register, with a dual function also as Constant Generator. It is named R2, SR or CG1.
- Constant Generator, named CG2.

Register R3 is exclusively used as a constant generator supplying instruction constants, and is not used for data storage. Its function is normally transparent to the user. This is explained in Chap. 4.

The PC and SP registers always point to an even address and have the least significant bit hardwired to 0. A particular feature in the MSP430 family is that these two registers can be used as operands in instructions, allowing programmers to develop applications with simpler software algorithms.

**Status Register** The SR register has the common flags of Carry (C), Zero (Z), Sign or Negative (N), overflow (V) and general interrupt enable (GIE). It contains in addition a set of bits, CPUOFF, OSCOFF, SCG1 and SCG0, used to configure the CPU, oscillator and power mode operations. These bits are explained in later sections and chapters. The bit distribution in the SR is shown in Fig. 3.5.

**Arithmetic-Logic Unit:** The MSP430 CPU ALU has a 16-bit operand capacity; the CPUX has 16- or 20-bit operand capacity. It handles the arithmetic operations of addition with and without carry, decimal addition with carry (BCD addition—see Sect. 2.7.3), subtraction with and without carry. These operations affect the overflow, zero, negative, and carry flags. The logical operations AND and XOR affect the flags, but other ones do not, like the Bit Set (OR) or Bit Clear.

Like many microcontrollers, the MSP430 ALU does not handle multiplications or divisions, which may be programmed. Several models contain a *hardware multiplier* peripheral for faster operation.

## 3.4 System Buses

Memory and I/O devices are accessed by the CPU through the system buses. A bus is simply a group of lines that perform a similar function. Each line carries a bit of information and the group of bits may be interpreted as a whole. The system buses are grouped in three clases: *address*, *data*, and *control* buses. These are described next.

### 3.4.1 Data Bus

The set of lines carrying data and instructions to or from the CPU is called the *data bus*. A *read* operation occurs when information is being transferred *into* the CPU. A data bus transfer *out from* the CPU into memory or into a peripheral device, is called a *write* operation. Note that the designation of a transfer on the data bus as read or write is always made with respect to the CPU. This convention holds for every system component. Data bus lines are generally bi-directional because the same set of lines allows us to carry information to or from the CPU. One transfer of information is referred to as *data bus transaction*.

| b15 | | b9 | b8 | b7 | b6 | b5 | b4 | b3 | b2 | b1 | b0 |
|-----|---|----|----|------|------|------------|------------|-----|---|---|---|
| Reserved | | | V | SCG1 | SCG0 | OSC OFF | CPU OFF | GIE | N | Z | C |

**Fig. 3.5** MSP430 status register

The number of lines in the data bus determines the maximum *data width* the CPU can handle in a single transaction; wider data transfers are possible, but require multiple data bus transactions. For example, an 8-bit data bus can transfer at most one byte (or two nibbles) in a single transaction, and a 16-bit transaction would require two data bus transactions. Similarly, a 16-bit data bus would be able to transfer at most, two bytes per transaction; transferring more than 16 bits would require multiple transactions.

## 3.4.2 Address Bus

The CPU interacts with only one memory register or peripheral device at a time. Each register, either in memory or a peripheral device, is uniquely identified with an identifier called *address*. The set of lines transporting this address information form the *address bus*. These lines are usually unidirectional and coming out from the CPU. Addresses are usually named in hexadecimal notation.

The width of the address bus determines the size of the largest memory space that the CPU can address. An address bus of $m$ bits will be able to address at most $2^m$ different memory locations, which are referred to by hex notation. For example, with a 16-bit address bus, the CPU can access up to $2^{16} = 64$ K locations named 0x0000, 0x0001, ..., 0xFFFF. Notice that the bits of the address bus lines work as a group, called *address word*, and are not considered meaningful individually.

**Example 3.3** *Determine how many different memory locations can be accessed and the address range (i.e., initial and final addresses) in hex notation with an address bus of (a) 12 bits, and (b) 22 bits. Justify your answer.*

*Solution: (a) For 12 bits, there are $2^{12} = 2^2 \times 2^{10} = 4$ K different locations that can be addressed. In binary terms, we think of them as the integer representations of 0000 0000 0000B to 1111 1111 1111B, which in hex notation become 0x000 to 0xFFF. (b) Working similarly, for 22 bits there are $2^{22} = 2^2 \times 2^{20} = 4$ M locations. Yet, since there are 22 bits and hex integers represent only four bits, the largest number with the two most significant bits is 3, so the address range is 0x000000 to 0x3FFFFF.*

**Example 3.4** *A certain system has a memory size of 32 K memory words. What is the minimum number of lines required for the address bus?*

*In powers of 2, 32 $K = 2^5 \times 2^{10} = 2^{15}$. Therefore, the address bus must contain at least 15 lines, one per bit. (The answer could have been also expressed as $n = \log_2(32 * 1024) = 5 + 10 = 15$).*

### 3.4.3  Control Bus

The *control bus* groups all the lines carrying the signals that regulate the system activity. Unlike the address and data buses lines which are usually interpreted as a group (address or data), the control bus signals usually work and are interpreted separately. Control signals include those used to indicate whether the CPU is performing a read or write access, those that synchronize transfers saying when the transaction begins and ends, those requesting services to the CPU, and other tasks. Most control lines are unidirectional and enter or leave the CPU, depending on their function. The number and function of the lines in a control bus will vary depending on the CPU architecture and capabilities.

## 3.5  Memory Organization

The memory subsystem stores instructions and data. Memory consists of a large number of hardware components which can store one bit each. These bits are organized in $n$-bit words, working as a register, usually referred to as cell or location. The contents of cells is a basic unit of information called *memory word*. In addition, each memory location is identified by a unique identifier, its *memory address*, which is used by the CPU to either read or write over the memory word stored at the location. In general, a memory unit consisting of $m$ cells of $n$ bits each is referred to as an $m \times n$ memory; for $n = 1$ and $n = 8$ it is customary to indicate a number followed by b or B, respectively. Thus, we speak of 1Mb (one Mega Bits) and 1 MB (one Mega Bytes) memories to refer to $1M \times 1$ and $1M \times 8$ cases.

Usually, addresses are sequentially numbered as illustrated in Fig. 3.6. However, in a specific MCU model some addresses may not be present. The example in the figure shows a memory module of 64 K cells, each storing an 8-bit word (byte), forming a 64 kilo-byte (64 KB) memory. In the illustration, for example, the cell at address

**Fig. 3.6** Memory structure

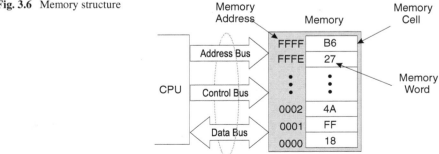

0FFFEh contains the value 27 h. The CPU uses the address bus to select only the cell with which it will interact. The interaction with the contents is realized through the data bus. In a write operation, the CPU modifies the information contained in the cell, while in a read operation it retrieves this word without changing the contents. The CPU uses control bus signals to determine the type of operation to be realized, as well the direction in which the data bus will operate—remember that the data bus lines are bidirectional.

**Remark on notation**: Hereafter, to simplify writing, memory addresses and contents will be written with notations such as [*address*] = *contents*, or *address*: *contents*. In figures, the contents of memory is shown always in hexadecimal notation without suffix or prefix. Taking as reference Fig. 3.6 for an example, [0002h] = 0x4A.

## 3.5.1 Memory Types

Hardware memory is classified according to two main criteria: *storage permanence* and *write ability*. The first criterion refers to the ability of memory to hold its bits after these have been written. Write ability, on the other hand, refers on how easily the contents of memory can be written by the embedded system itself. All memory is readable, since otherwise it would be useless.

From the storage permanence point of view, the two basic subcategories are the *nonvolatile* and *volatile* groups. The first group encompasses those memories that can hold the data after power is no longer supplied. Volatile memory, on the other hand, looses its contents when power is removed.

The nonvolatile category includes the different *read only memory* (ROM) structures as well as *Ferro electric RAM* (FRAM or FeRAM). Another special one is the nonvolatile RAM (NVRAM) which is in fact a volatile memory with a battery backup; for this reason, NVRAMs are not used in microcontrollers. The volatile group includes the *static RAM* (SRAM) and the *dynamic RAM* (DRAM).

From the write ability point of view, memory is classified into *write/read* or *in-system programmable* and *read only* memories. The first group refers to those memories that can be written to by the processor in the embedded system using the memory. Most volatile memories, SRAM and DRAM, can be written to during program execution, and therefore these memory sections are useful for temporary data or data that will be generated or modified by the program. The FRAM memory is non-volatile, but the writing speed is faster than the DRAM. Hence, microcontrollers with FRAM memory are very convenient in this aspect.

Most in-system programmable nonvolatile memories are written only when loading the program, but not during execution. One reason is the writing speed, too slow for the program. Another aspect of Flash memory and EEROM is the fact that writing requires higher voltages than the ones used during the program execution, thereby consuming power and requiring special power electronics hardware to increase

| Storage | Memory | In-system Writable | Comments |
|---|---|---|---|
| Nonvolatile | Masked ROM | No | Non programmable |
| | OTPROM | No | One time programmable with programming device |
| | EPROM | No | Erasable and programmable with external device |
| | EEPROM | Yes | Slow to erase/write. Not advisable to write during program execution. Requires higher voltage. |
| | Flash | Yes | Similar to EEPROM |
| | FRAM | Yes | Fast to write at low voltage |
| Volatile | Static RAM | Yes | Fastest to write/read |
| | DRAM | Yes | Fast to write/read |

**Fig. 3.7** Memory types

voltage levels. The FRAM, on the other side, is writable at the program execution voltage levels.

The FRAM, is both nonvolatile and writable with speeds comparable to DRAMS and at operational voltage levels. Moreover, unlike the Flash or DRAM, consumes power only during writing and reading operations, making it a very low power device. One disadvantage at this moment is that the temperature ranges of operation are limited, so their application must be in environments where temperature is not so variable. Further research is being done to solve this limitation. Table in Fig. 3.7 summarizes this discussion.

As a marginal note, for historical reasons, in the embedded community volatile read/write memories are referred to as RAM (random access memories)[7] while the ROM term is used for non volatile memory. This convention is hereafter adopted unless specific details are needed.

## 3.5.2 Data Address: Little and Big Endian Conventions

Most hardware memory words nowadays are one byte length. As explained above, each memory word has an address attached to it, referred to as its *physical address*, which is encoded by the group of the MCU address bus bits. Memory blocks one

---

[7] RAM, an acronym for "Random Access Memory", is a term coined in the 1950s to refer to solid state memories in which data could be accessed randomly, as opposed to other memory devices such as magnetic tapes and discs. Under this definition, semiconductor ROM devices are random access ones also, but the DRAM is not. But the terms are already stuck and understood by the embedded community.

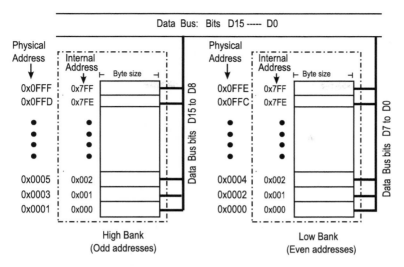

**Fig. 3.8** Example of bank connection

byte length are called *banks*. For data buses wider than a byte, two or more banks are needed to connect all the data bus lines, as illustrated in Fig. 3.8 for a 16-bit data bus. In this figure, physical address refers to the address seen by the CPU with the address bus, while the internal addresses are the addresses proper to the bank. In the case of 8-bit data bus MCU's, these two sets may be identical for an example like this one.

Here, two 2 KB hardware memory banks are connected to the 16-bit data bus forming a 4 KB memory segment starting at 0x0000. A *memory segment* is a set of memory words with continuous addresses. One of the banks, called *low bank* has only even addresses attached to its memory words while the other one, the *high bank* has odd addresses. For that reason, they are also known as *even address bank* and *odd address bank*, respectively. The data bus bits D15, D14, ..., D1, D0 are divided in two byte groups, which are connected to the banks.

On the other hand, instructions and data may be longer than a byte, which means that they cannot be stored in one memory cell, but need to be broken into one byte pieces, which are stored in consecutive memory locations. In this case, use the terms *instruction address* and *data address* using the lowest of the set of physical addresses containing the bytes, with the understanding that this number encompasses all the physical addresses of the set. For example, if the data is 23AFh with address F804h, than the physical addresses covered in this case are F804h and F805h, one per byte. Similarly, if the instruction of three words "40Bf 003A 0120" has the address F8AAh, this means that the physical addresses covered are F8AAh to F8AFh.

In particular, if there is a reference to the size of the data, the terms become *word address* for 16-bit long data, *double word address* for 32-bit long data, and so on. Thus, a word address includes two physical addresses and a double word address covers four physical addresses. The physical addresses of the individual

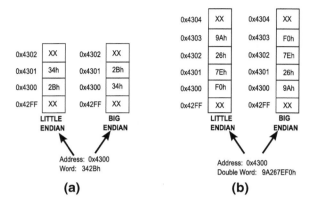

**Fig. 3.9** Big and Little Endian conventions for (**a**) a word ([4300h] = 3428h]) and (**b**) a double-word ([4300h] = [9A267EF0h])

bytes comprising the word or double word depend on the convention used to store them: *big endian* or *little endian*.

**Big Endian**: In the big endian convention, data is stored with the most significant byte in the lowest address and the least significant byte in the highest address.

**Little Endian**: In the little endian convention, data is stored with the least significant byte in the lowest address and the most significant byte in the highest address.

**Second Remark on Notation**: Hereafter, all data addresses and contents will also be written using the same notation as for memory contents introduced on (3.5), that is, [*address*] = *contents*, or *address*: *contents*. For example, 3840:23A2 is for the address 3840h of the word 23A2h. The same information is written as [3840h] = 23A2h. The addresses for individual bytes will follow depending on the endian convention used by the specific microcontroller.

Figure 3.9 illustrates the endian conventions for word (16-bit word) and double word (32-bit word) cases. The XX in this figure are "don't cares", irrelevant for the example.

It is convenient to have word-sized and double word-sized data at even addresses when the data bus is 16-bits wide. Similarly, if the data bus has 32 bits, then we should have double word-sized data at addresses multiples of 4. The reason may be understood taking a closer look at Fig. 3.8. The physical addresses of both banks differ only by bit A0 of the address bus: it is 0 for all bytes in the low bank, and 1 for those in the high bank. Since these are 2K cells, we can use A11 A10 A9 ...A2 A1 to access the internal cells in the banks. Thus, two consecutive physical addresses with the smaller one being even will catch in both banks cells with identical internal addresses. For example, the physical addresses pair 0x0301 − 0x0300 (0000 0011 0000 0001 and 0000 0011 0000 0000 in binary) will access the same internal address in both banks: 001 1000 0000, or 180h. Hence, when the word address is even, a single data bus transaction is needed for any transfer, since the bytes in two contiguous physical addresses corresponding to the two data bytes are accessed simultaneously. This is not possible if the word address is odd.

Associated with the data address, the term *word boundary*, refers to an even address, and *double word boundaries* which refers to addresses which are multiple of 4; in hex notation, the latter means any number ending in 0, 4, 8, C.

By the way, the connection of banks illustrated in Fig. 3.8 corresponds to a little endian convention; for a big endian case, the high bank has the even addresses.

**Example 3.5** *The debugger of a certain microcontroller presents memory information in chunks of words in the list form shown below, where the first column is the address of the word in the second column. Following the debuggers' conventions, all numbers are in hex system. If more than one word is on the line, the address is for the first word only. Assuming that all data is effectively 16-bit wide, break the information into bytes with the respective address, (a) assuming little endian convention, and (b) assuming big endian convention.*

F81E: E0F2 0041 0021
F824: 403F 5000
F828: 831F

**Solution:** *According to instructions, F81E is the address of word E0F2, F820 that of the second word 0041 and so on. With this in mind, we can break the memory information in byte chunks as follows:*

|          | Little endian |          | Big endian |          |
|----------|---------------|----------|------------|----------|
| F81E:F2  | F824: 3F      | F81E: E0 | F824: 40   |          |
| F81F: E0 | F825: 40      | F81F: F2 | F825: 3F   |          |
| F820: 41 | F826: 00      | F820: 00 | F826: 50   |          |
| F821: 00 | F827: 50      | F821: 41 | F827: 00   |          |
| F822: 21 | F828: 1F      | F822: 00 | F828: 83   |          |
| F823: 00 | F829: 83      | F823: 21 | F829: 1F   |          |

## 3.5.3 Program and Data Memory

The previous section focused on address aspects of memory. With respect to the contents, the structure of a microcomputer includes two differentiable types of memory depending on the kind of information they store: *Program Memory* and *Data Memory*.

Program memory, as inferred by its name, refers to the portion of memory that stores the system programs in a form directly accessible by the CPU. A *program* is a logical sequence of instructions that describe the functionality of a computer system.

In embedded systems, programs are fixed and must always be accessible to the CPU to allow the system to operate correctly. Thus, when power is removed from the system and later restored, programs must still be there to allow the system to function properly. This implies that program is usually stored in nonvolatile memory. Program memory capacity in typical embedded systems is in the order of kilo-words. Some microcontroller models may run a program from the RAM portion while others enforce using only ROM sections.

**Fig. 3.10** Topological difference between (**a**) von Neumann and (**b**) Harvard architectures

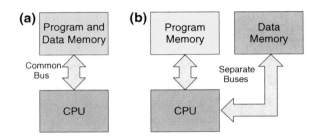

Data memory is used for storing variables and data expected to change during program execution. Therefore, this type of memory should allow for easily modifying its contents. Most embedded systems implement data memory in RAM (Random-Access Memory). Since data memory is meant to hold temporary information, like operation results or measurements for making decisions, the volatility of RAM is not an inconvenience for system functionality. The average amount of RAM needed in most embedded applications is relatively small. For this reason it is quite common to find embedded systems with data memory capacity measured in only a few hundreds words.

Particularly important data which cannot be lost when the system is de-energized, must be stored in the ROM section. Some microcontroller families may access these data directly from there, although they cannot change data during program execution. Other models always copy the data into the RAM section and enforce data manipulation only from these volatile segments.

### 3.5.4 Von Neumann and Harvard Architectures

Program and data memories may share the same system buses or not, depending on the MCU architecture. Systems with a single set of buses for accessing both programs and data are said to have a *Von Neumann* architecture or *Princeton* architecture.[8]

An alternate organization is offered by the *Harvard Architecture*. This topology has physically separate address spaces for programs and data, and therefore uses separate buses for accessing each. Data and address buses may be of different width for both subsystems.

Figure 3.10 graphically depicts the topological differences between a Von Neumann and a Harvard architecture.

Numerous arguments can be brought in favor of either of these architectural styles. Both are present in embedded systems. Texas Instruments MSP430 series uses a Von Neumann architecture while Microchip PIC and Intel's 8051 utilize Harvard archic-

---

[8] In the early days of computing, much of the work that defined the von Neumann architecture was developed by Princeton professor John von Neumann, after whom its named. Due to this affiliation, the von Neumann architecture is also called Princeton architecture.

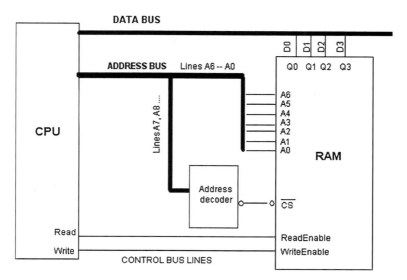

**Fig. 3.11** CPU to memory connection: a conceptual scheme

tectures. Without any preference in this respect, for most aspects of our discussion we will assume a Von Neumann architecture.

The IR and PC register widths will depend on the architecture. In Von Neumman models, they have the same widths as the other CPU registers that may hold addresses, while in Harvard architecture models they are independent of the other registers, since the buses do not have to be of the same size.

### 3.5.5 Memory and CPU Data Exchange: An Example

To illustrate the process by which the CPU and memory interact, let us utilize a simplified example of a static RAM memory interface. Figure 3.11 shows a conceptual connection scheme between the CPU and this RAM.

The RAM terminals can be divided in three groups: (a) the data Input/Output terminals Q0, Q1, ...; (b) The (internal) address terminals A0, A1, ...used to select a specific word cell inside the block; and (c) the select (CS) and control terminals (ReadEnable and WriteEnable) used to operate with the memory. The CS terminal is used to activate the block, i.e. make it accessible, while the other two determine the type of operation to be performed.

The Data Bus lines are connected to the data Input/Output terminals. Internal to the block, these terminals are connected via three state buffers. These buffers allow to set the direction of data flow (read or write) and also set high impedance to disconnect from the Data Bus when the RAM is not activated. The read and write transactions

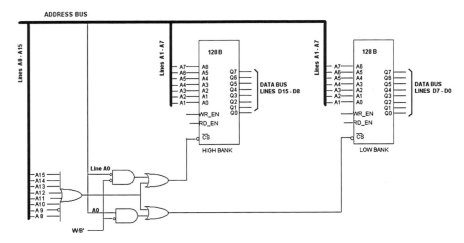

**Fig. 3.12**  256B memory bank example

are controlled by the CPU with signals that go through the control bus. The figure is by no means exhaustive and other control lines may intervene.

The $n$ address bus lines leaving the CPU are separated in two groups: a set of lines that are directly connected to the memory block internal address lines, and another set to an decoder used to activate the memory block. In the example figure, lines A1 to A7 connect directly to the decoder selectors in the RAM while the rest of the address bus lines go to a decoder that will activate the RAM block. The address of a memory location is then the word formed with the address bus bits that activate the RAM and those selecting the word line for the internal location.

**Example 3.6**  *Figure 3.12 illustrates an example of the connection illustrated by Fig. 3.8. It shows two 128B memory modules connected to a 16-bit data bus and driven by a 16-bit address bus and a control signal* **W/B'**. *Since each block has only eight I/O terminals, two of them are required to cover the data bus lines. The control signal* **W/B'** *indicates if the CPU is reading or writing a word (W/B' = 1) or a byte (W/B' = 0). Determine the range of addresses for this memory bank, and what addresses are for each block.*

*Solution: The blocks are activated when a low signal appears at the $\overline{CS}$ terminal, which are connected to OR outputs of the encoder subsystem. If the output of the eight input OR is high, the system is disconnected from the data bus. Hence, we need A15 = A14 = A13 = A12 = A11 = A10 = 0, A9 = 1, and A8 = 0 to ensure that the blocks can be activated. The address lines A7 to A1 activate the internal address lines of the activated RAM block.*

*Now, if signal W/B' = 1 then both blocks are enabled, irrespectively of A0's value. If W/B' = 0 then A0 = 0 activates the low bank RAM and A0 = 1 the high bank RAM. Therefore, the range of addresses covered by the memory bank is as follows:*

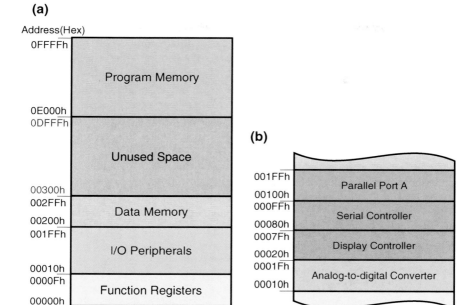

**Fig. 3.13** Example memory map for a microcomputer with a 16-bit address bus. **a** Global memory gap, **b** Partial memory map

*Range: From 0000 0010 0000 0000B to 0000 0010 1111 1111B, i.e., 0200h to 02FFh.*

*Low bank: Activated when A0 = 0, meaning even addresses in the range, that is: 0200h, 0202h, 0204h ..., 02FEh.*

*High bank: Activated when A0 = 1, meaning odd addresses in range: 0201h, 0203, ..., 02FFh.*

### 3.5.6 Memory Map

A *memory map* is a model representation of the usage given to the addressable space of a microprocessor based system. It is an important tool for program planning and for selecting the convenient microcontroller for our application. As implied by its name, the memory map of a microcomputer provides the location in memory of important system addresses. Figure 3.13a illustrates a pictorial representation of the memory map of an example computer system with a 16-bit address bus, with an addressable memory space of 64 K words for a Von Neumann model. In the map, memory is organized as a single flat array.

In this particular case, I/O device addresses and function registers are also mapped as part of the same memory array (see Sect. 3.6). This example is just one of the possible ways of arranging the address space of a microcomputer. In this particular example, the first 16 memory words (addresses 0h through 0Fh) are allocated for function registers. The next 496 locations are assigned to input-output peripheral devices, and only 256 words are reserved for data memory (addresses from 0200h through 02FFh). Program memory is located at the end of the addressable space with 8 K words in addresses from 0E000h through 0FFFFh. Addresses from 00300h through 0DFFFh correspond to unused memory space, that could be used for system expansion.

Memory maps can be global or partial. A *global memory map* depicts the entire addressable space, as illustrated in Fig. 3.13a. A *partial map* provides detail of only a portion of the addressable space, allowing for more insight in that portion of the global map. Figure 3.13b illustrates a partial map of a possible distribution of the I/O peripheral addresses.

### 3.5.7 MSP430 Memory Organization

The MSP430 has a Von-Neumann architecture with a little endian data address organization. Since it also works with an I/O mapped architecture, a concept discussed later in Sect. 3.6, the memory address space is shared with special function registers (SFRs), peripherals and ports. The address bus width depends on the microcontroller model. All models are based on the original 16-bit address bus with an address space of 64 K bytes, called simply the MSP430 architecture. The extended MSP430X architecture has a 20-bit address bus with an address space of 1 M byte. However, the first 64 K addresses in this case have the same map to provide complete compatibility. Figure 3.14 shows the basic global memory map for both models. Models with a relatively large RAM capacity, like those of the family 5xx may be somewhat different. See the device-specific data sheets for specific global and partial memory maps.

The amount of RAM and Flash or ROM depends on the model. RAM memory, which may start with only 128 bytes of capacity, usually starts at address 0200h, and ends depending on the model. Similarly, in the 16-bit model, the Flash/ROM memory ends at address 0FFFFh but the start depends on the capacity. Later models may differ because of capacity needs, but are nevertheless compatible. That is, although RAM may not start at 0200h, this address is still contained in the RAM section so programs written for other models may run.

The interrupt vector table, which is explained later, is mapped into the upper 16 words of Flash/ROM address space, with the highest priority interrupt vector at the highest Flash/ROM word address (0FFFEh). Some models have the possibility of other interrupts besides the basic sixteen original ones, and provide larger tables. See specific data sheets for more information. In addition, some models contain specialized modules such as Direct Memory Access (DMA), flash controllers, or RAM controllers, to optimize power and access time.

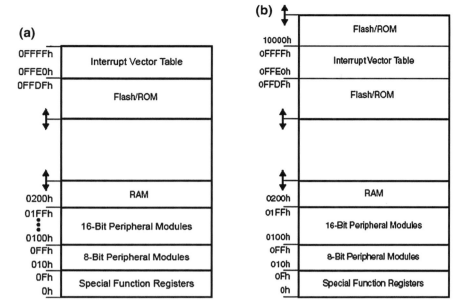

**Fig. 3.14** MSP430 global memory maps (*Courtesy of Texas Instruments, Inc*). **a** 16-bit address bus memory map, **b** 20-bit address bus memory map

**Data and Program Memory** Basically, we should expect to have the program in the flash/ROM section and data in the RAM section. This is the way in which assemblers organize memory unless otherwise instructed. The MSP430 has however an important feature. Since only one 16-bit memory data bus is used for any memory access, programs can be run from the RAM section and data accessed from the Flash/ROM section. This unique feature gives the MSP430 an advantage over some microcontrollers, because the data tables do not have to be copied to RAM for usage. Instruction fetches from program memory are always 16-bit accesses, whereas data memory can be accessed using word (16-bit) or byte (8-bit) instructions.

**Important Remark:** Attempts to access non existent addresses (vacant space) result in errors that reset the system.

## 3.6  I/O Subsystem Organization

The I/O subsystem is composed by all the devices (peripherals) connected to the system buses, other than memory and CPU. The I/O designation is collectively given to devices that serve as either input, output, or both in a microprocessor-based system, and also includes special registers or devices used to manage the system operation without external signals. Like in the read/write convention, the CPU is used as the reference to designate a device as either input or output. An *input transaction* moves information into the CPU from a peripheral device making it an input peripheral. *Output transactions* send information out from the CPU to external devices, making these output devices. Examples of input devices include switches and keyboards, bar-code readers, position encoders, and analog-to-digital converters. Output devices include LEDs, displays, buzzers, motor interfaces, and digital-to-analog converters. Some devices can perform both input and output transactions. Representative examples include communication interfaces such as serial interfaces, bi-directional parallel port adapters, and mass storage devices such as flash memory cards and hard disk interfaces.

Without giving an exhaustive list, common peripherals to be found in most embedded microcomputer systems include those listed below. Detailed discussion of these and other peripherals is found in Chaps. 6 and 7.

**Timers** These peripherals can be programmed for any use prescribing time intervals. For example, to measure time intervals between two events, generate events at specified time intervals, or generate signals at a specified frequency, as it is the case of *pulse width modulators*, and many others.

**Watchdog Timer (WDT)** This is a special type of timer, used as a safety device. It resets if it does not receive a signal generated by the program every X time units, a feature useful in several applications. It may also be configured to generate interrupt signals by itself at regular intervals of time.

**Communication interfaces** Used to exchange information with another IC or system. They use different protocols such as serial peripheral interface (SPI), universal serial bus (USB), and many others.

**Analog-to-Digital Converter (ADC)** Very common since many input variables from the real world vary continuously, that is, they are analog.

**Digital-to-Analog Converter (DAC)** It performs an opposite function to the ADC, delivering analog output signals.

**Development peripherals** These are used during development to download the program into the MCU and for debugging. They include the monitor, background debugger, and an embedded emulator.

The I/O subsystem organization resembles the organization of memory. Each I/O device requires an I/O interface to communicate with the CPU, as illustrated in Fig. 3.15. The interface serves as a bridge between the I/O device and the system buses. Each I/O interface contains one or several registers that allow for exchanging data, control, and status information between the CPU and the device. As a consequence, as far as it concerns the CPU, the I/O device "looks" very much alike hardware memory, with I/O interface registers connected to data bus and selected with the address bus. In the case illustrated in Fig. 3.15, the interface contains eight registers, accessible at addresses 0100h through 0107h.

An I/O interface, like in the case of memory system, may also contain address decoders to decode the addresses assigned to each particular device, buffers, latches, and drivers, depending on the application.

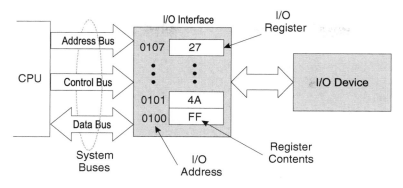

**Fig. 3.15**  Structure of an input/output (I/O) interface

Microprocessors like the Intel 80x86 family and Zilog Z80 have separate address spaces for memory and I/O devices, requiring specific input and output instructions to access peripherals in such a space. This strategy, called *Standard I/O* or *I/O-mapped I/O*, requires special instructions and control signals to indicate when an address is intended to the memory or to the I/O system.

An alternative is to include the I/O devices inside the memory space; this scheme is named *memory-mapped I/O*. This is normal since I/O registers are seen just as other memory locations by the CPU. This has become the dominant scheme in embedded systems, making I/O and memory locations equally accessible from a programmer's point of view. The memory map illustrated in Fig. 3.13 shows an example of memory assignments to the I/O peripherals (00000h to 001FFh) with more details in the partial map (b).

### Other Registers and Peripherals

Besides the I/O peripherals, the system may include other registers and devices which are not intended to connect to the external world or to be used as memory. This set includes special function registers, timers, and other devices whose function are related to the functionality of the system, power management, and so on. The CPU communicates with these registers and devices just as it does with the rest of the system, using the system buses. Figure 3.13 illustrates inclusion of special function registers in the memory map.

## *3.6.1 Anatomy of an I/O Interface*

The most general view of an I/O interface includes lines to connect to the address, data, and control buses of the system, I/O device connection lines, and a set of internal

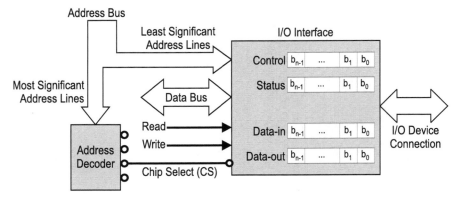

**Fig. 3.16**  Anatomy of an input/output (I/O) interface

registers. Figure 3.16 illustrates such an organization. A few least significant address lines are used to select the internal interface registers. The upper address lines are usually connected to an off-interface address decoder to generate select signals for multiple devices. Data lines in the designated data bus width (8- or 16-bit in most cases) are used to carry data to and from the internal registers. At least two control lines are used to indicate read (input) or write (output) operations and synchronize transfers. The device side includes dedicated lines to connect with the actual I/O device. The number and functions of such lines change with the device itself.

Internal registers in the I/O interface may be read-only, write-only, or write-and-read type from the CPU viewpoint, depending on the register type, the interface model, and the MCU model. Please refer to the appropriate user guide or data sheet. Three types of internal registers can be found in any interface:

**Control Registers**: Allow for configuring the operation of the device and the interface itself. One or several control registers can be provided depending on the complexity of the interface. Sometimes this type of register is called Mode or Configuration Register.

**Status Register**: Allow for inquiring about the device and interface status. Flags inside these registers indicate specific conditions such as device ready, error, or other condition.

**Data Registers**: Allow for exchanging data with the device itself. Unidirectional devices might have only one data register (Data-in for input devices or Data-out for output devices). Bi-directional I/O interfaces include both types.

Further and more detailed discussion on IO interfaces is found in Chaps. 6 and 7. Let us for the moment use the following examples to illustrate the above concepts.

**Example 3.7** *Consider a hypothetical printer interface as an example of an output device. This interface contains three 8-bit internal registers: control, status, and Data-out registers. The control register allows the CPU to control how the printer operates while the status register contains information about the status of the printer.*

**Table 3.3** Internal addresses for sample printer interface

| A1 | A0 | Register |
|---|---|---|
| 0 | 0 | Control |
| 0 | 1 | Status |
| 1 | 0 | Data-out |
| 1 | 1 | Not used |

*The data-out register is an 8-bit write-only location, accepting the characters to be printed.*

*Since there are three registers, at least two address lines, A1 and A0, are needed to internally select each of them. Assume the internal addresses are assigned as shown in Table 3.3.*

*The address bus interface in this case will connect the two least significant address lines: A1 and A0 to the respective address signals in the printer interface. An external address decoder uses lines A15 to A2 to set the interface base address at 0100h. Thus, the addresses of the control, status, and Data-out registers will then be 0100h, 0101h, and 0102h, respectively. Address 0103h will be unused since that address is not assigned to any register and cannot be redirected because the select signal is controlled by address lines A2, A3, ... A15. Figure 3.17 shows a diagram of how the interface would connect to the CPU buses and to the printer.*

*Let's assume the status and control register bits have the meaning shown in Fig. 3.18.*

*Using address 0100h, the CPU may then write on the control register a word that tells the printer how the character will be printed (normal, bold, etc.) or if the paper should be ejected, and so on. If a character is to be printed, the CPU then writes on the Data-out register, with address 0102h, the code of the desired character. Reading the status register, on memory address 0101h, the CPU can check if the printer is ready to receive a character, or if it has problems and what type of problem, and so on.*

*Thus, by issuing the corresponding transfer instructions, the operation of this printer interface can be easily managed by the CPU and have a document printed by sending one character at a time.*

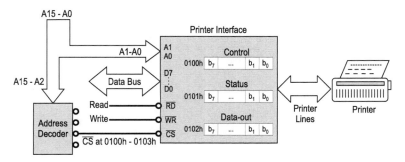

**Fig. 3.17** Connection of printer interface to CPU buses and printer itself

Let us now consider the I/O port registers for the MSP430 microcontrollers as an actual interface example of interest for us. In general, MCU I/O ports work as blocks, half blocks, or single lines. For example, if a port has eight bits, all eight bits could work as input or output terminals. Sometimes they can be configured part as output and part as input. MSP430 I/O ports have all pins independently configurable. Hence, even if all the set is associated with one register, the individual bits in the register may be working each one independent of the others. The following example gives an introduction to the digital I/O ports of the MSP30 architecture. A detailed discussion of I/O ports is provided in Chap. 6.

Ports in the MSP430 are named as P$x$, where $x$ may go from 1 to 10. The pins in a port (P$x$) are identified as P$x.n$, where $n$ goes from 0 to 7. Not all ten ports are present in all MSP430 microcontrollers.

**Example 3.8** *The MSP430 I/O port pins work independently. They can function either as input or output pins. That flexibility makes necessary at least three registers associated to the port: one for configuring the flow of information, one for input data, and one for output data. In addition, to optimize resources, many MCU pins are shared by two or more internal devices. This makes sense because it is very rare to use all of them in one application. Hence, one more register is necessary to select between the I/O or the other modules. The registers are listed next and their application as indicated in Fig. 3.19a.*

**Direction Register (PxDIR):** *Selects in or out direction function for pin, with 1 for output direction and 0 for input direction.*

**Input Register (PxIN):** *This is a read-only register. The value changes automatically when the input itself changes.*

**Output Register (PxOUT):** *to write signal to output. This is a read-and-write register.*

**Function Select Register (PxSEL):** *Used to select between I/O port or peripheral module function. With PxSEL.n = 0, pin Px.n operates as an I/O pin port; with PxSEL.n = 1, as a module pin.*

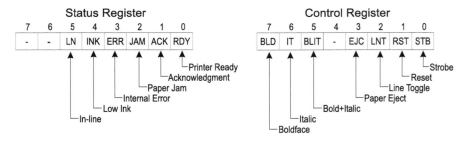

**Fig. 3.18** Status and control registers in printer interface example

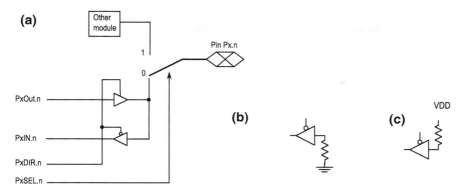

**Fig. 3.19** Basic IO Pin hardware configuration: **a** Basic I/O register's functions; **b** pull-down resistor for inputs; **c** pull-up resistors

*Let's say for example, pins 7 to 3 of port 6 are to be used as output pins, and pin 2 for operating the module, then we should use appropriate CPU instructions to write* 0xF8 *in the P6DIR register and* 04h *in the P6SEL register.*

*Since the input port goes through a three-state buffer, it is not advisable to leave it floating when the pin operates in input mode. It is necessary to connect a pull-up or a pull-down resistor, as shown in* Figs. 3.19b, c.

*Additional configuration registers might be included as part of an I/O port. Examples include registers to configure interrupt capabilities in the port, or to use internal pull-up/pull-down resistors, and other functions.* Section 8.1 *in* Chap. 6 *provides a detailed discussion of I/O port capabilities in MSP430 microcontrollers.*

### 3.6.2 Parallel Versus Serial I/O Interfaces

In microcomputer-on-a-chip systems, most peripherals are connected to the data bus via a parallel interface, i.e., all bits composing a single word are communicated simultaneously, requiring one wire per bit. But I/O ports interacting with devices external to the system, may connect via parallel or serial interfaces. The ports are then referred to as *parallel I/O ports* and *serial I/O ports*. Serial interfaces require only one wire to transfer the information, sending one bit at a time.

Serial interfaces were the first to be in use for inter-system communications, especially the RS-232 used for devices requiring long distance connections. A reason for preferring serial lines was the economy resulting of using only one wire (plus ground). However, in early computer systems serial ports were slow, and for short length, fast connections, parallel interfaces were preferred. Thus in early computer systems, most printers, mass storage devices, and I/O sybsystems made broad use of parallel ports. Protocols like the ISA/EISA/PCI Buses, GPIB, SCSI, and Centronics[9] are just a few examples of the myriad of parallel communications standards that flourished in that era.

On the other side, parallel connections have their own problems as speeds get higher and higher. In modern technology for example, wires can electromagnetically interfere with each other; also, the timing of signals must be the same and this becomes difficult with faster connections so now the pendulum is swinging back toward highly-optimized serial channels even for short length interconnections. Improvements to the hardware and software process of dividing, labeling, and reassembling packets have led to much faster serial connections. Nowadays we encounter printers, scanners, hard disks, GPS receivers, display devices, and many other peripherals that reside relatively close to the CPU using serial channels such as USB, FireWire, and SATA. All forms of wireless communication are serial. Even for board-level interconnections, standards like SPI, $I^2C$, and at a higher note, PCIe, to mention just a few examples, are representative of the trend serializing most types of inter- and intra-system communications.

### 3.6.3  I/O and CPU Interaction

Two major operations are involved when operating with the I/O subsystem: one is data transfer, i.e., sending or receiving data numbers, and the other is timing or synchronizing. In the first category, input and output port registers are the major players.

Input ports always transfers data toward the CPU. They may be of the *buffered input* or the *latched input* type. In the first case, they do not hold input data and the CPU can only read input present at that instant. Latched inputs do hold the data until this is read by the CPU, after which they are usually cleared and ready for another input.

Output port registers may be write-only or read-and-write registers, depending on the hardware design. The output state is in most cases latched, allowing to hold the output data at the pins until the next output operation is executed.

Timing or synchronizing is necessary when the nature of the peripheral or external device is such that interaction with the MCU must wait until the device is ready. For example, hard printing is a slower operation than electronic transfer. Hence, printers usually receive the information in a buffer, from which the data is processed. A printer cannot receive more data from the MCU once its buffer is full. Another example are data converters, which must complete conversion before delivering the result to the MCU.

Devices use a *flag* to indicate their readiness to receive or deliver data. A flag is a flip flop output that is set or cleared by the device when it is ready to communicate

---

[9] Centronics, although widely used in printers, never actually reached an official standard certification.

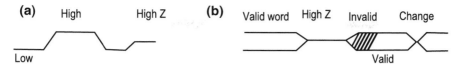

**Fig. 3.20**  Definitions of timing concepts. **a** Single signal timing convention, **b** Bus timing convention

with the MCU. Using this flag, two methods of synchronizing I/O devices are used: *polling* and *interrupt*.

Polling tests or polls the flag repeatedly to determine when the device is ready. The interrupt method continues with the normal program run and reacts to the flag when the device is ready. When a system is programmed so that it will only react to interrupts, it is said to be *interrupt driven*. Polling and interrupts are explained in more detail in later sections.

### 3.6.4 Timing Diagrams

Data transfer or communication between the CPU and other hardware components uses signals sent through the system buses. As explained earlier, data bus lines are used to transfer data, the address bus to select which memory cell or other piece of hardware the CPU works with, and signals from the control bus dictate what and how the transfer takes place. A bus requires a protocol to work, that is, a set of rules describing how to transfer data over them. This protocol includes information about the times required for each step in the process, to allow settling of states, delays and so on, as well as the order in which the signals are activated. The most common way to define the protocols is with *timing diagrams*, whose basic definitions are shown in Fig. 3.20.

In a timing diagram time flows from left to right. For single bits, two states are defined: high or low. Tri-state lines also include a third state called *high impedance* or "High Z" for short. State changes occur within a finite time greater than zero, as denoted by the oblique lines in Fig. 3.20. For buses, where several bits are to be considered, valid words are represented by parallel lines at high and low levels. High impedance states for the bus are represented by a line that is neither high or low.

Figure 3.21 shows a simplified example of time diagram protocols for write and read transfers. In this particular example, only two bus control signals are assumed to intervene: $\overline{enable}$, an active low signal which activates the memory block and marks the transfer time, and $read/\overline{write}$ which indicates the transfer direction. When $read/\overline{write}$ is high, it enables reading, that is, transfer from memory to the Data Bus to be retrieved by the CPU; and when low, enables writing from Data Bus into memory. Actual timing diagrams provided in manuals are more elaborated and

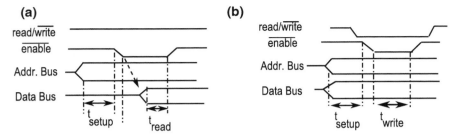

**Fig. 3.21** Example of simplified write and read timings. **a** Read timing diagram, **b** Write timing diagram

may include several signals from the control bus. Chapter 6 deals with more detail on this topic.

### 3.6.5 MSP430 I/O Subsystem and Peripherals

The I/O and peripheral subsystem of the MSP430 family of microcontrollers is memory mapped, occupying addresses 0x0000 to 0x01FF. The operation of the different MSP430 members is controlled mainly by the information stored in special function registers (SFRs) located in the lower region of the address space, 0x0000 to 0x000F. The SFRs enable interrupts, provide information about the status of interrupt flags, and define operating modes of the peripherals. The ability of disabling peripherals not needed during an operation is a low power feature since the total current consumption can be reduced.

There is a wide selection of MSP430 microcontrollers models, even within one generation, with diverse offerings in peripherals. A partial list of peripherals available in different models include internal and external oscillator options, 16-bit timers, hardware for pulse width modulation (PWM), watchdog timer, USART, SPI, I2C, 10/12-bit ADCs, and brownout reset circuitry. Some family members include less common features such as analog comparators, operational amplifiers, programmable on-chip op-amps for signal conditioning, 12-bit DACs, LCD drivers, hardware multipliers, 16-bit sigma-delta ADCs, and multiple DMA channels. The user must see the specific device data sheet for available peripherals and SFRs. Peripherals are described later in this book.

**Peripheral Map**

Peripheral modules are mapped into the address space. SFRs occupy the space from 0h to 0Fh, followed by the 8-bit peripheral modules from 010h to 0FFh. These modules should be accessed with byte instructions. Read access of byte modules using word instructions results in unpredictable data in the high byte. If word data is written to a byte module only the low byte is written into the peripheral register, ignoring the high byte.

The address space from 0100 to 01FFh is reserved for 16-bit peripheral modules.

### I/O Ports and Pinout

MSP430 microcontroller models, as most microcontrollers, usually have more peripherals than necessary for a given application. Therefore, to minimize resources, several modules may share pins. Figure 3.22 shows package pinout descriptions for two models.

Inset (a) shows the MSP430G2331 pinout. We see here that, with the exception of pin 5, which has only one port connection to bit 3 of port 1, P1.3, all other port connections have more than one functions. It becomes therefore necessary to configure the hardware so as to include the selection of the desired function. This is yet another control register to be included in the port interface.

In the MSP430 families '1xx to '4xx, I/O ports are 8-bit wide, and are numbered P0 to P8; not all ports are available in a chip. Later families '5xx and '6xx have 8-bit ports P1 to P11, and may combine pairs to create 16-bit I/O ports named PA to PE. For example, P1 and P2 together can formed one 16-bit port PA. The reader should consult the specific device data sheet for further information.

## 3.7  CPU Instruction Set

The architecture and operation of a CPU are intimately related to how its instructions are organized in the *Instruction Set* and how instructions access data operands, the *Addressing Modes*. Instructions for the CPU are stored in memory just as any other word. What makes it an instruction is the fact that it is going to be decoded by the CPU during the instruction cycle, to find out what to do.

This section focuses on the software component of the CPU, that is, the instructions and the addressing modes. As such, the focus is now turned to the programmer's model of the CPU while describing this software side of the architecture. Before going into these details, let us first introduce useful notation to describe instructions, operations, and their programmer's model.

**Fig. 3.22**  Package pinout of two MSP430 microcontroller models: **a** MSPG2231 Pin description, **b** MSP430x1xx Pin description (*Courtesy of Texas Instruments, Inc.*)

## 3.7.1 Register Transfer Notation

For the programmer, the detailed description of hardware is not the main interest. The concepts more important to know are what registers the CPU provide for data transactions, what are the memory and I/O maps of the device, the instruction set, and so on. Hardware details become important when fine tuning the program, optimizing, debugging, and so on.

It is important for the programmer to have a notation available for operations in MCU environments, independent of the specific MCU architecture but taking into consideration the features of the systems. One such notation is the *register transfer notation* (RTN).

After executing an instruction, the contents of a register or cells in memory may be written upon with a new datum. In this case, the register or cell being modified is called *destination*. The *source* causing the change at destination may be a datum being transferred (copied) or the result of an operation. This process is denoted in abstract form as

$$destination \leftarrow source \quad \text{or} \quad source \rightarrow destination \qquad (3.1)$$

In this book we adhere to the left hand notation. The notation in programmer's model for the different operands that may be used in RTN are as follows:

1. Constants: These are expressed by their value or by a predefined constant name, for example 24, 0xF230, MyConstant. Constants cannot be used in destination.
2. Registers: These are referred to by their name. If it is in abstract form without reference to a particular CPU, it is customary to use Rn, where n is a number or letters.
3. Memory and I/O: These are referred to by the address in parenthesis, as in (0x345E), which means "The data in memory at address 0x345E." Notice that it is data address, not physical address. If the address information is contained in register Rn we write (Rn), meaning by that "The data in memory at address given by Rn". We also say that the register *points* to the data.

Operations are allowed in address expressions, as in (Rn + 24h). Addressing modes can also be applied to RTN expressions, if there is no conflict in using them. Alternate notation for memory addresses is using the letter M before the parenthesis, as in M(0x345E), M(Rn + Offset).

When the size of an operand is not clear from notation, it must be explicitly stated at least once, to avoid ambiguity. (0240h), the datum at address 0240h, may be a byte, a 16-bit word, or a double word. If the size is not clear by the context, we write for example byte(0240h) or word(0240h).

Observe that Rn and (Rn) are distinct operands. The first one refers to the contents of register Rn, and also defines the data size. The second refers to the data in memory or I/O register whose address is the contents of Rn and does not specify the data size.

If the destination operand appears in the source too, the datum to be used in the source is that before the operation, while in the destination goes the result of the operation.

**Example 3.9** *The following transactions illustrate RTN for memory operands. For these examples, let us assume word-size data at addresses before each transaction as* [0246h] = 32AFh *and* [028C] = 1B82h. *Moreover, let us assume little endian storage.*

| RTN | Meaning | Result |
|---|---|---|
| (0246h) ← 028Ch | Store 028Ch at data address 0246h | [0246h] = 028Ch |
| Word(0246h) ← (028Ch) | Copy at memory location 0246h the word at memory address 028Ch | [0246h] = IB82h |
| Byte(0246h) ← (028Ch) | Copy at memory location 0246h the byte at memory address 028Ch | [0246h] = 3282h |
| (0246h) ← (0246h) + 3847h | Add 3847h to the current contents of memory location 0246h | [0246h] = 6AF6h |
| Byte(028Dh) ← (028Dh) − (0247h) | Subtract the byte memory location 0247h from the byte at memory address 028Dh | [028Ch] = E982h |

*Observe that in the second, third, and fifth transactions the data size had to be explicitly given, at least once, because otherwise the transaction become ambiguous.*

The next example combines registers and memory locations.

**Example 3.10** *Assume two 16-bit registers, R6 and R7, with contents* R6 = 4AB2h *and* R7 = 354Fh, *respectively. Moreover, assume words at addresses* [4AB2h] = 02ACh, [4C26h] = 94DFh *and* [4AB8h] = 3F2Ch. *Assume little endian convention when necessary. The following examples illustrate more RTN expressions:*

| RTN | Meaning | Result |
|---|---|---|
| R6 ← 3FA0h | Load R6 with 3FA0h | R6 = 3FA0h |
| R7 ← R6 | Copy in R7 the current contents of R6 | R7 = 4AB2h |
| R7 ← (R6) | Copy in R7 the word whose memory address is stored in R6. Also: Copy in R7 the word in memory pointed at by R6. | R7 = 02ACh |
| byte(4C26h)← byte(4C26h) (R6) | − Subtract from the byte at address 4C26h the byte whose address is given by R6 | [4C26h] = 9433h |
| (R6) ← (R6) + R7 | Add the contents of R7 to the word pointed at by R6. | [4AB2h] = 37FBh |
| R7 ← (R6 + 6h) | Copy in R7 the memory word whose address is given by R6 + 6h | R7 = 3F2Ch |
| byte(R6) ← (R6) − 0x92 | subtract 0x92 from the current memory byte at address given by contents of R6. | [4AB2h] = 021Ah |

Sometimes we may simplify notation by simply stating in words the objective. For example, when comparing register contents arithmetically or testing register bits, the destination is the status register flags, but we may simply state the comparison or testing being done, and leave the SR implicit.

### 3.7.2 Machine Language and Assembly Instructions

As explained in Sect. 3.3.1, an instruction cycle starts by fetching a word from memory using the address stored at the Program Counter. This word is the *instruction word*. The complete instruction may consist of more than one word, but only the leading one is the instruction word. Depending on the number of words comprising an instruction, as well as its type, the CU will require one or more instruction cycles to complete the operation.

Since instructions are found in memory and loaded in registers, it is obvious that they consist of 0's and 1's. We say that they are encoded in *machine language*.

#### Machine Language Instructions

The bits of an instruction word are grouped by fields. Usually, the most significant one is the *operating code* (OpCode) which defines the operation to be performed. Other fields contain information about the *operands* and the *addressing mode*. The addressing mode is the way in which an operand defines the datum. For example, in an RTN expression we may have Rn or (Rn). In both cases the operand is Rn, but the datum is found in different places.

Following the OpCode, there might be information for zero, one or two operands, depending on the instruction. Some MCU models include up to three operands. This case is not considered in this book. The operands, together with addressing mode and other fields, provide information about data needed to execute the operation. Figure 3.23 shows four CPU instructions in the MSP430 machine language together with the operations done shown in RTN.

Notice that the suffix "h" is not shown for the words in hex notation. This is not an accident. *By default, it is customary to consider the base for machine language and addresses as hexadecimal.*

#### Assembly Language Instructions

To avoid using binary or hexadecimal representation for instruction words, scientists and engineers devised *high level* and *assembly* languages. Java, C, and Basic belong to the first group. Instructions in assembly language, the first to appear, on the other side, are the same as machine language. That is, each assembly instruction is associated with one, and only one, machine language instruction. But assembly language

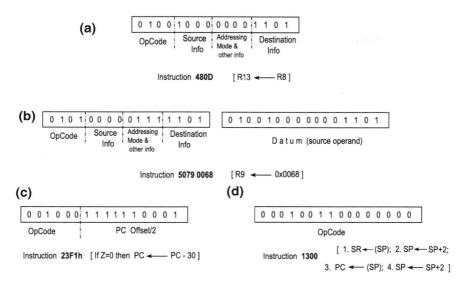

**Fig. 3.23** Four machine language instructions for the MSP430

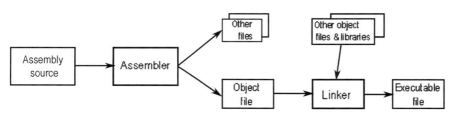

**Fig. 3.24** Basic assembly process

targets a more human-like notation. The left table below shows the MSP430 assembly translations for the machine language instructions of Fig. 3.23. The table on the right shows examples of machine and assembly encoding for Freescale HC11.

| Machine language | Assembly language |
|---|---|
| 480D | mov R8,R13 |
| 5079 006B | add.b #0x6B,R9 |
| 23F1 | jnz 0x3E2 |
| 1300 | reti |

**Basic assembly process.** The basic principles of the assembly process are illustrated in Fig. 3.24.

| Machine language | Assembly language |
|---|---|
| C8 34 | EORB #$34 |
| C0 07 | LDD #7 |
| 7E 10 00 | JMP $1000 |
| 5C | INCB |

Programmers write the program using a text processor to generate a *source file*. This source contains the assembly instructions as well as *directives*, instructions, *comments*, and *macro directives*. Directives are used to control the assembly process and are not executed by the CPU. Macro directives allow to group a set of assembly instructions into only one line; they serve to simplify the source from the human point of view.

The source file is then fed to an *assembler* which expands the macros and translates assembly instructions into machine language generating an object file and other files useful for debugging. *Interpreters* are a special type of assemblers that interpret assembly instructions in a program one line at a time.[10]

The object file is sent to the *linker*, which combines if necessary the file with previously assembled object files and libraries to generate the final *executable file*. This is the one to be actually downloaded onto the microcontroller memory.

Directives allow to organize the memory allocation, to define constants, labels, and introduce other features which facilitate the programmers work. Two pass assemblers run more than once over the source text to properly identify labels and constants.

Assembly instructions syntax depends on the microcontroller family, but the underlying principles are similar: a name for the OpCode, called **Mnemonics**, and **Operands**, specified using special syntax for addressing modes. Similarly, each assembler has its own directives, but again the underlying principles are very similar.

Chapter 4 is dedicated to assembly language programming for the MSP430 family. We focus in one family to have a real world reference, but the principles are similar for other MCU families, with the corresponding changes in syntax and adaptation to the underlying hardware.

### 3.7.3 Types of Instructions

At the most basic level, the instruction set of any CPU is composed of three types of instructions: data transfer, arithmetic-logic, and program control. The number and format of specific machine code instructions in each of these groups can change from one CPU architecture to another. However, in any case we expect a basic set of operations that serve as the basis for developing assemblers and high-level language compilers.

---

[10] An interesting online interpreter for the MSP430 instructions is found at http://mspgcc. sourceforge.net/assemble.html.

## Data Transfer Instructions

Data transfer instructions, as implied by their name, move data information from one source to a destination. In fact, they only copy the source in the destination without modifying the source. The exception to this behavior is the instruction swap.

The execution of data transfer instructions involves the BIL and registers. Flags in the status register SR are usually not affected by these instructions. Examples of common data transfer operations/instructions include:

- move or load: Copies data from a source to a destination operand. In RTN, *dest* ← *src*.
- swap: Exchanges the contents of two operands. In RTN, *dest* ↔ *src*.
- Stack operations push, and pop or pull: These are explained in Sect. 3.7.4
- Port operations in and out: to read and write data from I/O ports in I/O mapped architectures.

In early processors based on a load-store architecture, data transfer instructions always involve a special data register called *accumulator*. Nowadays some families continue with this tradition.

Port operations are required in I/O mapped architectures. I/O memory mapped devices use the same data transfer instructions for memory and I/O port operands.

The stack operations always involve the stack pointer. The importance of these operations makes it worth a special section, Sect. 3.7.4, and will be explained there.

## Arithmetic-logic Instructions

Arithmetic-logic instructions are those devoted to perform arithmetic and/or logic operations with data. They also include other operations with the contents of a register or memory word. Their execution uses the arithmetic logic unit, registers, and the BIL. Instructions of this type usually affect the flags in the SR.

Most MCUs, especially RISC type, involve at most two operands in the instructions. Therefore, this type of instructions are of the format

$$destination \leftarrow (Destination Operand \oslash Source Operand) \qquad (3.2)$$

where $\oslash$ is an operation. Typical expected operations are addition, subtraction, and logic bitwise operations such as AND, OR, and XOR. Multiplication and division are not supported by all ALU's.

The actual set supported by an MCU depends on the architecture design of the CPU and ALU. Those operations not supported by it, usually have to be realized with software when necessary.

**Compare and Test Operations:** These two types of instructions realize an operation, usually subtraction or AND, without modifying any operand. The objective is to affect flags in the status register, usually to make decisions.

**Rotate and Shift Operations:** These type of instructions are not of the format shown in (3.2). Instead, they displace the bits internal to the destination operand.

Load-store architectures restrict all arithmetic-logic instructions to use the accumulator as one of the source operands and always as destination for storing the result. This limitation is overcome in RISC architectures and orthogonal architectural designs where virtually any GPR or memory location can be used as source or destination.

**Bitwise Operations** When the operation in (3.2) is of the logic type, very often it is bitwise. That is, the execution is realized bit-by-bit without one affecting the other, in the form

$$dest\_bit(0) \leftarrow (dest\_bit(0) \oslash src\_bit(0), dest\_bit(1) \leftarrow (dest\_bit(1) \oslash src\_bit(1), \ldots$$

Bitwise logic instructions allow us to set, clear or modify specific bits in an operand without affecting others, by exploiting the logic properties of the operation in the way shown in Table 3.4 and illustrated in Fig. 3.25. Also, we may test if a particular bit is 1 or 0.

**Program Control Instructions**

Program control instructions allow us to modify the default flow of execution in a program. By default, the address of the instruction to be fetched next follows right after the one being currently executed. This address is automatically loaded into the PC after the decode phase. When a program control instruction is executed, the contents of the PC may be changed so the following instruction to be fetched is not the default one. There are three types of program control instructions:

**Unconditional Jump** An unconditional jump always changes the program control flow to an address as indicated by the instruction. The action is $PC \leftarrow NewAddress$.

**Conditional Jump** The PC contents will be changed only if some condition of the status flags is met. That is : "If flag $X = n$, then $PC \leftarrow NewAddress$". Table 3.5 illustrates the basic conditional jumps using mnemonics for the notation.

**Subroutine Calls and Returns** These instructions allow the programmer to transfer program flow to and from special sections of code called subroutines which are outside the memory segment of the main code. The modus operandi of these instructions is explained in Sect.3.7.4.

Jumps are also called Branches in literature. Table 3.5 does not show all conditional jumps, only the basic ones depending on the common flags. Other conditional jumps use combination or operations with flags. We can think of conditional jumps as the instructions that will be associated to flowchart decision symbol, as illustrated in Fig. 3.26.

Common uses of the `jnz` instruction are in wait or delay loops and in iteration loops where a process has to be repeated N times. This is illustrated in Fig. 3.27.

**Table 3.4**  Logic properties and applications

| | |
|---|---|
| 0.AND.X = 0;<br>1.AND.X = X | To **clear** specific bits in destination, the binary expression of the source has 0 at the bit positions to clear and 1 elsewhere (Fig. 3.25a) |
| 0.OR.X = X;<br>1.OR.X = 1 | To **set** specific bits in destination, the binary expression of the source has 1 at the bit positions to set and 0 elsewhere (Fig. 3.25b) |
| 0.XOR.X = X;<br>1.XOR.X = X̄ | To **toggle or invert** specific bits in destination, the binary expression of the source has 1 at the bit positions to invert and 0 elsewhere (Fig. 3.25c) |

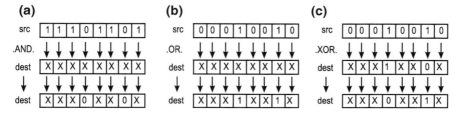

**Fig. 3.25**  Using logic properties to work with bits 1 and 5 only

**Table 3.5**  Conditional jumps

| Mnemonics | Meaning | Mnemonics | Meaning |
|---|---|---|---|
| jz | Jump if zero (Z = 1) | jn | Jump if negative (N = 1) |
| jnz | Jump if not zero (Z = 0) | jp | Jump if positive (N = 0) |
| jc | Jump if carry (C = 1) | jv | Jump if overflow (V = 1) |
| jnc | Jump if no carry (C = 0) | jnv | Jump if not overflow (V = 0) |

Notice in the pseudo code presented the use of the *label*. A label is attached to an instruction and it can be used in a jump instruction to indicate the address of which instruction should result in the PC if the branch is executed. We will go over labels later again.

### 3.7.4 The Stack and the Stack Pointer

The stack, a specialized volatile memory segment used for temporarily storing data, is managed with the *Stack Pointer* (SP), both by programming or by the CPU itself.

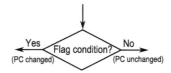

**Fig. 3.26**  Decision symbol associated to conditional jumps

To store data in this segment using a stack transaction is *to push*. Retrieving a datum from the stack is to *pop*, or to *pull*. Any stack transaction makes implicit use of the SP, which points to the address where the push or pop is to be realized.

The term "stack" was coined in analogy with tray stacks, which operate in a natural Last-In-First-Out (LIFO) fashion: the item on the top is the last one put there, and the first to be pulled out from the stack. For this reason, the contents of the stack pointer is usually called *Top-of-stack* (TOS). After each operation, the SP is updated with the new TOS.

In most MCU architectures, the stack is user or compiler defined, and the stack usually grows *downward*, not upwards.[11] This means that the SP contents diminishes each time a new item is pushed onto the stack and increases when an item is pulled. The SP usually starts pushing at the top address of RAM, opposite to usual memory operations which start storing at the bottom. Considering this, it makes sense for the stack to grow downward.

On the other hand, the analogy with the term "top of stack" address should be handled with care, since the memory address will depend on whether we are pushing or popping a datum. Let us illustrate the stack process with an analogy to a rack for DVDs, using Fig. 3.28. Each shelf has a number attached to it: 10, 9, ..., 0. Let these be the shelves' addresses.

In (a), five DVD's have been stored, starting at shelf 10. The last one in is occupying shelf 6. If another DVD is pushed, the TOS address to consider is shelf 5, where this DVD will go. If another item is pushed afterwards, it will go to TOS = 4. Now, if we pull or pop the last item in, TOS = 6 as illustrated in (b). After this pop, the new "last-in" item is in address 7, which becomes the TOS for popping.

Hereafter, we will simply use the term "top of stack" (TOS) for any case, unless it becomes necessary to differentiate. Since SP is precisely related to the TOS address, we refer to its contents as the TOS. The user, or the compiler, may define the stack segment by initializing the SP register.

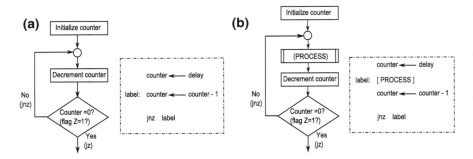

**Fig. 3.27** Flow diagram and instruction skeleton associated to (a) delay loops and (b) iteration loops

---

[11] Just like the stacks of most graphic calculators and computer monitors, where new items appear at the bottom and not at the top.

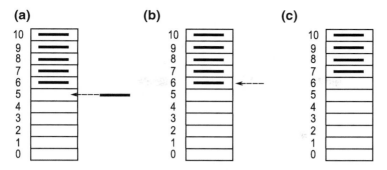

**Fig. 3.28** Illustrating the stack operation. **a** TOS = 5 for pushing, **b** TOS = 6 for pulling, **c** After pulling, new TOS = 7

Depending on the MCU, the SP will contain the TOS for pushing or the TOS for pulling. The case will determine how the push and pop operations are physically performed, and also how the programmer may read or change contents of previously pushed items without going into a stack transaction.

CASE A: Pushing when SP points at TOS: The operations are performed as follows:
1. Pushing a source:

$$(SP) \leftarrow src$$

$$SP \leftarrow SP - N$$

That is, it first stores the source and then updates SP.

2. Pulling into a destination:

$$SP \leftarrow SP + N$$

$$dst \leftarrow (SP)$$

That is, it first updates SP, and then retrieves the data.

CASE B: Popping with SP pointing at TOS. The operations are performed as follows:
1. When pushing a source:

$$SP \leftarrow SP - N$$

$$(SP) \leftarrow src$$

That is, it first updates SP and then stores the data.

2. When pulling into destination:

$$dst \leftarrow (SP)$$

$$SP \leftarrow SP + N$$

That is, it first retrieves the data and then updates the source.

The value of N is 1, 2 or 4, depending on the MCU model. For 4 and 8-bit MCU's, N = 1. For 16-bit MCU's, N = 2. For 32-bit MCU's, N may be fixed to 4, or else N = 2 or 4, depending on the length of the manipulated data. Let us illustrate the stack operation with two examples. The user should consult the MCU's documentation to know which case applies to the system being used. The MSP430, in particular, uses the TOS for pop.

**Example 3.11** *This example illustrates the stack process in the 16-bit MSP430 family using Fig. 3.29. In this family, register SP points to the last item pushed onto the stack (next to be pulled). SP updating is done by adding/subtracting 2 to its contents.*

*For simplicity, memory contents are expressed without the "h" suffix and refer to word data, occupying two physical byte space locations. Word addresses are even.*

*Figure 3.29a illustrates the push operation, with initial contents for registers SP and R9 being 027Eh and 165Ah, respectively (Fig. 3.29a.1). Notice that SP is pointing to the address 027Eh, which is the TOS. The push operation is executed in two steps:*

*Step 1: Update TOS: SP ← (SP − 2), i.e., SP becomes 027Eh-2 = 027Ch*

*Step2: Copy the contents of R9 at the TOS, (SP) ← R9.*

*The result is shown in Fig. 3.29a.2.*

*Now consider the pop operation illustrated by (b), with similar initial conditions, as shown in Fig. 3.29b.1. The pop operation is executed in two steps:*

*Step 1: Copy the contents of the TOS onto R9, R9 ← (SP)*

*Step2: Update TOS: SP ← (SP + 2), i.e., SP = 027Eh+2 = 0280h*

*The result is shown in Fig. 3.29b.2.*

**Example 3.12** *Now let us illustrate the push and pop operations for the 8 bit HC11 microcontroller, where SP points to the address for the new item. Now N = 1 for SP updating; the register being used in the example is named Accumulator A (ACCA). Fig. 3.30 illustrate the operations.*

*Figure 3.30a illustrates the push operation. Assume that before the operation the contents for registers SP and Accumulator A are 027Eh and 5Ah, respectively (Fig. 3.30a.1). Notice that SP is pointing to the address 027Eh, which is the TOS. The* **push operation** *is executed in two steps:*

*Step 1: Copy the contents of Accumulator A at the TOS, (SP) ← ACCA.*

*Step2: Update TOS: SP ← (SP − 1), i.e., SP becomes = 027Eh − 1 = 027Dh*

*The result is shown in Fig. 3.30a.2.*

*The pop operation illustrated by (b), with similar initial conditions, as shown in Fig. 3.30b.1. The pop operation is executed in two steps:*

*Step2: Update TOS: SP ← (SP + 1), i.e., SP = 027Eh + 1 = 027Fh*

*Step 1: Copy the contents of the TOS onto Accumulator A, ACCA ← (SP)*

*The result is shown in Fig. 3.30b.2.*

Observe in both examples that when a push is done, the original contents at the memory location is lost but this does not happen when a pop occurs. However, the

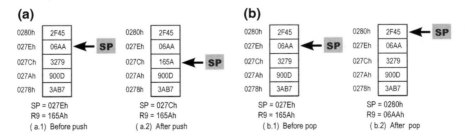

**Fig. 3.29** The stack, SP register. **a** Push operation, **b** Pop operation

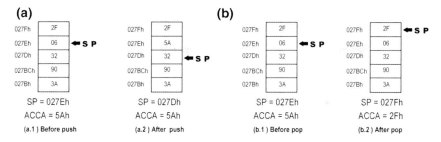

**Fig. 3.30** Another example: SP register. **a** Push operation, **b** Pop operation

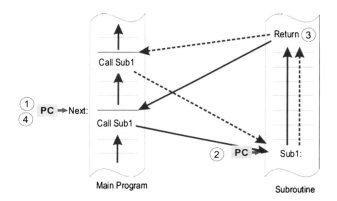

**Fig. 3.31** Invoking a subroutine

same data cannot be retrieved twice with a stack operation, because of the dependence on the pointer SP, which no longer refers to that location. Usually, other operations are available for this end.

Hereafter, irrespectively of how SP operates, push and pop operations will be denoted in RTN as "(TOS)← source" and "dest ← (TOS)", respectively.

**Stack and Subroutines** Even if the user does not utilize push and pop operations explicitly, these may occur transparent to the programmer when dealing with subroutines and interrupts, which involves a special kind of subroutine called Interrupt Service Routine (ISR). Subroutines are program modules logically separated and independent of the main program, which are invoked to do a specific task and then continue the main program with the instruction following the call or interrupt. The process is illustrated in Fig. 3.31 where a subroutine Sub1 is called from Main Program.

When a program invokes (calls) a subroutine, the current contents of the PC is pushed onto the stack. Observe that just before executing the call instruction, the PC contains the address of the next instruction to be fetched, the one right after the call. Therefore, by pushing the PC, the CPU is effectively storing in memory the address of the instruction following the call. Returning from a subroutine causes the

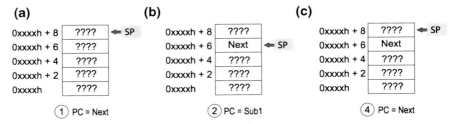

**Fig. 3.32** Stack pointer use in the invocation and return of subroutines. **a** Stack before call, **b** Stack after call, **c** Stack after return

PC to be popped from the stack. This way program execution can always resume at the instruction next to where the subroutine was invoked, no matter where in the program the invocation occurs. Figure 3.32 illustrates the stack operations for the previous figure. Just before the call, the SP is pointing at location 0xxxxh+8 as illustrated by inset (a). The process is then as follows:

Step 1:  After the CALL instruction is fetched and decoded, but still not executed, the PC will point to the next instruction, at address Next.

Step 2:  The execution of the invocation results in the current PC value being pushed onto the stack and loaded with the start address of Sub1, as depicted in Fig. 3.32b. Now the instruction to be fetched is the first one in the subroutine.

Step 3:  The subroutine instructions are executed in sequence up to the Return Instruction.

Step 4:  When the Return instruction is executed, the PC is popped, restoring the value Next, allowing the program execution to continue in the main program at the instruction following the call. At this point, SP will be back at address 0xxxxh+8 as shown in Fig. 3.32c.

Manipulating the SP in a program that uses subroutines requires caution to avoid loosing the return addresses. Chapter 4 provides a more detailed explanation of the use of the stack and push an pop operations.

### 3.7.5 Addressing Modes

Addressing modes can be defined as the way in which an operand is specified within an instruction so as to indicate where to find the data with which the operation is executed. The addressing mode is denoted using a specific syntax format, proper of the microcontroller family. Instructions with implicit operands are said to use *Implicit Addressing Modes*.

The machine language instruction contains information about the addressing mode which, when decoded, directs the sequence of operations in the CU to achieve the desired results. However, in some cases a strict disassembly of the machine language instruction, out of the context of a program, can yield a different result of what we could expect. There are several reasons for this. For example, there are modes used in the instruction fetching process which should not be of concern to the programmer, but are necessary to consider in the hardware design. For our purposes, we focus on the addressing modes used in assembly language.

Now, there are two different type of data we should consider. One associated with program flow instructions, like jumps, in which the data information refers to how the PC contents is to be modified. The second type is the one used in transfers and general type of instructions manipulating data. We discuss first this second group.

In general, the data to be used or stored in a transfer or in an arithmetic or logic instruction can be located in only one of the following possible places:

1. It may be explicitly given,
2. It may be stored in a CPU register,
3. It may be stored at a memory location, or
4. It may be stored in an I/O port or peripheral register.

In memory I/O mapped systems, the fourth and third options are equivalent. The syntax used in the addressing mode in an assembly language instruction tells the assembler which of these cases apply. Unfortunately, the syntax is not standard for the different microcontroller families, so the user must learn which one applies to the machine being used. Also, the names given to similar addressing modes is not completely standard, although this one tends to be more uniform. In what follows, the syntax used by the MSP430 family, which is shared by many other families, will be used. But the reader must bear in mind that it is absolutely necessary to look at the particular list of the family being used.

For the four cases listed above for the data location, the first two are almost standard for the different processors: The most basic types of addressing modes supported by almost all processors are listed next.

**Immediate Mode—syntax: #Number**. In this mode, the value of the operand Number is the datum.[12] Immediate mode is reserved only for source operands, since a number cannot be changed by an operation.

**Register—Syntax: Rn**. The operand is the CPU register Rn and the datum is contained in Rn.

When the datum is found in memory, we need to indicate the address. The information or mode is *absolute* if the address is given, and *relative* if it is expressed as an offset with respect to a given address. The basic modes are the following

**Absolute or Direct Mode—syntax: Number**. "Number" is the address where the datum is located. This mode is sometimes referred to as *symbolic mode*.

**Indirect—syntax: @Rn**. The datum is found in the memory location whose address is given by the contents of register Rn. We say that the register points at the datum. This mode is also called *Indirect Register Mode*.

---

[12] It receives the name "immediate" because at the machine level, the word containing the number goes immediately after the instruction word.

**Table 3.6**  Addressing modes examples

| Assembly | RTN | Comment |
|----------|-----|---------|
| mov src,dest | dest ← src | Copy or load source to destination |
| add src,dest | dest ← dest + src | Add source to destination |
| sub src,dest | dest ← dest − src | Subtract source from destination |
| and src,dest | dest ← dest .AND. src | Bitwise AND source to destination |
| xor src,dest | dest ← dest .XOR. src | Bitwise XOR source to destination |
| cmp src,dest | dest − src | Compare does not affect dest., only flags |

**Indexed—syntax**: **X(Rn)**. This mode specifies the datum address as the sum of number X address plus the contents of register Rn. X is sometimes called *base* and Rn is said to be used as an *index*; for this reason this mode is also named *base indexed mode*.

Although the basic concepts may the same, remember that neither the nomenclature nor the syntax are standard. Let us look now at the jumps.

Jump instructions can only have an address as an operand. Normally, but again not always, the machine language form uses what is called *Relative addressing mode*, which indicates an offset with respect to the current PC contents. The new PC contents is the sum of the current PC contents and a two's complement offset, (PC ← PC + offset). Hence, the maximum jump is limited by the number of bits used in the offset. In some processors, unconditional jumps use an absolute mode also in the machine language form.

To make programmers' life easier, the assembly form uses the absolute or direct mode only and the assembler calculates the needed offset. When using two-pass assemblers, labels can be used. A label is a symbolic name given by the programmer to an address. The RTN codes of Fig. 3.27 use labels.[13]

Two-pass assemblers allow programmers to use custom defined constants, including labels, thereby making the programmer's task easier. They are called "two-pass" because they need to go over the text more than once to identify and translate the constants that have been defined.

Other addressing modes can also be supported by a particular processor, which can enhance the flexibility of the instruction set. Also, remember that the syntax mentioned above is not standard. A discussion of the addressing modes supported by the MSP430 is offered in Chap. 4. The following examples illustrate the above addressing modes. For these examples we use some MSP430 assembly language instructions, shown in Table 3.6. The examples show the syntax for the source and destination to identify where the data is.

**Example 3.13** *This example illustrates the addressing modes mentioned before with the help of Fig. 3.33. This figure shows the contents of the 16-bit registers R6, R8*

---

[13] Some interpreters, which work line by line, require the offset to be directly specified, while others accept the address and calculate the offset.

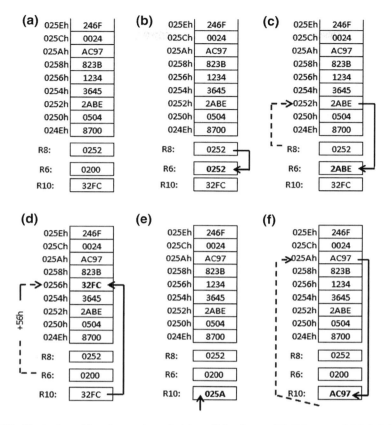

**Fig. 3.33** Illustrating addressing modes. **a** Intial condition, **b** mov R8, R6, **c** mov @R8, R6, **d** mov R10, 56h(R6), **e** mov #0x025A, R10, **f** mov 0x025A, R10

*and R10, and a memory segment of a 64 KB address space. Memory is shown in word format, that is, only even addresses are presented and contents are shown for a complete 16-bit word, four hex digits. Contents of registers and memory are in hex notation without suffix.*

*Part (a) shows the contents before any instruction. Parts (b) to (f) illustrate the addressing mode using data transfer instruction in format* mov source, destination. *The solid arrows show the transfer and the dashed lines the address pointers; the result is highlighted in each case. Now let us look into details.*

*(b)* mov R8, R6, *R6 ← R8 in RTN, means "copy the contents of register R8 onto Register R6". R6 = 0252h after this instruction.*

*(c)* mov @R8, R6, *R6 ←(R8) in RTN, means "copy the memory data word whose address is given by register R8, onto Register R6". Since R8 = 0252h, we search for address 0252h in the memory space where we see (0252h) = 2ABEh. Therefore, R6 = 2ABEh after this instruction.*

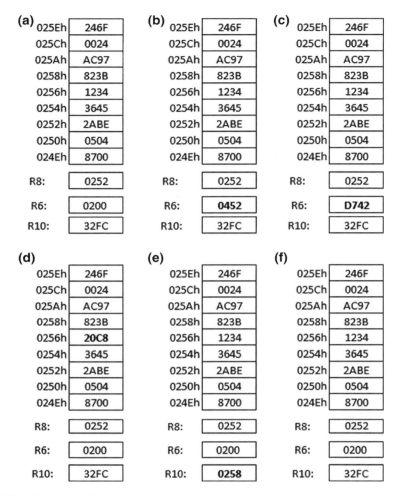

**Fig. 3.34** Illustrating addressing modes with arithmetic-logic instructions

*(d)* mov R10,56h(R6), *or* (R6 + 56h) ←R10 *in RTN, means "copy the contents of register R10 onto the memory location whose address is found by adding* 56h *to the contents of Register R6". Since* R10 = 32FCh, *then* (R6 + 56h) = (0256h) = 32FCh *after this instruction.*

*(e)* mov #0x025A, R10, *or* R10 ← 0x025A *in RTN, means "Register R10 is loaded with number 025Ah" Then* R10 = 025Ah *after this instruction.*

*(f)* mov 0x025A, R10, *or* R10 ← (025Ah) *in RTN, means "Copy at Register R10 the data in memory whose address is 025Ah" Looking at the memory space, we find* (025Ah) = AC97h. *Then* R10 = AC97h *after this instruction.*

Let us now work with other instructions, combined with different addressing modes. This is done in the following example.

**Example 3.14** *Using the same set of data shown in Fig. 3.33a, let us now work another set of operations, as shown in Fig. 3.34. As before, contents of registers and memory are in hex notation and inset (a) shows the contents before any of the other instructions.*

(b) add  R8,R6, R6 ← R6 + R8 *in RTN, means "add the contents of registers R8 and R6, and store the result in Register R6". Since* 0200h + 0252h = 0452h, *then* R6 = 0452h *after this instruction.*

(c) sub  @R8,R6, R6 ← R6 − (R8) *in RTN, means "subtract from the contents of R6 the data in memory located at the address provided by register R8, and store result in Register R6". Since* R8 = 0252h, *we search for address 0252h in the memory space where we see* [0252h] = 2ABEh. *Now* 0200h − 2ABEh = D742h. *Therefore,* R6 = D742h *after this instruction.*

(d) xor  R10,56h(R6), (R6 + 56h) ← R10.xor.(R6 + 56h) *in RTN, means "perform a bitwise XOR operation with the contents of register R10 and the content of the memory location whose address is found by adding 56h to the contents of register R6, and store the result in the same memory location". Since* R10 = 32FCh *and* (R6 + 56h) = (0256h) = 1234h *then the performed operation is as shown below. Notice the toggling effect according to the source.*

$$0001001000110100 \text{ .xor.}$$
$$\underline{0011001011111100} =$$
$$0010000011001000$$

so (*R6* + 56*h*) = 20*C*8*h* after this instruction.

(e) and  #0x025A,  R10, R10 ← 0x025A .and. R10 *in RTN, means "the contents of Register R10 is bitwise AND-ded with number 025Ah, the result being stored in R10" Since* R10 = 32FCh, *the operation is*

$$0000001001011010 \text{.and.}$$
$$\underline{0011001011111100} =$$
$$0000001001011000$$

*so* R10 = 0258h *after this instruction.*

(f) cmp  0x025A,  R10, *will take the difference between R10 and the data word at memory address* 0x025A, *that is* R10 − (0x025A), *without affecting any of the operands, just the flags. Since* (025Ah) = AC97h *and* R10 = 32FC, *and subtraction is done by two's complement addition, then* R10 − (0x025A) ⇒ 32FC + 5369 = 8665, *with the flags being then* C = 0, Z = 0, N = 1, V = 1.

## *3.7.6  Orthogonal CPU's*

A CPU is said to be *orthogonal* if all its registers and addressing modes can be used as operands, except for the immediate mode as a destination. Very few microcontrollers accept the immediate mode as destination operand, which make sense only for compare and testing purposes but not for cases where the destination should store a result.

Orthogonality has several advantages, among which the ability to write compact codes. Many designers prevent however the use of special registers as operands, either source or destination, to prevent unexpected results. Orthogonality requires the programmer to use some operands like the Program Counter register with great care.

## 3.8  MSP430 Instructions and Addressing Modes

The MSP430 CPU supports 27 machine language *core instructions*, that is, instructions that have unique opcodes decoded by the CPU. The length of the machine language instructions are one to three words. There are 24 additional *emulated instructions* supported by assemblers, introducing particular and more popular mnemonics for easier program writing and reading. For example instruction "`inv R5`" is emulated by the core instruction "`xor #0xFFFF,R5`". It is easier for the programmer to write the former when intending to invert the contents of R5. In a certain sense, emulated instructions may be considered as standard macros.

The MSP430 supports seven addressing modes, including the ones mentioned before, albeit with a slightly different notation. The instructions and addressing modes for the MSP430 are discussed in Chap. 4, which is devoted to assembly programming.

Although advertised as orthogonal, MSP430 microcontrollers are not fully orthogonal since two of its non immediate addressing modes are invalid for destination. This is a small price paid by the MSP430 designers to maintain all instruction words of equal length.

## 3.9  Introduction to Interrupts

The topic of interrupts and resets involve both hardware and software subjects, but it is also closely related to how a CPU operates. For this reason it becomes important to provide an introductory discussion on the subject at this point. An in-depth discussion of the hardware and software components for supporting a reset are offered in Chap. 6, Sect. 6.5. Moreover, Chap. 7 offers a comprehensive discussion of hardware and software components required for handling interrupts in general.

## 3.9.1 Device Services: Polling Versus Interrupts

The CPU in a microcontroller or microprocessor is a sequential machine. The control unit sequentially fetches, decodes, and executes instructions from the program memory according to the stored program. This implies that a single CPU can only be executing one instruction at any given time. When it comes to a CPU servicing its peripherals, the program execution needs to be taken to the specific set of instructions that perform the task associated to servicing each device. Take for example the case of serving a keypad.

A keypad can be visualized at its most basic level as an array of switches, one per key, where software gives meaning to each key. The switches are organized in a way that each key depressed yields a different code. For every keystroke, the CPU needs to retrieve the associated key code. The action of retrieving the code of each depressed key and passing them to a program for its interpretation, is what we call *servicing the keypad*. Like the keypad in this example, the CPU might serve many other peripherals, like a display by passing it the characters or data to be displayed, or a communication channel by receiving or sending characters that make up a message, etc.

When it comes to the CPU serving a device, one of two different approaches can be followed: *service by polling* or *service by interrupts*. Let's look at each approach with some detail.

### Service by Polling

In service by polling, the CPU continuously interrogates or polls the status of the device to determine if service is needed. When the service conditions are identified, the device is served. This action can be exemplified with a hypothetical case of real life where you will act like a polling CPU:

*Assume you are given the task of answering a phone to take messages. You don't know when calls will arrive, and you are given a phone that has no ringer, no vibrator, or any other means of knowing that a call has arrived. Your only choice for not missing a single call is by periodically picking-up the phone, placing it to your ear and asking "Hello! Anybody there?" hoping someone will be in the other side of the line. This would be quite an annoying job, particularly if you had other things to do. However, if you don't want to miss a single call you'll have to put everything else aside and devote yourself to continuously perform the polling sequence: pick-up the phone, bring it to your ear, and hope someone is on the line. Since the line must be available for calls to enter (sorry, no call-waiting service), you have to hang-up and repeat the sequence over and over to catch every incoming call and taking the messages. What a waste of time!* Well, that is polling.

## Service by Interrupts

When a peripheral is served by interrupts, it sends a signal to the CPU notifying of its need. This signal is called an *interrupt request* or IRQ. The CPU might be busy performing other tasks, or even in sleep mode if there were no other tasks to perform, and when the interrupt request arrives, if enabled, the CPU suspends the task it might be performing to execute the specific set of instructions needed to serve the device. This event is what we call an *interrupt*. The set of instructions executed to serve the device upon an IRQ form what is called the *interrupt service routine* or ISR.

We can bring this interrupting capability to our phone example above.

*Let's assume that in this case your phone has a ringer. While expecting to receive incoming calls, now you can rely on the ringer to let you know that a call has arrived. In the mean time, while you wait for calls to arrive you are free to perform other tasks. You could even get a nap if there were nothing else to do. When a call arrives, the ringer sounds and you suspend whatever task you are doing to pick-up the phone, now with the certainty that a caller is in the other end of the line to take his or her message. The ring sound acts like an interrupt request to you. A much more efficient way to take the messages.*

Interrupts can be used to serve different tasks. The following are just a few simple examples that illustrate the concept:

- A system that toggles an LED when a push-button is depressed. The push-button interface can be configured to trigger an interrupt request to the CPU when the push-button is depressed, having an associated ISR that executes the code that turns the LED on or off.
- A message arriving at a communication port can have the port interface configured to trigger an interrupt request to the CPU so that the ISR executes the program that receives, stores, and/or interprets the message.
- A voltage monitor in a battery operated system might be interfaced to trigger an interrupt when a low-voltage condition is detected. The ISR might be programmed to save the system state and shut it down or to warn the user about the situation.

## Servicing Efficiency: Polling or Interrupts?

There are innumerable events that can be configured to be served by interrupts. Note that any or all of them could be served by polling, but that would require having the CPU tied to interrogating (polling) the corresponding interface to determine when service is needed. This would result in a very inefficient use of the CPU.

To give an idea of how inefficient polling results, let's analyze one of the examples above. Let's assume the CPU serving the push-button and LED example is running at a clock frequency of 1 MHz and executes one instruction every four clock cycles. This is a very conservative scenario as today's processors con run at frequencies well over 1 GHz. This implies the CPU will roughly execute 250,000 instructions every second, or one instruction every four microseconds. To know if the push-button has been depressed the CPU only needs to execute two instructions: one to read the I/O pin associated to the push-button and a conditional jump instruction to make the decision: eight microseconds in total.

Consider a user repeatedly depressing the push button. Let's say the user is really fast, and able to push the key ten times per second. This is once every 100 ms. If we have the CPU polling this key interface to determine when the button was depressed, in 100 ms the processor would have checked it 12,500 times to find it was depressed only once. The efficiency in the CPU usage under these circumstances would be 1/12500 x 100 = 0.008 %. This implies that 99.992 % of the CPU cycles were wasted checking for a push-button that was not depressed. Servicing the same event with an interrupt would require only one action: toggling the LED when the push-button was depressed, which can be performed with a single instruction. This yields an efficiency of nearly 100 % and tons of CPU cycles saved. The CPU can use the rest of the time for attending other requests, executing tasks in the main program, or just saving power in a low-power sleep mode.

Despite this illustrative example, and all said about polling and interrupts, it deserves to mention that not in every situation using interrupts will be better than polling. There are situations where interrupts will not serve the purpose and polling will be the best alternative. Interrupts suffer from latency and are susceptible to false triggering. So in applications where the interrupt response lag might be too long or where noise might render interrupts unusable, polling becomes the first alternative to serving peripherals. Chapter 7 discusses these aspects in more detail.

## 3.9.2 Interrupt Servicing Events

In general, a CPU might support two different types of interrupt requests: maskable and non-maskable.

**Maskable Interrupt Requests**: A type of request that can be masked meaning that they can be disabled by clearing the CPU's global interrupt enable (GIE) flag in the processor status word (PSW). When the GIE is clear, the CPU will not honor any maskable interrupt request. This gives the programmer an ability of deciding when the CPU is to accept or not interrupt requests. Maskable interrupts are the most common type of interrupts managed in embedded systems. In most cases they are referred to as simply *interrupts*.

**Non-maskable Interrupt Requests (NMI)**: Cannot be masked by clearing the GIE in the CPU and are therefore always served when they arrive at the CPU. This type of interrupt requests are reserved for system critical events than cannot be made to wait. We'll refer to them as *NMIs*.

Although different CPU architectures might have different ways of serving interrupts, there are several steps that are common to any processor. Once an interrupt request is accepted, and assuming the processor was in the middle of executing an arbitrary instruction in a program, the fundamental steps taking place in the CPU to serve an interrupt include the following:

**Step 1** Finish the instruction being executed. The CPU never truncates instructions. Interrupt requests are always served between instructions.

**Step 2** Save the current program counter (PC) value and, in most CPUs, the status register (SR), onto the stack. Some processors might save other registers as well.

**Step 3** Clear the global interrupt enable flag. This action, performed by most CPUs, disables interrupting an ISR under execution. This condition could be overridden, but is generally not recommended.

**Step 4** Load the program counter (PC) with the address of the ISR to be executed. The way this step is accomplished changes with the interrupt handling style, but the net result is always the same: getting into the PC the address of the first instruction of the ISR to be executed.

**Step 5** Execute the corresponding ISR. The ISR ends when the special instruction *interrupt return* (IRET or RETI) is executed.

**Step 6** Restore the program counter and any other register that was automatically saved onto the stack in Step 2. This action is the result of executing the interrupt return instruction. Note that by restoring the status register, the interrupts become unmasked. Restoring the PC causes resuming the interrupted program at the next instruction where it was left.

Supporting interrupts in an application fundamentally needs four basic requirements:

1. A means for saving the program counter and status register, as indicated above in Step 2. This capability is provided by having a property allocated stack area in your program.
2. Having an interrupt service routine designed to render the service needed by the interrupt requesting device. This is the code executed in Step 5 above.
3. A means for the CPU locating the ISR corresponding to a particular request. This takes special meaning when the CPU can be interrupted by multiple devices, each having a different ISR. Usually this provision also deals with resolving any conflict that might develop if multiple devices place simultaneous requests to the CPU. This allows accomplishing Step 4 in the list above.
4. Enabling the interrupts in the system, otherwise they would not be served. By default, when the CPU is powered-up, interrupts are masked, so explicit software action is needed to unmask them. Enabling interrupts requires two levels of action: one at the CPU with the GIE flag, and another enabling the device(s) that will issue interrupt requests. This requirement will make possible to start the interrupt cycle at Step 1 in the sequence above.

Satisfying these four basic requirements will need additional details about how the CPU is designed to support interrupt processing. Chapter 7 provides a detailed discussion of the subject of interrupts.

### 3.9.3 What is Reset?

A *reset* is an asynchronous signal that when fed to an embedded system causes the CPU and most of the sequential peripherals to start from a predefined, known state. This state is specified in the device's user's guide or data sheet. In a CPU, the reset state allows for fetching the first instruction to be executed by the CPU after a power-up.

A reset occurs when power is applied for the first time to the system, or when some event that might compromise the integrity of the system occurs. These events include a power on reset (POR), Brown-out resets (BOR), and others.

A reset may also be generated by a hardware event, like for example the expiration of a watchdog timer, or through the assertion of the CPU reset pin, caused either by a user or an external hardware component.

A reset can be viewed as a special kind of interrupt. Upon a reset, the program counter (PC) is loaded with the address of the first instruction to be executed by the CPU, starting the fetch-decode-execute cycle. Section 6.5 in Chap. 6 provides a detailed discussion of how a reset is configured and programmed.

## 3.10 Notes on Program Design

Before leaving the general aspects in this chapter, let us discuss some directives on program design, especially considering some aspects pertaining embedded systems.

A program is a logical sequence of instructions, associated data values and compiler directives written down to carry out an algorithm to perform a specific task using a computer or embedded system hardware.

Programming should not be left to chance. The steps in creating a program include design and planning of the algorithm, documenting the different phases, encoding into a specific programming language using appropriate syntax rules, testing and debugging. Despite our efforts, many programs contain bugs and require testing and maintenance.

*Bugs* are mistakes that the programmer has introduced in the program, either of syntax or logical natures. *Testing* is done to prove the existence of bugs, while *debugging* consists in fixing the bugs. Poor programming practice not only has higher probability of producing more bugs, but also makes it almost impossible to debug without incurring in enormous costs.

Discipline in programming and documenting, from the first steps in design and planning to the last steps in devising tests and debugging procedures is absolutely necessary for embedded system designers. Consider that once the system is in work, it will have limited human interaction.

As an embedded system programmer, try to follow a minimum set of goals when programming:

**Write the shortest possible program**. A short program reduces use of resources, it is easier to maintain and modify, as well as to understand. If the planning yields a long program, try to break it in short identifiable subtasks (see top–down design below).

**Evaluate shortcuts before implementing them**. To write a short program never, never, try to introduce shortcuts or tricks which may seriously affect the other qualities of the program. It is better to sacrifice length than understanding.

**Write programs easy to understand**. Spend some time assigning resources, giving appropriate names to constants and variables, documenting the program, and so on. Make it a habit to follow similar structures for similar tasks whenever possible. The time spent in making the program understandable pays itself soon.

**Write programs easy to modify**. You will soon discover that different tasks can be achieved by simple changes in code when the design is clear. This easiness reflects in lower costs and faster development of future projects.

**Document your program as you work on it**. The importance of documentation can never be overemphasized. Without it, programs are difficult to read, modify, reuse and debug. Moreover, documenting after finishing the program usually yields a lower quality documentation and higher cost.

**Create your own catalog of codes for common tasks**. Hundreds    of    different projects require similar steps to be followed, subtasks to be accomplished and so on. For example, configuring the I/O ports and the peripherals, setting modes of operations are common to most projects. Having a catalog of subroutines, macros and particular codes will accelerate developments. Do not reinvent simple tested codes.

**Program for lowest power consumption**. Always try to consume the lowest as possible choosing the mode suitable for your application. Prefer interrupts to polling, consider putting to sleep the system or the unused components when possible, study power profiles if available, etc.

**Know your hardware system**. Don't forget that embedded systems require a good match between software and hardware. You have to program for the hardware and connections you have available. Hardware and program design very often goes in parallel.

**Avoid using unimplemented bits of memory or registers**. Future hardware revisions or enhancements can make your program useless if you do not keep this goal in mind. Also, moving modules to new memory locations have less or no effect on the correctness of the program.

**If possible, use a single resource for a single purpose**. For example, if you have decided to use register R15 as a counter, don't use it for other purpose unless you're forced to it by lack of resources. Plan for any multiple use in advance.

Of course one might look for more goals that a program must pursue. When solving one problem try to identify any new goal. This will certainly help in the future.

# 3.11 The TI MSP430 Microcontroller Family

Previous sections have introduced characteristics of the MPS430 family. The rest of the chapter is devoted to this particular set of microcontrollers.

The MSP430 family of microcontrollers developed by Texas Instruments targets the market of battery-powered, portable embedded applications. Since its beginning, it has been designed with an architecture optimized for low-power consumption.

The first members of the family were introduced in early 1992, with the series MSP430x3xx, targeted specifically at low-power metering equipment. Since then, new series have been added, with every new addition offering improved characteristics over its predecessor: the 1xx in 1996, the 4xx in 2002, the 2xx in 2005 and the 5xx/6xx in 2008 and 2009. The most recent family introduced in 2010 is the ferroelectric memory family, which incorporates the ferroelectric memory technology, making the full memory space both writable and non volatile. There are over one hundred different configurations for MSP430 MCUs. New members maintain downward software compatibility, keeping up with their original philosophy of being an architecture tailored for ultra-low-power embedded applications.

Aside from some early EPROM versions in the 3xx series and high volume mask ROM devices in the 4xx series, all family members are in-system programmable via a JTAG port or a built in bootstrap loader (BSL) using RS-232 or USB ports.

The 3xx and 1xx generations were limited to a 16 bit address space. Later generations include what is called '430X' instructions that allow a 20 bit address space. Most applications however work fine with the 64 K memory space, and we will work with this space. With the exception of a glance at the CPUX and the characteristics specific to the extended 20 bit address space, which are explained in Appendix D.

## 3.11.1 General Features

All MSP430 family members are developed around a 16-bit RISC CPU with a Von-Neumann architecture. The assortment of peripherals and features in each MSP430 device varies from one series to another, and within a series from one family member to another. Common features to devices include:

- Standard 16-bit Architecture: All devices share the same core 16-bit architecture and instruction set. The 20-bit CPUX registers are also based on this architecture.
- Different Ultra Low-power Operation modes: The devices can operate with nominal supply voltages from 1.8 to 5.5 V. The nominal operating current at 1 MHz ranges from 0.1 to 400 $\mu$A, depending on the supply voltage. The wake-up time from standby mode is 6 $\mu$S.
- Flexible and powerful processing capabilities: The programmer's model provides seven source-address modes and four destination-address modes with 27 core instructions. Extensive interrupt capability with prioritized, nested interrupts with

unlimited depth level. A large register file with RAM execution capability, table processing, and hex-to-decimal conversion modes.

- Extensive, memory-mapped peripheral set including: a 14-bit SAR A/D converter, multiple timers and PWM capability, slope A/D conversion (all devices); integrated USART, LCD driver, Watchdog Timer, multiple I/O lines with interrupt capability, programmable oscillator, 32-kHz crystal oscillator (all devices).
- Versatile device options include:

  - Masked ROM
  - OTP and EEPROM models with wide temperature range of applications
  - Models with ferroelectric memory
  - Up to 64 K addressing space for the MSP430 and 1M for the MSP430X
  - Memory mixes to support all types of applications.

- JTAG/debugger element, used for debugging embedded systems directly on the chip.

**Low-power Philosophy**

The MSP430 was designed since the beginning targeting low power consumption. This means not only using low power low voltage IC designs for the hardware, but also configuring the operation to different power modes, depending on application needs, and the possibility of turning off everything, including peripherals, while not in use. This feature is very useful for event driven designs, in particular for portable equipment.

The MSP430 can function in six different power modes, named Active, and the low power modes LPM0 to LPM4, where the CPU is off and can only be awaken by an interrupt or a reset. The modes are configurable via signals (bits) SCG1, SCG0, OSCOFF, and CPUOFF of the status register, which control the clock operations, as explained later in Chap. 7.

**Oscillators and Clocks**

The clock generator in MSP430 devices is designed to support low power consumption. It includes three or four clock sources, depending on the family and the specific member, that generate three internal clock signals for the best balance of performance and low power consumption. The clock sources include:

**LFXT1CLK**: A low-frequency/high-frequency oscillator that can be used with low-frequency watch crystals, namely external clock sources of 32,768-Hz, standard crystals, resonators, or with external clock sources up to 8 MHz.

**XT2CLK**: Optional high-frequency oscillator that can be used with standard crystals, resonators, or external clock sources in the 400-kHz to 16-MHz range.

**DCOCLK**: Internal digitally controlled oscillator (DCO).

**VLOCLK**: Internal very low power, low frequency oscillator with a 12 kHz typical frequency.

The three clock signals available from the basic clock are the auxiliary clock (ACLK), sub-main clock (SMCLK), and the master clock (MCLK) used for the feeding CPU the clock. They are software selectable to be fed from the available clock sources in the chip via software controlled configuration registers. Chapter 6 discusses in detail the MSP430 clock system.

### Mixed Signal Product Philosophy

This philosophy is reflected in the very a acronym MSP. A vast number of applications fall in the category of signal processing, and nowadays most are of the mixed-signal type, requiring both analog and digital hardware. The analog components used in the analog signal-chain design include analog-to-digital converters (ADC), analog comparators, operational amplifiers, and so on.

Most MSP430 models include some of these analog signal-chain components allowing the designer to reduce power consumption, component count, and space. This mixed-signal design of the MSP430 series is another attractive feature for the user. Analog signal-chain components in the MSP430 are discussed in Chap. 7.

## *3.11.2 Naming Conventions*

The MSP430 microcontrollers follow a more or less ordered sequence of fields in naming the parts. "MSP" stands for *Mixed Signal Processor*, emphasizing the fact that the analog and digital aspects of signal processing are included in the design and applications. The "430" refers to the MCU plattform.[14] The rest of the name is composed of several fields, some of which are optional. The name format is

MSP430{Device Memory}{Generation}{Family}{Series and Device Number}

... [A][Temperature Range][Packaging][Distribution Format][Additional Feature]

where fields between square brackets ([]) are optional. The definition of the fields are as follows:

**Device Memory** Which indicates the type of memory and specialized application:

Memory Type

- C = ROM
- F = Flash

---

[14] Some people say that the number stems from the date April 30, associated somehow to the development of the first prototype. We mention this reference without actually knowing its truthness.

- FR = FRAM
- G = Flash (value line)
- L = Non-volatile memory not included

**Specialized Application**

- FG = Flash Medical
- CG = ROM Medical
- FE = Flash Energy Metering
- FW = Flash Electronic Flow Meters
- AFE = Analog Front End
- BT = Pre-programmed with Bluetooth
- BQ = Contactless Power

**Generation** Refers to the series generation number, which does not necessarily refers to the order in which they were released.

- 1 = up to 8 MHz
- 2 = up to 16 MHz
- 3 = Legacy series, OTP
- 4 = up to 16 MHz with LCD
- 5 = up to 25 MHz
- 6 = up to 25 MHz with LCD
- 0 = Low Voltage Series

**Family** it refers to a family within the generation series type.
**Series and Device Number** within the given family
   On the other hand, the optional fields are:
**A**, which means a revision model
**Temperature Range**, which comes with the following option:

- S = 0 °C to 50 °C
- I = −40 °C to 85 °C
- T = −40 °C to 105 °C

**Packaging** A group of three letters describing the type of packaging for the device, as defined in http://www.ti.com/sc/docs/psheets/type/type.html
**Distribution Format**

- T = Small 7-in reel
- R = Large 11-in reel
- No marking = Tube or tray

**Additional Features:**

- −QI = Automotive qualified
- −EP = Enhanced product (−40 °C to 125 °C)
- −HT = Extreme high temperature (−55°C to 150°C)

## 3.12 Development Tools

Several tools are available for working with the MSP430. Below we mention some of them as well as internet sites where they are available or where more information can be found.

### 3.12.1 Internet Sites

Some important internet URLs where we can find tools are

**CCS** http://processors.wiki.ti.com/index.php/Download_CCS
**IAR** http://supp.iar.com/Download/SW/?item=EW430-KS4
**naken430** http://www.mikekohn.net/micro/naken430asm_msp430_assembler.php
**pds-430** http://www.phyton.com/htdocs/tools_430/tools_mca430.shtml
**cdk4msp** http://cdk4msp.sourceforge.net/ver
**mspgcc** http://mspgcc.sourceforge.net/
**mcc-430** http://www.mikekohn.net/micro/naken430asm_msp430_assembler.php

Some of these links include several of these features in the same package. Table 3.7 summarizes the availability of several tools described below.

### 3.12.2 Programming and Debugging Tools

Among the programming and debugging tools we have MSP430 simulators, C compilers and assemblers, linkers, and real time operating systems (RTOS). Simulators are of great help whenever a real system is not available or when it is not feasible to put to work dozens, hundreds, or even more systems at the same time. They also provide a tool for testing designs before actual loading it into the hardware system.

Several Real-Time operating systems have been developed around MSP430 systems. The RTOS' are good for systems with several tasks which are required to be

**Table 3.7**  Programming and debugging tools for MSP430

| Site | Simulator | C compiler | Assembler | Linker |
|------|-----------|------------|-----------|--------|
| CCS | X | X | X | X |
| IAR | X | X | X | X |
| mspgcc |   | X | X | X |
| naken | X |   | X | X |
| pds-430 | X | X | X |   |
| cdk4msp | X |   |   |   |
| mcc-430 |   | X |   |   |

completed within a predetermined time line. The following is a non-exclusive list of sites with RTOS' for MSP430 systems:

**TinyOS** http://www.tinyos.net/
**FreeRTOS** http://www.freertos.org/index.html
http://www.freertos.org/portmspgcc.html
**PowerPac** http://www.iar.com/website1/1.0.1.0/964/1/
**scmRTOS** http://scmrtos.sourceforge.net/ScmRTOS
**ecos** http://www.ecoscentric.com/index.shtml

Additional information about programming and debugging tools can be found at http://processors.wiki.ti.com/index.php/.

### 3.12.3 Development Tools

TI application engineers as well as third parties have designed several development tools which are readily available for the beginner, as well as for the intermediate and experienced users. These tools consist of emulators, evaluation kits, and device programmers. There are also a number of available tools specific for a very handy MSP430 development kit known as the eZ430.

**Emulators**

An emulator is a machine acting as if it were another "emulated" machine, by running the code of the latter. Though the result is probably a slower "version" of the original machine, this allows the user to examine the system behavior in a more complete fashion as opposed to simulation. The CCS and IAR sites mentioned before provide emulators. Others can be found at

**FET** http://focus.ti.com/docs/toolsw/folders/print/msp-fet430uif.html
**Project430** http://www.testech-elect.com/phyton/pice-msp430.htm

**Evaluation Kits**

Evaluation kits are relative small subsystems in a board which include a full featured specimen of an MCU (or other ICs). These kits typically expose several pins of the MCU that will allow external connections or the study of particular features.

The evaluation kits presented below typically reflects the complexity of the system and/or the availability or lack of additional components such as LCDs, speakers, battery connections, microphones, etc.

**MSP430-EXP430F5438** Includes: Dot-matrix LCD, 3-axis accelerometer, microphone, audio output, USB communication, joystick, 2 buttons, 2 LEDs

**MSP430-EXP430FG4618** Includes: LCD, capacitive sensitive input, audio output, buzzer, RS-232 communication, 2 buttons, 3 LED

**MSP430-EXP430G2** [15] Includes: Integrated flash emulation tool, 14/20-pin DIP target socket, 2 buttons, 2 LEDs, and PCB connectors. It is compatible with several 14-pin and 20-pin models.

### Device Programmers

In case you need to program several MSP430 MCUs at the same time there are at least two device programmers available, namely the MSPGANG430 and the GangPro430:

**MSPGANG430** Connection type: Serial Programs 8 devices at a time. It works with PC or as standalone.

**GangPro430** Connection type: USB Program 6 devices at a time via JTAG, Spy Bi-Wire, and BSL. Fast programming time.

### Tools for eZ430

Probably the most handy of all the MSP430 tools available for development are the eZ430 tools. These tools can go for as low as $20.00 (eZ430-F2013) and could be very handy. Available tools include the EX430-RF2560, eZ430-Chronos, eZ430-F2013, and others.

## 3.12.4 Other Resources

One of the best source of information for this family of MCU's can be found at the official TI's page for the MSP430: www.msp430.com. Here the reader can find several resources such as application notes, code examples, open source libraries, white papers, guides and books, and MCU selection wizards.

## 3.13 Summary

- A microcomputer system includes a Central Processing Unit (CPU), Data and Program Memory blocks, and Peripheral or Input/Output (I/O) subsystem, interconnected by system buses. Another set of components provide the necessary power and timing synchronization for system operation.

---

[15] Price: $4.30 (Really!).

- The Data Bus is used to transfer instructions to, and data to and from the CPU. Through the Address Bus, the CPU selects the location with which interaction takes place. The Control Bus includes all signals that regulate the system activity.
- The CPU fetches instructions from the program memory, decodes them and executes them accordingly. Hence, the CPU performs the basic operations of data transfer as well as the arithmetical and logical operations with data.
- Memory is organized as a set of cells or locations, characterized by the contents, i.e. the memory word, and the address. The program memory is stored in the ROM section, and the data memory in the RAM section. Von Neumann architecture systems include the program and data memories in the same space; the Harvard architecture systems use separate memory spaces and buses for program and data.
- Data wider than memory words utilize two or more of these for storage. The address of a datum is that of the lower byte. In little endian convention the least significant byte is stored there, while in big endian convention the most significant byte goes to that byte.
- A memory map is a representation of the usage given to the addressable space of a microprocessor based system.
- The minimal hardware components of the CPU are: Arithmetic Logic Unit (ALU), Control Unit (CU), a set of Registers, and a Bus Interface Logic (BIL). They are connected by an internal system of buses.
- The CPU registers provide temporary storage for operands, addresses, and control information. Some specialized registers for specific use are the Instruction Register (IR), the Program Counter (PC), the Stack Pointer (SP), and the Status Register (SR).

## 3.14 Problems

3.1 Answer the following questions.

    a. The main components of a microcomputer system are _____ .

    b. The acronym CPU means _____ .

    c. Which are the three system buses?

    d. What is a bus?

    e. What is a data bus transaction?

    f. What is a read operation? What is a write operation?

    g. How many data transactions are needed if the data consists of four bytes and the Data Bus width is 16 bits?

    h. If the address bus has N lines, what is the maximum number of addresses that the CPU may reach?

    i. What is the difference between physical address and data address?

    j. What is the difference between Harvard and a Von Neummann architectures?

    k. What is the difference between a microcontroller and a microprocessor?

l. If the address of the double word 2AF30456h is 0402h, what are the physical addresses of the individual bytes in a little endian memory system?

m. Repeat the previous question for a big endian system.

3.2 Figure 3.6 shows an example of a memory structure for a memory space of 64 KB, assuming that the address bus has 16-bits, so each address word will point to the respective cell in this space. Assume now a 20-bit address bus and an 8-bit data bus, together with several 64 KB memory modules similar to the one in the figure, with each module being independently activated or deactivated. The modules can all share 16 lines of the address bus, say A15 to A0, but only the activated module would be accessed by the data bus. The four additional lines A19 to A16 of the address bus could then be used to activate the appropriate module.

a. What is the maximum number of modules that can be used?

b. If all the possible memory modules were used, what memory size would be available?

c. Assume that only four memory modules are used. What is the effective size of the address space being used and what happens when the address words bits go from 0000xxx...xxB to 1111xxx...xxB?

d. It is said that each memory cell has a unique address. Considering the above item, is this an absolute truth? That is, is it possible for a memory cell to have more than one physical address? Justify your answer.

3.3 Figure 3.12 shows two memory banks being used for connection with a 16-bit data bus, one bank set for the even numbered addresses and the other for the odd numbered addresses.

a. What would happen if the least significant address line from the microprocessor is connected to the least significant address line in both banks?

b. How would you solve the problem presented in (a)?

3.4 The following data was obtained from the program memory of a particular microprocessor:

4031 0280 40B2 5A80 0120 D3D2 C022

E3D2 0021 403F 5000 831F 23FE 3FF9

Assuming the first address is 0xF800, could you order the data by physical addresses using the little endian convention? Repeat for the big endian convention.

3.5 Design the following decoders a. 2 to 4 b. 3 to 8 c. 4 to 16

3.6 If a particular microprocessor has only one data bus but has an area designated for program memory and another area designated for data memory, is it a Von Neumann or a Harvard architecture?

3.7 Show how you would connect eight 1 KB memory modules using a 3 to 8 decoder with a chip enable terminal. Assume each memory module has ten input address lines, eight data lines, and one chip enable terminal. Also assume the address bus coming from the microprocessor is a 16-bit address bus.

3.8  Consider a 1024B memory bank composed of two 512B memory modules: one for the low byte and one for the high byte. A signal from the microprocessor, R/W' is used to distinguish between a memory read and a memory write. The address bus from the microprocessor is 16-bit wide.

   a. Determine the address lines that will be connected to each of the memory modules.
   b. Determine the address lines that should be used to enable the chip select, an active low input signal for each of the memory modules.
   c. Design the decoder needed to enable the memory modules if the memory space for these modules begin at address 512.

3.9  Answer the following questions.

   a. What is the characteristic of an I/O memory mapped system?
   b. What other scheme exists beside the I/O memory mapped system for dealing with the I/O subsystem?
   c. Name four common peripherals to be found in most embedded systems.
   d. What is the main function of the watchdog timer?
   e. What is the difference for the CPU between an interface register with an I/O device and a memory location?

3.10  The following expressions are given in register transfer notation. All registers are 16-bit wide and memory addresses point to 16-bit word data. Before each expression, it is assumed that register and memory contents are as follows: R8 = 4286h; R9 = 32F4h; [4286h] = 3AC5h; [027Ch] = 90EEh.
Complete the following table filling up the column of results. Write your results using hex notation.

| Transaction | Result |
|---|---|
| R9 ← R8 | |
| R9 ← (R8) | |
| (R8) ← R9 | |
| R9 ← R9 + R8 | |
| (027Ch) ← R8 | |
| (R8) ← (R8) + 013Fh | |
| R8 ← 2468h | |
| (037Ch) ← (027Ch)–(4286h) | |

# Chapter 4
# Assembly Language Programming

Programming provides us with the flexibility to use the same hardware core for different applications. A device whose performance can be modified through some sort of instructions is called a *programmable* or *configurable device*. Most I/O ports and peripherals are examples of these blocks; their configuration should be part of our program design.

Assembly programming for the MSP430 microcontrollers is introduced in this chapter. We limit ourselves to the CPU case; CPUX instructions are briefly discussed in Appendix D. The assembly language instructions are particular to the MSP430 family, but the techniques and principles apply to all microcontroller families with the appropriate adaptations. The material should enable you to proceed to more advanced topics on your own by consulting manuals, examples available elsewhere[1] and, very important, by practicing programming.

The two most popular assemblers for the MSP430 are the IAR assembler by Softbaugh and Code Composer Studio (CCS) from Texas Instruments. They both have free limited versions to download from the Internet, and also come included in evaluation boards. In this text we use the IAR assembler since its directives and structure are more along line of most assemblers. It also has no limit in the memory size for assembly language code. Appendix C briefly discusses the CCS assembler.

## 4.1 An Overview of Programming Levels

Programs are written using a *programming language* with specific rules of syntax.These languages are found in three main levels:

a) *Machine language*,
b) *Assembly language*, and
c) *Level language*

---

[1] Texas Instruments has many code examples available for MSP430 at http://msp430.com.

M. Jiménez et al., *Introduction to Embedded Systems*,
DOI: 10.1007/978-1-4614-3143-5_4,
© Springer Science+Business Media New York 2014

To illustrate the differences among them, let us consider the code of Laboratory 1[2] in the three levels. The program toggles on and off an LED connected to pin 0 of port 1 (P1.0) of the microcontroller, as illustrated in Fig. 4.1. Toggling is achieved by changing the voltage level at the pin.

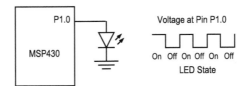

**Fig. 4.1**  Simplified hardware connection diagram for LED in board

The algorithm to achieve this goal is described by the flowcharts of Fig. 4.2, where (a) illustrates the process in its general terms, and (b) expands the different steps in more detailed actions.

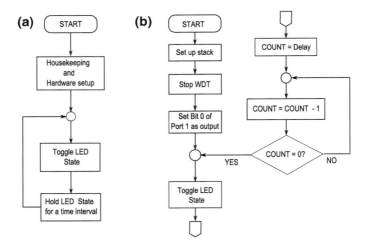

**Fig. 4.2**  Flow diagram for the blinking LED: **a** General Concept; **b** Expanded flow diagram

### 4.1.1 High-Level Code

Figure 4.3 shows a code in C language which executes the desired task. Readers with some proficiency in C will understand this *source code* without problem and associate it with the flow graph. The compiler takes care of the "set up stack" step.

---

[2] This hands on exercise is adapted from TI with permission

Listing 4.1: C Language Listing

```
1   #include    <msp430x12x.h>
2   void main(void)
3   { WDTCTL = WDTPW + WDTHOLD; /* Stop watchdog timer */
4     P1DIR |= 0x01;          /* Set P1.0 to output direction */
5     for (;;)
6     { unsigned int i;
7       P1OUT ^= 0x01;        /* Toggle P1.0 using ex-or */
8       i = 50000;            /* Delay */
9       do (i--);
10      while (i != 0);
11    }
12  }
```

**Fig. 4.3** C language code for LED toggling

Some differences with respect to C program sources for general purpose computers are noticeable. For example, hardware characteristics now require some attention. Despite this, high-level programming is more independent of the hardware structure, closer to our way of thinking, and easier to grasp. These are some of many reasons why this level is preferred for most applications. Chapter 5 introduces C programming for the MSP430.

A drawback is that this language is not understood by the machine. We need tools, called *compilers* and *interpreters*, to make the appropriate translation. The final machine product depends on the compiler used. Engineers also developed software tools called *debuggers* to help correct programming errors, usually called *bugs*; correcting them is called *debugging*.

## 4.1.2 Machine Language

Since digital systems only "understand" zeros (0) and ones (1), this is how the *executable code*, i.e., the set of instructions executed by the CPU, is presented to the system. Although possible, it is difficult to associate the machine code to an original high level source.

A program written in zeros and ones is a *binary machine language* program. The example in Listing 1.2 of Fig. 4.4 is for the flow graph of Fig. 4.2. Here, each line represents an instruction. The hex notation version in Listing 1.3, called *hex machine language*, is the form in which debuggers present machine language to users and also the syntax with which embedded programmers deal with this level. Obviously, machine language is not user friendly and hence the need of other programming levels.

|   | Listing 4.2: Binary Machine Language | Listing 4.3: Hex |
|---|---|---|
| 1 | 0100000000110001 0000001100000000 | 4031 0300 |
| 2 | 0100000010110010 0101101010000000 0000000100100000 | 40B2 5A80 0120 |
| 3 | 1101001111010010 0000000000100010 | D3D2 0022 |
| 4 | 1110001111010010 0000000000100001 | E3D2 0021 |
| 5 | 0100000000111111 1100001101010000 | 403F C350 |
| 6 | 1000001100011111 | 831F |
| 7 | 0010001111111110 | 23FE |
| 8 | 0011111111111001 | 3FF9 |

**Fig. 4.4** Executable machine language code for the example of Fig. 4.1

|   | Listing 4.4: Assembly Version | | Hex Machine Lang. |
|---|---|---|---|
| 1 | mov.w | #0x300,SP | 4031 0300 |
| 2 | mov.w | #0x5A80,&0x0120 | 40B2 5A80 0120 |
| 3 | bis.b | #001,&0x0022 | D3D2 0022 |
| 4 | xor.b | #001,&0x0021 | E3D2 0021 |
| 5 | mov.w | #0xC350,R15 | 403F C350 |
| 6 | dec.w | R15 | 831F |
| 7 | jnz | 0x3FC | 23FE |
| 8 | jmp | 0x3F2 | 3FF9 |

**Fig. 4.5** Assembly version for Fig. 4.4—Hex machine version shown for comparison

Instructions for the MSP430 CPU consist of one, two or three 16-bit words. The leading one is the *instruction word*, with information relative to the instruction and operands. Machine language for the MSP430CPU is considered in Appendix B.

## 4.1.3 Assembly Language Code

The first step toward a friendlier syntax was the *assembly language*, compiled with *assemblers*. Each instruction is now a machine instruction encoded in a more "human like" form. Translating a machine instruction into its assembly form is to *disassembly* and the software tool for this task is a *disassembler*. Figure 4.5 shows the assembly and hex versions of the executable code. Lines in both listings correspond one to one. Thus, "mov.w #0x300,SP" is the assembly for hex "4031 0300".

### Basic MSP430 Assembly Instruction Format

As illustrated, the very basic format of MSP430 instructions contains two components: the *mnemonic* and the *operands*. The mnemonic is a name given to the machine language opcode, and it is by extension sometimes also called opcode.[3] Mnemonics

---

[3] Strictly speaking, the opcode is part of the machine language code only and the mnemonic is the assembly language name for the opcode.

have a suffix ".b" or ".w" to differentiate byte and word size operands, as explained later in this chapter. No suffix is by default equivalent to ".w".

Operands are written in an appropriate MSP430 addressing mode, as introduced in Chap. 3. Operands are usually called *source* (src) and *destination* (dest). The instruction may have two, one or no operands using one of the format of expressions (4.1)–(4.3). The only core instruction without operand is reti, "Return from Interrupt"; all the other zero operand instructions are emulated. The operand in (4.2) may be a source or a destination, depending on the instruction.

$$\textbf{Mnemonic} \quad src, \, dest \tag{4.1}$$

$$\textbf{Mnemonic} \quad operand \tag{4.2}$$

$$\textbf{Mnemonic} \tag{4.3}$$

## 4.2 Assembly Programming: First Pass

Although simpler to read than machine form, the code of Fig. 4.5, which is a pure assembly listing, still has unfriendly notes. For example, the user needs the memory map to identify addresses 0x0022 and 0x0021, and knowledge of machine instruction lengths to know how many bytes the PC should jump. A friendlier version is found in Fig. 4.6, written with IAR assembler syntax. This is part of a complete *source file*. Each line in the source file is a *source statement*.

| | Listing 4.5: Assembly Code | Hex Code |
|---|---|---|
| 1 | `;Constants Declarations` | |
| 2 | `#include   "msp430g2231.h"` | |
| 3 | `LED        EQU    0x0001   ; LED at P1.0` | |
| 4 | `DELAY      EQU    50000    ;` | |
| 5 | `#define    COUNTER R15` | |
| 6 | `;- - - - - - - - - - - - - - - - - - - -` | |
| 7 | `            ORG 0F800h      ;Start Code` | |
| 8 | `;- - - - - - - - - - - - - - - - - - - -` | |
| 9 | `RESET      mov.w   #300h, SP     ; Set stack` | `4031 0300` |
| 10 | `StopWDT    mov.w   #WDTPW+WDTHOLD,&WDTCTL` | `40B2 5A80 0120` |
| 11 | `                               ; Stop WDT` | |
| 12 | `SetupP1    bis.b   #001h,&P1DIR   ;P1.0 output` | `D3D2 0022` |
| 13 | `        ;` | |
| 14 | `Mainloop   xor.b   #LED,&P1OUT   ; Toggle P1.0` | `E3D2 0021` |
| 15 | `Wait       mov.w   #DELAY,COUNTER` | `403F C350` |
| 16 | `                          ;Load Delay to Counter` | |
| 17 | `L1         dec.w   COUNTER       ; wait` | `831F` |
| 18 | `           jnz     L1       ; Delay over?` | `23FE` |
| 19 | `           jmp     Mainloop      ; Again` | `3FF9` |

**Fig. 4.6** Assembly language code for Fig. 4.4

**Fig. 4.7** Example of a list file line

Now assembly instructions have the format illustrated by (4.4); only the original instruction fields, mnemonics and operands, are compulsory. Mnemonics cannot start on the first column.

$$\underbrace{\text{SetupP1}}_{\text{Label}} \quad \underbrace{\text{bis.b}}_{\text{Mnemonics}} \quad \underbrace{\text{\#001h, \&P1DIR}}_{\text{Operands}} \quad \underbrace{\text{; P1.0 Output}}_{\text{Comment}} \qquad (4.4)$$

In fact, every source statement has the same format, all fields being in general optional. The mnemonics may be a machine instruction mnemonics or a directive. Operands must always go after a mnemonic.

Observe that the machine language and instruction mnemonics have not changed. But now there are new features in the listing that make it easier to read. Namely, we find *comments*, *labels*, *symbolic names*, and *directives*.

On the first column we can only have a comment (starting with the semicolon), a label, a C-type preprocessor directive, or a blank.

Assembling is done with a *two pass assembler*, which runs over the source code more than once to resolve constants and labels. The assembler receives the source file and outputs an *object file*, as well as other optional files. In particular, the list file is very helpful. For each source file statement which generates code or memory contents, the list file shows the address and contents. In the case of an instruction, this contents is the machine language instruction in hex notation, word size data. An example of a line of a list file is illustrated by Fig. 4.7.

### 4.2.1 Directives

Directives are for the assembler only. They do not translate into machine code or data to be loaded into the microcontroller memory. They serve to organize the program, create labels and symbolic names, and so on. Directives depend on the assembler, but instructions belong to the CPU architecture and are independent of the assembler. We will work with the IAR assembler.[4]

**Comments:** Anything on a line after and including a semicolon (;) is a comment. It is ignored by the assembler and serves as a documentation tool and to explain the purpose of an instruction. Remark how comments relate the code with the flow graph of Fig. 4.2b.

---

[4] For a more in depth treatment than the introductory level provided in this book, the reader may consult the document "MSP430 IAR: Reference Guide for Texas Instruments' MSP430 Microcontroller Family", published by IAR Systems Inc. and downloadable from the Internet.

**Symbolic Names:** These are names given to numbers, expressions, hardware devices, and so on. Labels fall in this category. The values and equivalences for these names are assigned during assembly and are thus called *assembly time constants*. The names can be used by the programmer as if they were already known before assembling.

**Labels:** These always start on the first column. They may contain alphanumeric characters, the underscore "_", and the dollar sign "$"; they may end with a colon (:) but never start with a number. The nature of this constant value will depend on the directive used with the label or if it associated to an instruction. In the latter case, the label takes the value of the memory address of the instruction word.

**Directives EQU, #define, #include and ORG:** The code example of Fig. 4.6 introduces some directives. IAR directives are usually written with all-caps letters. Those like "#define", "#include", and others are C-type pre-processor directives, so called because they come from C language and maintain its syntax. The directives included in the example are the following:

- EQU and #define which are used to define assembly time constants and some symbolic names, valid in the module, for numeric values, hardware pieces, and so on. EQU cannot be used with registers or hardware devices.
- #include, which serves to append external files. In this case, and almost always done, a header file where more assembly time constants are defined.
- ORG will tell the assembler the address in memory where the items following the directive will be stored.

The formats for EQU and #define are

```
LABEL   EQU <<Value or expression>>
```

and

```
#define <<Symbolic Name>> Value or expression or register
```

EQU is an example of a value assignment directive. Another one is SET. For #include, the format is

```
#include "filename"
```

MSP430 header files contain definitions for constants to be used. Since all MSP430 microcontrollers have the same addresses for similar peripherals and ports, related constants are available for MSP430 programmers in the *header files*. Therefore, this is a common line in IAR source files, with filename usually of the type *msp430xxxx.h*, where the xxxx field corresponds to either a model or a family. In the example of Fig. 4.6 we see #include "msp430g3221.h". More recent IAR versions allow *msp430.h* only, and the selection is made by the assembler among the different files available in a library. All the symbolic names that appear in the listing of Fig. 4.6 which were not defined explicitly with EQU or #define, are defined in the header file.

The format for the ORG directive is

```
ORG expr
```

where `expr` is the physical address where the program location counter will point to store the next items going to memory. The ORG directive requires the user to know the memory map, or at least the address of interest, of the microcontroller being used.

**Example 4.1** *Even if the source file is not yet complete, let us look at what happens when the assembly code of Fig. 4.6 is processed.*

*Labels* DELAY *and* LED, *as well as the symbolic name* COUNTER *are identified as* 50000 (=C350h), 1 (=01h), *and* R15, *respectively. All other symbolic names found, such as* P1DIR *and* P1OUT, *are defined in the header file. For other labels, the value will be generated, and the* ORG *directive allows us to understand the process.*

ORG 0F800h *is used just before the beginning of the source code. Therefore the first machine language instruction* 4031 0300 *following the directive goes to that address,* 0F800h. *From there on, every instruction will be stored continuously one after the other. The result can be appreciated with the list file, as partially illustrated below:*

```
                              ORG      0F800h           ;
F800 4031 0300        RESET    mov.w    #300h,SP         ; Set stack
F804 40B2 5A80 0120   StopWDT  mov.w    #WDTPW+WDTHOLD,&WDTCTL ; Stop WDT
F80A D3D2 0022                 bis.b    #LED,&P1DIR      ; P1.0 output
F80E - - - - -                 - - - -
       - -                     - -
```

*The assembly listing has the label* RESET *for this first instruction, so* RESET *becomes* 0F800h. *Similarly, the label* StopWDT *takes the value* F804h, *which is the address where the instruction is stored. Every label attached to an instruction will be then assigned the value of the corresponding instruction address.*

Using directive ORG requires from the user knowledge of the MCU memory map. In particular, the programmer needs to be aware where the executable code usually goes, whose starting address is machine dependent. RAM memory, used for variables and other uses, most often starts at 0200h. Other places need also to be known. Again, some names are defined in the header file to help the programmer in this task.

## Standard Symbolic Constants and Header Files

Even though the assembly instructions work only with integer constants, the definition of symbolic constants is for all practical matters "a must" in programming, and very important in assembly programming. To simplify readability and comprehension, as well as portability, symbolic names for the addresses of the peripheral registers, and for bits or groups of bits within registers are standard. Texas Instruments has generated *header files* files with these constant names. These are written in C language code, but shared by C compilers and assemblers. In the IAR assembler, the files are included in the source code with the directive `#include`.

**Fig. 4.8** Information for watchdog timer control register (WDTCTL) (*Courtesy of Texas Instruments, Inc.*)

| bit15 | | | | | | | bit8 |
|---|---|---|---|---|---|---|---|
| Bits 15-8 normally 69h;  WDTPW= 5Ah for writing | | | | | | | |

| bit7 | bit6 | bit5 | bit4 | bit3 | bit2 | bit1 | bit 0 |
|---|---|---|---|---|---|---|---|
| WDTHOLD | WDTNMIES | WDTNMI | WDTTMSEL | WDTCNTCL | WDTSSEL | WDTIS1 | WDTIS0 |

Fortunately for us, the user guides and data sheets for the different series and models use these symbolic names when explaining the peripheral registers, so the programmer does not need to know the details of the header files. But when necessary, these files are usually available in the IDE.[5]

Let us illustrate the naming conventions using the Watchdog Timer control register, presented in the user guides as shown in Fig. 4.8. The address is also given in the user guide. The meanings of the bit names are summarized in Table 4.1.

Names associated to single bits are usually declared indicating the action asserted when the bit is set (as in WTDHOLD for Watchdog Timer Hold) when this is possible, or to the type of control that it is exerted, as in WDTNMIES for Watchdog Timer Non Maskable Interrupt Edge Selection (0 for falling edge, 1 for rising edge).

With these definitions, "`mov #WDTPW+WDTHOLD,&WDTCTL`" is equivalent to "`mov #0x5A80,&0x0120`", and is used to stop (hold) the WDT operation. Remember that the first byte needs to be 5Ah in order to make changes to this register. Therefore, in the particular case of this register, every change must include WDTPWD in the immediate source operand.

For a group of bits, referred to as a set with a symbolic name ending as x, there are two general conventions. It is up to the reader to verify if both are used in the header file or only one of them. One is when x is substituted with the format "_N", then N represents the decimal value in the combination of the respective bits.

**Table 4.1**  Symbolic constants associated to watchdog timer control register (WDTCTL)

| Name | Number | Comment |
|---|---|---|
| WTDCTL | 0x0120 | Register address |
| WTDPW | 0x5A00 | Required to make changes |
| WTDHOLD | 0x0080 | When set, stops WDT |
| WDTNMIES | 0x0040 | selects the interrupt edge for the NMI interrupt when WDTNMI = 1 |
| WDTNMI | 0x0020 | Selects pin $\overline{RST}$/NMI function: 0 for Reset, 1 for NMI |
| WDTTMSEL | 0x0010 | WDT working mode: 0 for WDT, 1 for interval timer |
| WDTCNTCL | 0x0008 | Resets or clears the counter when set |
| WDTSSEL | 0x0004 | WDT clock source select: 0 for SMCLK, 1 for ACLK |
| WDTISx | (bits 1, 0) | WDT interval select. (Explanation below) |

---

[5] Unfortunately, now and then TI edits the files and changes names. If one of the errors shown during assembly indicates unknown definition, first check spelling. If this is not the source of error, check for an updated header file.

**Table 4.2**  Symbolic constants associated to watchdog timer control register (WDTCTL)

| Name | b1-b0 value | Using bits | Time interval |
|------|-------------|------------|---------------|
| WTDIS_0 | 0x00 | None | Watchdog clock source/32768 |
| WTDIS_1 | 0x01 | WTDIS0 | Watchdog clock source/8192 |
| WTDIS_2 | 0x02 | WTDIS1 | Watchdog clock source/512 |
| WTDIS_3 | 0x03 | WTDIS1+WTDIS0 | Watchdog clock source/64 |

In the other convention, x may be substituted by the bit number generating as many names as bits are in the group. Both conventions are illustrated for the group WDTISx in Table 4.2 in the first format. Remember that by default, the bits are 0.

The numbers 32768, 8192, 512, and 64 are associated to a frequency of 32.768 kHz for a crystal resonator. With this clock frequency, the time intervals are of 1000, 250, 16 and 1.9 ms. For other frequencies, these intervals will be different.

With these definitions,

`mov #WDTPW+WDTTMSEL+WDTCNTCL+WDTSSEL+WDTIS0,&WDTCTL`

is equivalent to `mov #561Dh,&0120h` and means: "Use the WDT as a normal timer, with the ACLK as its source and reseting every $f_{ACLK}/8192$ s". In particular, if we are using a watch crystal for the frequency source, the instruction is for using the WDT as a 250 ms interval timer. Every 250 ms, the interrupt flag is set. To avoid long operands, the complete sum is defined as another constant, "WDT_ADLY_250".

The naming principle illustrated is similarly applied to all registers, addresses for interrupt vectors, or even individual bits and other important combinations. The reader should look at different source examples and user guides to grasp the general idea.

### 4.2.2 Labels in Instructions

The above example gives more insight into labels. Although labels are optional, a good assembly programming practice follows some unwritten rules. Always use a label for

- Entry statement of the main code and of an Interrupt Service Routine (ISR). The label takes the value of the reset vector or interrupt vector.
- Entry statement of a subroutine. The label can be used to call the subroutine using it in immediate addressing mode in the **call** instruction, for example `call #Label`.
- Instruction to which reference is made, for example for a jump.

Labels are also useful as highlights for instructions, even if no reference is needed. Take a look at the listing in Fig. 4.6.

**Example 4.2** *Continuing with the same listing used in example 4.1, the reader can verify that the label* Mainloop *will be equal to* 0xF80E. *The label is used for jump instructions and can be used as any other integer constant with addressing modes as illustrated next.*

mov.w #Mainloop,R6 *(immediate mode for Mainloop) yields* R6 = F80Eh.
mov.w Mainloop,R6 *(direct or symbolic mode for Mainloop) yields* R6 = 403F.
call #Mainloop *(immediate mode for Mainloop) calls a subroutine with entry line at address* 0xF80E. *Hence, the CPU pushes the PC onto the top of the stack (TOS) and then loads it with* 0xF80E.
call Mainloop *(direct or symbolic mode for Mainloop) will push the PC onto the TOS and load it with* 0x403F. *In this particular example, there will be an error since the values at PC must be even. The CPU will reset.*

On the other hand, operands in jump instructions work differently depending on whether they are in a source being compiled or in an interpreter or single line assembler. Labels only apply to compilation with two-pass assemblers. Now, only the address of the target instruction is valid, where the address is given explicitly either as a label or a number, as illustrated in (a) and (b) below. Other modes are illegal, as illustrated in (c). Notice that the syntax is the same as a direct one but does not have the same meaning.

(a) *Instruction* jmp Mainloop *will load the PC with* F80E, *making* mov.w #50000,R15 *the next instruction to fetch.*
(b) *Instruction* jmp 0xF80E *has the same effect but requires the user to know the address of the target instruction.*
(c) *The instructions* jmp #Mainloop *or* jmp & Mainloop *are illegal.*

## 4.2.3 Source Executable Code

The executable code is the set of CPU executable source statements, namely, the instructions. It is the most important part of the source file. It consists of the main code and, optionally, subroutines and Interrupt Service Routines. Important characteristics to consider are

1. *Label for Entry line*: Names like "RESET", "INIT", "START" are common, but any one will do. This label serves to load the reset vector.
2. *Housekeeping and peripherals configuration*: This task includes

   • stack initialization,
   • configuration of the watchdog timer,
   • configuration of peripherals and I/O ports as needed

3. *Main routine or algorithms*: The instructions for the intended program.

Unlike programs written to be run on computers, microcontrollers programs usually do not have a stop or ending instruction. Here, they are terminated with either an

**Fig. 4.9** Initializing SP. **a**
At address just after RAM, **b**
First pushed word occupies
top of RAM

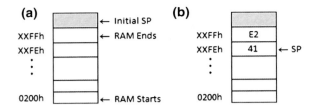

infinite loop or with the CPU off, waiting for an interrupt. Instruction `jmp $`, repeats itself indefinitely. This is useful for debugging and for interrupt driven designs.

**Housekeeping and configuration:** In the stack definition, the SP register is usually initialized with the *next* even value after the RAM, so the first pushed item occupies the highest word address in RAM, as illustrated in Fig. 4.9. RAM always begins at location 0200h in MSP430 microcontrollers, and its last address depends on the model. Thus, in the absolute code, for a 128 byte RAM the last address will be 0x027F so SP is initialized with 0x0280.

Strictly speaking, if the program does not include subroutines or interrupts, and neither includes push or pop instruction, then the stack pointer needs not to be initialized. Our example belongs to this kind of programs. However, it is a good practice to always include this initialization.

If the main code ends with an infinite loop, the Watch Dog Timer (WDT) should be stopped or else it will restart the CPU after some time. Hence, it is common to stop the WDT. This is usually done with the statement

$$\texttt{mov.w \#WDTPW+WDTHOLD,\&WDTCTL} \qquad (4.5)$$

as illustrated in our examples. It is recommended to stop the WDT whenever we have relative long loops in the code. Other WDT configurations are possible.

Finally, an embedded designer must be aware of how hardware connections interact with the CPU. This interaction takes place with internal peripherals and external hardware through I/O ports. Hence the need to configure ports and peripherals. The LED example given so far configures pin P1.0 of port 1 as an output, by setting bit 0 of the byte sized register named P1DIR.

Although a more detailed discussion on the peripherals and ports is provided in Chap. 7, there is an important point to consider with respect to ports. By default, port pins are selected as input pins at power up. Hence, if the pin is being used as input, there is no need to explicitly configure it at the beginning. Yet, if it is not being used, be sure the pin is not a floating input, or several problems can arise due to false signals and noise. Either connect a pull-up or pull-down resistor as illustrated in Fig. 4.10; some models have internal software configurable resistors. The resistor values are usually in the 30–56 kΩ. This practice should always be used when using input devices such as switches or pushbuttons.

If not using resistors, configure the pin as an output. In short, it is a good practice to *configure unused port pins as outputs to avoid hardware hazards*. In an example

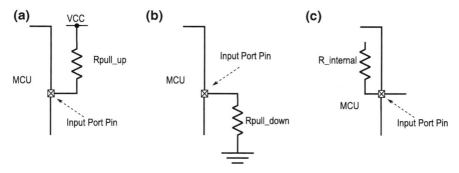

**Fig. 4.10** Input pins with **a** pull-up resistor, **b** pull-down resistor and **c** internal software configurable resistor

like the one used here, it would have been better to use `mov #0xFF, P1DIR`, since no pin is used as input.

**Main Algorithm:** After housekeeping and hardware configuration, we may start writing the instructions for the intended task. If interrupts are being enabled, the `eint` should be introduced after the hardware configuration, to avoid unwanted interruptions.

Since most, if not all, embedded systems work continuously and independently of a user intervention, many systems do not include a particular instruction to "stop" the program. MSP430 microcontrollers fall in this category. Instead, two common methods to "terminate" the main code are an infinite loop using a unconditional jump or setting the system in a proper low power mode, also turning the CPU off.

### 4.2.4 Source File: First Pass

Let us end this discussion about the assembly process with a first look at the layout of the full source file that includes the code considered up to now. Remember that the source file is the one that the programmer writes, with all necessary comments, instructions, directives, etc. to feed to the assembler. Figure 4.11 shows an *absolute source file* for our example. It is called absolute because all the memory allocations are given explicitly by the programmer with the directive ORG.

We recognize in this figure the part we have been working with. Namely, constant declarations and executable code. In addition we have a *documentation*, and the *reset vector allocation*. This last group must be in any assembly source file.

This example introduces two new directives: END and DW, equivalent to DC16.

END directive is compulsory to finish the source file, since the assembler does not read anything after this directive.

DW stands for "define word", and it is equivalent to DC16, which stands for "define 16-bit word". This directive tells the assembler to store in memory the 16-bit words,

```
 1  ;********************************************************;
 2      MSP430G2xx1 Demo - Software Toggle P1.0
 3  ;
 4  ;Description: Toggle P1.0 by xor'ing P1.0 inside
 5  ;   of a software loop.
 6  ;       ACLK = n/a, MCLK = SMCLK = default DCO
 7  ;
 8  ;                   MSP430G2xx1
 9  ;               -----------------
10  ;           /|\|                 XIN|-
11  ;            | |                    |
12  ;            --|RST              XOUT|-
13  ;              |                     |
14  ;              |                 P1.0|-->LED
15  ;
16  ;   Based on code written by D. Dang
17  ;   Texas Instruments Inc.
18  ;   October 2010
19  ;   Built with IAR Embedded Workbench Version: 5.10
20  ;********************************************************
21  #include   "msp430g2231.h"          ; standard constants
22  LED         EQU   01h                ;LED at pin P1.0
23  DELAY       EQU   50000
24  #define     COUNTER R15               ; R15 as counter
25  ;- - - - - - - - - - - - - - - - - - - - - - - - -
26              ORG   0F800h             ; Program Reset
27  ;- - - - - - - - - - - - - - - - - - - - - - - - -
28  RESET       mov.w   #0300h,SP         ; Initialize stack
29  StopWDT     mov.w   #WDTPW+WDTHOLD,&WDTCTL ; Stop WDT
30  SetupP1     bis.b   #001h,&P1DIR      ; P1.0 output
31                                        ;
32  Mainloop    xor.b   #LED,&P1OUT       ; Toggle LED
33  Wait        mov.w   #DELAY,COUNTER    ; Delay to counter
34  L1          dec.w   COUNTER           ; Decrement counter
35              jnz     L1                ; Delay over?
36              jmp     Mainloop          ; Again
37                                        ;
38  ;- - - - - - - - - - - - - - - - - - - - - -;
39              Interrupt Vectors
40  ;- - - - - - - - - - - - - - - - - - - - - - -;
41              ORG   0FFFEh             ; Address for
42              DW    RESET              ; RESET Vector
43              END
```

Documentation

Constants
Declaration

Absolute directive

Executable
Code

Reset vector
allocation

**Fig. 4.11** An absolute IAR listing for blinking LED (*Courtesy Texas Instruments Inc.*)

separated by a comma, that follow the directive. The storage starts at the memory address at which the assembler's *program location counter* (PLC) is pointing at that moment. It is convenient to have it in this case pointing at an even address.

**Reset Vector Allocation:** It is absolutely necessary to allocate the reset vector at address 0xFFFE. Remember that the reset vector is the address of the first instruction to be fetched. When the MCU is powered up or reset, the control unit makes the PC register be loaded with the word found at the mentioned address. We tell the assembler to allocate the reset vector with the lines.

```
ORG  0xFFFE
DW <<ResetVectorLabel>>
```

In our example, and very often, the label used is "RESET", but any valid name will do. Header files usually contain the definition of RESET_VECTOR as 0xFFFE, so we can use ORG RESET_VECTOR if this is the case. We will be back at this topic later.

### 4.2.5 Why Assembly?

High level languages increase productivity and are easier for working with complex algorithms. Why then should we bother studying assembly language?[6]

Assembly instructions are in an isomorphic relation with machine language instructions and thus intimately related to the embedded system. This provides many advantages among which we have: lower power consumption, the least memory usage and fastest operation, as well as lower costs and easier maintenance, desirable features in embedded applications, in particular portable ones. Hence, assembly language is a good choice for short to medium programs, where complexity is still manageable and we can keep good control on the machine performance. Typical uses include device drivers, and low-level embedded and real-time systems.

Another advantage appears when debugging, since programs are always stored in machine language form, no matter the original source program. Access to the system memory through debuggers provides invaluable room for maintenance and improvement. For these reasons, assembly language is a good choice for direct hardware manipulation, access to specialized processor instructions, or to address critical performance issues.

High level code cross compilers work by first translating the code into an assembly language one, giving us the opportunity to optimize the ultimate executable code applying our assembly language skills. For compilers not going through the assembly translation, we always have the possibility of disassembling the machine language to obtain the assembly version.

On the down side, vocabulary in assembly language is very limited and thus the program generally needs more lines, yielding slower code writing and higher costs in development. Some assemblers include high level features to mitigate this disadvantage, in particular for standard assembly code formats. Anyway, we should be clear that for complex algorithms and large codes, high–level languages are definitely preferred.

To close this discussion, let us point out that in embedded systems it is very often highly desirable to mix high-level and assembly instructions. This feature is very useful when we need to use the microcontroller with precise timing and special instruction sequences.

## 4.3 Assembly Instruction Characteristics

Using the source in Fig. 4.11 as an initial template, we are now in a position to look at the MSP430 instructions using the assembler. Of course, any other assembler or interpreter can be used to test our understanding of the instructions and partial listings.

---

[6] An excellent and more in depth discussion about this topic can be found at http://en.wikipedia.org/wiki/Assembly_language. We encourage the reader to go to this site and also look several of the references listed therein.

**Table 4.3** MSP430 addressing modes

| Addressing mode | Syntax | Comment |
|---|---|---|
| Immediate mode | #X | Data is X |
| Register Mode | Rn | Data is Register Rn Contents |
| *Data in Memory:* | | |
| Indexed mode | X(Rn) | Address is X + Rn contents |
| Indirect mode | @Rn | Address is Register Rn contents |
| Indirect autoincrement | @Rn+ | Address is Register Rn contents as before. Now, Rn is incremented after execution by 2 for word instructions and by 1 for byte ones. |
| Direct mode | X | Address is X |
| Absolute mode | &X | Address is X |

**Addressing Modes:**

Remember that operands in an instructions must be written with an appropriate addressing mode syntax. For instructions other than jumps, the seven modes used by the MSP430 are recalled and summarized in Table 4.3. The immediate, the indirect and autoincrement modes are not valid as destination operands.

**Core and Emulated Instructions**

The MSP430 architecture has twenty-seven *hardwired* core instructions, i.e., each one with a specific OpCode in machine language syntax. In addition, assemblers support twenty-four emulated instructions, with mnemonics easier to remember. For example, to "**inv**ert" all bits in R5, "inv R5" is easier to recognize than the equivalent core instruction "xor #0xFFFF,R5." There is no penalty for the use of emulated instructions.

Tables 4.4 and 4.5 list the complete sets of MSP430 core and emulated instructions, respectively.

## 4.3.1 Instruction Operands

Excepting for the jump instructions, an operand in an instruction is either

- a CPU register name, or
- an integer constant, or
- a character enclosed in single quotes ("), which is equivalent to its ascii value, or
- a valid user defined constant

Registers PC, SP and SR may also be referred to as R0, R1 and R2, respectively. Integer constants may be in

- decimal notation, without any suffix or prefix; only decimal numbers can have the minus sign ("–") attached;

**Table 4.4** Core MSP430 instructions

| Type | Instruction | Description | V | N | Z | C |
|------|-------------|-------------|---|---|---|---|
| Data | `mov src,dest` | Loads destination with source | - | - | - | -[1] |
| Transfer | `push src` | Pushes source onto top of stack | - | - | - | - |
| | `swpb dest` | Swap bytes in destination word | - | - | - | - |
| | `add src,dest` | Adds source to destination | * | * | * | * |
| | `addc src,dest` | Adds source and carry to destination | * | * | * | * |
| | `sub src,dest` | Adds $\overline{source}+1$ to destination | * | * | * | * |
| Arithmetic | | ( subtract source from destination) | | | | |
| | `subc src,dest` | Adds $\overline{source}+CF$ to destination | * | * | * | * |
| | | ( subtract with borrow) | | | | |
| | `dadd src,dest` | Adds source and carry to destination in Decimal (BCD) form[2] | * | * | * | * |
| | `cmp src,dest` | $dest-source$, but only affects flags[3] | * | * | * | * |
| | `sxt dest` | Sign extend LSB to 16-bit word | 0 | * | * | * |
| | `and src,dest` | "AND"s source to destination bitwise | 0 | * | * | * |
| | `xor src,dest` | "XOR"s source to destination bitwise | * | * | * | * |
| | `bit src,dest` | Like and, but only affects flags[4] | 0 | * | * | * |
| Logic | `bic src,dest` | Resets bits in destination | - | - | - | - |
| and bit | `bis src,dest` | Sets bits in destination. | - | - | - | - |
| management | `rra dest` | Roll bits to right arithmetically, i.e., $B_n \rightarrow B_{(n-1)}\ldots B_1 \rightarrow B_0 \rightarrow C$ | 0 | * | * | * |
| | `rrc dest` | Roll destinations to right through Carry, $C \rightarrow B_n \rightarrow B_{(n-1)}\ldots B_1 \rightarrow B_0 \rightarrow C$ | * | * | * | * |
| | `jz/jeq label` | Jump if zero/equal (Z = 1) | - | - | - | - |
| | `jnz/jne label` | Jump not zero/equal (Z = 0) | - | - | - | - |
| | `jc/jhe label` | Jump if carry (C = 1) – if higher or equal–– ($\geq$, for unsigned numbers) | - | - | - | - |
| | `jnc/jlo label` | Jump if not carry (C = 0)– if lower,– – ($<$, for unsigned numbers) | - | - | - | - |
| Program | `jn label` | Jump if negative (N = 1) | - | - | - | - |
| Flow | `jge label` | Jump if $V=N$ ($\geq$, for signed numbers) | - | - | - | - |
| | `jl label` | Jump if V$\neq$N (if $<$, signed numbers) | - | - | - | - |
| | `jmp label` | Jump to label unconditionally | - | - | - | - |
| | `call dest` | Call subroutine at destination | - | - | - | - |
| | `reti` | Return from interrupt | - | - | - | - |

[1]: For Flags: − means there is no effect; * there is an effect; "0", flag is reset.
[2]: Result is irrelevant if operands are not in format BCD
[3]: Used to compare numbers, usually followed by a conditional jump
[4]: Used to test if bits are set, usually followed by a conditional jump

- binary notation with suffix b, as in 00101101b, or in the form b'00101101';
- hex notation with suffix h, as in 23FEh or prefix 0x as in 0x23FE, or prefix h as in h'23FE'. Suffixed numbers must never start with a letter;
- octal notation with suffix q, as in 372q, or prefix q as in q'372'.

**Table 4.5** Emulated instructions in the MSP430

| Type | Instruction | Description | Core Inst. |
|------|-------------|-------------|------------|
| Data Transfer | pop dest | Loads destination from TOS | mov @SP+,dest |
| | adc dest | Add carry to destination | addc #0,dest |
| | dadc src,dest | Decimal add Carry to destination | addc #0,dest |
| | dec dest | Decrement destination | sub #1,dest |
| Arithmetic | decd dest | Decrement destination twice | sub #2,dest |
| | inc dest | Increment destination | add #1,dest |
| | incd dest | Increment destination twice | add #2,dest |
| | sbc dest | Subtract Carry from destination | subc #0,dest |
| | tst dest | Test destination | cmp #0,dest |
| | inv dest | Invert bits in destination | xor #0FFFFh,dest |
| | rla dest | Roll (shift) bits to left | add dest,dest |
| Logic | rlc dest | Roll bits left through carry | addc dest,dest |
| and bit | clr dest | Clear destination | mov #0,dest |
| Management | clrc | Clear carry flag | bic #1,SR |
| | clrz | Clear zero flag | bic #2,SR |
| | clrn | Clear negative flag | bic #4,SR |
| | setc | Clear carry flag | bis #1,SR |
| | setz | Clear zero flag | bis #2,SR |
| | setn | Clear negative flag | bis #4,SR |
| | br dest | Branch to destination | mov dest,PC |
| Program | dint | Disable interrupts | bic #8,SR |
| Flow | eint | Enable interrupts | bis #8,SR |
| | nop | no operation | mov R3,R3 |
| | ret | Return from subroutine | mov @SP+,PC |

### Operands for Jump Instructions

In MSP430, in a source program to be compiled by an assembler, the operand is direct, giving the address for the jump. This value is usually given as a labe, as in jmp mainloop, since it is very improbable that the programmer knows the address before compiling.

However, for line assemblers it must be an even signed number within the limits ±512. This is half the offset to be added to the PC register, and hence represents the number of bytes in program memory to be jumped to, forward or backward. Lines 7 and 8 in Fig. 4.5, which appear as jn 0x3FC and jmp 0x3F2 could have also be written as jn-4 and jmp-14, respectively.

## 4.3.2  Word and Byte Instructions

With some exceptions, instruction operands may be either byte-size or word-size. Word and byte instructions are differentiated with a suffix **.w** or **.b**, respectively. The default when no suffix is included is a word instruction.

In byte instructions, when the datum is in memory, the source or destination refers specifically to the byte in the cell. Addresses can be even or odd, with no particular problem. In addition, in the indirect autoincrement addressing mode @Rn+, register Rn is automatically incremented by 1 when working with byte instructions.

Now, when data is in a register, since the register itself is 16-bit wide, the situation is a little different. A source operand in register mode in byte operations points only to the least significant byte. When a destination, the result goes to the least significant byte and the most significant byte is cleared. These remarks are illustrated in Fig. 4.12.

Observe that the most significant byte of a register is not available with a byte instruction. Therefore, to access or change only this part of the register, swpb Rn may be utilized before/after the byte instructions.

The following example illustrates the operation of byte and word instructions.

**Example 4.3** *The following table shows examples of individual and sequence of instructions for byte and word operations. All numbers in the table are in hex notation without suffix or prefix. MSB(Rn) amd LSB(Rn) mean, respectively, most and least significant byte of Rn. Before each instruction or sequence, contents are*

R5 = 03DAh, R6 = 0226h, R15 = BAF4h, [03DAh] = 2B40h, [03DCh] = 4580h and [0226] = F35Ah.

*Since the MSP430 uses little endian method, [03DAh] = 40h, [03DBh] = 2Bh and so on. In byte operations with memory, only the byte is taken. Hence @R5, @R5+, 0(R6) and 1(R6) in the instructions point only to the byte, not the word. This was indicated only once in the RTN of the second column to eliminate any ambiguity.*

## 4.3.3  Constant Generators

Operands in register mode yield faster execution and require less program memory. Any non-register operand data, like a number in immediate mode or an address,

**Fig. 4.12**  Byte operations with data in register: **a** Register as source only, **b** Register as destination (Hx and Gx stand for generic hex digits.)

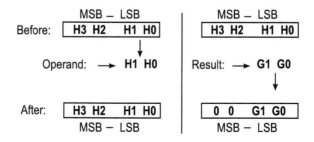

generates one word in machine language, as it can be verified in listing of Fig. 4.5. The immediate mode always generates the word immediately after the instruction word[7] whenever two words are appended. Since some immediate values are very often used, CPU designers usually include one or two registers hardwired to them in such a way that when the programmer write the corresponding immediate operand, the actual machine code uses a register mode expression.

| Instruction(s) | Register notation | After |
|---|---|---|
| mov.w R5,R6 | R6 ← R5 | R5 = 03DA, R6 = 03DA |
| mov.b R5,R6 | LSB(R6) ← LSB(R5)<br>MSB(R6) ← 00h | R5 = 03DA, R6 = 00DA |
| mov @R5,R6 | R6 ← (R5) | R5 = 03DA, R6 = 2B40,<br>[03DA] = 2B40 |
| mov.b @R5,R6 | MSB(R6)←0,<br>LSB(R6)←(R5) | R5 = 03DA, R6 = 0040<br>[03DA] = 2B40 |
| mov @R5,0(R6) | (R6) ← (R5) | R5 = 03DA, R6 = 0226<br>[03DA] = 2B40,<br>[0226] = 2B40 |
| mov.b @R5,1(R6) | Byte(R6 + 1) ← Byte(R5) | R5 = 03DA, R6 = 0226<br>[03DA] = 2B40,<br>[0226] = 405A |
| *Sequence:* | | |
| mov @R5+,R6 | R6 ← (R5);<br>R5 ← R5 + 2 | R5 = 03DC, R6 = 2*B*40,<br>[03DA] = 2B40, [03DC] = 4580 |
| mov @R5+,R15 | R15 ← (R5)<br>R5 ← R5 + 2 | R5 = 03DE, R15 = 4580,<br>[03DA] = 2B40,<br>[03DC] = 4580 |
| *Sequence:* | | |
| mov.b @R5+,R6 | MSB(R6)←0, LSB(R6)←(R5);<br>R5 ← R5 + 1 | R5 = 03DB, R6 = 0040<br>[03DA] = 2B40 |
| mov.b @R5+,R15 | MSB(R15)←0, LSB(R15)←(R5);<br>R5 ← R5 + 1 | R5 = 03DC, R15 = 002B<br>[03DA] = 2B40 |
| mov.b R6,&0227 | (0227)←LSB(R6) | R6 = 0226, [0226] = 265A |

MSP430 designers adopted a similar philosophy, with interesting variations. Both R2 and R3 are used as *constant generators* for several constants, as indicated below:

- R3 for immediate values 0, 1, 2 and −1 (0xFFFF)
- R2 for immediate values 4 and 8, and for absolute value 0.

Some bits in the instruction word identify the individual cases, as explained in Appendix B.2. Notice that R3 is not a general purpose register but may be used as an operand. It does not store any result as a destination, and as an explicit source is equivalent to #0. On the other hand, when R2 is explicitly mentioned as an operand, it refers to the SR register.

---

[7] Hence the name immediate mode.

**Example 4.4** *To illustrate the constant generation, consider lines 1–6 in listings of Fig. 4.5. They all have immediate mode sources, since instruction* dec.w R15 *emulates* sub.w #1,R15. *Let us now look at the following table:*

| Non-constant generation | | Constant generation | |
|---|---|---|---|
| Assembly Instruction | Machine language | Assembly Instruction | Machine language |
| mov.w #0x300,SP | 4031 **0300** | bis.b #001,& 0x0022 | D3D2 0022 |
| mov.w #0x5A8,& 0x0120 | 40B2 **5A80** 0120 | xor.b #001,& 0x0021 | E3D2 0021 |
| mov.w #0xC350,R15 | 403F **C350** | sub.w #001,R15 | 831F |

*Immediate values of instructions on the left hand column do not belong to the constant generation set. Therefore, they are explicitly included in the machine instruction after the instruction word (highlighted with bold fonts). On the right column, the source "# 1" is generated by register* R3. *The highlighted nibble 3 in the machine language instruction word refers to register R3, and there is no extra word for the immediate value.*

*Therefore, the constant generation property of R3 has saved one word (two memory locations) in memory, as well as power while achieving a faster execution.*

**Arithmetic and Logic Operations**

As illustrated in the previous examples, assemblers interpret logic and arithmetic expressions used in operands. Table 4.6 shows some valid operators for these expressions. Operations follow rules of precedence, with a smaller group number having more preference. In a same group, precedence is from left to right, except for unary operators which apply exclusively to the operand following them.

# 4.4 Details on MSP430 Instruction Set

MSP430 instructions may be classified into four groups:

1. Data transfer instructions;
2. Arithmetic instructions;
3. Logic instructions;
4. Program flow instructions.

Tables 4.4 and 4.5 indicate which instructions fall in each group. Before going into the assembly process, let us look quickly into the different groups.

**Table 4.6** Valid Operators in expressions listed by precedence order

| Group | Operator | Meaning |
|-------|----------|---------|
| 1 | + | Unary plus symbol |
|   | − | Unary minus symbol |
|   | ~ | 1s complement |
|   |   | Logical NOT |
| 2 | * | Multiplication |
|   | / | Division |
|   | % | Modulo |
| 3 | + | Addition |
|   | − | Subtraction |
| 4 | << | Shift left |
|   | >> | Shift right |
| 5 | < | Less than |
|   | <= | Less than or equal to |
|   | > | Greater than |
|   | >= | Greater than or equal to |
| 6 | = | Equal to |
|   | != | Not equal to |
| 7 | & | AND |
|   | ^ | XOR |
|   | \|\| | OR |

## 4.4.1 Data Transfer Instructions

The three core and one emulated MSP430 data transfer instructions are: **mov, push, swpb** and **pop**. They operate as indicated in the tables and do not affect any flag. The **mov** instruction is the most often used in assembly programs. In RTN notation:

```
mov src,dest  dest ← src      swpb dest  LSB(dest) ↔ MSB(dest)
push src      (SP) ← src      pop dest   dest ← (SP)
              SP ← SP-2                  SP ← SP+2
```

**Push and Pop:** The instructions `push` and `pop`, being common to all CPU's, have been explained in Chap. 3. In the MSP430, all SP updates are always by steps of two, and SP always points to an even address. These instructions require careful management to avoid bugs because of mishandling of these operations. Between the push operation of the datum and the pop operation to retrieve <u>the same</u> datum, either we have an equal number of push and pop operations, or else we should manipulate the SP register properly. Also, retrieving must be done in the reverse order of pushing. The following example illustrates this point.

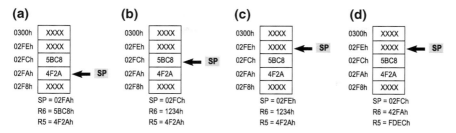

**Fig. 4.13** Illustration of **push** and **pop** in Example 4.5. **a** After push operations, **b** Just before pop R6 in correct sequence, **c** After pop R6 in correct sequence, **d** After pop R6 in incorrect sequence

**Example 4.5** *Assume* R6 = 5BC8h, R5 = 4F2AhSP = 02FE *before the following two sequences:*

*Correct Sequence:*

```
FIRST:  push R6        ;save R6
        push R5
        mov #1234h,R6
        mov #0xFEDC,R5
        pop R5         ;recover R5
LAST:   pop  R6        ;recover R6
```

*Incorrect Sequence:*

```
FIRST:  push R6        ;save R6
        push R5        ;save R5
        mov #1234h,R6
        mov #0xFEDC,R5
LAST:   pop  R6        ;recover R6
```

*Figure 4.13 illustrates the differences. Case* (a) *shows the* SP *value just after the* push R5 *operation, valid for both sequences. Case* (b) *corresponds to the first case. Notice that the original R5 was retrieved and R6 is about to be recovered too after the pop operation, as seen in* (c). *This is the situation for an equal number of push and pop operations, in a correct sequence, after* push R6; *notice that* SP *ends up with the same original value. An unequal number of push and pops case, without proper manipulation of* SP *is illustrated by* ( d ). *Notice that neither* R5 *nor* R6 *were recovered, and* SP *ends at a different address. Also, if we add* pop R5 *after* pop R6, *the registers will not be loaded with their original values.*

**Swap Bytes:** The swap-bytes instruction (**swpb**) exchanges the most and least significant bytes of the destination word, usually a register. For example, if R5 = $34FAh$, then `swpb R5` yields R5 = $FA34h$. This is handy because the programmer has direct access only to the least significant byte of registers in byte operations.

## 4.4.2 Arithmetic Instructions

For easiness, the arithmetic instructions are repeated in Table 4.7 with their RTN equivalent. These operations are performed by the ALU and have effect on flags C, Z, N and V. The "normal" effect mentioned in the table is defined as follows:

**Carry flag SR(0):** C = 1 carry occurs, C = 0 if no carry occurs;
**Zero flag SR(1):** Z = 1 if result is cleared; Z = 0 otherwise;
**Negative or Sign flag SR(2):** N reflects the MSB of the result (Bit 7 for byte operations, Bit 15 for word operations);
**Overflow flag SR(8):** V = 1 if one of the following occurs:

1. addition of two equally signed number produces a number of the opposite sign
2. if the remainder in subtraction when subtrahend and minuend are of different sign, has the sign of the subtrahend.

Flags N and V make sense for signed numbers only. For unsigned numbers, overflow is flagged by the Carry flag. Yet, remember that interpretation of flags is a task that belongs to the programmer, not to the controller. Notation SR(X) refers to the position of the bit in the status register. Emulated instructions affect flags according to the core equivalent.

The subtraction operations, with and without borrow, are actually realized as two's complement addition. Thus, "`sub src, dest`" executes "dest + NOT src + 1". In MSP430, carry C = 0 denotes the presence of a borrow while C = 1 indicates that no borrow was needed. See p. XXX for a reminder. Notice that subtraction with borrow has two mnemonics, so the programmer may use whichever makes more sense for the program.

The MSP430 family ALU supports BCD addition, also called decimal addition (see Sect. 2.7 .3) with instruction `dadd`, although involving the C flag. This instruction assumes that the operands are unsigned integers encoded in BCD. Therefore, the result of `dadd src,dest` is meaningless if any of the operands is not in this code. After the `dadd`, flag V is undefined and C = 1 if result is greater than 9999 for word operations or 99 for byte operations. Since the decimal addition also includes the carry, the programmer must secure that the carry flag is 0 before addition whenever necessary.

The sign extension instruction `sxt` affects the most significant byte of the word destination. If bit 7 is 0, then the MSB becomes 0x00; if the bit is 1, the MSB becomes 0xFF. In other words, it extends an 8-bit signed number, the LSB of the destination, into a 16-bit one. This is a useful instruction when processing data from 8-bit ports delivering signed data, as illustrated by the following sequence:

**Table 4.7** Arithmetic instructions for the MSP430

| Core Instructions | | | |
|---|---|---|---|
| Mnemonics & Operands | Description | Flags | Comments |
| add src,dest | dest← src + dest | Normal | Add src to dest |
| addc src,dest | dest← src + dest + C | Normal | Add with Carry |
| dadd src,dest | BCD algorithm used in | Special[a] | Decimal version of addc. |
| | dest← src + dest + C | | Data in BCD format |
| sub src,dest | dest ← dest + .NOT.src + 1 | Normal | Subtract src from dest. $(dest \leftarrow dest - src)$[b] |
| subc src,dest sbb src,dest | dest ← dest + .NOT.src + C | Normal | Subtract with borrow $(dest \leftarrow dest - src + C)$[b] |
| cmp src,dest | dest + .NOT.src + 1 | Normal | Only affect flags. |
| sxt dest | MSB← FFh×(Bit 7) | Special[c] | Word operand only. Signed LSB extended to 16 bit. |

| Emulated | | | |
|---|---|---|---|
| Mnemonics & Operands | Description | Emulated Instruction | Comments |
| adc dest | dest ← dest + carry | addc #0,dest | Add carry to dest. |
| dadc dest | BCD version for adc | dadd #0,dest | |
| inc dest | dest ← dest + 1 | add #1,dest | |
| incd dest | dest ← dest + 2 | add #2,dest | |
| dec dest | dest ← dest − 1 | sub #1,dest | |
| decd dest | dest ← dest − 2 | sub #2,dest | |
| tst dest | dest ← dest − 0 | cmp #0,dest | To test for sign or zero |

[a] C = 1 if result > 99 for bytes or result > 9999 for words; V is undefined
[b] Borrow needed if C = 0; Borrow = $\overline{Carry}$
[c] C = NOTZ, V = 0

```
mov.b &P1IN,R14    ;Read input data
sxt    R14         ;Sign extend for processing
```

Notice that multiplication and division are not supported by the MSP430 ALU. Some models include a hardware multiplier. We shall deal with this peripheral in later sections.

Let us illustrate some operations with core instructions, since emulated ones function similarly.

**Example 4.6** *Each case in the tables below is independent of others. For all cases, contents of registers and memory before the instruction are as follows, unless otherwise indicated. All contents and addresses are in hex notation without suffix or prefix:*

$$R5 = 35DA, \ R6 = EF26, \ R7 = 5469, \ R8 = 0268,$$
$$[0268] = 364A, \ [026A] = 2FD1, \ [03BC] = 1087$$

dummy

| Instruction | Operation | Results | Flage |  |  |  |
|---|---|---|---|---|---|---|
|  |  |  | C | Z | N | V |
| *Addition:* |  |  |  |  |  |  |
| add R5,R6 or |  |  |  |  |  |  |
| add.w R5,R6 | 35DAh+EF26h=12500h | R6=2500 | 1 | 0 | 0 | 0 |
| add.b #0x26,R5 | 26h + 0DAh = 100h | R5=0000 | 1 | 1 | 0 | 0 |
| *Decimal addition:* |  |  |  |  |  |  |
| A. Assuming carry C=0. |  |  |  |  |  |  |
| dadd.b #0x96,R6 | 0 + 96 + 26 = 122 | R6=0022 | 1 | 0 | 0 | X |
| B. Assuming carry C=1. |  |  |  |  |  |  |
| dadd R7,&0x03BC or |  |  |  |  |  |  |
| dadd.w R7,&0x03BC | 1+5469+1087 = 6557 | [03BC]=6557 | 0 | 0 | 0 | X |
| Instruction | Operation | Results | C | Z | N | V |
| *Subtraction, which actually uses two's complement addition* |  |  |  |  |  |  |
| sub 2(R8),R6 or |  |  |  |  |  |  |
| sub.w 2(R8),R6 | EF26h+D02Eh+1h = 19F55h | R6=9F55 | 1 | 0 | 1 | 0 |
| sub.b #67,R5 | 0DAh + 0BCh+1h = 197h | R5=0097 | 1 | 0 | 1 | 0 |
| *Sign extention* |  |  |  |  |  |  |
| sxt R5 | Bit7=1: MSB(R5)←FFh | R5=FFDA | 1 | 0 | 1 | 0 |
| sxt R6 | Bit7=0: MSB(R6)←FFh | R6=0026 | 0 | 0 | 0 | 0 |
| *Compare* |  |  |  |  |  |  |
| cmp R6,R7 or |  |  |  |  |  |  |
| cmp.w R6,R7 | 5469h+10D9h+1 = 6543h | No change | 0 | 0 | 0 | 0 |

Instructions addc, sbb, and dadd, involve the carry/borrow in addition and subtraction. There are situations in which this property becomes very useful. One such example is for addition or subtraction of data larger than the operand limitations. This is analogous to the hardware use of two or more adders, as illustrated by Fig. 4.14. The following example illustrates a software code for the same objective.

**Example 4.7** *To add* 36,297,659+2,382,878, *16-bit data cannot be used. However, we may combine registers to represent them and utilize more than one instruction to operate with them. The operation in hex equivalent numbers is*

$$36297659 + 2382878 = 38680537 \Rightarrow 0x0229DBBB + 0x00245C1E = 0x024E37D9$$

*Since more than two bytes are required for each number, let us use registers by pairs, with the combination R10-R11 for the first operand and for the result, and R14-R15 for the second operand. The following sequence of instructions realize the operation:*

```
mov  #0xDBBB,R11  ;LSW of first term
```

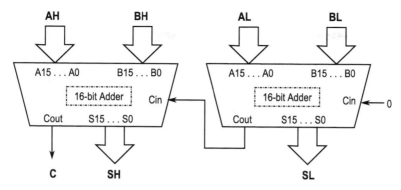

**Fig. 4.14**   Adding 32-bit words A and B, where A = AH AL and B = BH BL.

```
mov    #0x0229,R10    ;MSW of first term
mov    #0x5C1E,R15    ;LSW of second term
mov    #0x0024,R14    ;MSW of second term
add    R15,R11        ;add LSW's, R11 = 3729h, C = 1
addc   R14,R10        ;add MSW's and C, R10 = 024Eh, C = 0
```

*We could also work with the decimal addition, encoding with BCD:*

```
mov    #0x7659,R11    ;LSW of first term, BCD
mov    #0x3629,R10    ;MSW of first term, BCD
mov    #0x2878,R15    ;LSW of second term, BCD
mov    #0x0238,R11    ;MSW of second term, BCD
clrc                  ;Clear Carry flag
dadd   R15,R11        ;add LSW's, R11 = 0537h in BCD, C = 1
dadd   R14,R10        ;add MSW's and C; R10 = 3868h in BCD, C = 0
```

### 4.4.3  Logic and Register Control Instructions

The logic and register control instructions are shown in Table 4.8, together with the effect on flags. Emulated instructions affect flags according to the respective equivalent core instruction. Note that some of these are in fact arithmetic instructions. The "normal" effect on flags for logic operations is defined as follows:

**Carry flag SR(0):** C is the opposite of Z (C = NOTZ);
**Zero flag SR(1):** Z = 1 if result is cleared; Z = 0 otherwise;
**Negative Flag SR(2):** N reflects the MSB of result ;
**Overflow Flag SR(8):** V = 0.

#### Bitwise Logic Operations and Manipulation

Logic instructions are bitwise operations, which means that the operation is done bit-by-bit without any reference to other bits in the word. We can therefore target

**Table 4.8**  Logic and register control core instructions for the MSP430

| Core Instructions | | | |
|---|---|---|---|
| Mnemonics & Operands | Description | Flags | Comments |
| and src,dest | dest ← src.AND.dest | Normal | Bitwise AND |
| xor src,dest | dest ← src.XOR.dest | See Note * | Bitwise XOR |
| bic src,dest | dest ← (.NOT.src).AND.dest | Not affected | Clear bits in dest with mask src. |
| bis src,dest | dest ← src.OR.dest | Not affected | Set bits in dest with mask src. |
| bit src,dest | src.AND.dest | Normal | Test bits in dest with mask src. Only affects flags |
| rra dest | $b_n \to b_{n-1} \to \cdots$ $\cdots \to b_0 \to C$ | C←LSB | Roll dest right arithmetically. |
| rrc dest | $C_{old} \to b_n \to \cdots$ $\cdots \to b_0 \to C_{new}$ | C←LSB | Rotate dest right logically through C. |
| Emulated Instructions | | | |
| Mnemonics & Operands | Description | Emulated Instruction | Comments |
| inv dest | $bit(h) \leftarrow .NOT.bit(h)$ | xor #0xFFFF,dest | Inverts bits in dest. |
| rla dest | $C \leftarrow b_n \leftarrow \cdots$ $\cdots \leftarrow b_0 \leftarrow 0$ | add dest,dest | Roll dest left. |
| rlc dest | $C_{new} \leftarrow b_n \leftarrow \cdots$ $\cdots \leftarrow b_0 \leftarrow C_{old}$ | addc dest,dest | Rotate dest left through Carry. |
| clr dest | dest ← 0 | mov #0,dest | Clears destination. |
| clrc | C ← 0 | bic #1,SR | Clears Carry flag. |
| clrn | N ← 0 | bic #4,SR | Clears Sign flag. |
| clrz | Z ← 0 | bic #2,SR | Clears Zero flag. |
| setc | C ← 1 | bis #1,SR | Sets Carry flag. |
| setn | N ← 1 | bis #4,SR | Sets Sign flag. |
| setz | Z ← 1 | bis #2,SR | Sets Zero flag. |

C = NOT(Z), N reflects MSB, V = 1 if both operands are negative

specific bits in data without affecting the others. For this purpose, we use *masks*. A mask X is a source whose binary expression has 1's in the target bit positions and 0's elsewhere. The properties of the logic operations which we use to achieve this manipulation are the following:

$$X + A = \begin{cases} A & X = 0 \text{ Leaves A unchanged} \\ 1 & X = 1 \text{ Forces a set} \end{cases} \tag{4.6}$$

$$X \oplus A = \begin{cases} A & X = 0 \text{ Leaves A unchanged} \\ \bar{A} & X = 1 \text{ Toggles or inverts A} \end{cases} \tag{4.7}$$

$$X \cdot A = \begin{cases} 0 & X = 0 \ \text{Forces a clear or reset} \\ A & X = 1 \ \text{Leaves A unchanged} \end{cases} \tag{4.8}$$

$$\bar{X} \cdot A = \begin{cases} 0 & X = 1 \ \text{Forces a clear or reset} \\ A & X = 0 \ \text{Leaves A unchanged} \end{cases} \tag{4.9}$$

Using these operations, we can therefore target individual bits for manipulation:

**Clear bits** Instruction (**bic**) uses (4.9) to *clear* the bits selected by the mask, i.e. make the bits equal to 0.

**Set bits** Instruction (**bis**) uses (4.6) to *set* the bits according to the mask, i.e., force them to be 1.

**Toggling** Instruction **xor** uses (4.7) to *toggle* bit values, i.e. inverts the current values of the target bits.

**Bit testing** Instruction (**bit**) uses (4.8) to *test* if at least one of the target bits is set, i.e., equal to 1.

From (4.8), instruction **and** can also be used to *clear* bits. However, now the source should have 0's at the bit positions to be cleared. We illustrate these operations with an example.

**Example 4.8** *Assume contents of the registers and memory* **before** *any instruction as*

R12 = 25A3h = 0010010110100011, R15 = 8B94h = 1000101110010100,

[25A5h] = 6Ch = 01101100

*(a)* **AND** *and* **BIT TEST**

*Instructions:* and R15,R12 *or*    and.w R15,R12 *and* bit R15,R12 *or* bit.w R15,R12

| *Operation:* | *Flags:* C = 1Z = 0N = 0V = 0 |
|---|---|
| 0010 0101 1010 0011 (R12) AND | |
| <u>1000 1011 1001 0100</u> (R15) = | and R15,R12 *yields* |
| 0000 0001 1000 0000 | R12 = 0180    *but* |
| | bit R15,R12 *leaves* R12 *unchanged* |

*Instructions:* and.b 2(R12),R15  *and* bit.b 2(R12),R15

| *Operation:* | *Flags:* C = 1Z = 0N = 0V = 0 |
|---|---|
| 0110 1100 (Memory) AND | |
| <u>1001 0100</u> (LowByteR15) = | and.b 2(R12),R15 *yields* |
| 0000 0100 (new Low Byte R15) | R15 = 0004,    *but* |
| | bit.b 2(R12),R15 *leaves* R15 *unchanged*. |

*(b)* **BIT CLEAR (BIC)**

*Instruction:* `bis R15,R12` *or* `bis.w R15,R12`

| | |
|---|---|
| *Operation:* | *New Contents:* R12 = 2423 |
| 0010 0101 1010 0011 (R12) AND | *Flags:* not affected |
| 0111 0100 0110 1011 ($\overline{R15}$) = | |
| 0010 0100 0010 0011 (new R12) | |

*Instruction:* `bic.b 2(R12),R15`

| | |
|---|---|
| *Operation:* | *New Contents:* R15 = 0090 |
| 1001 0011 ($\overline{Memory}$) AND | *Flags:* not affected |
| 1001 0100 (LowByteR15) = | |
| 1001 0000 (new Low Byte R15) | |

*(c)* **BIT SET (BIS)**

*Instruction:* `bis R15,R12` *or* `bis.w R15,R12`

| | |
|---|---|
| *Operation:* | *New Contents:* R12 = AFB7 |
| 1000 1011 1001 0100 (R15) OR | |
| 0010 0101 1010 0011 (R12) = | *Flags:* not affected. |
| 1010 1111 1011 0111 | |

*(d)* **XOR**

*Instruction:* `xor.b #0x75,R15`

| | |
|---|---|
| *Operation:* | *New Contents:* R15 = 00E1 |
| 0111 0101 (0x75) XOR | *Flags:* C = 1Z = 0N = 1V = 0 |
| 1001 0100 (LowByteR15) = | |
| 1110 0001 | |

*(e)* **Invert** *(emulated instruction)*

*Instruction:* `inv.b &0x25A5` *equivalent to* `xor.b 0xFF,&0x25A5` *yields* [25A5h] = 10010011 = 93, *with flags:* C = 1Z = 0N = 1V = 0
*Instruction:* `inv R15` *equivalent to* `xor 0FFFFh,R15` *yields R15 = 0111 0100 0110 1011 = 746B with flags:* C = 1Z = 0N = 0V = 0

**Remark on BIT instruction:** When using instruction `bit`, the result is Z = 0 and C = 1 iff at least one of the target bits in datum is 1. An interesting and often applied case is when the set consists of only one bit. If this bit is 0, then Z = 1, C = 0; if the bit is 1, then Z = 0, C = 1. Hence, this instruction "transfers" or "copies" the tested bit onto the carry flag and the inverted bit onto the zero flag. This characteristic is useful to retrieve a bit or its complement from a word or byte.

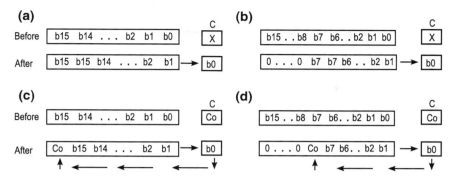

**Fig. 4.15** Right arithmetic rolling (shifting): **a** rra.w dest and **b** rra.b dest; Right rotation through carry: **c** rrc.w dest and **b** rrc.b dest

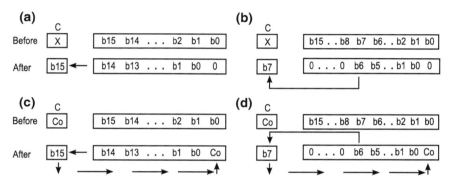

**Fig. 4.16** Left "arithmetic" rolling: **a** rla.w dest and **b** rla.b dest; Left rotation through carry: **c** rlc.w dest and **b** rlc.b dest

### Register Control Instructions

Although the logical operations may be considered as register control instructions, their bitwise operating characteristic makes them more appropriate for targeting specific bits inside the register. Other register control instructions are of the shift and rotating type. These instructions can be visualized on Figs. 4.15 and 4.16. Recall that in the MSP430, left rolling and rotations are actually emulated instructions. The following example illustrates these operations.

**Example 4.9** *Each case below is independent of others. For all cases, contents of register R5 before instruction is:* R5 = 8EF5 = 1000 1110 1111 0101

(*a*) **Right shift/rotations**
*Instruction:* rra R5 *or* rra.w R5
*Instruction:* rra.b R5
*Instructions:* clc *followed by* rrc R5 *or* rrc.w R5
(*b*) **Left Shifts/Rotations**

*Operation*: 1000 1110 1111 0101 $\overset{rra}{\rightarrow}$ 1100 0111 0111 1010 (LSB 1⇒ C)
*New Contents*: R5 = C77A    *Flags*: C = 1Z = 0N = 1V = 0

$$\overset{Higher Byte \quad rrahere}{}$$
*Operation*:1000 1110 1111 0101 $\overset{rra}{\rightarrow}$ $\overbrace{00000000}$ $\overbrace{11111010}$ (LSB 1⇒ C)
*New Contents*: R5 = 00FA    *Flags*: C = 1Z = 0N = 1V = 0

*Operation*: C = 0 *and then* 1000 1110 1111 0101 $\overset{rra}{\rightarrow}$ 0100 0111 0111 1010 1⇒ CF
*New Contents*: R5 = 477A    *Flags*: C = 1 Z = 0 N = 0 V = 0

*Instruction:* rla  R5   *or*   rla.w  R5, *equivalent to* add  R5,R5.

*Operation*:1000 1110 1111 0101 *yields* C←1 0001 1101 1110 1010
*New Contents*: R5 = 1DEA   *Flags*: C = 1 Z = 0 N = 0 V = 1

*Instruction:* rla.b  R5, *equivalent to* add.b  R5,R5

*Operation*: 1000 1110 1111 0101 ⇒ 0000 0000 1110 1010, C←1
*New Contents*: R5 = 00EA   *Flags*: C = 1Z = 0N = 1V = 0

*Instructions:* setc *followed by*   rlc  R5   *or*   rlc.w  R5, *equivalent to* addc R5,R5.

*Operation*: C = 1 *and then* 1000 1110 1111 0101 ⇒ 0001 1101 1110 1011 with 1⇒ C
*New Contents*: R5 = 1DEB   *Flags*:C = 1Z = 0N = 0V = 1

Notice that the flags for left rolls result from the core instructions, which are additions.

**Division and Mulitiplication by 2** Right rotation arithmetically, rra, may be interpreted as the division of a signed number by two in the sense

$$dividend = divisor \times quotient + residue$$

with the residue always nonnegative and less than the absolute value of the divisor (2). The carry holds the residue.

Right rotation through carry, on the other hand, may also be interpreted as a division by 2 for unsigned numbers, provided the carry is initially 0 (use clc before rotating).

Since left shifts are emulated by addition of destination with itself, they may be interpreted as unsigned multiplications by 2. Specifically, rla A may be interpreted as 2×destination, and rlc  A as $2 \times A +$ carry. For decimal, BCD, dadd A,A may be interpreted as $2 \times A +$ carry in decimal system.

Let us use the above remarks in the following example.

**Example 4.10**  *Give an arithmetic interpretation to the instructions of the previous example.*

*(a) Right rotations*

*For* rra R5*: The signed equivalents for the numbers involved are* 8EF5h $\rightarrow$ $-28,939$*; signed* C77Ah $\rightarrow$ $-14,470$*. The carry flag has the residue 1, so* $-28,939 = -14,470 \times 2 + 1$*, corresponding to* $-28939 \div 2$*.*

*For* rra.b R5*: The signed equivalents for the numbers involved are* F5h $\rightarrow$ $-11$*; signed* FAh $\rightarrow$ $-6$*. The carry flag has the residue 1, so* $-11 = -6 \times 2 + 1$*, corresponding to* $-11 \div 2$*.*

*For* clc *followed by* rrc R5*, the division is for unsigned numbers. Since* 8EF5h $\rightarrow$ $36,597$ *and* $477Ah \rightarrow 18,298$*. The carry flag has the residue, and* $36,597 = 18,298 \times 2 + 1$*, which means a division by 2.*

*The unsigned division occurs if a 0 is forced as a most significant bit. Therefore, preceding* rrc R5 *by* setc *does not result in a division.*

*(b) Left shifts: Since they are emulated by addition and addition with carry,* rla *and* rlc *may be interpreted, respectively, as 2x and 2x+C, including the Carry as the most significant bit for interpretations. The result is valid for both unsigned and two's complement signed encodings.*

Up to this point, we have illustrated the instructions as isolated steps. Let us look at another example, using left shift for binary to BCD conversion.

**Example 4.11  Binary to BCD conversion:** *Let us convert an 8-bit binary encoding into its equivalent BCD encoding. Since the largest decimal equivalent of an 8-bit number is 255, one 16-bit register suffices for the result.*

*Recall that*

$$\underbrace{b_7 b_6 \cdots b_1 b_0}_{\text{binary}} = \underbrace{b_7 2^7 + b_6 2^6 + \cdots + b_1 2^1 + b_0}_{\text{Decimal}}$$

*A little algebraic exercise shows that the right hand decimal expression can be transformed into the so called nested multiplication form*

$$((((((( 2 \times 0 + b_7) \times 2 + b_6) \times 2 + b_5) \times 2 + b_4) \times 2 + b_3) \times 2 + b_2) \times 2 + b_1) \times 2 + b_0$$

*Therefore, the power expansion in decimal form can be calculated by eight iterations of operations of the form* 2X+C*, as expressed with the following sequence:*

$$A_1 = 2 \times 0 + b_7$$
$$A_2 = 2 \times A_1 + b_6$$
$$A_3 = 2 \times A_2 + b_5$$
$$A_4 = 2 \times A_3 + b_4$$
$$\vdots$$

**(a)**

| | R7 (in hex) | C | LSB of R6 (in binary) |
|---|---|---|---|
| Initial State: | 0 0 0 0 | x | 1 1 0 1 0 1 0 1 |
| rla.b R6 | 0 0 0 0 | 1 | 1 0 1 0 1 0 1 0 |
| dadd R7,R7 | 0 0 0 1 | 0 | 1 0 1 0 1 0 1 0 |
| rla.b R6 | 0 0 0 1 | 1 | 0 1 0 1 0 1 0 0 |
| dadd R7,R7 | 0 0 0 3 | 0 | 0 1 0 1 0 1 0 0 |
| rla.b R6 | 0 0 0 3 | 0 | 1 0 1 0 1 0 0 0 |
| dadd R7,R7 | 0 0 0 6 | 0 | 1 0 1 0 1 0 0 0 |
| rla.b R6 | 0 0 0 6 | 1 | 0 1 0 1 0 0 0 0 |
| dadd R7,R7 | 0 0 1 3 | 0 | 0 1 0 1 0 0 0 0 |
| rla.b R6 | 0 0 1 3 | 0 | 1 0 1 0 0 0 0 0 |
| dadd R7,R7 | 0 0 2 6 | 0 | 1 0 1 0 0 0 0 0 |
| rla.b R6 | 0 0 2 6 | 1 | 0 1 0 0 0 0 0 0 |
| dadd R7,R7 | 0 0 5 3 | 0 | 0 1 0 0 0 0 0 0 |
| rla.b R6 | 0 0 5 3 | 0 | 1 0 0 0 0 0 0 0 |
| dadd R7,R7 | 0 1 0 6 | 0 | 1 0 0 0 0 0 0 0 |
| rla.b R6 | 0 1 0 6 | 1 | 0 0 0 0 0 0 0 0 |
| dadd R7,R7 | 0 2 1 3 | 0 | 0 0 0 0 0 0 0 0 |

**(b)**

```
NUMB:  mov.b  #0xD5,R6
       mov    #0,R7
       mov    #8,R15
LOOP:  rla.b  R6
       dadd   R7,R7
       dec    R15
       jnz    LOOP
```

**Fig. 4.17** Illustrating the binary to BCD conversion (11010101b → 213). **a** Sequence of steps in loop excluding counter; **b** assembly code for the loop

$$A_8 = Result = 2 \times A_7 + b_0$$

*The decimal 2X+C can be realized with* dadd, *provided that the carry flag is the appropriate bit. The bits can be extracted into the carry by left rollings using* rla.b. *Assuming that the result is written in register R7 and the number of interest is stored as the least significant byte of R6, a pseudo code for the above sequence is:*

Initialize R7 = 0
Iterate 8 times:
  Roll left LSB of R6 (rla.b R6)
  2×R7+C in decimal (dadd R7)

*Figure 4.17 illustrates as an example the conversion of* 11010101 *and the code for executing the algorithm.*

### 4.4.4 Program Flow Instructions

These instructions are listed in Table 4.9. The set can be divided in three groups: miscellaneous, jump, and subroutine handling instructions. The third group is discussed

**Table 4.9** Program flow instructions for the MSP430

| Core Instructions | | |
|---|---|---|
| Mnemonics & Operands | Description | Comments |
| call dest | Push PC and PC ← dest | Subroutine Call |
| jmp label | PC ← label | Unconditional jump (goto) |
| jc label | If C = 1, then PC ← label | "Jump if carry" |
| (or jhs label) | | "Jump if higher than or same as" |
| jnc label | If C = 0, then PC ← label | "Jump if no carry" |
| (or jlo label) | | "Jump if lower than" |
| jge label | If N = V, then PC ← label | "Jump if greater than or equal to" |
| jl label | If N ≠ V, then PC ← label | "Jump if less than" |
| jn label | If N = 1, then PC ← #label | "Jump if negative" |
| jnz label | If Z = 0, then PC ← label | "Jump if not zero" |
| (or jne label) | | "Jump if not equal" |
| jz label | If Z = 1, then PC ← label | "Jump if zero" |
| (or jeq label) | | "Jump if equal" |
| reti | Pops SR and then Pops PC | Return from interrupt |
| Emulated Instructions | | |
| Mnemonics & Operands | Description | Emulated Instruction |
| br dest | Branch (go) to label | mov dest,PC |
| dint | Disable Interrupts (GIE = 0 in SR) | bic #8, SR |
| eint | Enable Interrupts (GIE = 1 in SR) | bis #8, SR |
| nop | No operation | mov R3, R3 |
| ret | Return from subroutine | mov @SP+, PC |

in Sect. 4.8. The two latter groups control the flow of the program by changing the contents of the PC register. Jumps are limited to a range between $-1024$ and $+1022$ (even number) memory addresses from current PC contents.

## Miscellaneous Program Flow Instructions

These instructions are emulated. Two are for interrupt handling: eint, to enable maskable interrupts, and dint, to disable them.

The third instruction is the "no operation", nop, emulated by mov R3,R3. It achieves nothing, but takes one instruction cycle to execute. It is used to introduce one cycle delay. Useful to synchronize operations with peripherals like the hardware multiplier. Enabling and disabling interrupts require one extra cycle for the hardware to actually configure. Therefore, these operations usually go together with instruction nop too, as illustrated next.

```
eint            ;enable interrupts
nop             ; wait for enabling

dint            ;disable interrupts
nop             ; wait for disabling
```

Similar strategies are suggested when enabling and disabling interrupt requests from peripherals.

### Unconditional Jump and Branch Instructions

Unconditional jumps are realized with the jump instruction `jmp label`, and the the branch instruction `br dest`, which is emulated by `mov dest,PC`. The emulated branch instruction may go to any even address in the full memory range. Here, the operand follows the normal rules of addressing modes.

In embedded systems it is normal to close the main executable code with an unconditional jump to continuously repeat the task for which the system is being used. Now, when debugging or testing codes, it is customary to use

$$jmp \ \$$$

to finish the code or introduce breaks. This instruction tells the CPU to repeat itself.[8]

### Conditional Jumps and Control Structures

Figure 3.27 introduced the delay and iteration loops, using the `jnz` instructions. Let us expand the discussion.

Structured programming utilizes control structures of the IF-THEN and IF-THEN-ELSE type, and loop structures like FOR loops, WHILE loops, etc. We briefly review these structures and their coding in the assembly language using conditional jumps.

A conditional jump tests if a flag or a relation among flags is set or not set. The operation is illustrated by Fig. 4.18 using the flowchart decision symbols, for a single flag/relation, and for compound AND/OR.

During program design, the programmer will use statements which give origin to the flag testing, like "A≥B?", "Hardware ON", "Conversion Finished", and so on. Alternative mnemonics reinforces this process. In particular, after comparing A and B with `comp B,A`, we prefer the mnemonics as shown in Table 4.10.

Remember that "A > B" is equivalent to "A ≥ B AND A ≠ B", or "NOT(A < B) OR A = B". Similarly, "A ≤ B" is equivalent to "A < B OR A = B". For these compound statements we may apply the structures (b) or (c) of Fig. 4.18.

---

[8] It is meaningless to end a real world application with this command, just waiting until battery dries up!

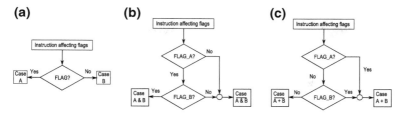

**Fig. 4.18** Illustration of conditional jump operation: **a** Single Flag/condition; **b** AND-compound statement; **c** OR-compound statement

**Table 4.10** Comparing A and B by A-B

| Case | Unsigned numbers | Signed numbers |
|------|------------------|----------------|
| A = B | jeq | jeq |
| A ≠ B | jne | jne |
| A ≥ B | jhs | jge |
| A < B | jlo | jl |

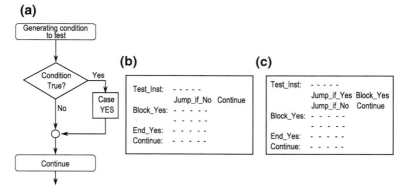

**Fig. 4.19** IF-Structure **a** Flowchart, **b** and **c** Assembly code examples

**IF-THEN and IF-THEN-ELSE structures** The flowchart for the simple IF-THEN construct is shown in Fig. 4.19a. Insets (b) and (c) show two assembly code formats to implement this flowchart. Other formats are of course possible. The flowchart for the simple IF-THEN construct is shown in Fig. 4.20a, together with two code examples in insets (b) and (c).

**Example 4.12** *Ten consecutive unsigned 16-bit numbers are stored in memory, with the first one at address NUMBERS. The objective is to add the first two numbers multiples of 4 and if the addition is not overflowed, store the result at address RESULT. Recall that a binary number is a multiple of 4 if the two least significant bits are 0.*

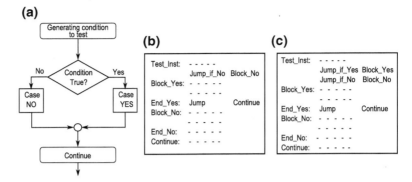

**Fig. 4.20** IF-ELSE Structure **a** Flow chart, **b** and **c** Assembly code examples

*The pseudo code in the left column may be programmed by the instructions of the right column:*

| | |
|---|---|
| Step 1. Initialize pointer = NUMBERS, SUM = 0, COUNTER = 2. | `Step1: mov #NUMBERS,R8`<br>`        mov #0,R9`<br>`        mov #2,R10` |
| Step 2. Read number & increment pointer. | `Read:  mov @R8+,R11` |
| Step 3. If number is not multiple of 4, go to step 2. | `Step3: tst #3,R11`<br>`        jnz Read` |
| Step 4. Add to SUM | `Step4: add R11,R9` |
| Step 5. Decrement COUNTER. | `Step5: dec R10` |
| Step 6. If COUNTER ≠ 0 go to step 2. | `Step6: jnz Read` |
| Step 7. Store Result. | `Step7: mov R9,&RESULT` |

**Example 4.13** *Working with unsigned numbers, assume we have a number A in register R5 and B > 5 in register R6 Our objective is to achieve the following:*

1. If $A < 5$, then $B \leftarrow B - A$

    *else if $A > 5$, then $B \leftarrow B + A$*
    else $B \leftarrow 3$.

2. Multiply B by 2.

*An assembly listing for this pseudo code is the following:*

```
test_inst:    cmp    #5,A          ; Test A-5
              jhs    NotLess       ; jump if A >= 5
Case_Less:    sub    A,B           ; if A < 5, B - A
              jmp    Dup1          ; endif
NotLess:      jz     Case_equal    ; else
```

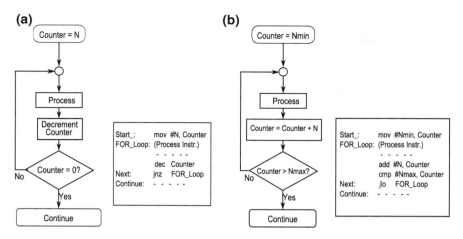

**Fig. 4.21** FOR-loop structure **a** To repeat N times a process, **b** General case

```
Case_Higher:  add   A,B        ; if A > 5, B + A
              jmp   Dupl       ; endif
Case_equal:   mov   #3,B       ; else if A = 5, B = 3
Dupl:         rla   B          ; Endif, 2B
```

**FOR-loops** Two flowcharts for FOR-Loops are shown in Fig. 4.21, together with code examples. That in (a) is very common and particularly useful when a process is to be repeated N times, as in several examples presented so far, including delay loops. In (b) we find the more traditional loop.

**WHILE and REPEAT loops** The principles for the WHILE-loop are illustrated in Fig. 4.22, and those for the REPEAT-UNTIL-Loop in Fig. 4.23. The loop process should include an instruction that affects the condition to be tested.

In a repeat loop, also known as *DO-UNTIL loop*, the process is executed at least once, irrespectively of the truth value of the condition to be stop, because this one is tested at the end of the loop. In a while loop, the condition is tested before going into the loop, so the loop process may not be executed at all.

### Example 4.14 A polling example
*Let us illustrate polling with an example: a red LED and a green LED are driven by pins P1.0 and P1.6 of port 1, respectively. An LED is on if the output at the pin is high. A pushbutton is connected at pin P1.3, provoking a low voltage when down and a high voltage when up. The objective of the code is to turn on the red LED with the green LED off while the button is kept down, and conversely when it is up. Figure 4.24 illustrates the flow chart and code for the infinite loop of the main code. This is only part of the complete source program.*

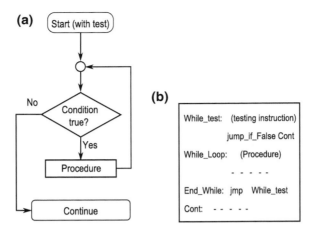

**Fig. 4.22** WHILE-Loop structure: **a** Generic pseudo code; **b** Flowchart; **c** Assembly structure

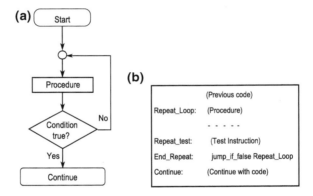

**Fig. 4.23** REPEAT-UNTIL-Loop structure: **a** Generic pseudo code; **b** Flowchart; **c** Assembly structure

## 4.5 Multiplication and Division

Two instructions absent from the MSP430 instruction set are those for multiplication and division. Traditionally, these two operations have not been supported by the CPU hardware. As mentioned before, successive right and left shifts/rotations can be used for multiplication or division by powers of 2. This property can be applied for multiplication by small factors, based on the binary power expansion of an unsigned number. For example, to multiply $A \times 10 = 2^3 A + 2A$, three left shifts plus an addition, and some extra instructions for temporary storage, will do it. There exist in the literature several general algorithms for multiplication and division. (See problems 8–10 for examples).

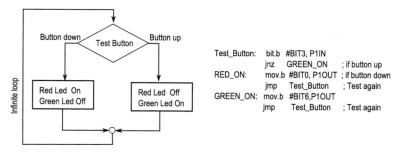

**Fig. 4.24** Polling a button down with LEDs: flowchart and code

## *4.5.1 16 Bit Hardware Multiplier*

Although not all MSP430 devices have a hardware multiplier for all devices, it exists in all families.[9] It is a memory mapped peripheral. In families '1xxx to '4xxx, the input registers to the multiplier are 16-bit wide but accept 8-bit data. Also, all registers have the same memory address. Families '4xxx and '5xxx/'6xxx have a 32-bit multiplier with more functions, but are software compatible with previous models. However, the source program must be written in symbolic name terms because addresses are not preserved in the new families. The 16-bit hardware multiplier is described next.

The hardware multiplier allows the following operations:

- Multiplication of unsigned 8-bit and 16-bit operands (MPY)
- Multiplication of signed 8-bit and 16-bit operands (MPYS)
- Multiply-and-accumulate function (MAC) using unsigned 8-bit and 16-bit operands
- Multiply-and-accumulate function (MAC) using signed 8-bit and 16-bit operands

Operand sizes may be mixed. The names in parentheses are the standard names (addresses) for the registers to be used. MAC multiplication actually carries an operation of the type

$$A \leftarrow A + B \times C$$

by adding the product to an accumulator. The other two types only work in the form $A \leftarrow B \times C$.

The multiplier registers are listed in Table 4.11. With the exception of SUMEXT, which is a read-only register, all others are read/write type.

The steps for multiplying are as follows:

1. Load first operand in the MPY, MPYS, or MAC register, depending on the operation. The register defines the operation.

---

[9] Most information in this section follows the application report slaa042, available from Texas Instruments website. Yet, this report has the old names for the registers: SUMLO and SUMHI for current RESLO and RESHI, respectively.

**Table 4.11**  16-bit hardware multiplier registers

| Register short form name | Address | Comment |
|---|---|---|
| MPY | 0130h | First operand, unsigned multiplication |
| MPYS | 0132h | First operand, signed multiplication |
| MAC | 0134h | First operand, unsigned MAC |
| MACS | 0136h | First operand, signed MAC |
| OP2 | 0138h | Second operand, any case |
| RESLO | 013Ah | Result low word |
| RESHI | 013Ch | Result lhi word |
| SUMEXT | 013Eh | Sign extension register for result |

2. Load the second operand in the OP2 register. As soon as this is done, multiplication is carried out.
3. For MAC operations, add a `nop` instruction to allow time for accumulation to be achieved. This step is optional in other multiplications but it is recommended as a rule.
4. The result is available in registers RESLO, RESHIGH and SUMEXT.

The delay mentioned in step 3 above for accessing the results is necessary for MAC operations and for special cases in other multiplications. It is therefore recommended to include it as a rule, easier to remember than to remember the cases.

**Example 4.15  Signed and Unsigned multiplications:**
*The largest unsigned multiplication with words is* FFFFh $\times$ FFFFh $=$ FFFE0001 *with no carry. Therefore, for all cases we have register* SUMEXT $=$ 0000h. *Let us now look at some operations:*
*Multiplying unsigned bytes* 0x34 *and* 0xB2, *will yield a 16-bit result to be kept in register* R5:

```
mov.b   #0x34,&MPY   ;load first factor
mov.b   #0xB2,&OP2   ; second operand
nop                  ; delay to get results
mov     &RESLO,R5    ; result to R5
```

*Multiplying unsigned words stored in memory at addresses* FIRSTFACTOR *and* SECONDFACTOR, *will yield a 32-bit result to be kept in memory at address* RESULT:

```
mov     &FIRSTFACTOR,&MPY   ; load first factor
mov     &SECONFACTOR,&OP2   ; second operand
nop                         ; delay to get results
mov     &RESLO,RESULT       ; Lower 16 bits of result
mov     &RESHI,RESULT+2     ; Higher 16 bits of result
```

*Let us now consider signed multiplication: the largest negative result is obtained with* $8000h \times 7FFFh = C0008000h$ *with* SUMEXT = FFFFh *since this register offers a 16-bit sign extension for the result. The largest positive results arises from* $8000h \times 8000h = 40000000h$ *with* SUMEXT = 0000h. *The availability of* SUMEXT *allows us to obtain signed results with 48, 64 or more bits. Let us now look at the previous examples, but this time for signed multiplication:*

```
mov.b   #0x34,&MPYS  ;load first factor for signed
                        multiplication
sxt     MPYS         ; sign extend MPYS register
mov.b   #0xB2,&OP2   ; second operand
sxt     OP2          ; sign extend OP2 register
nop                  ; delay to get results
mov     &RESLO,R5    ; low 16 bits of result to R5
mov     &RESHI,R6    ; high 16 bits result to R6
```

*The sign extension for the multiplier registers become necessary because the byte operation leaves the MSB cleared. The higher 16 bit storage is optional, depending on the needs. The following code shows the use of a 48 bit result.*

```
mov     &FIRSTFACTOR,&MPYS  ; load first factor
mov     &SECONFACTOR,&OP2   ; second operand
nop                         ; delay to get results
mov     &RESLO,RESULT       ; bits 0-15 of result
mov     &RESHI,RESULT+2     ; bits 16-31 of result
mov     &SUMEXT,RESULT+4    ; bits 32-47 of result
```

When using the unsigned multiply-and-accumulate function, the register pair RESHI-RESLO constitutes the accumulator to which the product will be added, and SUMEXT will be 0000h if the addition does not generate a carry, and 0001h if it does. In the signed case, it will contain the sign extension.

**Example 4.16** *Let us multiply two 2-dimensional vectors with word size unsigned numbers,*

$$[a1, a2] \times [b1, b2]^T = a1b1 + a2b2$$

*with the result to be placed in memory address* RESULT. *The vectors are stored in addresses* VECTOR1 *and* VECTOR2, *respectively.*

*To accomplish this task, we initialize the first product and then we multiply and accumulate the second one. We store the result including* SUMEXT *for the carry.*

```
mov     #VECTOR1,R4          ;initialize pointers
mov     #VECTOR2, R5
mov     @R4+,&MPY            ; initialize sum with
mov     @R5+,&OP2            ; first product
nop                          ; delay to get results
mov     @R4,&MAC             ; accumulate second product
```

```
mov     @R5,&OP2            ; to previous result
nop                         ; delay to get results
mov     &RESLO,RESULT       ; bits 0-15 of result
mov     &RESHI,RESULT+2     ; bits 16-31 of result
mov     &SUMEXT,RESULT+4    ; store carry
```

## 4.5.2 Remarks on Hardware Multiplier

The sum in a MAC operation does not generate an overflow flag. Hence, if it is necessary, this should be handled by software.

On the other hand, given that the multiplier, although a combinatorial system, takes time to accomplish multiplication, it is recommended to disable maskable interrupts during multiplication with a sequence such as

```
dint                ;disable interrupts
nop                 ;delay to settle
mov  #A,&MPY        ;multiplication
mov  #B,&OP2        ;AB
nop                 ;delay to settle
eint                ;restore interrupts
```

## 4.6 Macros

If your source program is rather long, chances are that it contains blocks of code that are repeated several times, either verbatim or with minor changes in operands and labels. To simplify this source encoding, assemblers allow us to create *macros*.

A macro is a set of instructions grouped under one user-defined mnemonic, used wherever the set of instructions should go. When compiling, the assembler substitutes this mnemonic by all the corresponding instructions and labels. We say that it *expands* the macro.

Notice that use of macros simplifies and shortens the user source program only, not the object files. Hence, if memory resources are limited and you must expand the macro several times, consider using subroutines instead.

On the other hand, in time intensive applications, the overhead in calling and returning from subroutines could become a problem, and macros offer good alternatives for a modular solution. At the end, a trade-off between using macros or subroutines should be evaluated.

A macro must be defined before it can be used. It can be defined either in the source file, in another file which can be copied or included, or in a macro library. It is possible to pass parameters to the macro, and to define local variables, especially labels, which the assembler will differentiate in the different expansion instances.

In the IAR assembler the macro definition starts with the statement

```
macroname MACRO [arg] [,arg]
```

and ends with the directive ENDM.

Here, "macroname" is the user-defined mnemonic for the macro. Optional arguments separated by commas are parameters to be passed and substituted by the assembler during expansion. Local variables are defined with the directive LOCAL within the macro definition.

**Example 4.17** *Assume that a red LED is connected to pin P1.0 of port 1, and a green LED to pin P1.6. (Let us define therefore for this connection* RLED  EQU  BIT0 *and* GLED  EQU  BIT6*). Now, the following macro is defined to turn-on an LED the number of times defined by* Times, *with a speed determined by the number defined with the parameter* DelayTime.

```
ledflash  MACRO LED, Times, Delaytime
          LOCAL Counting, Delay
          mov   #2*Times, R12          ;On-Off count
Counting: xor.b #LED,P1OUT             ;toggle LED
          mov   #DelayTime,R15         ;Load delay to counter
Delay:    dec   R15                    ; Delay Loop
          jnz   Delay                  ;until counter=0
          dec   R12                    ;repeat toggling
          jnz   Counting               ;
          ENDM                         ;end definition
```

*After this definition,* ledflash RLED, 20, 60000 *will cause the CPU to turn on and off the Red Led 20 times with a medium speed,* ledflash GLED, 10, 30000 *will work with the green led 10 times, twice the speed. Also,* ledflash GLED+RLED, 10, 50000 *will work with both leds.*

When using the hardware multiplier, it may be convenient to create macros to simplify and make the code easier to read and write with templates like

```
Mult_16   MACRO     FACT1,FACT2   ; unsigned word
                                    multiplication
          mov       FACT1,MPY
          mov       FACT2,OP2
          nop
          ENDM
```

Hence, Mult_16 R4,R5 becomes a convenient mnemonics for multiplication of registers R4 and R5.

The IAR assembler offers other techniques to simplify the source writing. The reader is encouraged to consult the IAR manual.

## 4.7 Assembly Programming: Second Pass

The source file layout was introduced in Sect. 4.2.4. The basic layout for a source file consists of (1) documentation, (2) Assembly time constants definitions, (3) Data and variables memory allocation, (4) Executable code, (5) reset and interrupt vector allocation, and (6) directive END.

The order is not necessary as mentioned, except for the ending directive. Also, only item(4), (5) and (6) are compulsory. When uploaded in the microcontroller memory, the executable code normally goes in the program or code memory section in the flash-ROM space, and data and variables to the RAM space, unless otherwise directed.[10] Every IAR source file ends with the directive END; anything after this directive is ignored by the assembler.

Let us now look at more directives and considerations about the source file.

### 4.7.1 Relocatable Source

Figure 4.11 was an example of an absolute source file, where the *program location counter* (PLC) used by the linker to load each instruction and data at the appropriate address, is controlled with the **ORG** directive. An alternative to this method is a *relocatable* source file, illustrated by Fig. 4.25 for the same task as before.

In relocatable codes we use *segment directives* to define or work memory segments, whose starting address is specified by the linker.

Absolute codes have the disadvantage that a program written for one model may not be suitable for another one because of a different memory map. Relocatable codes try to solve this problem and leave the storage work to the linker. The programmer must rely on the linker for doing a good work. Unfortunately, not all linkers are that reliable. Some assemblers, like the CCS from TI, only work relocatable codes.

**Segment directives** In relocatable codes the object file is organized by *relocatable sections*, managed with the use of *segment directives* . A section is a block of code or data that occupies contiguous space in the memory map. Each one has an associated *section program counter* (SPC), also called *section location counter* (SLC), which is initially set to 0.

The SLC functions within the relocatable section only, assigning an offset address with respect to the beginning of the section to the stored information. The actual physical address allocation is done by the linker during the compiling process. All sections are independently relocatable, which allows the user a more efficient use of memory.

IAR starts a relocatable section with directive **RSEG**, followed by the type or name of the section, as illustrated by Fig. 4.25. In the MSP430 models, some addresses or sections have names attached. An important one is RESET, for the reset vector

---

[10] The MSP430 CPU can run programs from the RAM space. This is not a feature of every microcontroller.

```
 1  ;********************************************************
 2  ;   MSP430G2xx1 Demo - Software Toggle P1.0
 3  ;   Description: Toggle P1.0 by xor'ing P1.0 inside
 4  ;   of a software loop.
 5  ;       ACLK = n/a, MCLK = SMCLK = default DCO
 6  ;
 7  ;                       MSP430G2xx1
 8  ;                   - - - - - - - -
 9  ;               /|\|                XIN|-
10  ;                | |                   |
11  ;                - -|RST           XOUT|-
12  ;                   |                   |
13  ;                   |            P1.0|- ->LED
14  ;
15  ;   Based on code written by D. Dang
16  ;   Texas Instruments Inc.
17  ;   October 2010
18  ;   Built with IAR Embedded Workbench Version: 5.10
19  ;********************************************************
20  #include    "msp430g2231.h"            ; standard constants
21  LED          EQU     01h               ;LED at pin P1.0
22  DELAY        EQU     50000
23  #define      COUNTER R15               ;using R15 as counter
24  ;- - - - - - - - - - - - - - - - - - - - - - - - - - -
25               RSEG    CSTACK            ; creates a stack in RAM
26               RSEG    CODE              ; Program goes to code
27  ;- - - - - - - - - - - - - - - - - - - - - - - - - - -
28  RESET        mov.w   #SFE(CSTACK),SP   ; Initialize stack
29  StopWDT      mov.w   #WDTPW+WDTHOLD,&WDTCTL  ; Stop WDT
30  SetupP1      bis.b   #001h,&P1DIR      ; P1.0 output
31                                         ;
32  Mainloop     xor.b   #LED,&P1OUT       ; Toggle P1.0
33  Wait         mov.w   #DELAY,COUNTER    ; Delay to R15
34  L1           dec.w   COUNTER           ; Decrement R15
35               jnz     L1                ; Delay over?
36               jmp     Mainloop          ; Again
37                                         ;
38  ;- - - - - - - - - - - - - - - - - - - - - - - - - - -
39  ;           Interrupt Vectors
40  ;- - - - - - - - - - - - - - - - - - - - - - - - - - -
41               RSEG    RESET             ; MSP430 segment for
42               DW      RESET             ; RESET Vector
43               END
```

Documentation

Constants
Declaration

Segment   direc-
tives

Executable
Code

Reset vector
allocation

**Fig. 4.25**  Relocatable IAR listing for blinking LED (*Courtesy Texas Instruments Inc.*)

allocation. Types of sections include CSTACK, CODE, DATA. Notice that RSEG CSTACK creates a stack at the top of the RAM; in this case, initialization of SP uses standard defined constant **SFE(CSTACK)**, whose value depends on the model.

**Example 4.18**  *Let us now see how the source code is assembled in relocatable mode. The difference with the previous example can be seen next:*

```
                    RSEG  CSTACK             ; creates stack
0000                RSEG  CODE        ; starts code section
0000  4031 0300     RESET   mov.w  #SFE(STACK),SP    ; Set stack
0004  40B2 5A80 0120  StopWDT mov.w  #WDTPW+WDTHOLD&WDTCTL ; Stop WDT
000A  D3D2 0022             bis.b  #001h,&P1DIR    ; P1.0 output
000E  - - - - -             - - - - -
        - -                   - -
                    RSEG  RESET             ; Allocates reset
0000  - - - -       DW    RESET             ; vector address
                    END
```

*We can see here the action of the section program counter. It only keeps track of the offset addresses within the section itself. Every time you change section, the respective SPC goes into effect from the last value where it was left. Labels do not*

**Table 4.12** Initialized Data definition or allocation IAR directives

| Directive | Use |
|---|---|
| DC8 or DB | Generates 8-bit constants, including strings. |
| DC16 or DW | Generates 16-bit constants. |
| DC32 or DL | Generates 32-bit constants. |
| DC64 | Generates 64-bit constants. |
| DF32 or DF | Generates 32-bit floating point constants. |
| DF64 | Generates 64-bit floating point constants. |
| .double | Generates 48-bit floating point constants in TI Format. |

Texas Instrument's 32-bit

*have yet any value assigned because the physical memory addresses are not known at this point.*

### 4.7.2 Data and Variable Memory Assignment and Allocation

In addition to the assembly-time constants, we also have variables and data. We can define *memory initialized constants* using a label as an address pointer with the format

```
LABEL <<Data-allocation-Directive>> data,[data,]
```

One or more data can be written on the same line, separated by commas. The Label in this case is the address value of the location for the first datum only. The IAR directives used in MSP430 assembly programs for this purpose are shown in Table 4.12.

If the constant is to be maintained fixed, and above all kept when the system is de-energized, then it is preferably stored in the flash ROM section. It can be defined in the code segment before the reset vector or after the executable code. On the other hand, if it is a value that in fact is going to be changed later during the program execution, we store it in the RAM section. We usually call it then *initialized variable*.

The following are examples for data allocation:

**Example 4.19** *The memory contents is all in hex notation without suffix h or pre-fix 0x. In all cases, label stands for the address value of the first byte. Directive* EVEN *aligns address to an even address, which is important for data larger than a byte. Data in memory is presented horizontally in byte-size information, so little endian convention should be observed. The IEEE floating point single precision equivalent for* −357.892 *is* 0xC3B2F22D *and the double precision equivalent is* 0xC0765E45A1CAC083.

It must be stressed that the label name points only to the address of the first byte, in all cases. In assembly language, the nature of the data is not defined by types, as it is the case in high level language. Thus, even though the user can define a floating number with the directive DF as shown in the example, the memory contents can be used as bytes, words or double words, without any reference to the origin. Thus,

```
Single DB 45,-14,0xB2,195
```

yields the same memory contents as before. This is an important point to remember, and of course a disadvantage of assembly language for structured programming.

For these types of definitions, the values are accessed with the labels in absolute or direct addressing mode. In the following example we illustrate the difference between the assembly-time constant and the memory initialized ones using the directive DW.

**Example 4.20** *The DELAY constant defined in the program examples is used to illustrate the differences between the constant definitions. In the left column below, the DELAY is defined as 50,000 using the* EQU *directive, while in the right column the value is stored in memory using the* DW *directive. In both columns, register* R15 *will be loaded with this value* 50000 = 0xC350. *However, in the first case we use immediate mode, while in the second case we can use either absolute or direct mode, because here DELAY has the value of the address where the datum is stored.*

```
DELAY EQU 50000 | DELAY DW 50000
...             | ...
mov #DELAY,R15  | mov DELAY,R15
                | or
                | mov &DELAY,R15
```

Memory locations can be assigned to constants using the appropriate *data definition or data allocation* directives using this format.

*Variables* are names given to those labels attached to memory assignment directives in the RAM section, pointing to addresses of memory data cells whose contents will be changed during the program execution. If the variable is initialized, we use one of the directives in Table 4.12. If we only separate memory segments, then we use the directives in Table 4.13 with the format

```
LABEL    DIRECTIVE    number-of-spaces-allocated
```

For example, "table DS8 0xA" reserves space for ten bytes, and "table DS16 0xA" for ten words.

**Example 4.21** *A pulse is sent to pin 2.5 to start a device, which will deliver ten signed byte size data, once at a time. The device sets a flag at pin 2.6 to indicate a datum is ready. The data is retrieved from Port 1 and stored in RAM as 16-bit sized data. The ten data are also added. The partial listing below achieves this objective. I/O ports configuration is not included.*

**Table 4.13** Initialized Data definition or allocation IAR directives

| Directive | Use |
|---|---|
| DS8 or DS | Allocates space for 8-bit data. |
| DS16 or DS 2 | Allocates space for 16-bit data. |
| DS32 or DS 4 | Allocates space for 32-bit data. |
| DS64 or DS 8 | Allocates space for 64-bit data. |
| .float | Allocates space for 48-bit floating point constants in TI Format. |

```
            ORG     0X0200
TABLE       DS16    0x0A            ; reserves 10 word spaces in RAM
SUM         DW      0               ; initialize addition

- - - - - - - -
;           The following goes somewhere in the code component
            mov     #0,R12          ; pointer to table start
            bis.b   #BIT5,&P2OUT    ; one cyle
            nop                     ; pulse
            bic.b   #BIT5,&P2OUT    ; to pin P2.5
            mov     #10,R15         ; data

POLL_lp:    bit.b   #BIT6,&P2IN     ; test until
            jz      POLL_lp         ; ready
            mov.b   &P1IN,R4        ; retrieve datum
            sxt     R4              ; extend to 16-bits
            mov     R4,TABLE(R12)   ; store datum
            incd    R12             ; point to next item
            add     R4,SUM          ; add datum
            dec     R15             ; retrieve next datum
            jnz     POLL_lp         ; until finish.
```

### 4.7.3 Interrupt and Reset Vector Allocation

The compulsory allocation of the the reset vector at address 0xFFFE was briefly discussed in Sect. 4.2. We also need to allocate interrupt vectors, so the CPU can go to the corresponding ISR when requested.

Each peripheral and I/O port with interrupt capability has an assigned address for its interrupt vector in the interrupt vector table. Table 4.14 shows the interrupt vector table for devices in the MSP430x20x3 family.

The format to allocate these vectors in an assembly program is the same as that for the reset vector, using the appropriate address from the interrupt vector table:

```
ORG   0xFFFE                    ;reset vector address
DW    <<ResetVectorLabel>>      ;Label used by user
ORG   <<Assigned address>>      ;address of ISR
DW    <<IntVectorLabel>>        ;Label used by user for ISR
```

**Table 4.14** Interrupt vector table for MSP430x20x3 devices.

| Vector address | Standard name | Source of interrupt | Interrupt flags | System interrupt |
|---|---|---|---|---|
| 0xFFE0 | – – | Unused | | |
| 0xFFE2 | – – | Unused | | |
| 0xFFE4 | PORT1_VECTOR | I/O Port 1[a] | P1IFG.0 to P1IFG.7 | Maskable |
| 0xFFE6 | PORT2_VECTOR | I/O Port 2[a] | P1IFG.6 to P1IFG.7 | Maskable |
| 0xFFE8 | USI_VECTOR | USI[a] | USIIFG, USISTTIFG | Maskable |
| 0xFFEA | SD16_VECTOR | SD16_A[a] | SD16CCTL0 SD16IFG and SD16OVIFG | Maskable |
| 0xFFEC | – – | Unused | | |
| 0xFFEE | – – | Unused | | |
| 0xFFF0 | TIMERA1_VECTOR | Timer_A2[a] | TACCR1 CCIFG and TAIFG | Maskable |
| 0xFFF2 | TIMERA0_VECTOR | Timer_A2 | TACCR0 CCIFG | Maskable |
| 0xFFF4 | WDT_VECTOR | Watchdog Timer+ | WDTIFG | Maskable |
| 0xFFF6 | – – | Unused | | |
| 0xFFF8 | – – | Unused | | |
| 0xFFFA | – – | Unused | | |
| 0xFFFC | NMI_VECTOR | NMI | NMIIFG | Non maskable |
| | | Oscillator Fault | OFIFG | |
| | | Flash memory access violation[a] | ACCVIFG | |
| 0xFFFE | RESET_VECTOR[a] | Power-Up | PORIFG | Reset |
| | | External Reset | RSTIFG | |
| | | Watchdog Timer+ | WDTIFG | |
| | | Flash Key Violation | KEYV | |
| | | PC Out of range[b] | | |

[a]Multiple source flags
[b]PC out of range if it tries to fetch instructions from addresses 0h to 0x1FF, or unused addresses

Notice that each address has a standard name, which is defined in the header file. Thus, instead of ORG 0xFFFE we can write ORG RESET_VECTOR. The following example works the LED toggling with an interrupt driven program. The LED toggles each time the pushbutton is pressed. Chapter 7 discusses in detail the subject of interrupts and the mechanisms used by the CPU to identify an interrupt request source.

**Example 4.22** *The listing in Fig. 4.26 shows an interrupt driven example for toggling the LED with a pushbutton; an internal resistor is used. The CPU is turned off and is awaken by the interrupt request that happens when the pushbutton at pin P1.3 is pressed. Observe the allocation of the reset and interrupt vectors at the end of the listing.*

```
 1  #include  "msp430g2231.h"
 2  ;- - - - - - - - - - - - - - - - - - - - - - - - - -
 3            ORG      0F800h                  ; Program Reset
 4  ;- - - - - - - - - - - - - - - - - - - - - - - - -
 5  RESET     mov.w    #0280h,SP               ; Initialize stac
 6  StopWDT   mov.w    #WDTPW+WDTHOLD,&WDTCTL  ;Stop WDT
 7  SetupP1   bis.b    #0xF7,&P1DIR            ;P1.0 output
 8                                             ;unused P1 pins as output
 9                                             ;Pin P1.3 input by default
10            bis.b    #0x4,&P1REN             ;P1.3 Resistor enabled
11            bis.b    #0x4,&P1OUT             ; as pullup resistor
12            bis.b    #0x4,&P1IE              ;enable interrupt at P1.3
13
14  SetupP2   bis.b    #0FFh,&P1DIR            ;Unused P2 pins as output
15
16  Mainloop
17            bis.w    #CPUOFF+GIE,SR          ;Turn off CPU, and
18                                             ; enable interrupts
19            nop                              ; Breakpoint for assembler
20            jmp      Mainloop                ; Again
21  ;- - - - - - - - - - - - - - - - - - - - - - - - -
22  PORT1_ISR ; Begin ISR
23  ;- - - - - - - - - - - - - - - - - - - - - - - - -
24            bic.b    #BIT3,&P1IF             ; Reset Interrupt Flag
25            xor.b    #0x04,&P1OUT            ; toggle LED
26            reti                             ; return from interrupt
27  ;----------------------------------------------------------------
28  ;         Interrupt Vectors
29  ;----------------------------------------------------------------
30            ORG      RESET_VECTOR            ; MSP430 RESET Vector
31            DW       RESET                   ;
32            ORG      PORT1_VECTOR            ; Port 1 Int. Vector
33            DW       PORT1_ISR
34            END
```

**Fig. 4.26**  Listing with an interrupt vector allocation

### 4.7.4  Source Executable Code

Let us add some comments on the executable code. Remember that this is the set of CPU instructions. We usually place this code in the flash-ROM section of the memory space. Let us focus now on some characteristics of the main algorithm.

**Main Algorithm:** After housekeeping and hardware configuration, we may start writing the instructions for the intended task. If interrupts are being enabled, the eint should be introduced after the hardware configuration, to avoid unwanted interruptions.

Since most, if not all, embedded systems work continuously and independently of a user's intervention, many systems do not include a particular instruction to "stop" the program. MSP430 microcontrollers fall in this category. Instead, two common methods to "terminate" the main code are with an infinite loop through an unconditional jump or sending the system into a low power mode, which turns the CPU off.

Examples of how to make an infinite jump have been provided in this chapter, like for example the main code loop in Fig. 4.6. Terminating the main program with a low-power mode requires a few additional considerations, like determining the events that will bring the CPU back into activity and configuring and activating their interrupts.

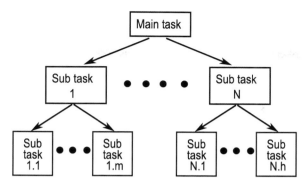

**Fig. 4.27** Modular programming concept

Low-power modes are extremely useful to reduce the power consumption in embedded designs. They can be achieved by configuring the SR bits CPUOFF, SCG1, SCG0 and OSCOFF, to produce one of four different CPU sleep modes with different power consumption rates, or the default active mode where the CPU and all enabled clocks and peripherals are active. When the CPU is powered-up, it begins operation in active mode.. Low-power modes need to be explicitly activated to take effect. Chapter 7 provides a detailed discussion of how to use low-power modes in MSP430 microcontrollers.

## 4.8 Modular and Interrupt Driven Programming

To simplify the design and debugging of large programs, it is a good practice to decompose it in short modules, called *subroutines* or *functions*. This is the so called *modular approach* illustrated by Fig. 4.27. We may include in this practice interrupt driven programming in which the particular Interrupt Service Routines (ISR) are intended to respond mainly to hardware interrupt requests, but may also obey to software requests; in this case they are called *software interrupts*. An interrupt driven program is one which exclusively responds to interrupt requests.

### 4.8.1 Subroutines

If a piece of code is used several times in a program, it may be worth writing a subroutine for it. It makes the main code easier to write and understand, and makes better use of resources. In addition, if it is the type of code that can be used in many different applications, including the subroutine in a library is very advantageous. We call this a *reusable function or subroutine*.

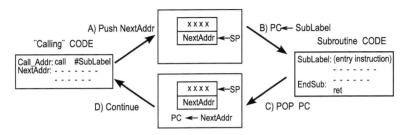

**Fig. 4.28**   The program flow after Call and ret instructions

In short, an assembly subroutine or function is a collection of instructions, a process by itself, separated from the main code and ending with the instruction ret, to return to the place in memory from which the subroutine was called. The executable subroutine code appears only once in memory and is invoked each time it is needed with the instruction  call  dest. The operand is the address of the entry line of the subroutine, and it will be loaded into the PC. Any valid addressing mode can be used for dest.

Subroutines should be as short as possible, and focus on doing one and only one task. It is common to think of them as reusable, since different programs may need the task programmed in this function. They can be included in the source file or in libraries.

The subroutine execution process is illustrated with Fig. 4.28 using the immediate mode for the destination operand. The call process takes five cycles and the return process seven cycles. Most of the time, this time expense is a very good investment when compared to the advantages. The steps shown are:

(a) The instruction **call** pushes the current contents of the PC register, which has the address of the next instruction (NextAddr), and then loads the program counter with the address SubLabel of the entry line of the subroutine.

(b) The fetched instruction in the following cycle is then the first one of the subroutine. The CPU continues with the sequence of instructions in the function, up to the last instruction **ret**, which means "return from subroutine".

(c) This emulated instruction pops back NextAddr into PC

(d) Hence, the next instruction to be fetched is the one after the call instruction in the main program.

This sequence is followed every time that the subroutine is invoked. Because of the "hidden" push and pop operations in this process, it is of extreme importance to verify that the register SP is pointing at the right address when the returning instruction is executed. This requires a good use of the push and pull instructions within the function, or additional manipulation of the SP before exiting the subroutine.

### Data Passing and Local Variables and Constants

Remark that the CPU and memory resources are shared between the main code and subroutines. Therefore, any register and memory location keeps the same contents before and after calling subroutines, except for the program counter. This is something to consider both for *passing data* and for local variables and constant values.

The process of passing data consists of providing a subroutine data information needed to realize the intended algorithm, and returning results from the subroutine to the calling module. Local variables and constants are those used exclusively in the subroutine, but not outside of it. When a subroutine is planned, it is necessary to define how both items will be managed, especially for reusable functions. We discuss these points next.

### Local Variables and Constants

Some recommendations and measures presented below may be skipped for subroutines proper to the program and not designed to be in a library, assuming that good care is taken to avoid difficulties.

- The fastest and simplest way to introduce local values or results is to use registers. Several designers suggest to use registers R4–R11, but you may proceed as necessary. In the subroutine code, start by pushing onto the stack all registers used locally and pop them in reverse order before returning.
- Assign a RAM segment for local variables. These may be overwritten outside the subroutine.
- Local constants not supposed to change, may be defined in ROM space. If they are only initial values, and may change, use a RAM address.
- The stack is a valid segment for local values if accompanied with appropriate manipulation of the SP register. For many designers, it is the best option after register use.

If you make a habit to use a specific set of registers for local variables and another set for passing data, these may or may not be pushed or popped, depending on the strategy needed.

**Example 4.23** *A subroutine reverses the bit order in register* R11—*data passed to function in* R11. *The routine returns as output the original* R11 *value and the reversed word in* R12—*results returned in* R11 *and* R12. *To iterate, we use* R15 *as a local counter or variable initialized with* 16, *the number of bits to be reversed. The following code does the work.*

```
INIT:     push R15      ; save R15 for local use
          push R11      ; To recover R11 at output
          mov #16,R15   ; load local counter
LOOP:     rla R11       ; msb to carry
```

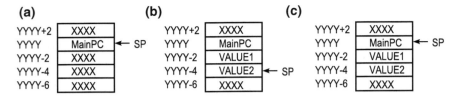

**Fig. 4.29** Illustration of stack for local constants

```
        rrc R12        ; push it at right on R7
        dec R15        ; if not finished
        jnz LOOP       ; go for next bit
        pop R11        ; recover original value
        pop R15        ; retrieve register used for local
                         value
_ENDS:  ret            ; return from subroutine
```

When using the stack for local variables, each time a word is pushed, it will be referred to by the address X(SP), X = 0, 2, 4, ..., in a reverse order to that being pushed. Before returning, the instruction add #2*n, SP, with $n$ being the number of local variables pushed, will return SP to the correct address to point to. This is illustrated by Fig. 4.29 for two local values VALUE1 and VALUE2.

Figure 4.29a shows the stack just after the call, with SP pointing at address YYYY to the returning address. Inset (b) shows the situation after pushing both local constants. Notice that VALUE2 is at address 0(SP) while VALUE1 at 2(SP). Finally, (c) shows the situation after the SP has been adjusted for return with add #4, SP. Let us change the subroutine of the previous example using the stack for the local variable. The listing would go then like this:

```
INIT:   push R11       ; To recover R11 at output
        push #16       ; load local counter
LOOP:   rla R11        ; msb to carry
        rrc R12        ; push it at right on R7
        dec 0(SP)      ; if not finished
        jnz LOOP       ; go for next bit
        add #2,SP      ; adjust SP
        pop R11        ; retrieve register used for local
                         value
_ENDS:  ret            ; return from subroutine
```

## Passing Data

Data passed to the subroutine or function, also called *parameters*, is most commonly transferred to and from a function

- By register
- By memory
- By stack

The preferred method is by register, since it is faster and does not use memory. The MSP430 has many general purpose registers, making this method quite feasible in most cases. The other two methods consume memory space and are slower, but are also an option. Using the stack may be dangerous if no precaution is taken, as explained below.

**Example 4.24** *Assume a data* VALUE *is to be passed to a subroutine starting at address* SUBR. *Assume that the first instruction in the function is to transfer the data to R7, followed by a byte exchange. Let us look at the three methods to pass the data.*

*(a) If data is passed using register R6, then the corresponding instructions in the main code and in the subroutine are as follows:*

```
Subroutine:                 Maincode:
SUBR: mov R6,R7             mov #VALUE,R6
swpb R7                     call #SUBR
```

*(b) If a memory address, say* SUBDATA *is used to transfer data to the subroutine, then the instructions are as follows:*

```
Subroutine:                 Maincode
SUBR: mov &SUBDATA,R7       mov #VALUE,&SUBDATA
swpb R7                     call #SUBR
```

*(c) The stack for passing data is not the best option unless carefully used, especially when the subroutine is used frequently. And don't forget that embedded microcontrollers usually run in infinite loops, or are not supposed to be turned off during lifetime, always reacting to an event. A frequent format for passing data with the stack is as shown next:*

```
Subroutine:                 Maincode:
SUBR: mov 2(SP),R7         push #VALUE
swpb R7                     call #SUBR
```

*The problem with this approach can be appreciated with Fig. 4.30. We see in inset (b) that after return one memory stack space (two bytes) is exhausted. Even if this is the only subroutine in the program the stack, and hence the RAM space, will be depleted after a certain moment. One way to avoid this phenomenon is to add the instruction* incd SP *to move the stack pointer up just after the call instruction. It is a small price to pay for not having the RAM depleted. This is illustrated in inset (c).*

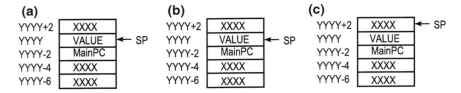

**Fig. 4.30**  Passing data with the stack. **a** Push value and call subroutine; **b** After return; **c** Saving memory after return

Passing data back to main code from subroutine may be done using similar procedures. Notice that the stack needs again special considerations, being that one reason why is not so popular for this task.

### 4.8.2  Placing Subroutines in Source File

Subroutines can be placed anywhere in the code section, as long as they end with `ret`. Although it is technically possible to include a subroutine within the main code, this is not at all recommended. The following example using polling illustrates placing of subroutines.

**Example 4.25**  *Each pin of port 1 of the MSP430 has an interrupt flag associated in the P1IFG register. The operation of this flag can be enabled with the Interrupt Enable Register (P1IE) and its operation configured with the Interupt Edge Select Register (P1IES). When bit x of P1IE is set, then bit x of the P1IFG register will be set every time that there is an appropriate voltage transition at pin P1.x of Port 1. The bit x of P1IES determines the type of transition: if 1 the transition is high to low, if 0 low to high. By default, all these registers are cleared. In this example we use P1IFG to do polling (Figs. 4.31, 4.32).*

*Our objective is as follows: A LED is connected to pin P1.0, while a push button is connected to P1.3. We want to toggle the LED by pressing and releasing the push button. Since this action on the push button will produce a pulse, the flag P1IFG.3 will be then set. We poll then this flag and whenever it is set we will call a subroutine that changes the LED state.*

*Two listing examples for this algorithm are shown in Figs. 4.3 and 4.32, with the subroutine placed at different positions with respect to the main code.*

**Subroutines in Libraries**

It is advisable to include reusable functions in a library with other functions of similar nature. In this case, the linker incorporates the object code of the function together with other codes, and gets the final version to be loaded onto the memory.

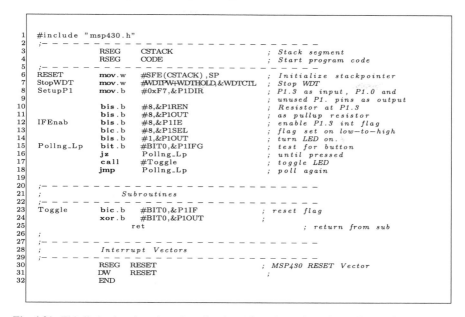

```
1   #include "msp430.h"
2   ;- - - - - - - - - - - - - - - - - - - - - - - -
3                   RSEG    CSTACK              ; Stack segment
4                   RSEG    CODE                ; Start program code
5   ;- - - - - - - - - - - - - - - - - - - - - - - -
6   RESET           mov.w   #SFE(CSTACK),SP     ; Initialize stackpointer
7   StopWDT         mov.w   #WDTPW+WDTHOLD,&WDTCTL ; Stop WDT
8   SetupP1         mov.b   #0xF7,&P1DIR        ; P1.3 as input, P1.0 and
9                                               ; unused P1. pins as output
10                  bis.b   #8,&P1REN           ; Resistor at P1.3
11                  bis.b   #8,&P1OUT           ; as pullup resistor
12  IFEnab          bis.b   #8,&P1IE            ; enable P1.3 int flag
13                  bic.b   #8,&P1SEL           ; flag set on low-to-high
14                  bis.b   #1,&P1OUT           ; turn LED on.
15  Pollng_Lp       bit.b   #BIT0,&P1IFG        ; test for button
16                  jz      Pollng_Lp           ; until pressed
17                  call    #Toggle             ; toggle LED
18                  jmp     Pollng_Lp           ; poll again
19
20  ;- - - - - - - - - - - - - - - - - - - - - - - -
21  ;                  Subroutines
22  ;- - - - - - - - - - - - - - - - - - - - - - - -
23  Toggle          bic.b   #BIT0,&P1IF         ; reset flag
24                  xor.b   #BIT0,&P1OUT        ;
25                  ret                         ; return from sub
26  ;
27  ;- - - - - - - - - - - - - - - - - - - - - - - -
28  ;               Interrupt Vectors
29  ;- - - - - - - - - - - - - - - - - - - - - - - -
30                  RSEG    RESET               ; MSP430 RESET Vector
31                  DW      RESET               ;
32                  END
```

**Fig. 4.31** This listing has the subroutine after the main code, and uses immediate addressing mode when calling

### 4.8.3 Resets, Interrupts, and Interrupt Service Routines

Resets and interrupts were introduced in Chap. 3, Sect. 3.9 and considered briefly in this chapter in previous sections, including the manipulation of the GIE flag and the allocation of interrupt vectors. We deal now with some additional aspects considerations in the assembly source. We start by looking at general issues and then go to some specifics.

**MSP430 Interrupt Process**

When the request for an interrupt is received, with flag GIE active for maskable interrupts to be enabled, the MSP430 processor responds with the following sequence, which is illustrated in Fig. 4.33:

1. It completes the instruction being currently executed, if any.
2. The PC and the SR registers are pushed, in that order, onto the stack. In this way, the state of the CPU is preserved, including the operating power mode before the interrupt.
3. The SR is cleared, except for the OSCOFF flag, which is left in the current state. This means that the system is taken into the active mode.

```
1   #include "msp430.h"
2   ;- - - - - - - - - - - - - - - - - - - - - - - - - - -
3               RSEG    CSTACK                  ; Stack segment
4               RSEG    CODE                    ; Start program code
5   ;- - - - - - - - - - - - - - - - - - - - - - - - - - -
6   ;                          Subroutines
7   ;- - - - - - - - - - - - - - - - - - - - - - - - - - -
8   Toggle      bic.b   #BIT0,&P1IF             ; reset flag
9               xor.b   #BIT0,&P1OUT            ; toggle LED state
10              ret                             ; return from sub
11  ;- - - - - - - - - - - - - - - - - - - - - - - - - - -
12  ;                          Main code
13  ;- - - - - - - - - - - - - - - - - - - - - - - - - - -
14  RESET       mov.w   #SFE(CSTACK),SP         ; Initialize stackpointer
15  StopWDT     mov.w   #WDTPW+WDTHOLD,&WDTCTL  ; Stop WDT
16  SetupP1     mov.b   #0xF7,&P1DIR            ; P1.3 as input, P1.0 and
17                                              ; unused P1 pins as output
18              bis.b   #8,&P1REN               ; Resistor at P1.3
19              bis.b   #8,&P1OUT               ; as pullup resistor
20  IFEnab      bis.b   #00001000b,&P1IE        ; enable P1.3 int flag
21              bic.b   #4,&P1SEL               ; flag set on low-to-high
22              bis.b   #1,&P1OUT               ; turn LED on.
23  Pollng_Lp   bit.b   #BIT0,&P1IFG            ; test for button
24              jz      Pollng_Lp               ; until pressed
25              call    #Toggle                 ; toggle LED
26              jmp     Pollng_Lp               ; poll again
27  ;- - - - - - - - - - - - - - - - - - - - - - - - - - -
28  ;             Interrupt Vectors
29  ;- - - - - - - - - - - - - - - - - - - - - - - - - - -
30              RSEG    RESET                   ; MSP430 RESET Vector
31              DW      RESET                   ;
32              END
```

**Fig. 4.32** This listing has the subroutine after the main code, and uses symbolic addressing mode when calling

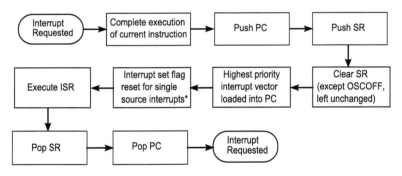

**Fig. 4.33** Sequence of events during interrupt processing

4. The highest priority interrupt vector is loaded into the PC. For single source interrupts, the interrupt flag is reset. Multiple sources interrupt flags should be reset by software, and priority established within the ISR.
5. The Interrupt Service Routine (ISR) is executed. When the reti instruction is executed, it will pop the SR and PC, returning to the place where the interrupt occurred, restoring also the previous controller state.

All the steps in the sequence above, just before the ISR execution, are automatic, transparent to the programmer in the sense that no code is require for them to take place. For this sequence to take place, we already identified in Sect. 3.9.2 four fundamental provisions that need to be made at the software level, for interrupts to work: Providing a stack, having an ISR, configuring the vector table, and enabling interrupts at the global and local levels. These issues are discussed in detail in Chap. 7. Here we remark them to put them in perspective within the context of designing assembly language programs.

**Remarks on Interrupt Service Routines**

Service routines are planned and written by the programmer. They contain the instructions that service the device that requested the interrupt. ISRs must end with the instruction `reti`, for "return from interrupt", which is responsible for pulling back the SR and PC registers from the stack to finish the interrupt sequence. When writing ISRs, the code shall be kept as short as possible. If a long processing is necessary, send most of it to the main code. Also, unless it is needed, avoid nested interrupts. The CPU, by default takes care of that. And lastly, do not forget the problem of data sharing. Define carefully what data is passed and any local variables or values.

Additional recommendations are discussed in Chap. 7. In particular Sect. 7.3 provides general considerations for ISRs and interrupt driven software from a broader design perspective.

## 4.9 Summary

- There are three levels for programming: High level, assembly level, and machine language programming. Compilers and assemblers are used to translate the first two into machine instructions, native to the CPU.
- MSP430 CPU machine language instructions consist of one to three 16-bit words. The leading word is the instruction word, in which a group of bits is the opcode and the rest of the bits determine the operands and addressing modes. The CPU has 27 core instructions.
- Assembly instructions are in a one to one correspondence with machine instructions and have as components mnemonics and operands. The mnemonics is the assembly name given to the opcode. The operands are stated in an addressing mode syntax.
- Two pass assemblers have directives that allow us to use labels, comments, and symbolic names to help us in writing and reading the source program. The format of an instruction is
  "`label mnemonics source,destination ;comment`,
  where the label and comment are optional. The operands depend on the instruction.
- The label in an instruction is a constant that takes, after compilation, the address of the instruction as its value.

- The MSP430 assemblers accept 24 emulated instructions with mnemonics easier to remember than their core counterpart because they are easier to associate to the operation. There is no penalty in using them.
- There are seven addressing modes for the MSP430: (1) register mode – Rn –, (2) immediate mode –#X –, (3) indexed mode – X(Rn) –, (4) indirect mode – @Rn –, (5) Indirect auto increment – @Rn+ –, (6) direct mode – X –, and (7) absolute mode – &X –. The immediate, indirect, and indirect autoincrement modes are not valid in the destination. Any register is valid as an operand.
- Most instructions may work with word size or byte size operands. This is indicated by adding a suffix to the mnemonics: ".w" for words, ".b" for bytes. Word size is the default.
- In byte operations with register mode, only the LSB is considered. When the register is the destination, the MSB is cleared.
- The source program must include allocation of reset vector and interrupt vectors for each ISR present in the code.
- The executable code in the source must include hardware configuration and stack pointer initialization. Interrupt enabling should be done only after the system is ready to work.
- The source program may be designed in absolute or relocatable form. In the first case, using the ORG directive, the user must know the memory map of the micro-controller being used.
- When placing the CPU in a low power mode, only a reset or an interrupt request will awaken the CPU. The ISR may do the service or else may return the PC to the main code with an appropriate strategy.
- Macros and subroutines help write shorter and more efficient codes.

## 4.10 Problems

4.1 Answer the following questions

    a. What are the programming levels used?
    b. What relationship exists between assembly and machine language?
    c. Describe the basic format of an assembly instruction with two and one operands.
    d. What is an assembler?
    e. What is a debugger and what is it used for?
    f. Why is R3 used as a constant generator and how does it function?
    g. What is a source program or source file?
    h. What is the difference between directives and instructions?
    i. What is the general format of a source statement?
    j. What is meant by "source executable code"?

4.2 In the MSP430 machine language, the instruction word with two operands is decoded in four nibbles, where the most significant nibble is the opcode.

For register modes in the operand, the least significant nibble is the number for the destination register, and the nibble next to the opcode is the source register. Consider now assembly and hex machine versions for the executable code in Fig. 4.5 and answer the following questions:

a. How many two operand instructions are there? (You have to identify emulated ones!)
b. For each two operand instruction, associate the mnemonic with the corresponding opcode.
c. In the format Rn, what value of n corresponds to register SP?
d. Does dec.w R15 have a source register? If affirmative, tell which one is it and why is it there. If negative, justify your answer.
e. What is the machine instruction word for mov.w #0xC350, R10 and dec R6?

4.3 Example 4.1 illustrates what value was assigned to label RESET as well as how the addresses for instructions are assigned once given the reset vector. Using the machine code of Fig. 4.6 as a reference, find the values that are assigned to each label of the source code.

4.4 Verify that the **jge** instruction works correctly by testing $N \oplus V = 0$ when there is an overflow.

4.5 Eleven byte sized signed numbers are stored in RAM, starting at address 0x200. The numbers are to be added and the result stored in the next even address available in RAM. Write a code to do this and include it in a program. Test your source with the set $-120, +38, -57, -110, +18, -97, +60, +85, -125, 78$, and $-1$.

4.6 Example 4.11 illustrates an algorithm to convert from binary to BCD, using the decimal addition. The example converts an 8-bit binary number. Modify the algorithm to convert 16-bit binary numbers, and store the result in memory.

4.7 Multiplication by 10 in binary can be done thinking in $A \times 10 = 2^3 A + 2A$ or $A \times 10 = (2^2 A + A) \times 2$. Write a subroutine to multiply a nibble by 10 and use it in nested multiplication of the power expansion form of a decimal number to convert a BCD number to its binary equivalent. Limit the numbers to 9999, so a 16-bit register can hold the result.

4.8 Neither multiplication nor division are supported by the MSP430 CPU ALU. For multiplication some microcontroller models include hardware multiplier peripherals which can be used; otherwise, multiplication must be realized by software.

Both multiplication and division of unsigned numbers can be realized by direct brute force methods. That is

**Multiplication:** To multiply AxB, add A times B.

**Division:** To divide A/B, subtract B from A as many times as possible until the difference is smaller than B. The difference is the residue R and the number of subtractions is the quotient Q.

a. Test your understanding by multiplying 11x3 and dividing 11/3 by hand with the algorithms.
b. If each factor in a multiplication is $n$ bits wide, what is the number of bits that we must reserve for the product?
c. Write pseudo codes or draw a flowchart for each operation, and encode them in assembly language. For the multiplication, assume both factors are bytes. For the division A/B, assume A is a 16-bit word and B a byte.
d. Test your code in the CPU (you may use the simulator).

4.9 The sum algorithm for multiplication is not efficient. There are several other multiplication algorithms available which are much better. One of them is an old one, used already in ancient times, and known as the *Russian Peasant Algorithm*. Two multiply AxB, the algorithm can be stated as follows:

**Step 1 IF** A is even, **THEN** initialize P = 0. **ELSE,** P = B.
**Step 2** While A$\geq$1 do
    **2.1** A$\leftarrow$A/2 and B$\leftarrow$2$\times$B
    **2.2 If** A is even **THEN** P$\leftarrow$P+B

In this algorithm, A/2 is the integer quotient, discarding the residue or fractional part.

a. Test your understanding of the algorithm by multiplying $34 \times 27$ by hand in decimal terms.
b. Multiplying two $n$-bit unsigned numbers, A and B, the result is a number P with at most $2n$ bits. With this in mind, write a pseudocode or draw a flowchart to implement the algorithm and implement it in assembly language for byte operands.
c. Repeat the above procedure for 16-bit operands.
d. Test your codes in the microcontroller (You may use the simulator).

4.10 For integer division M/D, D$\neq$0 and M being an $n$-bit word, yielding a quotient Q and remainder R, one algorithm that mimics long division is the following:

**Step 1**: Initialize Q = 0, R = 0
**Step 2**: **FOR** $j = n - 1$ to 0 **DO**
    **2.1** R$\leftarrow$ 2R + M($j$)
    **2.2 IF** R$\geq$D **THEN**
      a.     R $\leftarrow$ R $-$ D
      b.     Q($j$) = M($j$)
    **ENDIF**

In this algorithm, M($j$) means the bit of N in the $j$-th position.

a. Write an assembly code for the above algorithm and test it. Assume M is a 16-bit word.
b. Use your code as a function in a program to convert a 16-bit unsigned binary word into a BCD number.

# Chapter 5
# Embedded Programming Using C

In this chapter we present the basics of how to use C language to program embedded systems. More specifically, to program the Texas Instruments' MSP430 using the C language. Be aware that this is not a self-contained chapter in this topic or a substitute for a textbook on the subject. Being this an introductory chapter and limited in scope we are obliged to left aside several topics, among them for example programming for real time systems, a topic which would need a chapter by itself.

The C language evolved from BCPL (1967) and B (1970), both of which were type-less languages. C was developed as a *strongly-typed* language by Dennis Ritchie in 1972, and first implemented on a Digital Equipment Corporation PDP-11 [28]. C is the development language of the Unix and Linux operating systems and provides for *dynamic memory allocation* through the use of *pointers*.

A strongly-typed computer language is one in which a type must be declared for every variable before it can be used. Dynamic memory allocation refers to the task of requesting memory space allocation at run-time, when the process is running, and not during compilation or before the process begins to run. A *pointer* is the address of an object in memory. This object can be as simple as a variable or as complex as the most sophisticated structure.

## 5.1 High-Level Embedded Programming

High level languages arose to allow people to program without having to know all the hardware details of the computer. In doing so, programming became less cumbersome and faster to develop, so more people began to program. An important feature is that an instruction in a high level language would typically correspond to several instructions in machine code.

On the downside, by not being aware of the mapping of each of the high level language instructions into machine code and their corresponding use of the functional units, the programmer has in effect lost the ability to use the hardware in the *most efficient* way.

M. Jiménez et al., *Introduction to Embedded Systems*,
DOI: 10.1007/978-1-4614-3143-5_5,
© Springer Science+Business Media New York 2014

For most programmers, human-friendly characteristics of high level language for computer programming make it a "natural choice" for embedded programming as well. C language has features that make it attractive for this sort of applications. Perhaps the most important one is that it allows the programmer to manipulate memory locations with very little overhead added. Since input/output units can be treated as memory, C will also allow the programmer to control them directly.

In general, you use C language for embedded systems when one or more of the following reasons apply:

- tight control of each and every resource of the architecture is not necessary;
- efficiency in the use of resources such as memory and power are not a concern;
- ease of software reuse;
- company policy or user requirements;
- lack of knowledge of assembly language and of the particular architecture.

On the other hand, you use assembly language for embedded systems when one or more of the following reasons apply:

- it is imperative or highly desirable that resources be efficiently handled;
- there is an appropriate knowledge of the assembly language and of the particular architecture;
- company policy or user requirements.

In embedded systems, sometimes it becomes imperative to work both aspects. Hence, unlike what we could design as "typical" C programming, where the complete source is written in the same language, compilers for embedded systems provide methods to mix both languages. We explain this with greater depth in Sect. 5.4.

### 5.1.1 (C-Cross Compilers Versus C-Compilers) or (Remarks on Compilers)?

Recall that a *compiler* is a program developed to translate a high-level language source code into the machine code of the system of interest. For a variety of reasons, compilers are designed in different ways. Some produce machine code other than for the machine in which they are running. Others translate from one high level language into another high level language or from a high level language into assembly language. The term *cross-compiler* is used to define these type of translators. When the embedded CPU itself cannot host a compiler, or a compiler is simply not available to run on a particular CPU, a cross-compiler is used and run on other machine. We will be using the terms compiler and cross-compiler interchangeably, but the user has already been made aware of their differences.

Most compilers that have been developed for the MSP430 are in fact cross-compilers, yielding an assembly language code. One of the advantages comes when linking the compiled source with other files which were actually assembled.

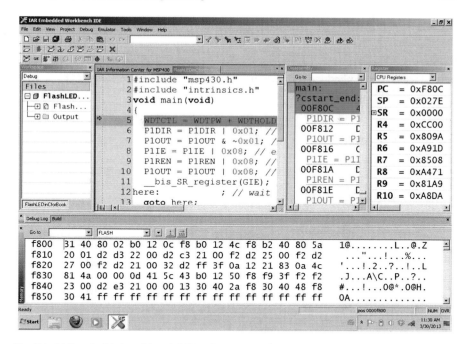

**Fig. 5.1**  IAR embedded workbench IDE: **a** Source code; **b** auxiliary tools and information windows *(Courtesy of IAR Systems AB)*

There are also *optimizing compilers*, i.e., compilers (or cross-compilers) in which code optimizing techniques are in place. These are compilers designed to help the code execute faster, and include techniques such as constant propagation, inter-procedural analysis, variable or register reduction, inline expansion, unreachable code elimination, loop permutation, etc.

We should emphasize that although the buzzword "optimization" is very attractive, there are times where it should be avoided, especially because the techniques involved may damage parts of the code or execution. The source must include the keyword `volatile` when declaring a variable which must not be subject to optimization.

In terms of software development, an Integrated Development System or IDE is an application that usually consists of an editor, a compiler, a linker, and a debugger, all within the same package. The IDE may also include other programs such as a graphical user interface or GUI, a spell checker, an auto-completion tool, a simulator, and many other features. Figure 5.1 shows a window of the IAR Embedded Workbench with several sub windows for the user information. IAR will be used for compilation of the codes presented here. This compiler also has the capability of performing code optimization.

## 5.2  C Program Structure and Design

We review briefly in this section the basic principles of C, introducing also some features related to embedded programming. The reader should consult another source for a more complete introduction to C language, and for the MSP430 programming, [29] is a good reference.

Basically, a C program source consists of preprocessor directives, definitions of global variables, and one or more functions. A function is a piece of code, a program block, which is referred to by a name. It is sometimes called procedure. Inside a function, local variables may be defined. A compulsory function is the one containing the principal code of the program, and must be called `main`. It may be the only function present.

Two source structures are shown below. They differ in the position of the `main` function with respect to the other functions. The left structure declares other functions before the main code, while in the right structure the main function goes before any other function definition. In this case, the names of the functions must be declared prior to the main function. The compiler will dictate which structure, or if a variation, should be used. There may also be variations on syntax. The reader must check this and other information directly in the compiler documentation.

Structure A:

```
preprocessor directives
global variables;
function_a
{
    local variables;
    program block;
}
function_b
{
    local variables;
    program block;
}
int main(void)
{
    local variables;
    program block;
}
```

Structure B:

```
preprocessor directives
global variables;
function_a, function_b;
int main(void)
{
    local variables;
    program block;
}
function_a
{
    local variables;
    program block;
}
function_b
{
    local variables;
    program block;
}
```

An example of the left structure is shown below in Example 5.1 The code is intended to illustrate the structure, not to be an optimal source. Further remarks are included in the example.

**Example 5.1**   A C language program structure example. *The following is a C program source that prints an integer value.*[1] *It contains a function* hello *which does not have local variables defined within its body, but it does have a formal parameter k of type* int *which actually acts as a local variable. Note that no semicolon is used after the function declarations. Comments are enclosed between /\* and \*/*

```
#include <stdio.h>
#define MAX 15
void hello(int k)      /* no semicolon (;) here! */
{
    printf("%d\n",k);
}
main( )    /* no semicolon (;) here either! */
{
    int i = MAX;
    hello(i);
}
```

The alternate program structure for the previous example is shown in Example 5.2.

**Example 5.2**   *The following source works as in the previous example. The change in order requires now a declaration. Note the use of the semicolon.*

```
#include <stdio.h>
#define MAX 15
void hello(int k);  /* a semicolon is used after a function
                       header */
main( )             /* no semicolon (;) after main( ) */
{
    int i = MAX;
    hello(i);
}
void hello(int k)   /* no semicolon (;) after the function
                       name */
{
    printf("%d",k);
}
```

The function declaration main, which must be in the source, tells the compiler, that the program execution starts there. It is therefore used by the compiler to define the reset vectors. Yet, the format for declaration is compiler dependent. Some variations found are main, main( ), and main(void). Many compilers require to declare the function with the type included.

*In the IAR compiler for MSP430 we usually write* void main(void)

---

[1]   The printf function makes sense only if you are printing to a screen monitor.

### 5.2.1 Preprocessor Directives

High level compilers very often include a *preprocessor*. This is a system that rewrites the source before the actual compiler runs over it. The source contains specific directives aimed at the preprocessor to do this task.

C language preprocessor directive statements begin with the special sharp character (#), and are usually written flushed to the left. Important preprocessing directives, among others, are #include , #define , and #pragma . For more directives, the reader may consult the compiler manual.

The directive include is generally used at the beginning of a source code; it lets the preprocessor know that we wish to include a file as part of our source code. These are usually header files. When the preprocessor finds this directive, it opens the specified file and inserts its content at the location of the directive. The formats used for this directives are

```
#include <filename>
#include "filename"
```

The <> syntax is mostly used for standard headers, like stdio.h, which contains the declaration for the printf function, among others. The other syntax is mostly used for user-defined or specific headers, for example, those pertaining to a particular microcontroller.

The syntax for the #define directive is

```
#define Symbolic_NAME expression
```

where "expression" may be as simple as a value or as complex as a piece of code. In general, it is said that the directive is used to define macros, being constants considered one special case. We will consider different applications of this directive as we encounter them.

The main objective of the define directive is for the compiler substitute the Symbolic_NAME by the corresponding expression every time it is found in the source. For example,

```
#define My_Ideas #include "my_header.h"
```

indicates that in the source "My_Ideas" is to be substituted by #include "my_header.h".

The #pragma directive is a compiler specific directive, and in many instances it instructs the compiler to use implementation-dependent features. Among other applications, with this directive we tell the compiler where to place interrupt vectors. Later examples will illustrate this feature.

### 5.2.2 Constants, Variables, and Pointers

As a minimum we expect to find in a function constants, variables, program statements, and comments. Any text on a line after two slashes (//) is a comment in the

IAR compiler. Also, comments may be enclosed between (/\*) and (\*/), as shown in the above examples. Let us now consider the other items.

## Compiler Time Constants

*Constants*, or more specifically, *compiling time constants* are introduced with the expression

```
#define CONST_NAME constant value or expression
```

By convention, the name of a constant is written in capital letters. In the codes of examples 5.1 and 5.2, `#define MAX 15` instructs the compiler to substitute string `MAX`, wherever it is found, with the constant value 15. We can perfectly use this defined constant in an expression like

```
#define MAX2 2*MAX
```

to create a new constant with value 30.

Notice that the compiler does not reserve memory space in the system for these constants. Many developers prefer the term "object-like macros" to "compiler-time constants" for the similarity in the definition format.

## Variables

We always think of a "variable" as a symbol for a quantity that is generally allowed to change during program execution. Thus, for example, a statement such as

```
P1OUT = P1OUT + 2;
```

is thought of as "add 2 to the variable P1OUT", so the variable P1OUT has a new value now.

When we declare a *variable*, what the compiler does is to reserve a place in memory for the variable, and the name is actually given to a memory location, i. e., an address. When the programmer thinks of the value of a variable, this value is actually the contents of the memory. Assembly language syntax is clearer in that sense. The above expression corresponds to `add #2, P1OUT`, a syntax that makes clear that P1OUT is a symbolic name—a compiling time constant—for a memory address.

However, here is precisely one of the advantages of high-level language: For all practical purposes, the programmer is manipulating a variable in the way he/she does it with Algebra, not thinking on memory contents at a given address!!

The format for variable declaration in C is

```
type var1name, var2name, ..., varxname;.
```

One or more variables, separated by commas, can be declared in one line. All them need to be of the same type. The declaration ends with a semicolon. Variables can also be arrays, like `a[9]`, either one dimensional or two dimensional.

By convention, user defined variables are generally written in lowercase letters, or a combination of lowercase and capital case letters for easier reading, as in
"dataValue". The compiler usually places variables in the main volatile RAM, especially if they are uninitialized. In the literature of most embedded system, including
the MSP430 family, capital letters assigned to symbolic names which are actually
declaring the addresses of I/O and peripheral registers with appropriate keywords
[30], as it was the case with the P1OUT used above.

Variables can be initialized, as i in our examples 5.1 and 5.2. Also, a local variable
to a function may be actually declared as an argument in the function declaration.
This was the case with k for the function hello( ) by writing the declaration
within the parenthesis. Variables which, like k, are defined as arguments for the call
to a function are named *formal parameter* to the function. The value or variable i
used during the call, which initializes the value for the formal parameter, is said to
be an *actual parameter*.

Variables in C have attributes: *name, type, scope, storage class*, and *storage duration*. We have mentioned the name. Types are explained later in this section.

The storage class of a variable may be auto, extern, static and
register. The storage class also specifies the storage duration which defines
when the variable is created and when it is destroyed. The storage duration can be
either *automatic* or *static*. An automatic storage duration indicates that a variable is
automatically created (or destroyed) when the program enters (or exits) the environment of code where it is declared. Unless specified otherwise, local variables in C
have automatic storage duration. The static storage duration is the type of storage
for global variables and those declared extern.

The scope of visibility, or simply the scope, defines and limits the parts of the code
where the variable can be accessed. The scope can be: *file scope, function scope*, or
*block scope*.

## Pointers

We have just said that by declaring a variable, the compiler reserves a location in
memory for it. For example, let us declare an initialized byte-size variable (see next
section) by

```
char myVar = 126
```

and assume that the value is stored in memory address $0 \times 0304$.

If we were writing an assembly language source, we would have written

```
myVar db 126
```

We can now use the addressing mode syntax to specify clearly what we want to
say. Notation #myVar (immediate mode) yields 0x0304, the address, while myVar
(absolute mode) or &myVar (symbolic mode) yields 126, the contents.

Well, the good news is that in C we have a similar convenient method to achieve
this differentiation, though with its own syntax: while myVar refers to the contents

**Fig. 5.2** Illustrating the
pointer with the address-of
operator (&)

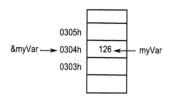

(126), &myVar has the value of the address 0x0304. It is said to be a *pointer*.[2] The
ampersand (&) in C before a variable name is called *address-of operator*. This is
illustrated in Fig. 5.2.

An asterisk (*) becomes an *at-address operator*, so that *(&myVar) is, effectively
the same as myVar. The &myVar value can only be assigned to pointers, which are
declared with an asterisk as follows:

```
int *ptr
```

This statement declares the pointer ptr and now a statement

```
ptr=&myVar
```

assigns the address of myVar to ptr. From now on, ptr can be used as an integer,
and *ptr as a pointer.

### Data Types

The type of a variable is responsible for the size of the memory location assigned to
hold its content and the format in which it will be held in memory. It is also a guide
for the compiler to detect programming errors when variables are used in invalid
'human' situations.[3] Not all C present in most cases. Focusing on the MSP430,
Table 5.1 shows the data types available for MSP430 C programs using the IAR
compiler.

### Operators

For your reference, the operators available as part of the C language are shown in
order of precedence, in Table 5.2. Note that the bitwise complement operator ~ has
a precedence of 2 whereas the bitwise or operator | has a precedence of 10 and the
assignment operator = has a precedence of 14. In C, the lower the precedence number

---

[2] It is said that the symbolic mode &N in MSP430 language was introduced because of the C
reference

[3] Types are not used in assembly language. This may cause valid "programming" expressions
but unwanted results because data is actually not differentiated. For example, the intention of the
programmer is different when defining a string 'AB' than when defining the number 16961. Yet,
the data in memory looks the same in bytes.

**Table 5.1**  C language data types

| Type | Size | Decimal range |
|------|------|---------------|
| *Integer variables* | | |
| char, signed char | 8 bits | −128 to 127 |
| unsigned char | 8 bits | 0 to 255 |
| | | ASCII values |
| int, signed int | 16 bits | −32, 768 to 32,767 |
| unsigned int | 16 bits | 0 to 65,535 |
| long, signed long | 32 bits | $-2^{31}$ to $2^{31} - 1$ |
| unsigned long | 32 bits | 0 to $2^{32} - 1$ |
| *Real variables* | | |
| float | 32 bits | IEEE single precision floats |
| double | 32 bits | same as float |
| *Miscellaneous* | | |
| void | | Generic type |
| pointer | 16 bits | Binary or Hex representations |

**Table 5.2**  C language operators

| Pre | Op | Descrip | Pre | Op | Descrip | Pre | Op | Descrip |
|-----|-----|---------|-----|-----|---------|-----|-----|---------|
| 1 | ( ) | parenth | 3 | / | div | 11 | && | and |
| 1 | [ ] | subscr | 3 | % | mod | 12 | \|\| | or |
| 1 | . | dir memb | 4 | + | add | 13 | ? : | cond |
| 1 | -> | ind memb | 4 | − | substr | 14 | = | assign |
| 2 | ++ | incr | 5 | « | shift l | 14 | += | assign |
| 2 | − | decr | 5 | » | shift r | 14 | −= | assign |
| 2 | * | deref | 6 | < | lt | 14 | *= | assign |
| 2 | & | ref | 6 | <= | le | 14 | %= | assign |
| 2 | ! | neg | 6 | > | gt | 14 | = | assign |
| 2 | ~ | bw comp | 6 | >= | ge | 14 | »= | assign |
| 2 | + | unary + | 7 | == | eq | 14 | «= | assign |
| 2 | − | unary - | 7 | != | ineq | 14 | &= | assign |
| 2 | sizeof | size | 8 | & | bw and | 14 | ≙ | assign |
| 2 | (cast) | cast | 9 | ^ | bw ex-or | 15 | , | comma |
| 3 | * | mult | 10 | \| | bw or | | | |

the higher the precedence of the operator. Some implementations of the C language may not support all the operators.

Remember that bitwise operations are very useful to manipulate individual bits within a word. We set bits with the OR operations, clear with AND together with complement, and toggle with XOR.

**Example 5.3**   *The following examples show common targets with logic bitwise operations in C:*

```
P1OUT |= BIT4;      // Set P1.4

P1OUT ^= BIT4;      // Toggle P1.4

P1OUT &= ~BIT4;     // Clear P1.4
```

## *5.2.3 Program Statements*

Program statements in C may include assignment statements, decision structures,
loop structures, functions, and pointers, among others. Assignment statements end
with semicolon (;). Notice however other uses of semicolon in the definitions below.

### Decision Structures

The main decision blocks are the if-structure type. We can distinguish three formats:
The basic if, the if-else, and the switch structures. These formats are shown
next:

```
BASIC IF:              IF-ELSE:                 SWITCH:


if (condition)         if (condition)           switch (condition)
{                      {                        {
    statements;            statements;          case 1:
}                      } else                       statements;
                       {                            break;
                           statements;
                       }                        case 2:
                                                    statements;
                                                    break;

                                                default:
                                                    statements;
                                                }
```

There are other structures available which the reader may consult elsewhere. The
structures are not unrelated. For example, the switch structure may be thought as a
larger if-else structure, and this one as the combination of very basic if statements in
just one block.

In the switch structure, it is very important to use the break; statement at the
end of each case, otherwise the program will flow into the next case and execute
the instructions included there. The break keyword causes the innermost enclosing

loop to be exited. The default label is used to specify what the program must do in case none of the alternatives specified before are satisfied.

Let us illustrate the structure with the following example.

**Example 5.4** *The following are to codes that increment odd integers by 1 and even integers by 2. First code is an if-else structure:*

```
if (a%2==1)
{
    a=a+1;
}
else
{
    a=a+2;
}
```

*The next code is a switch structure:*

```
switch (a%2)
{
    case 1:
                a=a+1;
                break;
    case 0:
                a=a+2;
                break;
    default:   /* not needed for this example */
}
```

**Loop Structures**

The loop structures available in the C language are the for, while, and do-while structures. One important point that the reader should remember is that in embedded applications it is common to have infinite loops running. Therefore, it is very important to plan the program before attempting to write it, checking that loops are correctly identified.

FOR Loop. The for-structure is defined with the following format:

```
for(cv=initial;final expression on cv;cv=cv + 1)
{
    statements;
}
```

The following are examples for this statement, using array variables. First a simple one:

**Example 5.5** `for (i=1;i<=N;i++)`
```
{
    cin>>array[i];
    array[i]=array[i]*array[i-1]+4;
}
```

Next, an example with nested loops, multiplying matrices:

**Example 5.6** *In this example an N-by-N matrix or array is updated with new values following the matrix multiplication rules.*

```
for (i=0;i<N;i++)
{
    for (j=0;j<N;j++)
    {
        for (k=0;k<N;k++)
        {
            c[i][j]=c[i][j]+a[i][k]*b[k][j];
        }
    }
}
```

An infinite loop, so common in embedded applications, can be obtained with a `for` structure with the declaration

```
for(;;)
```

This causes the statements within the block to repeat forever.

WHILE and DO-WHILE Loops. The structure for the `while`-structure is

```
cv = initial value;

while (condition)
{
    statements;
}
```

and for the `do-while` structure is

```
cv = initial value;

do {
    statements;
} while (condition);
```

Notice that the control variable `cv` is initialized before the loops. The WHILE loop is executed only if the variable satisfies the condition, while the other loop is executed at least once, irrespectively of the condition. Infinite loops are obtained by leaving the control variable unchanged with a value satisfying the condition.

**Functions**

The structure for a function definition in the C language is

```
return_type function_name(typed parameter list)
{
    local variable declarations;
    function block;
    return expression;
}
```

The parameters listed in the parenthesis are called *formal parameters* of the function. Technically, the type may be omitted, although it is a good programming practice to *always* explicitly specify the type of a parameter. When omitted, the default one is int. The formal parameters may be of different types.

Functions are basically sub routines. Data passing toward the function is done through parameters, and from the function with the return keyword in the expression. Let us look at at following example. For all practical purposes, the parameters become local variables within the function, with the exception of pointers, as explained later.

**Example 5.7**  *The following function returns the maximum of two real values.*

```
float maximize(float x, float y)
{
    /* no local variables except for x and y */
    if ( x > y)
        return x;
    else
        return y;
}
```

The values used in the actual call or instantiation of a function are called the *actual parameters* of the function. These values must be of the same type specified for the formal parameters, like for example in the statement

```
z = maximize(3.28, myData).
```

In this case, both z and myData should be of the float type.

Passing parameters to a function can be done *by-value* or *by-reference*. The example just mentioned is of the first group. When a parameter is passed by-value, the formal parameter is declared as a variable. In this case, only a copy of the parameter is used as input and the original value cannot be changed by the function.

On the other hand, when a value is passed by-reference, the formal arguments are the pointer type, for example as int *x. The input is actually the pointer of the value, that is the address of the original parameter. Now the original contents will be affected by the function. This method requires less memory than the other, but has the disadvantage of parameters not being local variables.

### 5.2.4 Struct

Let us finally look at two other types, `pointer` and `struct`. Although `struct` is not commonly used in microcontroller programming, it is worth a brief visit.

A *record* is a data structure whose members do not have to be of the same type. It is declared using the `struct` keyword as follows:

```
struct struct_specifier
{
    declarator_lists;
};
```

**Example 5.8** *An example of a record called* **date_of_year** *and its use is shown below. This record has an array member named month whose nine (12) elements are of type* char, *a member named* day_of_week *that points to a location in memory of type* char, *i.e. it is a pointer to an array or* string *of characters of arbitrary length, and another member named* day_of_month *of type* int. *After the record is declared, it is followed by the declaration of the variable* birthday *of the same type, as well as assignment statements needed to initialize the variable*

```
struct date_of_year
{
    char month[9];
    char *day_of_week;
    int day_of_month;
};
struct date_of_year birthdate;
strcpy(birthdate.month, December);
birthdate.day_of_week = Wednesday;
birthdate.day_of_month = 15;
```

*The structure of record is depicted in Fig. 5.3a, and the result for the variable in (b). We have used the function* strcpy, *which is declared in the header file* string.h.

## 5.3 Some Examples

Let us work some examples. The first two work a blinking led example. The second one, however, works it by polling for a push-button. A third example introduces handling of interrupts. This requires placing the interrupt vector and writing the ISR.

The programs are written using standard constants. If you need to fully understand the meaning, consult the user guide and look for the corresponding peripheral information, specifically the registers.

**(a)**

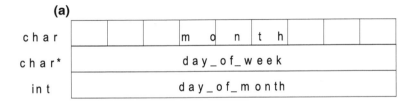

**(b)**

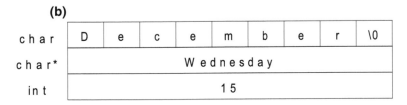

**Fig. 5.3** Structures `date_of_year` and `birthdate`

For example, take a look at the user guide for the 2xx family, slau144g [32]. In Chap. 10, you will see that WDTCTL is the short form for the Watchdog timer control register at address 0120h, which means that the variable is WDTCTL = 0120 h. On the other hand, looking at the description of this register, it is seen that WDTPW is placed covering the most significant byte, with the note "*Must be written as 05Ah*". Since it is the MSB of the register, it means that the constant WDTPW is 5600 h.

On the same register we see that WDTHOLD is bit 7, with a note "*rw-0*". This means that the bit is a read/write type, cleared on reset. The description tells that when the bit is 1, the watchdog timer is stopped. The constant is defined as 0080 h, corresponding to "bit7 = 1".

With this information, WDTPW | WDTHOLD yields 5680 h. Notice that WDTPW + WDTHOLD yield the same result. You can use any expression.

The same exercise can be done for other constants and variables.

**Example 5.9** *Let us start with a traditional blinking led operation. Assume that LED1 is connected to pin P1.0 in active high mode (it turns on when P1.0 outputs a high voltage, a 1) and LED2 to pin P1.3 in active low mode (it turns on when P1.3 outputs a low voltage, a 0). Starting with LED1 on and LED2 off, let us toggle both LEDs. The diodes are kept in the state for a period of time. We can use the following code:*

```
#include  <msp430x22x4.h>     //For device used
#define   LED1   BIT0  /LED at pin P1.0
#define   LED2   BIT3  /LED at pin P1.3

/* BIT0 and BIT3  are defined in header file  */
```

```
void  main(void)
{
        WDTCTL = WDTPW | WDTHOLD;      //Stop watchdog timer
        P1DIR = LED1 | LED2;           // Set output directions for
                                          pins
        P1OUT = LED1 | LED2;           // Initialize state for LEDs
                                       // LED1 on (high) and LED2 off
                                          (high too)
        for (;;)                       //infinite loop
        {
            unsigned int delay;        // to control led state
                                          duration
          delay=50000;
            do{delay--}                //Delay loop: decrement delay
            while(delay != 0);         // until it becomes 0
            P1OUT ^= LED1 | LED2;      //Toggle LEDs
      }                                   // close here infinite loop
}
```

Next example uses polling and two additional functions. To make it short, the second additional function is for illustration purposes only. A normal program would process results (for example, average, looking for maximum, and so on).

**Example 5.10** *An external parallel ADC is connected to port P1, and handshaking is done with port P2, bits P2.0 and P2.1. The ADC starts data conversion when P2.0 makes a low-to-high transition. New datum is ready when P2.1 is high. The objective of the code is to read and store 10 data in memory. Normal programs would do something else after data is collected. For illustration purposes, let the "something else" be sending a pulse to P2.3. After set up, the* main *will call in an infinite loop the two functions for reading and for sending the pulse. The flowcharts for the main and Read-and-Store function are shown in Fig. 5.4.*

*A more itemized pseudo code for our objectives is as follows:*

1. *Declare variables and functions*
2. *In main function:*

    (a) *Setup:*
        - *Stop watchdogtimer*
        - *P2.0, P2.2 and P2.3 as output*
    (b) *Mainloop (forever):*
        - *Clear outputs.*
        - *Call function for storing data.*
        - *Call function for something else*

3. *Function to Read and Store Data.*

    - *Request conversion*
    - *Wait for data*

**Fig. 5.4** Example 5.10: **a** Main Function; **b** Expanded flow diagram

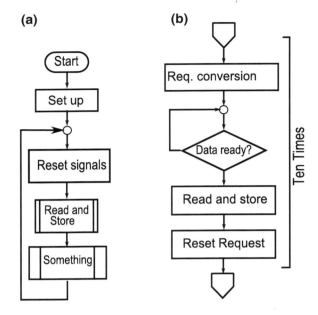

- *Store the data*
- *Prepare for new*

4. *Something Else*

  - *Send pulse to* P2.3

*The following piece of code satisfies the task requirements.*

```
#include  <msp430x22x4.h>          //For device used

unsigned int dataRead[10];         // Data in memory
void  StoreData(), SomethingElse(); // functions

void  main(void)
{
WDTCTL = WDTPW | WDTHOLD;      //Stop watchdog timer
P2DIR = BIT0  | BIT3;          // Set output directions for pins

for (;;)                       //infinite loop
    {
        P2OUT = 0;                 // Initialize with no outputs
        StoreData( );              // Call for data storage
        SomethingElse( );          // Call for other processing
    }
}

void StoreData (void)
```

```
    {
      volatile unsigned int i;   // not affected by optimization
      for(i=0; i<10; i++)
        {
        P2OUT |= BIT0;            // Request conversion
        while (P2IN & 0x08!=0x08){ } // wait for conversion
        a[i] = P1IN;
        p2OUT &= ~BIT0;           // prepare for new conversion.
        }

void SomethingElse (void)
    {
        volatile unsigned int i;  // not affected by optimization
        P2OUT |= BIT3;            // Set voltage at P2.3
        for(i=65000; i>0; i--);   // pulse width
        P2OUT &= ~BIT3;           // Complete pulse
    }
```

The following example is a demo for using Timer_A to toggle a LED. The code was written by A. Dannenberg from Texas Instruments [33].[4]

**Example 5.11** *Use Timer_A CCRx units ($x = 0, 1, 2$), and overflow to generate four independent timing intervals. TACCR0, TACCR1 and TACCR2 output units are selected with port pins P1.1, P1.2 and P1.3 in toggle mode. The pins toggles when the TACCRx register matches the TAR counter. Interrupts are also enabled with all TACCRx units, software loads offset to next interval only—as long as the interval offset is added to TACCRx, toggle rate is generated in hardware. Timer_A overflow ISR is used to toggle P1.0 with software. Proper use of the TAIV interrupt vector generator is demonstrated. An external 32 KHz watch crystal is used.*

$ACLK = TACLK = 32\,\text{kHz}, MCLK = SMCLK = default\,DCO\,1.2\,\text{MHz}$
*As coded and with* $TACLK = 32,768\,\text{Hz}$, *toggle rates are:* $P1.1 = TACCR0 = 32,768/(2*4) = 4,096\,\text{Hz}$ $P1.2 = TACCR1 = 32,768/(2*16) = 1,024\,\text{Hz}$ $P1.3 = TACCR2 = 32,768/(2*100) = 163.84\,\text{Hz}$ $P1.0 = overflow = 32,768/(2*65,536) = 0.25\,\text{Hz}$

```
;=========================================================================
;MSP430F22x4 Demo - Timer_A, Toggle P1.0-3, Cont. Mode ISR, 32kHz ACLK
;By A. Dannenberg - 04/2006
;Copyright (c) Texas Instruments, Inc.
;-------------------------------------------------------------------------
#include "msp430x22x4.h"

void main(void)
{
  WDTCTL = WDTPW + WDTHOLD;        // Stop WDT
  P1SEL |= 0x0E;                   // P1.1 - P1.3 option select
  P1DIR |= 0x0F;                   // P1.0 - P1.3 outputs
  TACCTL0 = OUTMOD_4 + CCIE;       // TACCR0 toggle, interrupt enabled
  TACCTL1 = OUTMOD_4 + CCIE;       // TACCR1 toggle, interrupt enabled
  TACCTL2 = OUTMOD_4 + CCIE;       // TACCR2 toggle, interrupt enabled
```

---

[4] See Appendix E.1 for terms of use.

```
    TACTL = TASSEL_1 + MC_2 + TAIE;   // ACLK, contmode, interrupt enabled

    while(1==1){ }                    // Wait for interrupt doing nothing
}

// Timer A0 interrupt service routine
#pragma vector=TIMERA0_VECTOR
_ _interrupt void Timer_A0(void)
{
    TACCR0 += 4;                      // Add Offset to TACCR0
}

// Timer_A3 Interrupt Vector (TAIV) handler
#pragma vector=TIMERA1_VECTOR
_ _interrupt void Timer_A1(void)
{
  switch (TAIV)         // Efficient switch-implementation
  {
    case  2: TACCR1 += 16;                    // Add Offset to TACCR1
             break;
    case  4: TACCR2 += 100;                   // Add Offset to TACCR2
             break;
    case 10: P1OUT ^= 0x01;                   // Timer_A3 overflow
             break;
  }
}
```

## 5.4  Mixing C and Assembly Code

Having introduced the reader to assembly language in Chap. 4, and after a review of fundamental concepts of the C language in the previous sections, let us now look at techniques that IARWE provides for mixing both languages, taking advantage of the best of both worlds.

The programmer may insert assembly language in the C code or mix both at the linking stage. Probably the best example for the latter method is when, writing code in one form, a particular subtask is already available and efficiently developed in the other language. It would take longer to produce the new code than to reuse the existing foreign one.

The most common examples of mixed codes in a source arise when the programmer needs a better control of CPU hardware while avoiding as much overhead as possible.

In this section we introduce the reader to the techniques for mixing both C and Assembly language in IAR Embedded Workbench (IAREW). This environment provides two ways to use assembly instructions in a C code. One is by *inlining* and the other one is using *intrinsic functions*. It is possible to create *mixed functions*. The reader is encouraged to consult [34] for a more detailed coverage of this topic.

## 5.4.1 Inlining Assembly Language Instructions

Inlining assembly code into a C program using IAREW is done with the asm keyword in the format

```
asm("assembly_instruction");
```

For example, asm("bis.b #002h,R4"); inserts the instruction in the program and directly works with the CPU register, something extremely difficult to do with C. There is no limit to the number of assembly language instructions inserted in this way. As seemingly simple and practical as it appears, this practice is discouraged in IAREW for the following reasons [34]:

1. The compiler is not aware of the effect of the assembly code. For this reason, registers and memory locations that are not restored within the sequence of inline assembler instructions might cause the rest of the code to not work properly.
2. As opposed to the Application Programming Interface used with high level code, inline assembly code sequences do not have a well-defined interface with the surrounding code. This makes it fragile and introduces the possibility of maintenance problems.
3. Optimizations will not take into account inline assembly code sequences.
4. With the exception of the data definition directives the rest of the assembler directives will cause errors.
5. There is no control over the alignment.
6. Variables with auto duration cannot be accessed.

Because of the above, intrinsic functions should be the preferred method of mixing both languages. If intrinsics are not available, then the method of mixed function explained in a later section could be used.

## 5.4.2 Incorporating with Intrinsic Functions

This is the preferred method to incorporate assembly code in C language programs. Intrinsic functions are codes that provide direct access to low-level processor operations and compile into inline code, either as a single instruction or as a short sequence of instructions. The compiler interfaces the sequence with variables and register allocation, and can optimize the functions that use these sequences.

Intrinsic functions are defined in the header file intrinsics.h, which must be included in the source code. Unfortunately, it seems that some of the modules have bugs, so the reader must verify the function before using it. Several examples of intrinsic functions in IAREW are shown in Table 5.3.

The following examples are part of Texas Instruments' (TI's) *MSP430F20xx,*
*MSP430F20xx Code Examples* packages available at Texas Instruments msp430.com

**Table 5.3**  Some IAREW intrinsics

| _ _bic_SR_register( ) | Clears bits in the SR register |
|---|---|
| _ _bis_SR_register( ) | Sets bits in the SR register |
| _ _delay_cycles | Provides cycle-accurate delay functionality |
| _ _disable_interrupt | Disables interrupts. Could use |
| _ _get_R4_register | Returns the value of the R4 register |
| _ _set_R4_register | Writes a specific value to the R4 register |
| _ _swap_bytes | Executes the SWPB instruction |
| _ _low_power_mode_n | Enters a MSP430 low power mode |

website.[5] Similar listings can be obtained for other members of the MSP430 family from TI's website. The reader is encouraged to browse through these and other codes to gain more insight on C programming of the MSP430 incorporating assembly language.

A nice features of these examples is that they use peripherals, interrupts and low power mode. The directive #pragma is also introduced. Here, it is used to load the interrupt vector in the interrupt vector table.

**Example 5.12** *The LED connected to pin P1.0 is toggled using* Timer_A *every* 50,000 *SMCLK clock cycles through an interrupt service routine. The intrinsic function* _ _BIS_SR_REGISTER *is used to enter low power mode and enable maskable interrupts [35]*

```
;=========================================================================
;MSP430F20xx Demo - Timer_A, Toggle P1.0, CCR0 Cont. Mode ISR, DCO SMCLK
;By M. Buccini and L. Westlund - 10/2005
;Copyright (c) Texas Instruments, Inc.
;-------------------------------------------------------------------------
#include <msp430x20x3.h>
#include <intrinsics.h>

void main(void)
{
  WDTCTL = WDTPW + WDTHOLD;                // Stop WDT
  P1DIR |= 0x01;                          // P1.0 output
  CCTL0 = CCIE;                           // CCR0 interrupt enabled
  CCR0 = 50000;
  TACTL = TASSEL_2 + MC_2;                // SMCLK, contmode
  _ _BIS_SR_REGISTER(LPM0_bits + GIE);    // Enter LPM0 w/ interrupt
}

/* Timer A0 interrupt service routine   */
  #pragma vector=TIMERA0_VECTOR
  _ _interrupt void Timer_A (void)
{
  P1OUT ^= 0x01;                          // Toggle P1.0
  CCR0 += 50000;                          // Add Offset to CCR0
}
```

---

[5] See Appendix E.1 for terms of use of these examples.

**Example 5.13** *This code by Westlund [36] uses the Analog to Digital Converter ADC10 incorporated in the the MSP430 model. A LED connected to pin 0 in port P1 will be lit if the voltage at terminal A1 is greater than 0.5\*AVcc, and will be turned off otherwise.*

*Looking at the MSP430 pinout, we find that terminal A1 shares the pin with bit 1 of port P1, or P1.1. we should therefore select the ADC. The code is as follows:*

```
; =======================================================================
;MSP430F20x2 Demo - ADC10, Sample A1, AVcc Ref, Set P1.0 if > 0.5*AVcc
;By L. Westlund - 05/2006
;Copyright (c) Texas Instruments, Inc.
; -----------------------------------------------------------------------
#include ''msp430x20x2.h''
#include ''intrinsics.h''

void main(void)
{

  WDTCTL = WDTPW + WDTHOLD;                 // Stop WDT
  ADC10CTL0 = ADC10SHT_2 + ADC10ON         // ADC10ON, interrupt enabled
              + ADC10IE;
  ADC10CTL1 = INCH_1;                       // input A1
  ADC10AE0 |= 0x02;                         // PA.1 ADC option select
  P1DIR |= 0x01;                            // P1.0 to output direction

  for (;;)
  {
    ADC10CTL0 |= ENC + ADC10SC;        // Sampling and conversion start
    _ _bis_SR_register(CPUOFF + GIE); // LPM0, ADC10_ISR will force exit

      if (ADC10MEM < 0x1FF)
          P1OUT &= ~0x01;  // Clear P1.0 LED off
      else
          P1OUT |= 0x01;      // Set P1.0 LED on
  }
}
// ADC10 interrupt service routine
#pragma vector=ADC10_VECTOR
_ _interrupt void ADC10_ISR(void)
{
    _ _bic_SR_register_on_exit(CPUOFF);// Clear CPUOFF bit from 0(SR)
}
```

*Note again the handling of the status register (SR) with the use of intrinsic functions* _ _bis_SR_register *and* _ _bic_SR_register_on_exit.

**Example 5.14** *This example introduces yet another of the MSP430 peripherals, the universal serial interface or USI [37]. The USI interrupt service routine toggles the LED connected to pin 0 in port P1. Since the auxiliary clock ACLK is chosen and divided by 128 and then the USI counter USICNT is loaded with 32, the USI interrupt flag USIIFG is set after a total of 4096 or 128\*32 pulses. USICNT is reloaded after toggling the LED inside the interrupt service routine.*

```
; =======================================================================
```

```
;MSP430F20xx Demo - USICNT Used as a One-Shot Timer Function, DCO SMCLK
;By M. Buccini and L. Westlund - 09/2005
;Copyright (c) Texas Instruments, Inc.
;-----------------------------------------------------------------------
#include ''msp430x20x3.h''
#include ''intrinsics.h''

void main(void)
{
  // Stop watchdog timer
  WDTCTL = WDTPW + WDTHOLD;
  BCSCTL3 |= LFXT1S_2;             // ACLK = VLO
  P1DIR |= 0x01;                   // Set P1.0 to output direction
  USICTL0 |= USIMST;               // Master mode
  USICTL1 |= USIIE;                // Counter interrupt, flag remains set
  USICKCTL = USIDIV_7 + USISSEL_1;  // /128 ACLK
  USICTL0 &= ~USISWRST;            // USI released for operation
  _ _BIS_SR_REGISTER(LPM3_bits + GIE); //Enter LPM3 w/ interrupt
}

/* USI interrupt service routine /* #pragma vector=USI_VECTOR
 _ _interrupt void universal_serial_interface(void) {
  P1OUT ^= 0x01;   // Toggle P1.0 using exclusive-OR
  USICNT = 0x1F;   // re-load counter
}
```

### 5.4.3 *Mixed Functions*

These are user defined functions to write modules in assembly language and call them from the C program. The objective is to reduce the negative impact in performance as compared with inlining assembly instructions. The procedure for doing this in the IAREW environment is outlined next.

The assembly language subroutine to be called from C must comply with the following conditions [34]:

1. Follow the calling convention.
2. Have the entry-point label PUBLIC.
3. Its declaration should be done before any call and it should be defined as external, to allow type checking and optional promotion of parameters, as in:
   extern int f1(void;) or
   extern int f1(int, int, int);

In order to comply with the above, you can create skeleton code in C, compile it, and study the assembler list file. To do this in the IAREW IDE, specify list options on file level an do the following:

- Make sure you select the file in the workspace window.
- Now choose `Project, Options`, and go to the C/C++ Compiler category and
  - select `Override inherited settings`
- On the List page
  - Deselect `Output list file`, and
  - select the `Output assembler file` option and its sub-option `Include source`.
- Make sure to specify a low level of optimization.

## 5.4.4 Linking Assembled Code with Compiled C Programs

Previous sections were focused on mixing C and assembly at the source level. There are times though when the assembly and C codes are separately assembled and compiled. In this case, for the C code to be able to access variables and call functions in assembly language and for the assembly code to access variables and call functions in C language, some conventions shown below and detailed in [29] must be observed. The definitions shown in Table 5.4 are also needed.

The caller program performs the following tasks when calling the callee or called function. The resulting memory allocation of these tasks is illustrated in Fig. 5.5:

1. The caller is responsible for making sure save-on-call registers across the call as needed.
2. If the parameter returned by the callee is a structure, then the caller allocates the space needed for the structure and passes in its first argument the address to the callee.

**Table 5.4**  Definitions

| | |
|---|---|
| Argument block | This is the part of the local frame used to pass arguments to other functions. As opposed to pushing the arguments onto the stack, they are passed by moving them into the argument block |
| Register save area | This is where the registers are saved within the local frame when the function is called and restored from when the function exits |
| Save-on-call registers | These are registers R11-R15. The caller must save them if needed because the callee will not preserve their contents |
| Save-on-entry registers | Registers R4-R10. The callee is responsible for preserving the contents of these registers by saving them in case it needs to modify their contents and restoring them before returning to the caller |

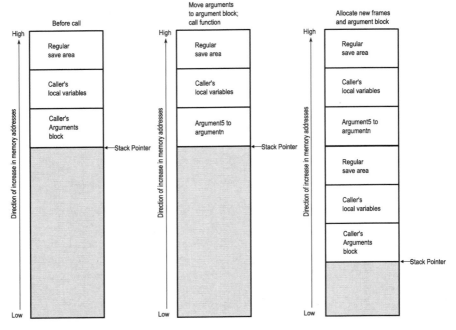

**Fig. 5.5**   Stack during a call to a function

3. The first arguments are placed by the caller in registers R12–R15. The remaining arguments are placed by the caller in the argument block in reverse order, placing the leftmost remaining argument at the lowest address so that it is placed at the top of the stack.
4. The caller program calls the callee function.

The callee performs the following tasks:

1. If the function can be called with a variable number of arguments, i.e. if it can be called with an ellipsis, and if they meet the following criteria, then the callee pushes these arguments onto the stack:

   (a) The argument is including the last explicitly declared argument or follows it.
   (b) The argument parameter passing is via a register.

2. The callee pushes the save-on-entry registers (R4–R10) values onto the stack. Additional registers may need to be preserved if an interrupt service function.
3. The callee allocates memory for the local variables and argument block by subtracting a constant from the SP.
4. The callee executes.
5. The value returned by the callee is placed in R12 (or R12 and R13).
6. If the callee returns a structure, it copies the structure to the memory block pointed to by the first argument, R12. In case the caller has no use for the return value, R12 is set to zero (0) to direct the callee not to copy the return structure.

7. The callee deallocates the frame and argument block allocated in step 3.
8. The callee restores all registers saved in step 2.
9. The callee returns.

### 5.4.5 Interrupt Function Support

If the appropriate steps are taken, then C programs can be interrupted and returned without causing harm to the C environment. If interrupts are needed by the systems, enabling them, e.g. using intrinsics, has no effect in the C environment. No arguments can be used in the interrupt service routine (ISR) declarations and they must return *void*, unless they are software interrupt routines. ISRs must preserve the registers they use or that are used by functions called by them.

The following should be remembered:

1. It cannot be assumed that the run-time environment is set up after a system reset interrupt. These means that local variables cannot be allocated nor can information on the run-time stack be saved. For example, the C boot routine creates the run-time environment using a function which resets the system, called c_int00 (or _c_int00). The run-time-support source library contains the source to this routine in a module named boot.c (or boot.asm).
2. If assembly language is used, then precede the name of a C interrupt with the appropriate linkname. For example, c_int00 should be referred to as _c_int00.

## 5.5 Using MSP430 C-Library Functions

There are many library functions available in the different C language implementations. We have already mentioned library functions, i.e. printf(), and strcpy(). Some of the most commonly used library functions are included as part of the stdio.h, math.h, and string.h header files. For example, the printf() and scanf() are included as part of the stdio.h header file. On the other hand, the pow(), log(), and floor() functions are part of the math.h header file and the strcpy(), strlen(), and strcat() are part of the string.h header file. The programmer needs to verify that the corresponding header file is included with the preprocessor directive #include before using the library function in the code.

IAR Embedded Workbench does a very good job in providing header files as part of their IDEs along with several run-time-support libraries. The reader is encouraged to review the libraries available for the C language implementation. Most of the information presented in this section can be found in greater detail in [34].

The reference also has guidelines for the user who may need to create new libraries of functions, or add functions to an existing library.

## 5.6 Summary

In this chapter we presented the basics of the C language for embedded programming. The most important features of the C language were presented such as its program structure, data types, constants, variable attributes, operators, loop and decision structures, arrays, records, pointers, and functions. We then proceeded to present those features of the C language that make it suitable for embedded programming, along with how to mix C and assembly language instructions, how to develop and modify a library. Several examples showing how to program the MSP430 using the C language were presented. The IAR Embedded Workbench IDE was used as the vehicle for developing programs in C and assembly language.

## 5.7 Problems

5.1 What is the name of the program modules in the C language?

5.2 What is the minimum number of functions that a C program must have? Identify them.

5.3 Can you explain why you can not use the exact same compiler on some specific PC to produce code to be run on a PC and also to produce code to be run on some embedded system?

5.4 What do we mean by "C is a strongly-typed language"?

5.5 What are the main parts of a C program?

5.6 Why is a compiler needed?

5.7 What type of code does a compiler produce?

5.8 What is a cross-compiler?

5.9 Name several preprocessor directives in the C language.

5.10 What is a header file?

5.11 What do you need to know in order to be able to use memory locations and I/O units in your C language programs?

5.12 Write a small C language program that will set the lower four bits of port 1 (P1) in the MSP430 as input bits and the upper four bits as output bits. Assume P1DIR is defined in the msp430xxxx.h file.

5.13 Write a C language program that will read 8 input bits from port 2 (P2), rotate them twice and send them out port 4 (P4) of the MSP430. Assume P2DIR, P4DIR, P2IN, P4OUT are defined in the msp430xxxx.h file. If you are using a microcontroller that does not have all these ports, e.g. the MSP430G2231, then use the ports it has available.

5.14 Angel and Carlos wanted to allow more time for the LEDS used to test the program in example 5.9 to be on and off. Thus, they increased the loop count from 50,000 to 75,000. When testing the program they found out, however, that the LED would now turn on and off at a faster rate. Explain what happened. Hint: Variables in the MSP430 are 16-bit long. Could you modify the program to do what Angel and Carlos want?

5.15 What is a "context switch"?. When can a context switch occur during the execution of a program?

# Chapter 6
# Fundamentals of Interfacing

Before a microcontroller can be put to work, the fundamental components that enable its functionality must be in place. These components are what we call the basic MCU interface. The first part of this chapter deals with the discussion of how to provide the fundamental requirements of power, clocking, reset, and bootstrapping needed to enable the functionality of any MCU. MSP430 devices have different levels of support for their basic interface. The supports level changes depend on the specific generation and family. In this chapter we discuss how these requirements are satisfied across all device generations.

## 6.1 Elements in a Basic MCU Interface

Any microprocessor, needs a basic interface to become usable. This interface includes four fundamental elements:

- A power source to feed power to the CPU and system peripherals
- A clock generator to synchronize the system operation
- A power-on reset circuit (POR), to take the system to its initial state
- A booting function, to initiate the system software execution

Microcontrollers frequently have several of the required elements embedded within their chip, while stand-alone CPUs require external components to provide all the elements for their basic interface. These components are pictorially illustrated in Fig. 6.1. The next sections provide insight into of each of these components.

M. Jiménez et al., *Introduction to Embedded Systems*,
DOI: 10.1007/978-1-4614-3143-5_6,
© Springer Science+Business Media New York 2014

**Fig. 6.1** Components in a
basic CPU interface: power
source, clock generator,
power-on-reset, and boot
sequence

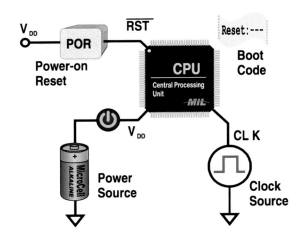

## 6.2 Processors' Power Sources

Regardless of the processor type, power sources need to be externally provided. They
can be implemented from different sources: batteries, wall connected AC power
outlet, or some forme of energy collecting or scavenging device. In every case,
selection and design criteria must be applied to properly feed power to the CPU and
its peripherals.

Power requirements of a microcontroller-based application are not too different
from those of any other type digital system. Basic requirements call for a steady
supply voltage and enough output current to feed, under worst case conditions, the
nominal system load. Such a load typically includes the CPU and its surrounding peripherals. The electrical characteristics provided in the data sheet of a device
($\mu$P, MCU, or peripheral) clearly indicate the allowed values for supply voltage and
regulation. It is of outmost importance to differentiate between *Absolute Maximum
Ratings* and *Recommended Operating Conditions* when searching for the appropriate
supply voltage level for a device.

The Absolute Maximum Ratings of a device specify levels of stresses that if
exceeded or used as operating conditions will affect reliability or cause permanent damage to the device. Therefore, we should never design for working at these
levels. The safe operating levels are those specified in the *Recommended Operating
Conditions* so these should be the choice for a good, reliable design, instead.

Table 6.1 shows the absolute maximum ratings of an MSP430G2231 microcontroller. Here we can observe the device's absolute limits for the maximum and minimum supply voltages along with other limiting parameters. For this device, setting
$V_{CC} = 4.1$ V would be a poor design choice, because although this value does not
exceed the absolute maximum, it would stress the device to the point where it would
easily fail due to overvoltage.

**Table 6.1** Absolute maximum ratings for MSP430G2231

| Parameter | Condition | Limits |
|---|---|---|
| Voltage applied at $V_{CC}$ to $V_{SS}$ | | $-0.3$ V to $4.1$ V |
| Voltage applied to any pin | | $0.3$ V to $V_{CC} + 0.3$ V |
| Diode current at any device pin | | $\pm 2$ mA |
| Storage temperature range, $T_{stg}$ | Unprogrammed device | $-55\,°$C to $150\,°$C |
| | Programmed device | $-40$ to $85\,°$C |

**Table 6.2** Recommended operating conditions for MSP430G2231

| Symb. | Parameter | Condition | Min. | Max. | Unit |
|---|---|---|---|---|---|
| $V_{CC}$ | Supply voltage | Program execution | 1.8 | 3.6 | V |
| | | Flash programming | 2.2 | 3.6 | |
| $V_{SS}$ | Supply voltage | | 0 | 0 | V |
| $T_A$ | Operating temp. | Free air | $-40$ | 85 | °C |
| $f_{clk}$ | Clock frequency | $V_{CC} = 1.8$ V, Duty cycle $= 50\,\% \pm 10\,\%$ | dc | 4.15 | MHz |
| | | $V_{CC} = 2.7$ V, Duty cycle $= 50\,\% \pm 10\,\%$ | dc | 12 | |
| | | $V_{CC} = 3.3$ V, Duty cycle $= 50\,\% \pm 10\,\%$ | dc | 16 | |

Table 6.2 shows the recommended supply levels for the same device. Note that the maximum recommended supply voltage value keeps a safe margin of 0.5 V below the absolute maximum. The ground level voltage specification can be also observed to remain at zero, independently from the value chosen for $V_{CC}$.

In the case of the MSP430G2231, there is no specification for a regulation requirement. This is a characteristic commonly found in microcontrollers. MCUs frequently accommodate internal references and regulators for on-chip components susceptible to $V_{CC}$ variations. For example, the MSP430G2231 embeds an internal 1.5 V voltage reference for its analog-to-digital converter and a programmable supply voltage supervisor that allows for selecting the minimum allowable operational voltage for the device, within the recommended range. This relaxes the regulation requirements of the unit. Nevertheless, it is always recommended to use a regulated power supply for whatever voltage level is selected for operating the chip. Processing speed, signal compatibility, power consumption, and system predictability are all dependent on the steadiness of the supply voltage. Moreover, older MCUs and microprocessors have less elaborated on-chip power supply modules. In such cases, the provision of a regulated supply becomes a must.

**Fig. 6.2** Structure of a typical MCU power supply

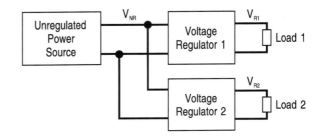

## 6.2.1 Power Supply Structure

Bare power sources, regardless of their type, are far from ideal and thus rarely meet the voltage steadiness requirement for digital circuits or microprocessor-based systems. For example, the output voltage of an AC to DC converter tends to change with load fluctuations or input voltage changes. Its outputs also carry ripple voltage fluctuations that worsen as the load current increases. Batteries tend to decrease their voltage level as they give off their stored charge. Less conventional sources such as solar cells or other forms of energy scavenging devices are even less stable in terms of output voltage levels. For these reasons, bare power sources are frequently referred to as *unregulated power sources.* Making them usable in digital systems require some form of power conditioning between them and their load.

A typical MCU power supply includes an unregulated power source and a voltage regulator to feed the load. If more than one voltage level were required, multiple regulators would be used, one for each different voltage level. This situation is illustrated in Fig. 6.2 for a system requiring two different voltage levels $V_{R1}$ and $V_{R2}$.

The unregulated source must be able of supplying the total power required by all loads plus the consumption of the regulators themselves. The requirements of each load must be met by the individual voltage regulators. To design the system power supply we need to determine the requirements of each load in terms of voltage level, peak current, and regulation.

The voltage level and regulation requirements are given by the microcontroller or MPU specifications. Each independent peripheral in the system will also have its own set of power specifications. Most devices allow in their specifications for a range of supply voltage values, easing the power supply selection process. However, once a specific voltage level is selected, it must be kept fixed at the selected value for proper system operation. Take as an example the case of the MSP430F149, for which the recommended operating conditions a supply voltage level goes from 1.8 to 3.6 V. If we choose to operate the MCU at 3.3 V, such a voltage level must be supplied through a regulated source maintaining it at the chosen value.

Microprocessors usually have more specific and restrictive requirements. For example, an OMAP-L137 requires a DVDD supply voltage of 3.3 V with a regulation of ±4.5 %. Such a requirement calls for a dedicated voltage regulator capable of maintaining the supply voltage level from 3.15 V to 3.45 V for all loading conditions.

**Fig. 6.3** Frequency versus supply voltage plot for an MSP430F149 MCU

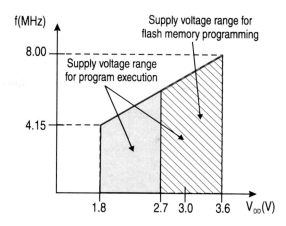

## 6.2.2 Power-Speed Tradeoff

One important consideration when choosing the supply voltage of a digital system is its power-delay trade-off: the lower the voltage, the smaller the power consumption but also the slower its processing speed. Power in digital circuits, particularly in those based on CMOS technologies, varies with the square of the supply voltage. Reducing the supply voltage of a CMOS circuit by a factor of two produces a reduction in power consumption by a factor of four, which results very attractive from a power consumption standpoint.

However, the designer must also consider that a reduction in the power supply level carries a reduction in the maximum frequency at which the system can operate. In microprocessor or microcontroller based systems, the maximum frequency of operation is almost directly proportional to the supply voltage. This situation is illustrated in Fig. 6.3 for an MSP430F149. It can be observed that when $V_{DD}$ is set to its minimum value ($V_{DD} = 1.8$ V, the maximum clock frequency is limited to 4.15 MHz. To operate the chip at its specified maximum frequency of 8.0 MHz, the supply voltage needs to be raised to 3.6 V.

The best recommendation for choosing the supply level of a microprocessor-based application is using the lowest voltage level that allows for the proper system operation.

## 6.2.3 Power Supply Capacity

Further consideration regards the current and voltage output capability of the chosen power source.

With this in mind, let us first look at the regulator requirements. The sustained current capacity of a regulator must supply the maximum sustained current of the

**Fig. 6.4** Non-regulated
source feeding a load through
a 3-terminal linear voltage
regulator

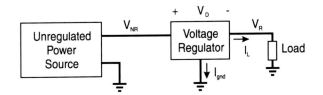

total load it feeds. Thus, given $n$ loads $\{L_1, L2, \ldots, L_n\}$ to be fed by a voltage regulator, each with nominal current $I_{Li}$, the regulator feeding them must be able to provide a current $I_R \geq \sum_{i=1}^{n} I_{Li}$. As a practical measure, the regulator capacity is chosen with a 15–20 % of overhead capacity.

On the other hand, a voltage regulator needs to maintain a minimum voltage drop between its input and output terminals and also requires a small amount of current to operate. The minimum voltage drop between the input and output terminals is called the regulator's *dropout voltage* $V_D$. This requirement imposes a minimum voltage level for the unregulated power source, as indicated in Eq. (6.1).

Figure 6.4 shows a block diagram of the connection of a non-regulated source to a load via a linear regulator. Here we can also observe the regulator's ground current, $I_{gnd}$. This current is necessary to bias the regulator's internal pass element to maintain a steady output voltage. From this connection, we see that the minimum voltage level VNR for the unregulated power source must satisfy

$$V_{NR} \geq (V_R + V_D) \qquad (6.1)$$

and the current capability for the supplied INR must satisfy

$$I_{NR} \geq (I_R + I_{gnd}) \qquad (6.2)$$

Finally the total power supplied by the non regulated source is $P_{NR} = V_{NR} \cdot (I_R + I_{gnd})$ and the power delivered to the load $P_R = V_R \cdot I_R$. We can therefore estimate the regulator's efficiency as $Eff = P_R/P_{NR} \times 100\,\%$. This tells us that we should select $V_{NR}$ as close as possible to its minimum value to reduce the power loss at the regulator.

### 6.2.3.1  Additional Considerations on Regulators

In general, when selecting a regulator, the designer can choose between linear regulators, either in their standard or low-dropout (LDO) versions, or switching regulators. Linear regulators use a pass device, typically one or more transistors, with a feedback network to maintain a steady output voltage. In a few words, the pass element is like an adjustable resistor that changes its value to keep a steady output voltage. In particular, standard linear regulators in their three-terminal packages are available in a wide range of input and output voltages and current ranges. They can provide regulated currents, as large as 10 A, with small $I_{gnd}$ requirements, typically less than 10 mA, and moderate dropout voltages (1.7 V $\leq V_D \leq$ 2.5 V). Among linear regulators, LDOs have the lowest $V_D$ requirement, typically between 0.1 and 0.7 V at the expense of increased $I_{gnd}$ (between 20 and 40 mA). On the downside, their maximum output current is in general limited, typically less than 1 A.

Another factor to consider when using linear regulators is their efficiency. The larger the difference $V_{NR} - V_R$, the lower the regulator efficiency. For example, neglecting $I_{gnd}$ and choosing a 9 V source to feed a constant load at 3.3 V through a linear regulator would yield a maximum efficiency of 36.67 %. Using a 4.5 V source instead, the efficiency would be increased to 73.3 %. The minimum value of $V_{NR}$ is actually limited by Eq. (6.1).

Switching regulators, also called DC-to-DC converts, operate by turning on and off the input voltage to the load, exerting control on the average voltage seen by the load through the duty cycle of the switching signal. Reactive elements like inductors and capacitors are used to smooth-out the output signal. This gives them great flexibility in producing output voltages that can be either higher or lower than the input voltage.

Switching regulators, are particularly attractive by their efficiency, frequently reaching values over 85 %. However, DC-to-DC converters tend to be a lot noisier than standard linear or LDO regulators. In fact, when the load includes noise sensitive components, it is common to place an LDO at the output of a DC-to-DC converter.

Example 6.1 evaluates two alternatives for feeding a hypothetical embedded design application.

**Example 6.1** *Consider using a standard 9 V alkaline battery for feeding a 3.3 V, 120 mA load via a linear voltage regulator uA78M33C. Estimate the approximated battery life and usage efficiency assuming a constant load current.*

*Solution: This regulator is rated at 3.3 V/500 mA, has a maximum bias current $I_{gnd}$ = 6 mA and a dropout voltage $V_D$ = 2.0 V. Using the Fig. 6.4 as reference, the total load current seen by the battery would be approximately $I_{NR} = I_L + I_{gnd}$ = 126 mA and the minimum required non-regulated input voltage to the regulator should be $V_{NR} \geq$ 5.3 V.*

*A typical 6LR61 9 V disposable alkaline battery provides a charge of approximately 480 mAh at a discharge rate of 100 mA, and drops to about 380 mAh at 300 mA load [38]. Both these values correspond to a minimum battery voltage of 4.8 V. Interpolating these values for the particular load and minimum voltage of this problem, yields a usable charge of approximately 447 mAh. The expected battery duration would be about:*

$$t_{bat} = \frac{Q_{bat}}{I_{NR}} = \frac{447\ mAh}{126\ mA} = 3.55\ h.$$

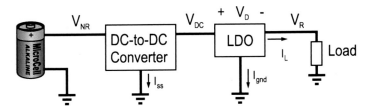

**Fig. 6.5** Power supply integrating a DC-to-DC converter and LDO

*In this case the efficiency at the regulator would be:*

$$Eff = \frac{V_R \times I_R}{V_{NR} \times (I_R + I_{gnd})} \times 100\,\% = \frac{3.3\,V \times 120\,mA}{9\,V \times (120+6)mA} = 34.92\,\%$$

*This means that over 65\% of the energy supplied by the battery would be lost as heat in the regulator.*

*A better design could use a DC-to-DC switching converter plus an LDO as illustrated in Fig. 6.5. A TPS54232 DC-to-DC converter can be configured to reduce the 9 V battery voltage to 3.45 V with 88\% efficiency. With an LDO like a TPS79933 operating with a nominal dropout of 150 mV and $I_{gnd} = 50\,\mu A$, its efficiency would be 96\%. The combined DC-to-DC–LDO efficiency would now be 85\%.*

*As the TPS54232 converter can accept input voltages down to 3.5 V, the usable battery charge is now estimated to 567 mAh. At 9 V the battery current will be approximately 52.3 mA, while at 3.5 V it will increase to about 134.5 mA. A rough battery life estimate based on a linear approximation with this current range yields 7.1 H of operation, nearly 7 h of continuous operation. This is almost twice the expected battery life. This estimate could be further improved if the load, assuming an embedded application, used some form of low-power mode.*

### 6.2.4 Power Supply Distribution Tips

Many problems arising in microprocessor-based designs are caused by noise in the power supply lines. *Power supply noise* can be introduced by multiple factors, being among the most notorious the switching nature of digital circuits. Every change of state in a digital circuit induces a change in the power supply current. This creates a time varying current demand on the power supply line, which acting upon the inherent impedance of power supply distribution lines and components such as DC converters and regulators, cause voltage fluctuations that account for most of the power supply noise manifestations we see in digital circuits.

The most nocive parasitic impedance component in power distribution lines is inductance. Unfortunately, inductance is unavoidable in power lines, since it is related precisely to the change of current, as we know from electromagnetism. In addition to this parasitic inductance, there are further contributions from the power devices. A linear, three terminal regulator contributes between $1\,\mu H$ and $2\,\mu H$ of output inductance to the power supply line. Switching regulators have much higher output parasitic inductance. Interconnections also contribute with about 20 nH of parasitic inductance per inch. When an active load is connected to such an impedance, the level of noise created will increase with the rate of change of the supply current, as dictated by Eq. (6.3). The rate of change of current in time increases with the circuit operating frequency, thus, the higher the frequency the more noisy the power supply lines would become.

$$v_L(t) = L\frac{di(t)}{dt} \tag{6.3}$$

Fortunately, most problems caused by power supply noise can be relatively easily mitigated by using simple measures when interconnecting and routing power supply distribution lines. The two most common techniques for reducing power supply noise are *Bypassing* and *Decoupling*.

**Bypassing Techniques**

Bypassing refers to the act of reducing a high frequency current flow in a circuit path by adding a shunting component that reacts to the target frequency. The most commonly used shunting devices in microprocessor-based designs are bypassing capacitors .

A bypass capacitor reduces the rate of change of the current circulating in the power line by providing a high-frequency, low impedance path to the varying load current. Two factors determine the effectiveness of a bypassing capacitor: size and location.

The size of a bypassing capacitor is selected in accordance to the frequency of the noise they are intended to attenuate. Assuming a supply voltage variation $\Delta V_{CC}$, a supply current $I_{CC}$, and a worst case transient time $\Delta T$ are known, the value for a bypass capacitor $C_b$ can be estimated with Eq. (6.4).

$$C_b = \frac{I_{CC}}{\Delta V_{CC}/\Delta T} \tag{6.4}$$

In most cases, a standard non-electrolytic capacitor between 1 and $0.01\,\mu F$ shall work. The smaller the value of $C_b$, the higher the frequency it will shunt. Due to the parasitic components of a real capacitor, namely its nominal capacitance $C_b$ and its equivalent series inductance (ESL), this formula is valid only through the resonant frequency $f_c$ of the capacitor, as illustrated in Fig. 6.6.

The best place for placing a bypassing capacitor is the one that provides the shortest return path to the high-frequency current component being shunted. Such a selection would ensure a small path inductance and therefore would minimize the transient effect it introduces. Based on this premise, the best place for a bypass capacitor to

**Fig. 6.6** Change of equivalent impedance of a bypass
capacitor as a function of
frequency

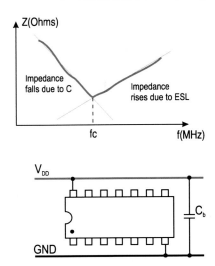

**Fig. 6.7** Placement of a
bypass capacitor as close
as possible to $V_{DD}$–GND
terminals

shunt the induced power supply noise of a packaged MCU or peripheral IC would
be the closest possible to the $V_{DD}$–GND terminals. Figure 6.7 illustrates a possible
bypass capacitor placement for a dual-in-line (DIP) package.

### Source Decoupling

Decoupling refers to the isolation of two circuits in the same path or connection
line. Power supply decoupling is usually achieved through the installation of low-
pass filters in strategic points of the power distribution line to reduce the strength of
high-frequency noise components from one side of the system, reaching potentially
sensitive components in the other side.

AC-to-DC converters, switching regulators, and DC-to-DC converters make particularly noisy sources. In such cases, adding a decoupling circuit at the power supply
output helps to isolate the power supply noise from the rest of the circuit. A practical
decoupling circuit can be made by adding a fairly large capacitor at the output of the
source. In most cases a 10–100 μF electrolytic capacitor shall suffice. If additional
filtering were needed, an LC filter at the power supply output could be used. The
electrolytic decoupling capacitor would also be large, as above, or even larger for
very noisy loads such as DC motors. The value of the inductor, although not critical, is also recommended to be fairly large, with typical values in the range from
10 to 100 mH. In circuits sensitive to power supply ringing, the decoupling circuit
could make the power supply underdamped. If this were an issue, a damping resistor
of value $R_{Damp} = 2\sqrt{\frac{L}{C}}$ in series with the decoupling capacitor might be added
for improved stability. Figure 6.8 shows a decoupling circuit including both, a filter
inductor and damping resistor next to an electrolytic decoupling capacitor.

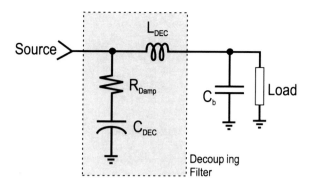

**Fig. 6.8** Typical topology of a decoupling LC filter including an optional damping resistor

## Analog Versus Digital Supply Lines

Microcontrollers that incorporate analog or mixed-signal circuitry such as embedded OpAmps, data converters (ADCs or DACs), analog comparators, and other functions, commonly provide separate, independent pins in their package for digital and analog supplies. This means separate lines for the digital supply ($DV_{DD}$) and analog supply ($AV_{DD}$), and their respective ground leads, $DGND$ and $AGND$.

The reason for such a separation is to allow for independently routing of the power lines for analog and digital components. The pulsating operation of digital circuits introduces considerable amounts of noise into their power supply lines. Separation of analog and digital supplies, and their corresponding ground connections, allows for reducing the amount of digital noise reaching the analog components. Although we talk about separate routings, the ground signals must have a common point. To minimize interference, ground lines for analog and digital must meet only at the power supply output. The same rule applies to common $V_{DD}$ lines. Figure 6.9 illustrates this type of connection. The case illustrated assumes a single $V_{DD}$/GND output at the power supply. Assuming this embedded system contains both, analog and digital circuitry, and the MCU provides for separated supply leads for analog and digital components, the connections shall route them separately.

To achieve such a noise isolation it is necessary to observe good circuit layout techniques. The list below includes several recommendations for making a good layout.

- Common points for both VDD and GND poles shall occur only at the power supply. If the system is enclosed, this is also a good point to ground the enclosure or chassis.
- If using wires for interconnections, physically place analog wiring separated from the digital wiring. Try to place each pair as close as possible to each other. This is, AVDD close to AGND, and DVDD close to DGND. If twisting each pair were possible, go for it. Whenever possible, use thick, sort wire connections to minimize the series impedance.

- When routing the interconnections on a printed circuit board (PBC), try to establish ground planes for both, analog and digital parts. Isolate the planes. Use wide traces for AVDD and DVDD. Route analog traces on the opposite side, right above the AGND plane. Do similarly for the digital traces and DGND plane.
- Use decoupling capacitors at the power supply output and bypassing capacitors at each package, in both, analog and digital sides.
- Try to locate the power supply block in the center of one of the sides of your board. This shall create better chances of separating one side for analog components and other for digital, at both sides of the power supply block.
- Do not surround analog components with digital or viceversa. This would make difficult to separate the routing of analog and digital components.

## 6.3  Clock Sources

The need for a clock source in microprocessor-based systems arises from the synchronous sequential nature of the digital logic making up most of the system components. The synchronous operation of the control unit's finite state machine (FSM) requires a periodic signal to yield precise transitions between the different states assumed by the CPU. Peripheral circuits also work driven by internal FSMs, requiring steady clock signals as well. Examples of such peripherals include timers, data converters, serial adapters, and displays, among others.

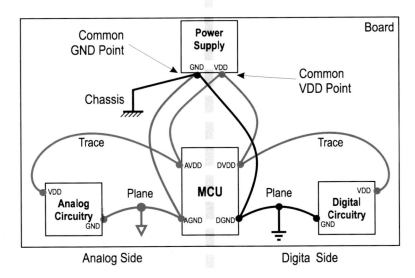

**Fig. 6.9**  Layout of analog and digital supply lines in a mixed-signal embedded board

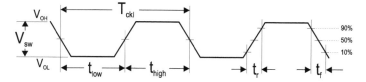

**Fig. 6.10** Clock signal timing specifications

In many cases, particularly in MCU-based designs, the clock sources for both the CPU and peripheral devices are derived from the same clock generator, which is designated for that reason *System Clock*.

## 6.3.1 Clock Parameters

The clock source for a microprocessor system must provide a steady square wave signal. In this context, steadiness implies fixed amplitude swing ($V_{sw}$), fixed frequency ($f_{clk} = \frac{1}{T_{clk}}$), constant duty cycle ($DC = \frac{t_{high}}{T_{clk}}$), and sharp edges denoted by small rising and falling times ($t_r$, $t_f$). Figure 6.10 illustrates the waveform parameters of a typical digital clock signal.

A minimum set of clock specifications for a microcontroller includes the clock swing, frequency ($f_{clk}$), duty cycle, and some measure of frequency stability.

The swing of a clock signal ($V_{sw} = V_{OH} - V_{OL}$) needs to match the levels established by the processor's logic levels. Given the processor supply voltage, the swing of the clock signal is expected to bring the low and high values to the safe margins established by $V_{IL}$ and $V_{IH}$.

The frequency, duty cycle, and stability are determined by the oscillator used to generate the clock signal. The value of these parameters must be chosen to satisfy the requirements specified in the processor's data sheet.

The clock generator is provided entirely outside the chip using dedicated oscillators with square wave output. In microcontrollers, although the use of external, dedicated clock generators is possible, the most common configurations have their oscillators residing either partially or in its totality within the chip. When an entirely on-chip oscillator is used, the voltage swing, the clock frequency, stability, and duty cycle are internally set. Such sources are very convenient since no external component are required to establish the system clock.

However, not every application will benefit from using such internal sources. Most MCUs use resonant RC circuits for these internal clock sources, which as a consequence suffer from limited bandwidth, and are sensitive to temperature and supply voltage variations affecting frequency stability. Taking into consideration such limitations, MCUs clock generators also allow for complementing their oscillators with external components that aid in overcoming the limitations of internal sources.

**Clock Frequency**

External components are able to establish the clock frequency, duty cycle, and stability, and therefore they can be chosen to satisfy the application expectations, always considering the physical limits specified in the processor's recommended operating conditions. It is also important to remember that the supply voltage level $V_{DD}$ at which the chip is operated imposes an additional limit to the maximum clock frequency at which the MCU can be operated. If we look back at Table 6.2, we would see that for an MSP430G2231, the recommended frequency range goes from DC up to 16 MHz, but the actual maximum frequency will depend on the supply level. For example, when operated with $V_{DD} = 2.7$ V, the maximum achievable frequency is only 12 MHz.

**Clock Duty Cycle**

The clock duty cycle defines the ratio of $t_{high}$ to the period of the clock signal, expressed in percent. A clock with frequency 1 MHz and duty cycle of 40 % will have a period of 1 μs, and a $t_{high} \cong 400$ ns. The duty cycle is determined by the topology of the oscillator circuit and the value of its components. In the case illustrated in Table 6.2 for the MSP430G2231 the duty cycle is specified at 50 %. This is a common requirement in many modern MCUs.

**Clock Stability**

The clock stability requirement specifies a statistical measure of the maximum allowable frequency fluctuations of a clock signal over a given time interval. This is necessary because clock signal parameters vary depending on factors that include the oscillator type, capacitive loading, oscillator circuit age, supply voltage, and temperature. In the case of the MSP430G2231, Table 6.2 specifies a maximum frequency deviation of $\pm 10$ % from the nominal target value. This means that if for example we set the clock frequency to 12 MHz, the maximum deviation from this value the processor could tolerate would be $\pm 120$ KHz. This is a particularly relaxed specification, which is convenient from a design flexibility standpoint. However, the actual acceptable deviation depends not only on physical specifications, but also on the intended application. This situation is illustrated in Example 6.2.

**Example 6.2** *Consider a 12 MHz clock signal that exhibits a $\pm 10$ % deviation from its nominal value. Two applications employing such a clock signal are analyzed: a dynamic display that sets a 60 Hz refresh ratio from this clock and a real-time clock slated to run uninterruptedly for weeks. Evaluate the impact of the clock accuracy on each system.*

*Solution: The impact of the clock deviation will be analyzed independently for each system.*

*Impact on the display system:* A 10% frequency deviation would translate into an equally proportional deviation in the refresh rate, implying that the actual rate could be 6 Hz off the target value. In the worst case, assuming a 10% frequency loss, the refresh ratio would be 54 Hz. Considering that for persistence of vision, the human eye only requires 24 Hz of refresh ratio to perceive motion, this 10% change of frequency can be deemed as negligible.

*Impact on the RTC:* A 10% deviation in frequency would cause the RTC to drift from the actual time at a rate of 6 s each minute or 2 h and 24 min per day. At the end of only one week, assuming a negative $\delta f_{CLK}$, the error would accumulate to 16.8 h, which would be totally unacceptable.

This simple example highlights the fact that the application itself is what ultimately defines the clock stability requirements.

Deviations from the nominal frequency in oscillator circuits can be caused by multiple factors. They can be generally grouped in factors affecting short or long term frequency stability. *Clock jitter* and *frequency drift* are two illustrative examples.

## Clock Jitter

Clock Jitter refers to the uncertainty in the periodicity of a clock signal, manifested as the randomness with which the signal passes through a predefined voltage threshold. Jitter is a complex phenomena in clock signals. Its origin can be related to both stochastic and deterministic factors that include internal noise, particularly thermal noise; power supply variations, loading conditions, and interference coupled from nearby circuits (crosstalk). Its accumulation can lead to data and synchronization errors in digital systems, specially in communications channels. When given as a clock specification, the *total jitter* defines the maximum accumulated period deviation caused by all jitter sources that a processor can tolerate in its clock input over a number of cycles. This is a common specification for microprocessor clock sources.

**Example 6.3** *Consider a microprocessor with a total system clock jitter specified at 150 ps under the JESD65B standard. This specification establishes that the total time deviation in the signal period resulting from the sum of all deterministic and random sources in the clock frequency over a minimum of $10^4$ cycles cannot exceed 150 ps. For some devices, the number of cycles specified in the JESD65B standard establishes a bare minimum. A commonly used number is $10^5$ cycles, but for some devices it can reach as much as $10^{12}$ cycles.*

## Clock Drift

Frequency Drift refers to the linear component of a systematic change in the frequency of an oscillator over time. Frequency drift is typically measured in parts-per-million (PPM). Drift differs from jitter in the sense that jitter is a measure per cycle count while drift refers to the accumulated frequency deviation in time. Aging affects the frequency of many types of oscillators causing a frequency drift that accumulates in time. For example, a 4 MHz oscillator with an age induced drift of 20 PPM per year will deviate its frequency by 80 Hz every year.

## 6.3.2 Choosing a Clock Source

The selection of the clock for a microcontroller application, although bounded by the specifications in the MCU data sheet, must be guided by the application requirements. The most important parameter to choose is the clock frequency. When selecting the clock frequency, the designer should look for the lowest frequency that allows for reliable and correct system operation. The source type, topology, and attributes are then decided to satisfy the rest of the parameters.

Some designers might feel tempted to assign the highest frequency allowed by the MCU, regardless of the system requirements. This could result in a poor design. The designer must always analyze the characteristics of the system at hand and take the appropriate decisions to avoid over-design. An overclocked system, besides running faster than needed will consume more energy, will dissipate more heat, and will be more expensive than necessary. Questions that should be answered in the process of selecting the right frequency and configuring the clock include:

- *What is the fastest event the system will need to handle?* The clock frequency must be set to assure that no event goes missed. The fastest process in your system would define the shortest time interval you'll need to keep-up with. Once you identify the frequency for such a process, all the other would fit. Consider for example the clock requirements for an embedded system handling the keystrokes of a typist on a keyboard. Assuming that no other event in this system would surpass the typist speed, a processor running in the KHz range would do fine. The requirements would not be the same for example for the controller of 100 Mbps embedded switch to handle internet traffic.
- *Has the value of $V_{DD}$ been assigned?* As indicated in the previous section, the value of $V_{DD}$ imposes a limit to the maximum clock frequency. Be aware of what this value is to know how high your processor can run. Sometimes the application speed requirement would dictate rising the $V_{DD}$ level or even switching to a faster processor.
- *What peripherals will share the same clock frequency?* In many cases, the peripherals, not CPU speed, will dictate the clock requirements. In control or signal processing applications, where precise sampling periods are required, the speed of the clock driving the timer to set the sampling period and the speed of the ADC are of outmost importance. Other typical considerations arise when a real a time clock or baud rate generator is to be driven by the system clock. The frequency chosen must be the one producing the least error for counting seconds or for obtaining the desired baud rate in the serial communications port.
- *How precise does the clock need to be?* Real time clocks running for extended periods of time (days, weeks, months, etc.) or fast asynchronous serial channels

(19,200 baud or faster) require clocks with small frequency drift and low jitter. Most systems designed to react at the speed of their human oriented user interfaces will probably have low clock precision requirements. The precision requirement along with the actual frequency dictated by the application become primary factors to decide whether the limited accuracy of an on-chip oscillator would suffice or if an external, more precise alternative would need to be considered.

- *What are the capabilities of the Clock System in my MCU?* Understanding the structure and features of the clock support logic in your MCU will certainly simplify the system clock design. Even in the case of using dedicated, external clock generators, knowing the capabilities of the clock generator would allow the designer to take advantage of embedded features that simplify obtaining the necessary clock signal(s).

Certain applications are easier to implement using specific frequency values. For example, a "sweet" frequency for real-time clock applications is 32.768 KHz. The reason for such a sweetness is that by successively dividing this frequency by two, yields a exactly 1.000 Hz with no decimal fractions. Moreover, being $32,768 = 2^{15}$ implies that the division to produce a 1 Hz signal can be readily made with a single 16-bit timer. Note that for example, using a 4 MHz frequency, would not yield the same result. The closest value you will get is 0.953674 Hz. Such an approximation would cause an error of 4.63 % per second in the timekeeping process, which quickly accumulates drifting the measured time from the actual value. Table 6.3 lists some of the most commonly used frequencies in digital systems.

## 6.3.3 Internal Versus External Clock Sources

When selecting the actual clock source for an MCU-based embedded application, the designer has multiple options to choose from. One of the first questions arising

**Table 6.3** Common frequencies for embedded systems applications

| Freq. (MHz) | Typical application |
| --- | --- |
| 0.032768 | Real-time clocks. Allows binary division to 1.0000 Hz ($2^{15} \times 1$ Hz) |
| 1.843200 | UART clock. ($16 \times 115,200$ baud or $96 \times 16 \times 1,200$ baud) |
| 2.457600 | UART clock. ($64 \times 38,400$ baud or $2,048 \times 1,200$ baud) |
| 3.276800 | Allows binary division to 100 Hz ($32,768 \times 100$ Hz, or $2^{15} \times 100$ Hz) |
| 3.579545 | NTSC M color subcarrier and DTMF generators |
| 3.686400 | UART clock ($2 \times 1.8432$ MHz) |
| 4.096000 | Allows binary division to 1 kHz ($2^{12} \times 1$ kHz) |
| 4.194304 | Real-time clocks, divides to 1 Hz signal ($2^{22} \times 1$ Hz) |
| 6.144000 | UART baud rates up to 38,400. |
| 6.553600 | Allows binary division to 100 Hz ($2^{16} \times 100$ Hz) |
| 7.372800 | UART clock ($4 \times 1.8432$ MHz) |
| 9.216000 | Allows integer division to 1,024 kHz ($2^{10}$) |
| 11.059200 | UART clock ($6 \times 1.8432$ MHz) |

is whether using an internal or external clock source. External clock sources are dedicated oscillators providing a square wave output with specific characteristics of frequency, duty cycle, and stability. These circuits use components such as quartz crystals or ceramic resonators that allow for establishing a base frequency that can be either divided or multiplied to produce virtually any frequency value required by a processor. External sources are more flexible than its internal counterparts in terms of providing a wider range of frequencies to choose from, providing larger driving capabilities, and high levels of accuracy. This flexibility comes at the expense of an increased number of board-level components, increased space, and higher associated cost.

Internal clock sources are common in microcontroller chips. Most contemporary MCUs provide internal oscillators that produce a square wave signal within the chip, requiring no external components at all. This kind of internal sources are convenient for keeping a low count of board-level components, reducing area and cost. However, these sources are frequently based on RC or ring oscillator topologies, that provide less stability and a limited set of frequency values to choose from. To overcome these limitations, the oscillators in MCUs also allow for using external components for completing their clock generators. This feature allows for using more stable references, like quartz crystal or ceramic resonators, or even dedicated external clock generators, at the expense of additional board-level components.

When deciding for an externally assisted configuration, the MCU clock generator might allow for using either RC assisted topologies, crystal-based circuits, or encapsulated external oscillators. RC-based topologies are easy to configure, cheap, and usually require only an external resistor or capacitor connected to the MCU clock inputs. The advantage of this option compared to using the internal MCU clock generator is the added flexibility in setting the clock frequency. However, this option will suffer of the same limitations of any other RC oscillator: limited bandwidth and frequency susceptibility to temperature and voltage changes.

The real advantage of using external components to configure the clock generator comes from the use of quartz crystals. This option allows taking advantage of the superior frequency stability of a crystal-based oscillator, plus the flexibility of a wider choice of frequency values, as the clock frequency is determined by the resonant frequency of the crystal.

### 6.3.4 Quartz Crystal Oscillators

Quartz is a piezoelectric material, meaning that the application of a controlled external voltage to a properly cut and shaped piece of crystal would deform it to a certain amount. If the external excitation were suddenly removed, the mechanical deformation would produce a voltage difference while it returns to its un-deformed shape. The duration and strength of this signal is a function of the crystal shape. This basic behavior is similar to that of a resistor-capacitor-inductor circuit (RCL) under a similar excitation, which implies the crystal has a very specific resonant frequency.

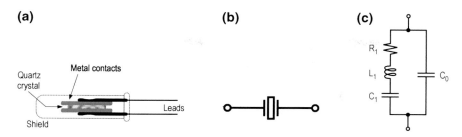

**Fig. 6.11** Quartz crystal structure (**a**), symbol (**b**), and equivalent circuit (**c**)

Figure 6.11 shows the structure, symbol representation, and equivalent circuit of a quartz crystal.

The equivalent circuit denotes the nature of the resonance: a series RCL branch that accounts for the piezoelectric property of the material and a parallel capacitor representing the capacitance of the metallic contacts on the crystal. Thus the circuit operates with two resonant frequencies, a series resonant frequency $f_s$ with $L_1 - C_1$ and a parallel resonance frequency $f_p$ with $L_1$ and the series combination of $C_0$ and $C_1$. The expressions for estimating the two resonant frequencies are given by Eqs. (6.5) and (6.6), respectively. Note that $f_s < f_p$. Also, since $C_0 \gg C_1$, their series combination will be close to $C_1$, making the difference $f_p - f_s$ small.

$$f_s = \frac{1}{2\pi\sqrt{L_1 \cdot C_1}}$$ (6.5)

$$f_p = \frac{1}{2\pi\sqrt{L_1 \frac{C_0 \cdot C_1}{C_0 + C_1}}}$$ (6.6)

Note that quartz crystals or ceramic resonators do not make oscillators by themselves. These devices need an excitation source and some form of positive feedback to make a sustainable oscillator circuit from them, as is also the case of plain RLC components. Microcontrollers that support external quartz crystal connections provide internal circuitry to complete the oscillator. In most practical applications, the quartz crystal is biased with a—parallel—voltage source which excites the equivalent resonant circuit, making the fundamental frequency of the crystal tend to $f_p$. In practice, the crystal's fundamental frequency does not reach $f_p$ due to limitations imposed by stray capacitances in the leads and the interconnection.

The two most commonly used oscillator topologies in MCUs are the Pierce and Colpitts configurations. Some processors are hardwired to use only one of them, while others accept being configured to operate with either of them through the connection of external components. Each one offer advantages and disadvantages that should be considered when an MCU allows for either of them to be used.

**Fig. 6.12** Simplified topology of a Pierce crystal
oscillator

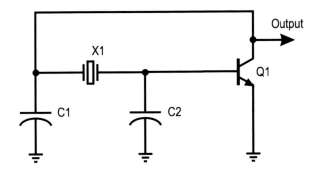

## Pierce Crystal Oscillators

A Pierce crystal oscillator is a series resonant mode circuit that places the crystal as part of the feedback path. Figure 6.12 illustrates the fundamental configuration of a Pierce crystal oscillator, excluding the bias network for clarity.

External capacitors $C_1$, $C_2$ and stray capacitances from the interconnect and the transistor determine the load capacitance seen by the crystal. The combination of these capacitors must closely match the equivalent capacitive load expected by the crystal to be tuned at its fundamental frequency. Moreover, these capacitors also establish the amount of feedback reaching the amplifier. Too little and the circuit might not oscillate, too much and the crystal gets overdriven, causing excess RF emissions, increased power dissipation, and physical wear in the crystal, accelerating aging. For these reasons, the selection of the values for $C_1$ and $C_2$ is critical.

The Pierce configuration is the most used topology in microcontrollers. A common configuration is illustrated in Fig. 6.13. This configuration includes two additional resistors: $R_S$, inserted to reduce the overdrive on the crystal; and $R_B$, which acts as feedback resistor for the inverting gate, stimulating oscillation. Some MCUs include these resistors on-chip, while others require them as external components. The Pierce topology offers the advantages of less sensitivity to noise and stray capacitances, and starts-up faster. It however, tends to consume more power and if the circuit might experiment accelerated aging the overdrive condition is not prevented.

## Colpitts Crystal Oscillator

A Colpitts crystal oscillator is a parallel resonator circuit that removes the crystal from the amplifier's feedback path. Figure 6.14 shows a circuit schematic of basic amplifier and feedback loop for this topology.

In the circuit, the series combination of capacitors $C_1$ *and* $C_2$ in parallel to the transistor input capacitance form the load capacitance seen by the crystal. Thus, the selection of the values of these capacitors needs to be carefully done to match the expected crystal load. This configuration produces a DC voltage over the crystal that contributes to aging. Figure 6.15 shows the external connection of a crystal to an

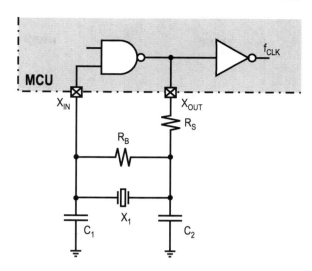

**Fig. 6.13** Crystal-based Pierce configuration for digital MCU clock generation

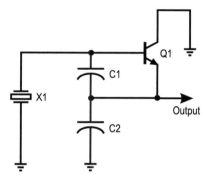

**Fig. 6.14** Simplified topology of a Colpitts crystal oscillator

MCU crystal port supporting a Colpitts configuration. Note the insertion of capacitor $C_{DC}$ with the purpose of blocking this DC component. This configuration is also more susceptible to noise and stray components since they appear across the crystal itself. One of the attractive features of the Colpitts oscillator is that when coupled with a branch to limit the output amplitude, it reduces the RF emissions and lowers the power consumption.

## Crystal Startup Time

One last note regarding crystal oscillators is about startup time. When initially energized, a crystal oscillator circuit will only produce noise. The frequency component of this noise that matches the phase condition for oscillation will be amplified and fed back to the input. As the oscillator operates with positive feedback and gain greater than one, the amplitude will grow until the amplifier gain is reduced to one by an

**Fig. 6.15** Crystal connection
to MCU pins for a Colpitts
configuration

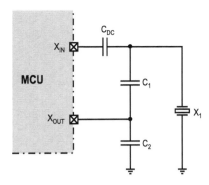

automatic gain control loop in the oscillator (not illustrated in the figures above).
The time elapsed from power-up until the oscillation amplitude reaches its nominal
value is called the *oscillator startup time*.

The startup time in a crystal oscillator can be anywhere from several hundred
milliseconds to several seconds. Factors influencing the actual startup time for a
particular crystal include the frequency value, quality factor $Q$, and load capacitance
seen by the crystal, among others. The lower the frequency of a crystal, the longer
its startup time. High $Q$-factor crystals or those having a large load capacitance will
also have a slow startup. In contrast, RC-based oscillators with their low $Q$-factors
exhibit a quick startup.

Startup times play an important role when the MCU is operated in a low-power
mode that shuts down the clock generator. Upon wake-up, the oscillator needs to be
restarted going through its startup process. As we will see later, there are ways to
reduce the startup time for cases like this. A note of caution here: it is possible to
reduce the startup time by overdriving the crystal. But be aware, this would increase
power consumption in the clock generator and would also accelerate the aging process
of the crystal.

## 6.4 The MSP430 System Clock

The system clock in MSP430 devices provides synchronization signals for the CPU
and embedded peripherals. The system is designed for providing low cost clocking
alternatives and supporting the processor's low-power modes.

Throughout the successive generations of MSP430 controllers launched by Texas
Instruments, the system clock topology has evolved, growing in the number of clocking options and their capabilities. Three fundamental topologies have been introduced
for the six generations of controllers in the MSP430 family. Each topology has been
used through the generations of MCUs as listed below.

- The Frequency Locked Loop (FLL) Clock Module used in MSP430x3xx devices and its "plus" version used in MSP430x4xx generation.
- The Basic Clock Module and its "plus" version used in generations MSP430x1xx and MSP430x2xx, respectively.
- The Unified Clock System (UCS), used in the fifth and sixth generation MSP430x5xx and MSP430x6xx devices.

The next sections provide insight into the evolution of these topologies, highlighting key features and differences among them.

## 6.4.1 The FLL and FLL+ Clock Modules

The first two generations of MSP430 devices, series x3xx and x4xx, both use the same basic clock generation design, the *Frequency Locked-Loop (FLL) Clock Module*. This fundamental design was introduced with x3xx devices, and an improved version included in x4xx devices. The discussion below provides details on these clock modules.

### The FLL Clock Module

The earliest generation of the MSP430 device family, the MSP430x3xx uses a *Frequency-Locked Loop* clock module as system clock generator. This clocking system consists of two oscillators: a Pierce crystal oscillator and a frequency stabilized, RC-based, Digitally Controlled Oscillator (DCO). The DCO is locked to a multiple of the crystal frequency, forming a frequency-locked loop (FLL). This clocking system fundamentally provides two clock signals: the *Main Clock* (MCLK), taken from the DCO output and an *Auxiliary Clock* signal (ACLK), taken out from the crystal oscillator. A buffered, software selectable output XBUF is also available, that provides as output either MCLK, ACLK, ACLK/2, or ACLK/4. The XBUF clock can also be turned off via software. The MCLK signal is used to drive the CPU, while ACLK and XBUF can be software selected to drive peripherals. Figure 6.16 shows a simplified diagram of the FLL Clock Module.

The crystal oscillator in the FLL module is designed to support an external 32.768 KHz quartz crystal with no need for additional external components. Load capacitors are internally provided and set to 12 pF. The oscillator can be turned off via control bit OscOff. Note that without an external crystal, there would be no ACLK signal, nor its derivatives through XBUF. ACLK can also be fed by an external square waveform, connected through pin Xin. In such a case the crystal oscillator would be bypassed and its internal circuitry deactivated.

The DCO is a software programmable, on-chip ring oscillator with RC-like characteristics. The DCO frequency is set via a multiplying factor $K$ obtained as $K = N + 1$, where $N$ is a value set via software. Upon a valid reset (PUC),

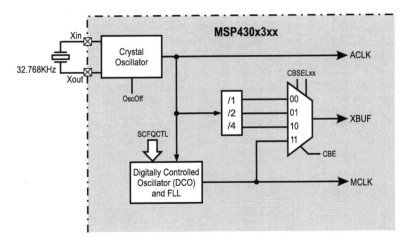

**Fig. 6.16** Simplified block diagram of FLL clock module in MSP430x3xx devices

$N$ is preset to 31 causing the system to begin operation at a frequency 32 times the crystal oscillator frequency. That is, if the external crystal frequency were 32.768 KHz, then, upon a PUC, MCLK would have a frequency of 1.048576 MHz. Under software control, the multiplying factor can be set to any value $1 \leq (N+1) \leq 128$ via the system clock frequency control register (SCFQCTL). The least significant seven bits in SCFQCTL allow setting $N$ to any value between 1 and 127. Note that as with any RC oscillator, the DCO frequency will vary with temperature and voltage. However, the inclusion of a FLL that tracks the crystal oscillator output also allows stabilizing the DCO frequency. The stability achieved brings the frequency variability to that of the external crystal connected to the chip. In addition, by using a digital FLL and DCO, the system clock exhibits a short startup time similar to that of an RC oscillator.

In the absence of the ACLK signal, an internal oscillator fault condition would be created. This fault could be caused by a crystal failure or by just the absence of an external crystal in the chip interface. This condition would trigger a non-maskable interrupt to the CPU if the oscillator-fault interrupt enable (OFIE) bit in the interrupt enable register 1 (IE1) was enabled. This event however, does not prevent the DCO from generating an unlocked frequency on the MCLK output. In fact, OFIE is by default clear upon PUC which allows the processor to operate without an external crystal. Take into consideration that under this condition the FLL would be unable to compensate for the inherent frequency variations of the RC oscillator.

### The FLL+ Clock Module

The second generation of MSP430, the MSP430x4xx, features an improved clock generator designated the *FLL+ Clock Module*. This improved module, in addition

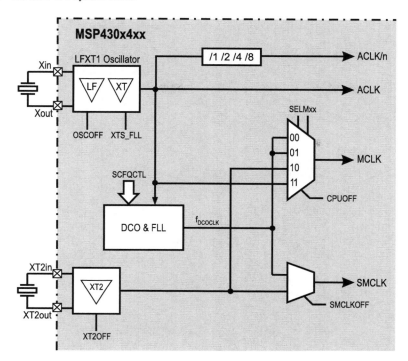

**Fig. 6.17**  Simplified block diagram of FLL+ clock module in MSP430x4xx devices

to LFXT1, DCO, accommodates an optional secondary oscillator, XT2, included in selected family members. The FLL+ module can provide four clock signals to the system: ACLK, ACLK/$n$ ($n = 1, 2, 4$, or 8), MCLK, and SMCLK. Figure 6.17 shows a simplified block diagram of the FLL+ clock module.

Oscillator LFXT1 is an improved version of the Pierce crystal oscillator of x3xx devices. It can now accept crystals of frequency higher than 32.768 KHz by including an additional high-speed feedback amplifier (XT) capable of operating with crystals in the range from 400 KHz to 8 MHz. Control bit XTS_FLL allows for choosing the correct amplifier for the kind of external crystal connected. The value of the internal load capacitors can now be selected via software to match the connected crystals, or optionally, externally provided. The LFXT1 oscillator feeds clock signal ACLK. As before, LFXT1 can be bypassed by an external clock fed via pin Xin. ACLK/n is a submultiple of ACLK software selectable to be divided by a factor of either 1, 2, 4, or 8.

The DCO is essentially the same as in the previous generation. As before, it operates in an FLL tracking a multiple of ACLK, and providing the same advantages of frequency stability and quick startup.

Oscillator XT2 is a new addition in this generation, but is present only in selected family members. It has the same capabilities as the high-frequency portion of LFXT1, except that it does not incorporate on-chip load capacitors, which must be externally provided.

The main clock (MCLK) can now be fed either from ACLK, the DCO, or XT2 (if present), and provides the same failsafe capabilities as in the previous generation. FLL+ allows for the CPU to be operated from the DCO without external crystals. Clock signal SMCLK (sub-main clock) replaces the old XBUF, and can now be fed from either the DCO or XT2 (if present).

## 6.4.2  The Basic Clock Module and Module+

Generations x1xx and x2xx featured a modified clock module with respect to the design included in earlier generations. This new clock generator is designated as *Basic Clock Module*. The basic clock module design is introduced in x1xx devices, and an improved version, the *Basic Clock Module+* featured in x2xx devices.

### The Basic Clock Module

The basic clock module in MSP430x1xx devices is a revised design of the FLL+ used in the earlier generation. This somewhat simpler, and yet improved version of the FLL+ still includes three oscillators (LFXT1, XT2, and DCO). However, the DCO is now redesigned to operate in open loop (no FLL), while offering a better frequency stability without the need of an external crystal. In addition, only three clock signals are made available: ACLK, MCLK, and SMCLK. Figure 6.18 shows a simplified block diagram of the basic clock module.

The structure of oscillator LFXT1 in this generation of the MSP430 retains most characteristics from its predecessor in series x4xx. It supports both low (32.768 KHz) and high frequency (450 KHz to 8 MHz) external crystals, selectable with the XTS control bit. Although the LF amplifier still includes on-chip load capacitors, this time fixed at 12 pF, its high-frequency counterpart, XT, does not. Load capacitors for high-frequency crystals must be externally provided. Bypass capability with an external clock is maintained.

Secondary oscillator XT2, available in pins XT2in/XT2out, maintains the same structure as before: high-frequency only, crystal-based, requires external load capacitors, and is available only on selected devices.

The major change of this generation system clock is in the DCO. The new DCO uses a current-controlled ring oscillator with RC-type characteristics. It is redesigned to sustain a stable programmed frequency output without the need of an external crystal and to provide an even faster startup time. This new DCO has eight frequency taps and does not include an FLL feedback loop. Note in Fig. 6.18 that now the DCO is independent from ACLK. The DCOCLK frequency now depends on three factors:

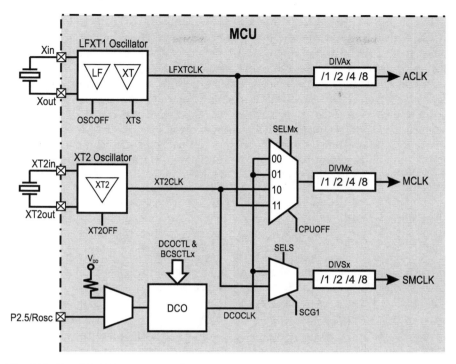

**Fig. 6.18** Simplified block diagram of basic clock module in MSP430x1xx devices

- A resistor-controlled current injection. Internally, the DCO incorporates a DC generator whose current output is controlled by either an internal or externally provided resistor. Control bit DCOR allows choosing either. Using an external resistor provides better tolerance to temperature and supply voltage changes.
- The value assigned to the DCO frequency select bits in the DCOCTL register. These three bits select which of the eight frequency taps in the DCO internal modulator will be used for the output frequency. Each successive tap has a frequency which is approximately 10 % higher than the previous one.
- The adjustment introduced by the modulator specified in the MODx bits of DCOCTL register. Each modulator can automatically adjust the selected frequency tap to maintain the clock frequency within bounds.

The specific combination of values for the modulator and DCO frequency varies for each desired frequency. The MSP430 header file of the specific device used in a particular application contains pre-defined values to accurately obtain a set of pre-calibrated frequencies with the DCO. The designer is referred to these values in the corresponding data sheets for a straightforward use of this resource in the basic clock module. General formulas for a wider range of frequencies are provided in the MSP430x1xxx user's guide.

The outputs from the three oscillators can be distributed through the clock signals ACLK, MCLK, and SMCLK as follows. The auxiliary clock ACLK can be fed exclusively from LFXT1. Remember that ACLK can be used to clock on-chip peripherals such as timers, baud rate generators, and data converters. The main clock (MCLK) has the most flexibility as it can be fed with either of the three oscillators. ACLK, as explained before, is the CPU clock and can be assigned to some peripherals as well. SMCLK, the sub-main clock can have as sources either the DCO or XT2, if present in the particular chip being used. All three clock signals can be divided via software control through respective frequency dividers located at the output of each selector, as illustrated in Fig. 6.18. Failsafe capabilities allow detecting failures in either of the high-frequency oscillators. There is no fault detection for LFXT1 in LF mode.

**The Basic Clock Module+**

The fourth generation of the MSP430, the MSP430x2xx, expands the basic clock module design by introducing several new features that include a fourth oscillator, and increasing the maximum clock frequency for the system. This new clock module, designated the *Basic Clock Module+*, inherits most features of the basic clock module. Figure 6.19 shows a simplified diagram of the basic clock module+.

The improvements can be summarized as follows:

- Addition of a new clock source, the very-low-power, low-frequency oscillator (VLO). With a typical frequency of approximately 12 KHz, the VLO is an internal RC type clock source that can replace the LFXTCLK in ACLK. Control bits LFXT1Sx allow for selecting between VLOCLK or LFXTCLK.
- Inclusion of minimum pulse filters (MPF) at the outputs of all clock generators to prevent spurious pulses propagating through the clock system. Observe the LPF blocks at the outputs of DCO, LFXT1, and XT2 oscillators.
- The internal resistor to set the DCO frequency is replaced with a current source, for added precision in the internal setting of the DCO DC signal.
- The high-frequency oscillators and DCO are now capable of operating up to 16 MHz. This is twice the maximum frequency achievable in the original basic clock module.
- Addition of a fault detection indicator for failure in the LFXT1 oscillator.
- Oscillator LTXF1 now has internal programmable load capacitors on-chip.

Not all these features are included in every x2xx device. All other remaining characteristics from the original basic clock module remain the same in these new devices.

## 6.4.3 The Unified Clock System

The last MSP430 generation reviewed, the MSP430x5xx/x6xx, uses the most comprehensive system clock designed so far for these controllers. The *Unified Clock*

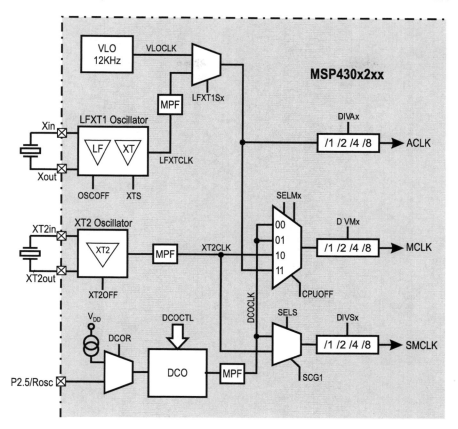

**Fig. 6.19** Simplified block diagram of basic clock module+ in MSP430x2xx devices

*System* (UCS) features functional attributes developed in previous generations now consolidated in one clock system. It incorporates five general oscillators, XT1, VLO, REFO, DCO, and XT2 that can be equally used for clocking the CPU or embedded peripherals through three main clock signals ACLK, MCLK, and SMCLK. A sixth dedicated oscillator, MODOSC and its output MODCLK, are reserved for clocking flash memory and certain peripherals. Any of the five general oscillators can be assigned to any of the clock outputs, as denoted in the simplified block diagram of the unified clock system shown in Fig. 6.20.

As in previous generations, oscillator XT1 maintains its capability of handling either low-frequency (LF) or optionally, high-frequency (HF) crystals through pins Xin, Xout. Control bit XTS for selecting the type of crystal connected. LF remains intended for 32.768 KHz crystals, while HF is now redesigned to accept crystals or clock sources in a range from 4 to 32 MHz. On-chip software selectable load capacitors make unnecessary to use additional external components to run from LF.

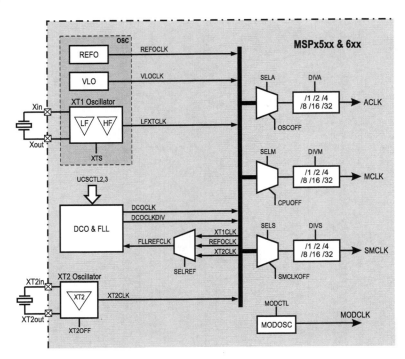

**Fig. 6.20** Simplified block diagram of the unified clock system in MSP430x5xx/x6xx devices

HF requires external capacitors. Xin maintains its capability to be bypassed with an external clock signal.

As in previous generations, optional oscillator XT2 runs with high-frequency crystals. Its frequency range is improved to match that of XT1-HF.

The very-low-power, low-frequency oscillator (VLO) introduced in the BCM+ is retained in this generation, now with a nominal frequency of 10 KHz. This RC-type oscillator is optimized for low power, with typical current requirements as low as 10 nA.

A new oscillator appearing in the UCS is the REFO, an internal trimmer low-frequency reference oscillator. REFO is trimmed to 32.768 KHz and intended to be used as a crystal free reference frequency for the DCO, although it can be used as any of the other oscillators.

The DCO in the UCS returns to its roots, with the ability to operate in a frequency-locked loop (FLL). In this revised version, the DCO frequency can be adjusted by software or stabilized by the FLL to a multiple of FLLREFCLK/$n$, for $n = 1, 2, 4, 8, 16$, or 32. The FLL reference clock, FLLREFCLK, can be software configured to be either XT1CLK, REFOCLK, or XT2CLK. The DCO frequencies DCOCLKDIV and DCO-CLK are configured through the UCS control registers UCSCTL2 and UCSCTL3, assigning factors $D$, $N + 1$ and $n$ to conform Eqs. (6.7) and (6.8).

**Table 6.4** Summary of clock configurations for MSP430 generations x1xx through x6xx

| MSP430 | Name | Oscillators | Clock signals | Max freq. (MHz) | DCO type |
|---|---|---|---|---|---|
| x3xx | FLL | XTAL, DCO | MCLK, ACLK,XBUF | 0.032 | FLL |
| x4xx | FLL+ | LFXT1, DCO, XT2 | MCLK, SMCLK, ACLK, ACLK/n | 8.000 | FLL |
| x1xx | BCM | LFXT1, DCO, XT2 | MCLK, ACLK, SMCLK | 8.000 | DCO |
| x2xx | BCM+ | LFXT1, DCO, XT2, VLO | MCLK, ACLK, SMCLK | 16.000 | DCO |
| x5xx x6xx | UCS | LFXT1, DCO, XT2, REFO, VLO, MODOSC | MCLK, ACLK SMCLK | 32.000 | FLL/DCO |

$$f_{DCOCLKDIV} = (N + 1) \times \left( \frac{f_{FLLREFCLK}}{n} \right) \qquad (6.7)$$

$$f_{DCOCLK} = D \times f_{DCOCLKDIV} \qquad (6.8)$$

When the FLL is disabled, achieved by deactivating either the internal DC generator or frequency integrator, the DCO continues to operate as it did in the basic clock system of series x1xx/x2xx. This implies that $f_{DCOCLK}$ is defined by the DCO frequency range selection and divider (DCORSEL) and the DCO modulator pattern bits (MOD). When enabled, the FLL is controlled by the DCO frequency tap and MOD specified in UCSCTL0. The specific frequency values that can be obtained depend on the particular device being programmed, as specified in the corresponding data sheets.

The clock distribution system allows assigning any of the six oscillator outputs REFOCLK, VLOCLK, LFXTCLK,XT2CLK, DCOCLK, or DCOCLKDIV to any of the clock signals ACLK, MCLK, or SMCLK. The clock signals are served divided by a factor $n = 1, 2, 4, 6, 8, 16, 32$, as specified in the corresponding DIVx control bits.

### 6.4.4 Using the MSP430 Clock

The clock options in the different generations of MSP430 can be summarized as illustrated in Table 6.4. There we can see that the low-frequency crystal oscillator and the DCO are a common denominator in all MSP430 generations. We can also see that the architecture has converged to providing MCLK, SMCLK, and ACLK as the clock signals for all the synchronization needs in the MCU.

Despite the number of different clocking alternatives provided by the MSP430, probably more than 90 % of the design needs can be satisfied with simple topologies and setup as illustrated in the examples below.

**Fig. 6.21** Hardware setup for internal clock generation in an MSP430F2274

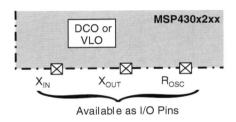

**Example 6.4** *Provide the hardware setup and initialization sequence to have the basic clock module in an MSP430F2274 operating on the internal DCO at 1 MHz with internal current source. Route ACLK from the VLO and turn off unused internal oscillators.*

*Solution: This configuration is one of the simplest that can make the MSP430 quickly usable. The hardware setup requires no external components, leaving all oscillator related pins (Xin, Xout, and Rosc) available for use as I/Os. Figure 6.21 shows the hardware setup for this configuration.*

*Note that although this example calls for using the DCO, the above hardware setup would also be valid if we were using the VLO as the internal clock.*

*To internally configure the oscillator, we first note that the default configuration upon power-up (PUC) has clocks MCLK and SMCLK fed by the DCO and running with its internal current source at a frequency of approximately 1.15 MHz. This is because upon PUC, the DCO is configured to use its internal current source (DCOR = 0) with DCOx = 3, MODx = 0, RSELx = 7. The auxiliary clock ACLK, is sourced from LFXT1, but in this case it would be inactive since there is no external crystal connected. Therefore, to avoid energy waste it would be necessary to deactivate LFXT1. In summary, all that is needed to satisfy the requirement is configuring ACLK to be fed from VLO and turn off LFXT1. The only action required for achieving both objectives would be setting LFXT1Sx = 10b. This would not only route ACLK from the VLO but will also disable LFXT1. This can be achieved by executing the instruction*

```
MOV.B #LFXT1S0,&BCSCTL3 ;Set ACLK feed from VLO and deactivate LFXT1
```

**Example 6.5** *Provide a hardware setup and initialization sequence to have the BCM+ in the MSP430F2274 operating from the DCO at 2.0 MHz. Use an external resistor for a temperature tolerant DCOCLK frequency. Assume no peripheral will use ACLK.*

*Solution: The hardware setup illustrated in Fig. 6.22 shows a connection to use an external resistor to set the DCO reference current. Such a setup reduces the temperature sensitivity of DCOCLK.*

*By default, after a PUC DCOCLK sources MCLK, LFXT1 is on, and DCOR is set to the internal current source. A PUC also makes DCOx = 3, MODx = 0, and RSELx = 7. These settings make $f_{DCO} \cong 1.15\,MHz$. To get $f_{DCO} = 2\,MHz$ with an external resistor we need to change the default configuration. Since LFXT1 is not going to be used, we should turn it off to save power. Consulting the MSP430F2274 data sheet (SLAS504F), the DCO characteristic with external resistor shows that an appropriate setting to get $f_{DCO} \cong 2.00\,MHz$, with $V_{DD} = 3.3\,V$ would be making*

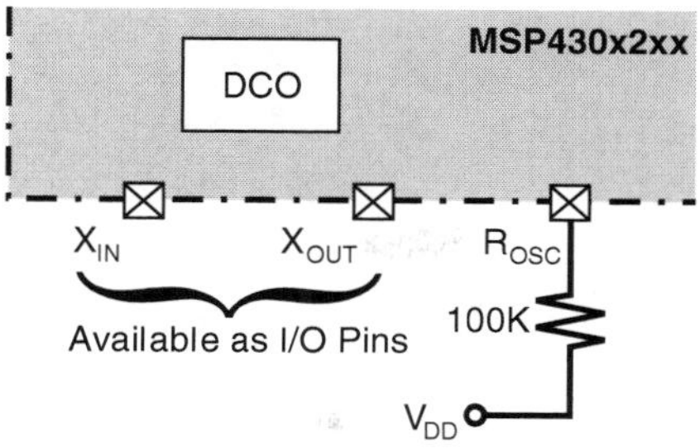

**Fig. 6.22** Hardware setup for resistor sourced DCO in an MSP430F2274

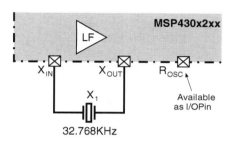

**Fig. 6.23** Hardware setup for clocking an MSP430F2274 with a low-frequency crystal

$R = 100\,K$, $RSELx = 4$, and leaving the default values on DCOx and MODx. The following code fragment accomplishes the objective:

```
; Assumes Vdd = 3.3 V and reset values in MODx and DCOx
BIS.B #DCOR,&BCSCTL2        ; Rosc set to external
BIC.B #RSEL1+RSEL0,&BCSCTL1 ; RSEL = 4
BIS.W #OSCOFF,SR            ; XTAL not used
```

**Example 6.6** *Use an external 32.768 KHz crystal to run the MSP430F2274 ACLK at the crystal frequency and the CPU at 8 MHz. Set SMCLK at 2 MHz.*

*Solution:* In this case, we can connect an external 32.768 KHz crystal through XTAL1 in LF mode, set the DCO to feed MCLK at 8 MHz, and run SMCLK from the DCO, dividing the output via DIVSx by a factor of 4. The hardware setup is quite simple, as illustrated in Fig. 6.23. Since no external resistor is being used, Rosc pin remains free to be used as I/O.

Upon power-up, XTAL1 in LF mode becomes the source of ACLK, so we can leave it with its default configuration. The DCO frequency desired in this case can be obtained by configuring DCOx, MODx, and RSELx. All we need is to step-up the DCO frequency from its default value of 1.15–8 MHz. This could be achieved with DCOx and MODx in their default values of three and zero, respectively, and changing RSELx to 13. With a 3.3 V supply voltage it would provide approximately 8.1 MHz. However, $f_{DCO} = 8$ MHz is one of the calibrated values for this chip, so we will instead resort to this setup. Calibrated values configure DCOx, MODx, and RSELx bits, clearing all other bits in DCOCTL and BCSCTL1. Note that this action will leave XT2 on, which for this example needs to be explicitly turned off. We'd

*also need to have SMCLK sourced from DCOCLK and divided by a factor of 4 to*
*get 2 MHz. After PUC, SELS = 0, which sets SMCLK = DCOCLK, so no change*
*needed there. DIVSx is by default 00b and needs to be changed to 10b. The code*
*fragment below would achieve the desired effect.*

```
; Assumes Vdd = 3.3 V and reset values in MODx, DCOx, and BCSCTL2
MOV.B &CALBC1_8 MHZ,&BCSCTL1 ; Set range for calibrated 8 MHz
MOV.B &CALDCO_8 MHZ,&DCOCTL  ; Set DCO step + modulation for 8 MHz
BIS.B #XT2OFF,&BCSCT1        ; XTAL2 not used
BIS.B #DIVS_2,&BCSCTL2       ; SMCLK divided /4 = 2 MHz
```

## 6.5 Power-On Reset

The sequential nature of the CPU's control unit, besides a clock, also requires a
RESET signal for taking its finite state machine (FSM) to its initial state upon
power-up. All sequential circuits have such a requirement. The reset signal in a
microprocessor-based system causes several actions in the CPU and its surrounding
logic. The specific list of actions changes from one processor to another, but there is
a list of tasks that are common to most architectures. These include:

- Loading the program counter register (PC) with the address of the first instruction to
  be executed. This causes execution of the first instruction in the booting sequence.
- Disabling the reception of maskable interrupts by the CPU.
- Clearing the status register (SR). The specific value loaded into the SR changes
  from one processor or MCU to another.
- Initializing some or all system peripherals (list changes for specific devices). For
  example many MCUs set all their I/O pins to input mode, timers are initialized to
  zero, and the default CPU operating mode is selected.
- Canceling any bus transaction in progress and returning control to the default bus
  master (details of bus Arbitration in Chap. 7).

The RESET signal in an embedded system is generated through a specialized
circuit called a *power-on reset circuit* or POR for short.

### 6.5.1 Power-On Reset Specifications

The specifications of a reset signal change from one chip manufacturer to another
and even for the same manufacturer, among device families. Despite this, there are a
few requirements than can be generalized to virtually any processor. These include
the assertion level of the reset pulse, and the minimum pulse width of the signal
to be accepted as a valid reset. Figure 6.24 illustrates a typical reset timing for a
generic CPU.

When power is applied to a system, the supply voltage ramps-up until reaching
its nominal value $V_{DD}$. During this period, the oscillator providing the clock signal

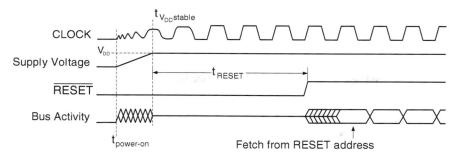

**Fig. 6.24** Typical CPU RESET timing

begins its startup time. The reset signal, in this example assumed to be asserted-low, must remain active at least for $t_{RESET}$ time after $V_{DD}$ has stabilized for the CPU recognizing it as a valid reset. Bus activity might show some noise, but usually the busses are held inactive until the reset signal is de-asserted. Upon RESET de-assertion, the CPU begins fetching instructions from program memory, starting at the reset address. The function of the POR circuit is to provide such a timing to the reset signal.

## 6.5.2 Practical POR Circuits

There are different alternatives for supplying a POR signal to a microprocessor or microcontroller. Many microcontrollers have their internal reset source, requiring minimal or no external components. Sometimes a custom-built circuit using discrete components, able of providing the minimal reset requirements would suffice. In other cases, the application requirements would call for dedicated reset supervisor ICs. We discuss these cases below.

### Embedded POR Circuits

Contemporary microcontrollers frequently include some form of on-chip POR circuit, reducing the count of external components needed to implement an application. When they do, they also provide the option of feeding an externally generated reset signal to the CPU. The MSP430 includes an on-chip POR circuit that requires no external components for operation. Section 6.7 provides a detailed discussion of the MSP430 reset structure.

**Fig. 6.25** A POR circuit
based on a one-shot topology

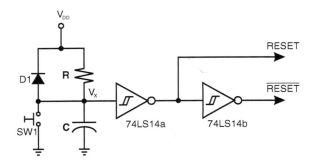

## One-Shot-Based POR Circuit

Most solutions for POR circuits are based on the structure of a monostable multi-vibrator, also known as a one-shot circuit. This circuit operates between two states: one stable and one unstable. The unstable state is entered when a perturbation to the stable state is triggered. The duration of the unstable state depends on the time constant of some energy storing component in the circuit. In most cases this is implemented with a simple RC circuit. When using a one-shot as a POR circuit, its time in the unstable state is used to produce the desired reset timing. Figure 6.25 shows a practical implementation of a power-on reset circuit based on a one-shot topology. Resistor R and capacitor C determine the circuit time constant that defines the reset pulse width.

Before power is applied to the circuit we assume capacitor C is discharged. At power-up, C begins to charge through R, with time constant RC. When Vx reaches the positive threshold of the Schmidt triggered inverter ($V_T^+$), this causes an abrupt change in the inverter's output. The second inverter provides the correct level of assertion for an active low reset signal. Figure 6.26 shows the waveforms in the critical circuit nodes. The width of the reset pulse is determined by the time constant RC, the supply level $V_{DD}$, and the positive threshold voltage of the inverter $V_T^+$, as given by (6.9).

$$t_{reset} = RC \cdot \ln\left(\frac{V_{DD}}{V_{DD} - V_T^+}\right) \tag{6.9}$$

Note that for a given requirement of $t_{reset}$, Eq. (6.9) is undetermined. What is commonly done to resolve the equation is to assign a suitable value of C and calculating the value of R. In most cases a capacitor around $1\,\mu F$ should work, as pulse widths for reset are at most in the range of milliseconds.

Figure 6.25 also features a couple of additional devices: D1 and SW1, which provide additional functionality to the basic POR circuit.

When the supply voltage drops below its nominal value, it might reach a level at which correct device functionality might not longer be guaranteed. Such a condition is called a *Brownout* event. Diode D1 in Fig. 6.25 acts as an elemental brownout detector in the POR circuit. It detects voltage drops in the supply line, triggering

**Fig. 6.26** Waveforms in the
critical nodes of the one-shot
based POR circuit

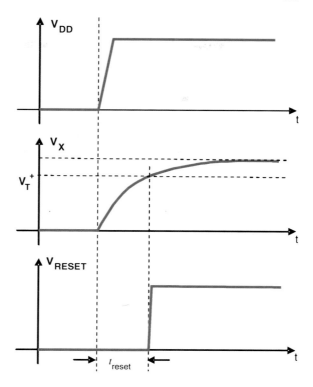

a POR if a threshold is reached. Its operation can be explained as follows: After a sufficiently long time of operation, Vx shall have reached $V_{DD}$, which would reverse bias D1. If $V_{DD}$ falls at least one diode drop below its nominal value, D1 would become forward biased, providing a low-impedance discharge path for C1. This would trigger the circuit into its unstable state, producing a valid reset pulse.

Sometimes it might result convenient to manually trigger a system reset without having to remove power from the system. For such situations, switch SW1 provides for manually discharging C1. Note that when SW1 is closed it creates a short-circuit to the capacitor terminals, making it possible to manually reset the system without the need of removing or dropping the supply voltage.

**Example 6.7** *A microcontroller with no internal POR logic requires an asserted-low reset pulse of at least 0.1 ms, applied to pin $\overline{RST}$ to accept a valid POR signal. Provide a suitable circuit to satisfy the chip requirements. Assume the MCU operates from a 3.3VDC power supply.*

*Solution: In this case, a solution could be provided with the circuit in Fig. 6.25. Here, we essentially need to determine the values of R and C to satisfy this require-ment. In this case we will use a 74HC14 inverter, which can be readily operated at 3.3 V and features $V_T^+ = 1.82\,V$. From Eq. (6.9), we readily obtain*

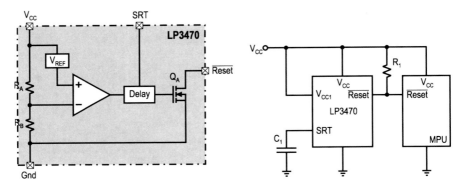

**Fig. 6.27**  Structure and typical application of LP3470 (*Courtesy of Texas Instruments, Inc.*)

$$t_{reset} = RC \cdot \ln \left( \frac{3.3\,V}{3.3\,V - 1.82\,V} \right)$$

$$t_{reset} = 0.80RC$$

*Using the specification of $t_{reset} \geq 100\,\mu s$ and assigning $C = 0.1\,\mu F$ (10% tolerance) yields $R = 1.24\,K\Omega$. In this case, we can choose a standard resistor of $1.5\,K\Omega$, 5%, which would produce an actual reset pulse width of $0.120\,ms$. This is a comfortable value that covers the component tolerance and still ensures a pulse wider than the minimum required.*

### Reset Supervisory Circuits

General purpose processors rarely provide on-chip POR circuitry. This function is typically served by an external, dedicated POR generator circuit. In most microprocessors the POR signal is served by a dedicated chip that also serves as power monitor and provides other supervisory functions that change from one processor to another.

An example of such a circuit is National Semiconductor's LP3470. This power reset supervisor has the capability of providing an externally programmable reset timeout period using an external capacitor. It also monitors the $V_{DD}$ line for undervoltages, triggering the reset signal if the supply level falls below a established threshold value. Figure 6.27 shows a functional block diagram and typical application of this circuit.

For additional details on how to configure the length of the reset pulse and the supply voltage threshold to trigger a POR, the reader is referred to the chip data sheets [40].

# 6.6 Bootstrap Sequence

The bootstrap sequence, frequently shorthanded as the *Boot Sequence* refers to the sequence of instructions that a CPU executes upon reset. What does it have to deal with boots and straps? The term bootstrap seems to be a metaphor of the phrase "pull oneself up by one's bootstraps" denoting those with the ability of helping themselves without external means. In a sense, that is what the bootstrap sequence does for a computing system: starting it when no other software is yet running. No basic microprocessor or MCU interface is complete without a properly designed boot sequence.

The boot sequence, when compared to the MCU interface elements described earlier in this chapter, can be perceived as an abstract component. After all, it refers to a piece of software. This software, however, requires a physical memory where it can reside and be accessed by the CPU. Thus, in this discussion the ability to provide a booting sequence presumes the existence in the system of a non-volatile program memory where such a code is stored and accessible to the CPU. Note however that we do not call the program memory an element of the basic interface. Just having memory in place does not imply that there is a valid boot sequence in the system. At this point, and for the rest of our discussion, we will assume that such a memory exists in the system, mapped in the right place within the processor's memory map. Section 3.5 provides a discussion on memory.

Specific tasks performed in an MCU boot sequence are influenced by multiple factors that include: specific processor used, system architecture, application, and even the programming style. However, it is possible to identify typical operations taking place in this sequence. These include:

- Identifying and initializing system peripherals. This includes configuring I/O ports, initializing timers, setting-up communication interfaces, etc.
- Setting-up the stack. In order to be usable, a stack needs to be initialized by setting the stack pointer register to point to the top of the memory area designated for stack operations.
- Initializing system-wide variables in memory. Many tasks in embedded software require global variables initialized to specific values for proper software operation. This is a task frequently handled in the boot sequence.
- Performing diagnostics and system integrity check-up. All systems at startup need to verify the integrity of its components to provide some degree of reliability. The boot sequence is the preferred place to run self-tests, particularly on system critical components.
- Loading an operating system or other type program. The boot sequence in these cases is also called the *boot loader*. In some systems this would load the kernel and other components of an operating system or the system run-time environment. In MCUs, for example, with the aid of a JTAG interface, user programs get uploaded into the MCU non-volatile memory.

## 6.6.1  Boot Sequence Operation

In a computer system the boot sequence begins to be executed when a valid reset signal is accepted by the CPU. Many processors also allow for critical system events other than a power-up detection to trigger the reset sequence as well. Examples include low supply voltage conditions, bus error faults, system integrity violations, and memory access faults, among others.

Another mechanism frequently provided in many systems is the ability to invoke software calls to the reset sequence from within programs. Regardless of the specific trigger, the execution of the boot sequence causes the CPU's control unit to load into the program counter (PC) the address where the first instruction resides. This address is called the reset address. Note that the advantage of starting from a known location in program memory is the main characteristic that makes the boot sequence so important for a computer system. It makes possible not only to start the system upon power-up, but also to recover the status of the system upon a runaway situation.

## 6.6.2  Reset Address Identification

There are two major modalities in which the reset address can be provided in MCUs: as part of a jump table or as a fixed vector also called *auto-vector*.

Systems using a jump table have a hardwired memory location where the CPU begins execution upon reset. A microcontroller using this type of scheme, upon a valid reset, loads the program counter (PC) with this physical address, making possible to begin the fetch-decode-execution cycle right from the reset location. This type of scheme would also apply to all system interrupts in the same way: a fixed address where execution begins when a qualified events triggers them.

The initial addresses for the reset code as well as for other system interrupts are located in a particular place in memory, each separated from the other by only a few memory words. Typical separations range from four to six bytes away depending on the device. This amount of memory is, in the best of the cases, enough for storing only one or two instructions, insufficient to accommodate a long list of tasks such as those we have identified in a typical reset sequence.

A way to handle this limitation is by placing at the reset address just an unconditional jump instruction that takes program control to the rest of the code. A similar arrangement is made for all other interrupts served by the system. When the entries for reset and the other events are provided, the memory locations for these events become populated with only unconditional jump instructions to the corresponding service routines. For this reason this memory area is called a jump table. MCUs like the classical MCS8051 from Intel and many of its derivatives use a jump table to serve reset and other interrupts. The pseudocode sequence below illustrates how a jump table-based reset address can be configured.

```
; ======================================================================
; Addresses RESET, EVENT1_ISR, etc. are declared in header file
; ----------------------------------------------------------------------
#include "headers.h"
; ----------------------------------------------------------------------
StartUp:    Instruction 1       ; Startup code begins here
            Instruction 2
            ...
Main:       Instruction a       ; Main program (infinite loop)
            Instruction b
            ...
            do_forever
; ----------------------------------------------------------------------
EVENT1_ISR: ---                 ; Service routine for Event 1
            ...
            IRET
; ----------------------------------------------------------------------
; Jump table begins below
; ----------------------------------------------------------------------
            ORG RESET_ADDRESS   ; Aligns linker to reset address
            JMP StartUp         ; Unconditional jump to reset code
            ORG INT_EVENT1      ; Aligns linker to event1 address
            JMP EVENT1_ISR      ; Unconditional jump to Event 1 ISR
            ---                 ; Other jump table entries
            ...
            END
; ======================================================================
```

Fixed vector systems have a hardwired memory location containing an address pointer to the location where the CPU is to begin execution upon reset. This entry is also called an *auto-vectored* entry. Microcontrollers using an auto-vectored scheme, upon a valid reset, load the PC with the contents of the fixed address location, defining in this way where the instruction fetch-decode-execute cycle must begin. Systems handling a power-up reset through this scheme, also handle interrupt servicing in the same manner. When all vectors are defined, the memory area where they are stored becomes a *fixed vector table*. This is a convenient mechanism for locating the reset code since no jump instruction needs to be executed.

MSP430 devices use an auto-vectored mechanism to support reset and interrupt events. The demo program below by A. Dannenberg illustrates a code organization to support a reset event in an MSP430 device [41].[1] The program toggles I/O port P1.0, producing a square wave signal that could be used to toggle on and off an external device like an LED or a buzzer. The waveform period is controlled with the value loaded into R15. Observe the declaration of the reset auto-vectored entry and the label identifying the startup code.

```
; ======================================================================
;MSP430F22x4 Toggle P1.0 Demo by A. Dannenberg. - 04/2006
;Copyright (c) Texas Instruments, Inc.
; ----------------------------------------------------------------------
```

---

[1] See Appendix E.1 for terms of use.

```
#include "msp430x22x4.h"                    ; Header file
;-------------------------------------------------------------------
            RSEG    CSTACK                  ; Stack declaration
            RSEG    CODE                    ; Code into flash memory
;-------------------------------------------------------------------
STARTUP     mov.w   #SFE(CSTACK),SP         ; Initialize stack pointer
            mov.w   #WDTPW+WDTHOLD,&WDTCTL  ; Stop watchdog timer
            bis.b   #001h,&P1DIR           ; Configure P1.0 as output

Main        xor.b   #001h,&P1OUT           ; Toggle P1.0
            mov.w   #050000,R15            ; Use R15 as delay counter
Repeat      dec.w   R15                    ; Decrement count
            jnz     Repeat                 ; While not zero repeat
            jmp     Main                   ; Do it again
;-------------------------------------------------------------------
            COMMON  INTVEC                 ; Auto-vector table
;-------------------------------------------------------------------
            ORG     RESET_VECTOR           ; Power-on reset auto-vector
            DW      STARTUP
            END
;===================================================================
```

The invocation to the reset sequence is very similar to that of processing an interrupt, except that there is no return address involved: the PC is loaded with the reset address and the status register is cleared.

Microprocessor-based systems in their vast majority handle interrupt events through a vectored system. In such a system, all interrupt events are issued through a single interrupt request pin. An interrupt acknowledgment signal from the interrupting device issues a pointer to an entry in a memory held vector table whose entries are the addresses of the service routines for the supported events. Even in these system it is common to find that a reset event is auto-vectored, requiring no interrupt acknowledgment signal. On example of such an arrangement is found in Intel 80x86 and Pentium devices. A more detailed discussion of vectored methods in microprocessor system is included in Sect. 7.1.

## 6.7  Reset Handling in MSP430 Devices

The MSP430 provides a flexible on-chip system initialization hardware in all its devices. This means that the required power-on-reset signal necessary to initialize the system can be internally generated without the need of external components. The internal reset circuitry also supports an externally triggered reset signal ($\overline{RST}$) through external pin $\overline{RST}/NMI$. As implied by its name, this pin is shared with the non-maskable interrupt (NMI) input. The default function of the $\overline{RST}/NMI$ pin upon power-up is for accepting an asserted-low reset signal. Although it could nominally be tied to $V_{DD}$ if not being used for either of its functions, adding an external RC (R $= 47$ K, C $= 10$ nF) is generally recommended to reduce the chance of false triggering due to power supply noise.

The functionality of $\overline{RST}/NMI$ can be changed via software from its default setting to become an NMI input. All generations of the MSP430, except x5xx/x6xx devices, contain in the Watchdog Timer Control Register (WDTCTL) the bits for configuring this pin. In series x5xx/x6xx the functionality is moved to a Special Function Register (SFR).

The reset circuitry in all MSP430 devices has a fundamental structure that provides for two internal reset signals: Power-on Reset (POR) and Power-up Clear (PUC). Figure 6.28 shows a simplified block diagram of the base reset structure.

POR is the MCU reset signal. This signal is generated when the device detects a power-up or brownout event in the main voltage supply line ($V_{CC}$) or when the external $\overline{RST}/NMI$ line, in its reset mode, is held low for at least $t_{reset}$ time. The minimum length of $t_{reset}$ is typically 2 μs for most family members, although the specification is left as a device dependent parameter. Another system event capable of triggering a POR in MSP430 devices equipped with a Supply Voltage Supervisor (SVS) is a low-voltage detection in the analog supply voltage line ($AV_{CC}$) or in a selected external voltage. An SVS circuit is a handy addition to enable monitoring subtle voltage variations in the supply line that could escape the brownout detection circuit and compromise the correct device functionality.

The PUC signal is like a soft reset for the device in the sense that it can be configured to be triggered by additional system events than those triggering a POR. A PUC is always triggered when a POR is triggered, but not viceversa. Conditions triggering a PUC include the expiration of the watchdog timer, a password violation when modifying the WDT control register or FLASH memory, and an invalid access caused by fetching instructions outside the valid range of program memory addresses (bus error).

### 6.7.1 Brownout Reset (BOR)

Every MSP430 device features a brownout reset circuit that triggers a POR signal when power is applied or removed from the supply voltage pin or when voltage fluctuations capable of compromising the integrity of internal registers or memory

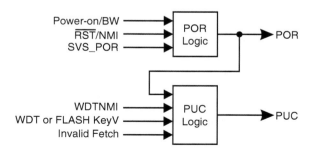

**Fig. 6.28** Basic power-on reset module in MSP430 MCUs

occur. The conditions for triggering a BOR event require the supply level $V_{CC}$ to cross the $V_{CC(start)}$ level. The BOR event remains active until $V_{CC}$ crosses the $V_{(B\_IT+)}$ threshold and lasts for at least $t_{(BOR)}$. The length $t_{(BOR)}$ is adaptively determined by the slope of the supply voltage fluctuation, being longer for a slow ramping $V_{CC}$ changes. Levels $V_{(B\_IT+)}$ and $V_{(B\_IT+)}$ define dual hysteresis thresholds that prevent successive BOR events unless the supply voltage drops below $V_{(B\_IT)}$. It deserves to mention that the level of $V_{(B_{IT})}$ is well above the minimum voltage level necessary to activate the POR circuit. Thus, it reacts earlier than the POR circuit upon power failures. Figure 6.29 shows the voltage levels leading to the triggering of a BOR event.

### 6.7.2  Initial Conditions

After a POR, the initial conditions in an MSP430 device are the following:

- The functionality of the $\overline{RST}/NMI$ pin is set to $\overline{RST}$.
- GPIO pins in all ports are configured as inputs.
- The processor status register (SR) is loaded with the reset value, which clears the V, N, Z, and C flags, disables all maskable interrupts (GIE = 0), and the CPU is set to active mode, cancelling any low-power mode previously set.
- The watchdog timer is activated in watchdog mode.

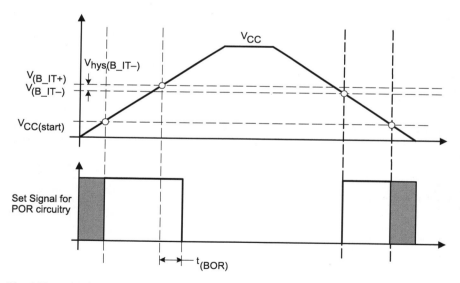

**Fig. 6.29**  Voltage levels and timing of an MSP430 BOR event (*Courtesy of Texas Instruments, Inc.*)

- The program counter (PC) is loaded with the address pointed by the reset vector (0FFFEh). Setting the reset vector contents to 0FFFFh, disables the device, entering a low power mode.
- Particular peripheral modules and registers are also affected, depending on the specific device being used.

Considering the particular conditions induced by the POR sequence, there are specific actions that must be performed by any user software running in the MSP430 upon reset. These include:

- Initializing the stack pointer (SP) to the top of the designated area in RAM.
- Deactivating or reconfiguring the watchdog timer to the mode required by the application.
- Configuring the peripheral modules necessary to support the particular application.
- Configuring desired low-power modes.
- Optionally, to cope with potential threatening situation, examining particular flags in the watchdog timer, oscillator fault, and/or flash memory status registers to ascertain the cause of the reset.

## 6.7.3 Reset Circuitry Evolution in MSP430 Devices

The fundamental MSP430 reset structure has exhibited several improvements as its design has evolved through the different device generations from series x3xx through x6xx.

Early MSP430x3xx had a basic system reset circuit with power-up/brownout detection, external RST/NMI support, and internal detection of watchdog timer expiration and key violations. POR could be triggered by a power-up/brownout or by the reset pin $\overline{RST}$ being driven low. PUC could be triggered by a POR or by a watchdog timer expiration or a WDT key violation.

As a measure to prevent accidental modifications to the on-chip flash memory, the MSP430 requires that any read or write access to the on-chip flash controller be word-size, containing in the upper byte the value "0A5h" as keyword. Failure to do so triggers a security key violation condition, denoted by the KEYV flag. This event triggers a PUC system reset. Accesses to the WDT control register are similarly protected. The failure to include the 0A5h keyword will cause a WDT key violation. Additional details on the operation of both the WDT and the on-chip flash memory are discussed in Chap. 7.

Series x4xx shared the same reset functionality as its predecessor, with the added functionality of including FLASH memory key violations among the sources for triggering a PUC.

The reset structure of MSP430 series x1xx and x2xx devices inherit the characteristics of x4xx devices plus a few additions. Series x1xx introduces the ability of configuring the SVS to trigger a POR. This improves device reliability in terms of a better ability to detect a wider range of potentially harmful power conditions.

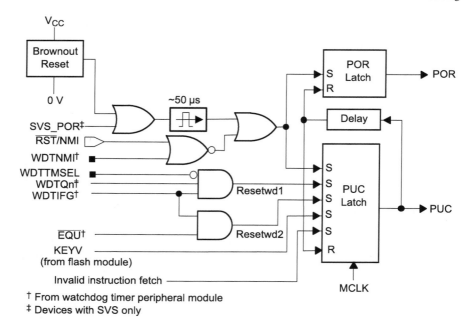

**Fig. 6.30**  Circuit schematic of POR and PUC logic in MSP430x2xx devices (*Courtesy of Texas Instruments, Inc.*)

The introduction of x2xx devices adds the ability to trigger a PUC by an invalid instruction fetch. Figure 6.30 shows a schematic of the logic governing the behavior of signals POR and PUC in MSP430x2xx devices.

The latest generations of MSP430 devices, series x5xx/x6xx feature an expanded reset and initialization circuit as part of the System Control Module (SYS). The reset logic now provides three internal signals: brownout reset (BOR), power-on reset (POR), and power-up clear (PUC). The BOR signal is separated from the POR to denote the expanded nature of a POR event in the MCU architecture.

The BOR signal is now the device reset, generated by either a power-up event/ brownout condition, a low pulse in the $\overline{RST}/NMI$ pin when in reset mode, a wake-up from the low-power modes LPM3.5 or LPM4.5, or a software triggered BOR event. Note that this generation of devices now supports specific software triggered BOR events. From the list of events triggering a BOR, only a power-up/brownout event cannot be masked through software-controlled enable flags. This gives an absolute highest priority to this event.

The POR signal now acquires a softer functionality, since all sources triggering it (excluding a BOR) can be software masked. A POR is triggered whenever a BOR goes off (not the opposite). In addition, voltage levels below the system voltage supervisor high and low levels can be enabled to trigger a POR, as well as a software triggered POR event.

In the MSP430x5xx/x6xx reset chain, PUC is the last signal link, being triggered whenever a POR is generated. As in previous generations, a PUC is also triggered by watchdog timer expiration or key violation events, by Flash memory violations, and an invalid fetch. In this new generation a PUC is also triggered by a password violation in the power management module.

## 6.8 Summary

This chapter provided concepts and applications of developing interfaces for microcontroller-based systems, from the basic interface necessary to start-up an MCU to specific interfaces that allow to integrate user interfaces, large loads, and motors. Section 6.1 introduced the fundamental components that make an MCU work: Power, clock, reset, and bootstrapping.

General purpose I/Os form the most important means for microcontrollers to interact with the external world. Using their Data-In/Data-Out registers, configuring their direction registers, and enabling their interrupt capabilities were the main subject of Sect. 8.1. Next, the fundamental considerations to interface switches and switch arrays, LEDs, LED arrays, numeric alphanumeric, and graphic LCDs, were discussed in detail in Sects. 8.3 and 8.4.

The subject of handling large loads, like heaters, solenoids, motors, lightning fixtures, etc. was discussed in Sects. 8.2 and 8.10. Every section featured how MSP430 devices support each of the base functions discussed, along with hardware and software examples illustrating their applicability.

## 6.9 Problems

6.1 Enumerate the elements of a basic microcontroller interface and indicate two major criteria for the selection of each.

6.2 Why should a designer ensure that an MCU in an embedded application does not work at its absolute maximum ratings?

6.3 An embedded system incorporates an MSP430F2274 clocked by a 4.096 MHz at 3.3 V, driving two seven-segment displays and two discrete LEDs. The seven-segment display draws 7.5 mA per segment at 1.8 V and the discrete LEDs operate with 10 mA each at 1.6 V. Three push-buttons are needed for completing the user interface. Design a suitable interface to connect all LEDs, seven-segments, and push-buttons. Estimate the circuit power requirements and recommend a non-regulated power supply and a suitable regulator.

6.4 Assume the system described in Problem 6.1 is to be fed from a 4.0 V, 2,400 mAh lithium battery. Estimate the expected battery life, assuming LEDs are driven dynamically at 30 % duty cycle. What would be the regulator efficiency? Determine the MCU thermal dissipation for the given loading conditions and verify if it is operating within a safe temperature range.

6.5 A crystal oscillator with a nominal frequency of 7.3728 MHz exhibits an aging induced drift of 12 PPM per year. If the system where it is being used has an accuracy requirement of 5 %, after how long shall the crystal be replaced to maintain compliance with frequency accuracy requirements?

6.6 Perform a comparative analysis of the advantages and disadvantages of using Colpitts versus Pierce oscillators in microcontroller-based systems. Identify at least one instance where one type would be more convenient than the other.

6.7 An MSP430F5438 is driven by an external crystal oscillator of 32.768 KHz. Devise an initialization sequence to produce a frequency of 11.0592 MHz in the ACLK line to drive a communication channel. Use the external crystal to stabilize the VLO generated frequency.

6.8 A certain microcontroller has an asserted-low Reset input that requires at least 25 clock cycles of assertion to be accepted as a valid reset input. Using the basic POR circuit illustrated in Fig. 6.25, determine suitable values of R and C, considering tolerances of 5 % in each to assure the circuit will always provide a valid reset when the MCU is running at 8 MHz.

6.9 An MSP430F2274 is operated from a 3.3 V power supply. What should be the power supply stability requirements to prevent that a normal voltage fluctuation in the supply voltage might trigger a brownout reset in the device? What would be the lowest safe voltage level at which the chip might operate?

6.10 Provide an initialization sequence for the MCU in Problem 6.1, and write a program that will make the system operate as a down counting stopwatch. Label the three keys as "b1", "b2", and "b3" and write a short program to display the key name on the seven-segment display while the corresponding key is depressed.

6.11 Devise an algorithm for joining a keypad scan function and a software debouncing function for the keypad such that a seven-segment display shows the character corresponding to the depressed key. Assume only single key depressing will be allowed.

6.12 Write a program to return on R5 the ASCII code corresponding to the depressed key in a 16-key keypad. Show how shall the keypad be interfaced to the MCU.

6.13 Provide a stand-alone program that will make a dedicated 12-key keypad scanner from a launchpad, placing in a predesignated memory location the ASCII code of the depressed key.

6.14 Design a dumb HEX terminal using a 16-key keypad and a 4x20 alphanumeric LCD on an MSP430F169. Provide a keypad buffer in the MCU with a capacity of 20 characters with a last-in first-out replacement policy for the data sent to the LCD.

6.15 A 120VAC fan is to be controlled by an MSP430F2273. Assuming the maximum fan current is 2 A, provide a solid-state relay interface using a Sharp S108T02. Verify both voltage and current compatibility in the control interface. Add a single push-button interface to the system and a status LED to be lit when the fan is ON. Provide a short program to toggle the fan with the push-button.

6.16  A standard Futaba S3003 servo-motor operated at 5.0 V moves between its two extreme positions 180° apart when fed a 50 Hz signal with 5–10 % duty cycle. Nominally the servo moves to its center position with a 7.5 % duty-cycle signal. Assuming the digital signal requires less than 1 mA when pulsed, and the motor consumes anywhere between 8 and 130 mA in a load range from idle to full torque. Provide a suitable interface to control the servo from an MSP430 Launchpad I/O pin and provide a software-delay based function that would allow controlling the shaft with a resolution of 0.5°.

6.17  A 12VDC, 1.5 A per phase, two-phase PM bipolar stepper motor with eight rotor poles is to be controlled from an MSP430F2274. Provide a discrete component H-bridge interface to control the motor with a GPIO port in the MSP and indicate the actual components to be used. Provide a safe to operate interface and a function accepting as parameters(via registers) the direction in which the stepper will move and the number of steps in each move. Assume the motor is to be operated in half-step mode. Determine the number of steps per revolution to be obtained in this interface.

6.18  Provide an enhanced interface for the motor in Problem 6.10 by using a DRV8811. Upgrade the driving software function to operate the motor in microstepping mode with eight microsteps per full step of the motor. Calculate the new step resolution of the motor.

# Chapter 7
# Embedded Peripherals

This chapter immerses readers into the array of peripherals typically found in a microcontroller, while also discussing the concepts that allow understanding how to use them in any microprocessor-based system. The chapter begins by discussing how to use interrupts and timers in microcontrollers as support peripherals for other devices. These are the two most important support peripherals in any microcontroller as they form the basis for real-time, reactive operation of embedded applications.

The subjects of flash memory storage and direct memory access are also discussed in this chapter, with special attention to its use as a peripheral supporting for low-power operation in embedded systems. The MSP430 is used as the platform to provide practical insight into the usage of these peripherals.

## 7.1 Fundamental Interrupt Concepts

Interrupts provide one of the most useful features in microprocessors. As it was explained earlier, in the introductory discussions in Chaps. 3 and 4, interrupts provide an efficient alternative to handle the service requests of peripheral devices and external events in a computer system. When compared to the alternative of polling, interrupt servicing allows a much more efficient use of the CPU. This fact was analyzed in Chap. 3, when a comparison was made of the efficiency of polling versus interrupts (Sect. 3.9).

In addition to the CPU usage efficiency advantage, using interrupt servicing also provides the following advantages:

- Compact and modular code: Interrupt service routines (ISR) induce code modularity and software reusability.
- Reduced energy consumption: As ISRs lead to less CPU cycles, this has a direct impact in the amount of energy consumed by the application, particular when combined with low-power modes.

M. Jiménez et al., *Introduction to Embedded Systems*,
DOI: 10.1007/978-1-4614-3143-5_7,
© Springer Science+Business Media New York 2014

- Faster response time: When many events and devices need to be served, well designed ISRs provide a quick response to the triggering event. Well coordinated priorities among sources will allow for a quick response with minimal processing overhead.

To have the ability of working with interrupts, the system designer must include both hardware and software considerations. Depending on the level of integration of the processor being used, the hardware components to support interrupts might need to be externally provided, as in the case of microprocessors, or can be found embedded in the chip as happens with most microcontrollers. In either case, once the hardware components are in place, dealing with interrupts requires their correct configuration and programming.

In this section we provide a discussion of fundamental concepts that help understanding how interrupts work in general. Specific treatment of the MSP430 interrupt structure is provided in Sect. 7.2, and guidelines and recommendations on how to design interrupt-based programs are the subject of Sect. 7.3.

### 7.1.1  Interrupts Triggers

Interrupts are mainly triggered by hardware events. Events such as a push-button depression, a threshold reached, and timer expiration, are few examples of events that might be configured to trigger interrupts. Once a hardware event has been configured to trigger interrupts, their occurrence is asynchronous and unpredictable. They can be triggered at any time. Thus, the software structure supporting the interrupts events, ISRs, variables, configurations, etc. must be written considering this fact.

**Interrupt Request Sensitivity**

The way an input pin in a processor detects an interrupt request can be either level sensitive or edge sensitive.

A level-sensitive interrupt input has an assertion-level that indicates an interrupt request is active. The assertion level can be either low or high. An active-low assertion level, the most commonly used, specifies an interrupt request by holding at a logic low level the interrupt request input to the processor. Asserted-high interrupt inputs will become activated by a high-level logic input.

Either type of assertion level requires the interrupt pin to remain asserted until an interrupt request is received or until service is rendered. This guarantees the request will be seen by the CPU when interrupt processing conditions are given. Some peripherals are designed to remove the interrupt request signal assertion upon access. For example, it is common for serial communication adapters to drop their receiver ready interrupt request when a read operation is made to their data-in buffer, but this is not a standard feature across all serial adapters. A designer might end-up working with one that needs to be explicitly turned-off from within the ISR. Therefore, the designer must verify in the device data what is the expected behavior of its interrupt request signal.

An edge-triggered interrupt input will respond to the transition detected in the interrupt request pin. The edge detection can be either with a rising or a falling edge of the input signal. To allow this functionality, the request needs to be latched, otherwise the request might go undetected.

Most modern microcontrollers allow for specifying the type of sensitivity desired for their external interrupt inputs. In these MCUs, a configuration register allows to choose between level- and edge-sensitive operation, the assertion level for the former, and the edge type for the latter. In the case of edge sensitivity some devices even allow triggering the interrupt request with an "any edge" option, meaning that either a rising or falling edge would trigger the request. The designer needs to verify the flexibility offered by the device at hand and ensure that the chosen configuration matches the output signal protocol from the requesting device.

### Software Interrupts

An ISR may also be called by software, in which case we talk of a *software interrupt*. Unlike hardware triggered interrupts, software interrupts are predictable and become part of the normal program sequence. Being so, the invocation to a software interrupt is equivalent to a function call.

Many embedded systems are designed to perform the solely task of reading and processing data from one or more peripherals whenever there is new data. This is most efficiently done by interrupts. When the MCU is programmed just to respond to interrupts, the system is said to be *interrupt driven*.

## 7.1.2 Maskable Versus Non-Maskable Interrupts

Earlier, in Chap. 3, two types of interrupt requests were identified: maskable and non-maskable interrupt requests. Let's revisit these types here to see a few additional details.

A maskable interrupt request is one that can be masked by the CPU through the global interrupt enable (GIE) flag to ignore it. This is done by clearing the GIE in the processor status register (PSW). This action makes the CPU ignore all interrupt request from all maskable sources.

The GIE flag is a useful feature that allows the CPU to choose when to accept or reject interrupt requests. A device placing a maskable interrupt request to the CPU knows that its request was served when the CPU executes its designated ISR. Some systems might involve an *Interrupt Acknowledgement Signal*, but that depends on the interrupt management style of the processor architecture.

The second type is a non-maskable interrupt (NMI) request. This type of request cannot be masked by the CPU and therefore always has to be served, even if the GIE flag is clear. This type of request is reserved only for system critical events whose service cannot be postponed under any circumstance. One example of such an event could be a low-battery voltage condition in a portable device. Upon a critical battery level that could be detected by a voltage comparator, an NMI would be triggered whose ISR would save the CPU state and any critical data held in registers and volatile memory to keep them from being lost and then shut the system down. Many battery operated laptops and cell phones nowadays use such a feature.

The vast majority of the interrupts served by a CPU are maskable interrupts. This is reflected by the fact the maskable interrupts are simply called *interrupts*. When a reference is made to a non-maskable interrupt then the reference is explicit calling it an NMI or non-maskable interrupt.

**Interrupt Masking and Polling Support**

Disabling or masking an interrupt request is a common and necessary situation in the management of interrupts. There are instances when the CPU might not be ready to manage interrupts, or simply it might not be appropriate or convenient to serve an interrupt. For example, just after a reset, before the system has been configured, the CPU cannot not accept interrupts if they are not configured yet. To this end, whenever the CPU is reset, its GIE is cleared and remains so until the program being executed explicitly enables it. Enabling the interrupt flag shall occur only after the whole interrupt structure has been configured.

Another instance when interrupts are disabled is when the CPU enters an ISR. Before begining the execution of an ISR, the GIE is cleared. This prevents other interrupts from interrupting an ongoing ISR. Upon returning from the ISR the GIE becomes re-enabled. The code within the ISR may include instructions to set the GIE flag, enabling the ISR interruption. However, this requires writing a code with re-entrance and possibly recursive capabilities. Most applications do not need such a prevision, and therefore in most cases it is recommended not to enable the global interrupt flag inside an ISR.

Besides disabling interrupts through the GIE flag, most hardware interfaces with the ability of triggering interrupt requests also include some mechanism for deactivating their interrupt capabilities. Consider for example, an interrupt managed push-button connected to the CPU via a GPIO port. The GPIO will include an interrupt enable flag (IEF) that allows to enable/disable its interrupt generating capability. Thus, for the push-button actually triggering an interrupt to the CPU, both the GIE in the CPU and the GPIO IEF in the push-button interface must be enabled.

Having interrupt enable flags in the interfaces themselves results convenient because they allow for individually enabling or disabling the interrupt generation ability of individual peripherals, as needed, without affecting the interrupt capability of other devices and their interfaces. Remember that when the GIE is disabled, all maskable interrupts become masked. Peripheral IEFs are frequently located within the interface control register(s).

Another feature closely related to an IEFs in device interfaces is the inclusion of status flags indicating when service requests (SRQ) have been placed. These flags are convenient for polled operation of the interface. Polled operation is possible by clearing the local IEF an interrogating the local SRQ flag.

Sometimes it becomes unavoidable to poll devices even when they are being operated by interrupt. This is a common situation arising when multiple events within a peripheral interface trigger a single, shared interrupt request signal. The activation of the common IRQ signal notifies the CPU about the need for servicing an event in that peripheral interface, but to determine which specific event triggered the request, the first action to be performed within the ISR is to poll the device's SRQ flags to determine what event caused the trigger. A typical scenario where this situation can be found is in general purpose I/O ports with interrupt generation capability. Frequently, the GPIO is designed to give each pin the ability generating service requests, while the entire port has a single request line to the CPU. An ISR serving such a port must first poll the pins SRQ flags to determine which one triggered the request, to then proceed providing the appropriate service.

### 7.1.3 Interrupt Service Sequence

The sequence of events taking place when an interrupt request is accepted by the CPU were outlined in Sect. 3.9.2. The same sequence is represented here, in a pictorial form in Fig. 7.1.

Steps 1 through 6, graphically denoted in the figure above are explained below. The stack and GIE operations are denoted in subgraphs (a), (b), and (c). A set GIE

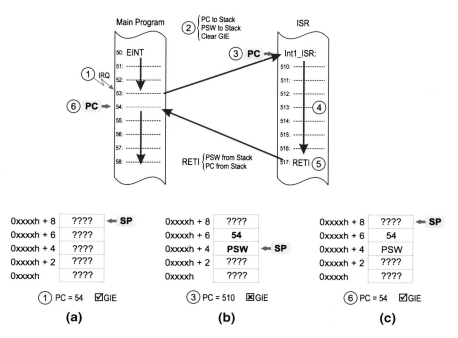

**Fig. 7.1** Sequence of events in the CPU upon accepting an interrupt request. **a** Stack before IRQ, **b** Stack after IRQ, **c** Stack after RETI

with is denoted with a checkmark, while a crossed-out GIE denotes it has been cleared.

1. An interrupt request arrives while a main program, with GIE flag set, is executing instruction 53. Instruction 53 completes execution.
2. The processor saves the current PC, pointing to instruction 54, and the status register (PSW), still with the GIE flag set, onto the stack. Next, the GIE flag is cleared.
3. The PC is loaded with the address of the first instruction of interrupt service routine of the device that issued the interrupt request (*How does this happen?*).
4. The ISR is executed, and finishes when a return from interrupt instruction (RETI) is encountered and executed.
5. The execution of the RETI instruction causes the PSW and PC to be restored from the stack.
6. The interrupted program is resumed, continuing from instruction the instruction pointed by the PC, in this case instruction 54.

Note the similitude between the interrupt processing sequence and that of calling a software function, illustrated in Fig. 3.31. This similarity is what makes interrupts to be sometimes regarded as hardware initiated function calls.

One aspect of the interrupt processing sequence that has yet to be explained is: How does the PC gets loaded with the correct ISR address? This question takes particular meaning when considering that there might be multiple interrupt capable devices in an embedded system, with equal number of ISRs. In complex systems there might be dozens of interrupt sources.

Another issue to consider is that having multiple devices issuing interrupt requests at arbitrary times, there is a high probability that two or more devices may simultaneously place service requests to the CPU. Recall that the CPU is a sequential machine and will only serve one device at a time.

Addressing these issues require having in place management mechanisms for (a) identifying the source of an interrupt request among many others that might coexist in the system to activate the correct ISR, and (b) resolving the conflict caused when multiple devices simultaneously place interrupt requests to the CPU. In the next sections, a discussion is made of how these issues are handled.

## 7.1.4 Interrupt Identification Methods

Throughout the evolution of microprocessors, different methods have been used to identify the source of an interrupt request in systems with multiple interrupt capable interfaces. In general, these methods, although having differences from a processor architecture to another, can be classified into one of three types: non-vectored, auto-vectored, or vectored systems.

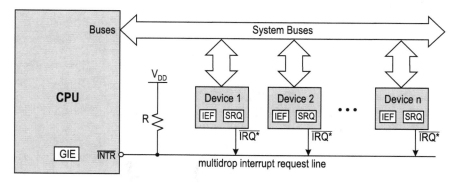

**Fig. 7.2** Device interfaces in a system managing non-vectored interrupts

## Non-vectored Interrupts

The most elemental way to serve interrupts in a system is by having all requesting devices placing their requests through a single, multidrop interrupt request line. When either of the devices places a request, the INTR line into the CPU becomes asserted, triggering a request. By just sensing the INTR line, the CPU will not be able to identify who placed the request, so within the ISR, the CPU proceeds to check the service request flags of all its peripherals to find out who placed the request. This is called a non-vectored interrupt system. Figure 7.2 illustrates a block diagram of a system managing non-vectored interrupts.

Each device interface has a an open collector or open drain request line tied to the processor $\overline{INTR}$ input. The activation of a request by a particular device interface will set its SRQ flag. If the interface's interrupt enable flag (IEF) is also set, its $\overline{IRQ*}$ will become asserted, placing a request to the CPU. Assuming the CPU has its GIE flag set, the interrupt will take place by loading the PC with a predefined address where the ISR is stored. The ISR address in non-vectored systems is usually fixed a certain location in program memory where the ISR must be stored in order to be executed. When any of the devices places a request, a single ISR is executed and code within the ISR polls the SRQ flags of each device interface to determine who placed the request. The absence of a hardware mechanism allowing the CPU automatically identifying who placed the service request is what gains this method the name of non-vectored.

A slightly improved mechanism is offered by some CPUs where instead of a fixed address for the ISR, a fixed location is specified where the address of the ISR is stored. Jumping to the ISR is achieved by loading the PC with the value stored at this location. This scheme makes more flexible the process of locating the ISR, but it still does not provide automated identification of the device who placed the request.

If multiple simultaneous requests were placed in this system, the polling order would determine what device would be served first. This establishes a priority service order in the system.

A non-vectored system, although simple, usually results in a delayed response due to the need of checking by software who placed a request at the beginning of the ISR.

**Vectored Interrupts**

A more efficient scheme to identify the source of an interrupt request is provided in a vectored system. In a vectored system, a hardware mechanism is provided that allows for automatic identification of the interrupt request source without having to poll the service request flags in the interfaces. The most common mechanism uses an interrupt acknowledgment signal (INTA), which, when received at the interrupt requesting interface causes the latter to send an identification code to the CPU. This ID code, which allows the CPU to determine the location of the device's ISR, is named a *vector*. A reserved space in memory holds the addresses associated to each interrupt vector in the system. This memory space is called the system vector table.

Vectored approaches are commonly used in microprocessors, where external interfaces issuing interrupts send their ID codes via the data bus when they receive the INTA signal. In most cases the vector ID is not the actual ISR address. Instead, the ID is a number that can be used to calculate the actual location of the ISR address. Intel processors from the 80x86 and successors use this scheme.

The process triggered in a vectored system by the reception and acceptance by the CPU of an interrupt request is called an *interrupt acknowledgment cycle*. Such a cycle begins with the CPU issuing an INTA signal, followed by the peripheral interface sending its vector ID, and then the CPU determining from the vector, the address that gets loaded into the PC to begin execution of the ISR. The whole interrupt acknowledgment cycle occurs automatically within the CPU control unit, with no instruction involved in this process. Sometimes this scheme is called a *full vectored system*.

**Auto-vectored Interrupts**

In microcontrollers, where multiple on-chip peripherals like timers, I/O ports, serial ports, data converters, etc., can issue interrupts to the CPU, a simpler mechanism called *auto-vectoring* is used. In an auto-vector system, each device has a fixed vector or a fixed address leading to its ISR. Thus, the CPU does not need to issue an INTA signal and the peripherals do not need to issue a vector. When any of the internal interrupt capable peripherals issues an interrupt request (assuming GIE is set), the MCU loads into the PC the address of the corresponding ISR. MCUs designed with fixed ISR addresses directly load the PC with the predefined ISR address, while MCUs with fixed vectors load the PC with the address stored at the corresponding vector entry from the vector table.

The MSP430 handles all interrupt requests using auto-vectors. This includes all maskable sources, non-maskable, and reset. The top portion of the 64K memory is reserved for storing ISR and reset addresses. Additional details for the MSP430 are discussed in Sect. 7.2.

## 7.1.5 Priority Handling

When several peripherals with interrupt capability coexist within the same system, there will be instances when two or more of them will simultaneously place service

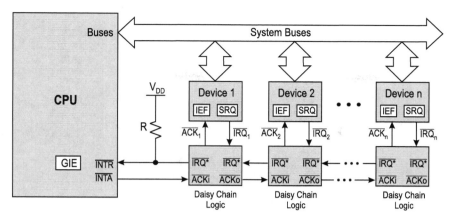

**Fig. 7.3** Device interfaces connected in a daisy chain

requests[1] to the CPU. When this situation arises, the CPU needs to have in place a way of prioritizing service. Recall that the CPU can handle only one task at a time.

There are multiple ways to provide for priority management in microprocessor-based systems, either via software or hardware.

When interrupts are served with non-vectored mechanisms, we saw earlier that the order in which service request flags are polled establishes a priority order. In such a system it becomes necessary to identify device priorities prior to deciding the polling order to ensure those with more pressing service needs are served first. The same concept applies when devices are served by polling.

Vectored and auto-vectored systems require hardware support for resolving priorities. Two methods are preferred: using *daisy chain-based arbitration*, or with an *interrupt controller*.

### Daisy Chain-based Arbitration

A daisy chain is perhaps the simplest hardware setup to provide priority arbitration in an embedded system. It assumes all devices place service requests through a common multidrop request line, and a common granting or acknowledgment signal is returned indicating the requested service has been accepted. This implies that each peripheral device in this scheme must have the capability of handling both, request and acknowledgment lines. Each peripheral device is connected to a link of the chain.

Each link in daisy chain contains a logic circuit that lets passing any device request to the CPU, but limits the acknowledgment signal to reach only up to the highest priority device interface that has placed a interrupt request to the CPU. Figure 7.3 shows an arrangement of device interfaces sharing a daisy chain scheme.

---

[1] Since interrupt service requests are served only at the end of each instruction, to consider two request simultaneous, they just need to arrive within the time frame of a single instruction execution.

**Fig. 7.4** Logic structure of a
daisy chain link

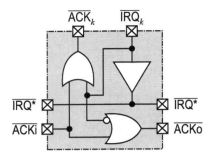

Due to the blocking action of link elements in the chain, priorities are hardwired as a function of the distance from the CPU ports: the closer the peripheral to the CPU, the higher the priority. Figure 7.4 shows the structure of a daisy chain link.

If two devices assert their request signals simultaneously, say devices 1 and 2, the interrupt request line into the CPU $\overline{\text{INTR}}$ becomes asserted, causing it to respond with $\overline{\text{INTA}}$ and clearing its GIE. However, the acknowledgment signal will only reach to Device 1. Note how the assertion of $\overline{\text{IRQ}_1}$ blocks the propagation of $\overline{\text{INTA}}$ down the chain. Thus device 1 will issue its vector ID through the systems buses and the CPU will execute ISR 1. Upon returning from ISR 1, the CPU re-enables the GIE, becoming able to serve another interrupt. Serving device 1 causes it to de-assert $\overline{\text{IRQ}_1}$. Since device 2 still has its $\overline{\text{IRQ}_2}$ asserted, the MCU will detect the interrupt request (from device 2). This time the $\overline{\text{INTA}}$ will reach device 2, allowing the execution of ISR 2.

A daisy chain is a simple solution to the priority arbitration problem, but has some limitations. First, priorities are hardwired, making difficult balancing services among peripheral devices. If a device close to the CPU places frequent requests, it could monopolize service, preventing devices down the chain to receive on-time CPU attention. Also, devices down-the-chain have a longer propagation delay to receive the $\overline{\text{INTA}}$ signal, which could delay their servicing. Another limitation is that a daisy chain will only work with level-sensitive interrupt requests. Additional latching hardware would be required to manage edge triggered requests.

If any of the above limitations becomes relevant in an application, the designer would need to resort to a different priority handling mechanism. Next section introduces how dedicated interrupt controllers work.

**Interrupt Controller-based Arbitration**

An interrupt controller is a special case of a single-purpose processor also called a *priority arbiter*. A priority arbiter is a programmable support peripheral, configured by the CPU at startup to allow for handling requests from multiple devices or events to a common resource. In the case of an interrupt priority arbiter, the common resource being accessed is the attention of the CPU.

Priority interrupt arbiters are the preferred choice for handling interrupts in vectored system as they reduce the hardware requirements of peripheral interfaces to comply with the signal protocol required by the interrupt acknowledgment cycle . An interrupt controller is placed between the CPU and the requesting devices,

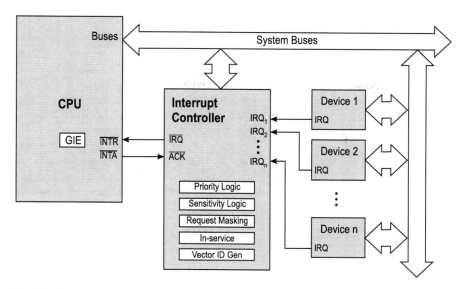

**Fig. 7.5** Interrupt management using a programmable interrupt controller

in such a way that individual interrupt requests from peripheral interfaces are received by the arbiter, and the arbiter, based on the configured management scheme, relays them to the CPU. This arrangement is illustrated in Fig. 7.5.

As an arbiter, the interrupt controller not only manages incoming requests from device interfaces allowing to establish a flexible priority management scheme, but also provides other functions that would otherwise require additional logic and software overhead in the ISR to be furnished. Some of the functions provided by an interrupt controller include:

- Signalization for the interrupt acknowledgement cycle: Upon receiving the $\overline{\text{INTA}}$ signal from the CPU, based on the configured priority scheme it selects the device to be served and issues its vector ID.
- Schemes to configure individual IRQ sensitivity: Such a scheme allows for selecting between level- or edge-sensitivity and the type of level or edge in each case. The interrupt controller would also include the latching logic necessary for edge sensitivity.
- A flexible priority management scheme: This adds flexibility in the management of priorities by allowing using either fixed or dynamic priority schemes. Dynamic schemes such as rotating priorities, where a device after being served goes to lowest priority, avoiding service monopolization, are possible.
- Service management registers: These include masking registers allowing disabling individual request lines and in-service indication registers that facilitate automated interrupt nesting.

The specific functions rendered by an interrupt controller change from one architecture to another. In microprocessors, programmable interrupt controllers are specifically designed to comply with host CPU architecture and protocols. Early systems had dedicated chips designed for interrupt control. A representative example of one of such devices is the Intel 8259A, a controller designed for early 80x86 processors, but its flexibility and versatility gained acceptance in a wide number of systems and applications. Modern microprocessor system rely on custom designed cores embedded into the processor or supporting chipsets.

In the arena of microcontrollers, interrupt handling also rely on custom designed hardware embedded into the MCU. The way these designs work internally result transparent to the programmer. The fundamental aspects to be known about a particular embedded interrupt control structure are its handing capabilities and programmer's model. A first step for a system designer successfully utilizing such resources is understanding them both by reading the manufacturer's documentation. The interrupt handling structure of MSP430 devices, discussed in the next section, is a representative example of this trend.

## 7.2 Interrupt Handling in the MSP430

The MSP430 is a microcontroller designed with a large and versatile interrupt management infrastructure. All its internal peripherals have the capability of being operated by interrupts, resulting in a system with a large number of hardware interrupt sources. As part of its ultra low-power philosophy, all these sources also have the ability of waking up the CPU from its featured low power modes. Among the list of internal devices capable of triggering interrupts in the MSP430 we find: general purpose I/O ports, data converters, serial interfaces, general purpose and watchdog timers, and system exceptions, to name just a few of the most common. A specific list of interrupt sources depends on the device generation, family, and the particular configuration of embedded peripherals in the chosen device.

MSP430 MCUs use a fixed priority management scheme resulting from using a daisy-chain arbitration structure, so peripherals closer to the CPU have a higher priority. As the list of embedded peripherals changes from one MSP430 model to another, knowing the exact priority level of a particular device requires consulting its data sheet.

Interrupt source identification in the MSP430 uses an auto-vector approach with vector locations fixed at the end of the memory map. Early MSP430 families allocated memory addresses from 0xFFE0 to 0xFFFF to form the *interrupt vector table*, including the address for the reset vector at 0xFFFE. Later families have extended the table to start at 0xFFC0, but they maintain backward compatibility with the arrangements provided for earlier generations.

When writing interrupt managed programs, the programmer needs to consult the data sheet of the particular MSP430 model being used in his or her application and then proceed to configure the vector entries of the interrupt sources used in their programs. The process of allocating vectors is discussed in Chap. 4, Sect. 4.7.3.

In general, an MSP430 microcontroller can handle three types of interrupt sources that include: system resets, non-maskable (NMI), and maskable interrupts. Below we discuss each of them.

## 7.2.1 System Resets

System resets, although been listed among interrupt sources, behave differently and serve a different purpose than conventional interrupts. The main similarity between resets and interrupts is that both change the normal course of execution of a program by loading into the PC the contents of their corresponding vector. However, a reset is quite different as it does not save the processor status or a return address, does not return, and has as a function initializing the state of the system. Thus the usage of interrupts and resets is not interchangeable.

Chapter 6 provides a detailed discussion of reset events in embedded systems. In particular, Sect. 6.7 discusses in detail the reset structure in MSP430 microcontrollers.

## 7.2.2 (Non)-Maskable Interrupts

(Non)-maskable interrupts in the MSP430 can be considered as a type of pseudo-NMI, based in our definition in Sect. 7.1.2 of an NMI. The reason is that although MSP430 NMIs cannot be masked by the GIE bit, they *can* be masked by the corresponding individual enable bits in their sources, namely NMIIE, OFIE, and ACCVIE (see sources below).

Whenever a (non)-maskable interrupt is accepted by the MSP430 CPU, *all* NMI enable bits are automatically reset, and program execution begins at the address pointed by the (non)-maskable interrupt vector, stored at address 0FFFCh. This single vector is shared by all NMI sources. (Non)-maskable interrupts in the MSP430 can be generated by any of the following sources:

- **An edge on the $\overline{\text{RST}}$/NMI pin when configured in NMI mode:** The function of the $\overline{\text{RST}}$/NMI pin after power-up is, by default, in reset mode. All MSP430 generations prior to series x5xx/x6xx, allow this functionality to be changed through the $\overline{\text{RST}}$/NMI bit in the watchdog control register WDTCTL. When in the NMI mode, with the NMI enable bit (NMIIE) is set, a signal edge in the $\overline{\text{RST}}$/NMI input will trigger an NMI interrupt. The triggering edge can be configured to rise or fall through the WDTNMIES bit. Triggering an NMI will also set the $\overline{\text{RST}}$/NMI flag NMIIFG.
- **The detection of an oscillator fault:**] By enabling the oscillator fault flag (OFIE) in the interrupt enable register 1 (IE1), an NMI can be triggered if a malfunction

is detected with the external crystal oscillator. As was explained in Sect. 6.4.1, the absence of signal from an external crystal switches the MCLK to the DCO and sets the OFIFG bit in the interrupt flag register 1 (IFG1). User software must consider this as an expected condition whenever the MSP430 is started. However, if an external crystal were present and after the due crystal startup period no signal were detected, then an enabled oscillator fault NMI can facilitate the implementation of counter measures. User software must check and clear the OIFG upon such an event.

• **An access violation to the flash memory:** To prevent corrupting the contents of the on-chip flash memory, the MSP430 forbids write accesses to the flash memory control register (FCTL1) during flash erase or write operations (WAIT = 0). Failure to observe this rule triggers an exception indicated by the access violation interrupt flag (ACCVIFG) in the flash memory control register FCTL3. If the ACCVIE flag is set, this event also triggers an NMI. User software action is required to clear ACCVIFG after it has been set.

As an NMI can be triggered by multiple sources, the code within the ISR needs to discriminate the activation source and take the appropriate remedial actions. When the NMI is triggered, it automatically clears the enable bits o all its sources, namely NMIIE, OFIE and ACCVIE, preventing interruption of the NMI ISR. Note that these flags are not restored by the RETI instruction and therefore user software needs to re-enable them if their functionality is to be restored. The corresponding interrupt triggering flags, NMIIFG, OFIFG, and ACCVIFG are not. Thus the NMI ISR must also reset the interrupt triggering flags. Figure 7.6 illustrates a recommended flowchart for planning a well-designed NMI handler.

### 7.2.3 Maskable Interrupts

Most of the interrupt sources in the MSP430 fall in the category of maskable interrupts, so all interrupt sources other than those listed in the NMI section fall in the maskable category.

All maskable interrupt sources in the MSP430 are disabled when the global interrupt enable (GIE) flag in the status register (SR) is cleared. In addition, each individual interrupt source has enable flags in their corresponding interfaces. This means that for any device interrupting the MSP430 CPU, both the GIE flag *and* the particular enable flag associated to the device must be enabled.

Interrupt requests in the MSP430 are checked between instructions, implying that interrupts are only served between instructions. When an interrupt is accepted by the MSP430 CPU, it pushes both the PC and the SR onto the stack in that order. Then, through its internal daisy chain logic it chooses the requesting device with highest priority. When the requesting device has a unique interrupt source, the CPU automatically clears its request; however, in devices featuring multiple sources of interrupt, the disabling must be part of the ISR. Next, the status register (SR) is cleared,

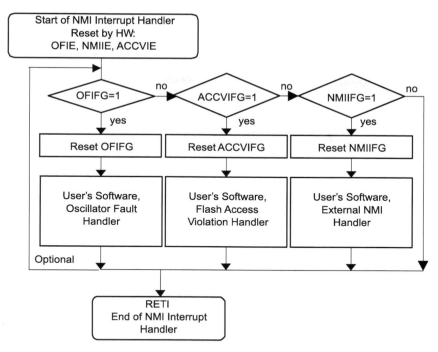

**Fig. 7.6** Recommended flowchart for a well-designed NMI handler (*Courtesy of Texas Instruments, Inc.*)

disabling the global interrupt enable flag, which prevents interrupting the ISR. Clearing the SR also cancels any previous low-power mode configured. At this point, the PC is loaded with the vector contents of the chosen interrupt source to begin executing its ISR. Upon executing the return from interrupt instruction (#RETI#), the MSP430 restores both SR and PC from the stack, re-enabling the GIE and restoring any low-power mode previous to the interrupt.

The specific list of interrupt sources in a particular MSP430 microcontroller and their priorities, as explained earlier, depends on the particular device chosen for an application. The designer needs to consult the data sheet of the particular device used in an application to have its specific vector and priority assignment. As an illustrative example, the interrupt vector table for an MSP430F2274 model is shown below in Table 7.1.

Note that unlike traditional microcontrollers, the MSP430 does not provide external dedicated interrupt request inputs. Instead, Ports 1 and 2 provide each interrupt capable GPIO inputs. Each pin in these ports can be configured to trigger an interrupt to the CPU, each having individual making and interrupt flags. All pins associated to each port share the same interrupt vector, resulting in up to 16 external interrupt pins with two separate vectors.

**Table 7.1**  MSP430F2274 interrupt vector table (*Courtesy of Texas Instruments, Inc.*)

| Interrupt source | Flag | Type | Address | Priority |
|---|---|---|---|---|
| Power-up<br>External reset<br>Watchdog<br>Flash key violation<br>PC out-of-range[1] | PORIFG<br>RSTIFG<br>WDTIFG<br>KEYV[2] | Reset | 0FFFEh | 31<br>highest |
| NMI<br>Oscillator fault<br>Flash access violation | NMIIFG<br>OFIFG<br>ACCVIFG[2][3] | NMI | 0FFFCh | 30 |
| Timer_B3 | TBCCR0<br>CCIFG[4] | Maskable | 0FFFAh | 29 |
| Timer_B3 | TBCCR1<br>TBCCR2<br>CCIFGs<br>TBIFG[2][4] | Maskable | 0FFF8h | 28 |
|  |  |  | 0FFF6h | 27 |
| Watchdog Timer | WDTIFG | Maskable | 0FFF4h | 26 |
| Timer_A3 | TACCR0<br>CCIFG[3] | Maskable | 0FFF2h | 25 |
| Timer_A3 | TACCR1<br>CCIFG<br>TACCR2<br>CCIFG<br>TAIFG[2][4] | Maskable | 0FFF0h | 24 |
| USCI_A0/USCI_B0 Rx | UCA0RXIFG<br>UCB0RXIFG[2] | Maskable | 0FFEEh | 23 |
| USCI_A0/USCI_B0 Tx | UCA0TXIFG<br>UCB0TXIFG[2] | Maskable | 0FFECh | 22 |
| ADC10 | ADC10IFG[4] | Maskable | 0FFEAh | 21 |
|  |  |  | 0FFE8h | 20 |
| I/O Port P2 (8 flags) | P2IFG.0 to<br>P2IFG.7[2][4] | Maskable | 0FFE6h | 19 |
| I/O Port P1 (8 flags) | P1IFG.0 to<br>P1IFG.7[2][4] | Maskable | 0FFE4h | 18 |
|  |  |  | 0FFE2h | 17 |
|  |  |  | 0FFE0h | 16 |
| (5) |  |  | 0FFDEh | 15 |
| (6) |  |  | 0FFDCh to<br>0FFC0h | 14 to 0,<br>lowest |

(1) A reset is generated if the CPU tries to fetch instructions from within the module register memory address range (0h to 01FFh) or from within unused address range. (2) Multiple source flags (3) (non)-maskable: the individual interrupt-enable bit can disable an interrupt event, but the general interrupt enable cannot. Nonmaskable: neither the individual nor the general interrupt-enable bit will disable an interrupt event. (4) Interrupt flags are located in the module. (5) This location is used as bootstrap loader security key (BSLSKEY). A 0AA55h at this location disables the BSL completely. A zero (0h) disables the erasure of the flash if an invalid password is supplied. (6) The interrupt vectors at addresses 0FFDCh to 0FFC0h are not used in this device and can be used for regular program code if necessary

## 7.3 Interrupt Software Design

When it comes to develop programs for interrupt-driven embedded systems, the adherence to good programming practices becomes of outmost importance. Interrupt-driven systems induce modular software structures that can be very agile, quickly responding to real-time external events. However, interrupt-driven systems are difficult to debug and tend to grow complex as the number of interrupt generating events increases. For these reasons, careful software planning and orderly code writing are essential.

### 7.3.1 Fundamental Interrupt Programming Requirements

Before interrupts can work in any application, there are four fundamental requirements that must always be met: stack allocation, vector entry setup, provision of an ISR, and interrupt enable. These requirements have been indicated earlier in Chaps. 3 and 4. Here we provide a more detailed discussion to each of them.

**1. Stack Allocation** The allocation of a stack area is fundamental for the proper functionality of any interrupt. The stack is used to store the program counter (PC), processor status (SR) and any other register saved upon interrupt acceptance. The stack allocation must be explicitly coded in assembly programs.[2] When allocating a stack, it is necessary to estimate the maximum stack space to be used by ISR invocations, function calls, temporary storage, and parameter passing. If nested functions or reentrant code is being used, nesting depth has to be controlled to avoid stack overflow. Each assembler has its own rules for stack declaration, allocation, and stack pointer management. In the case of MSP430 under the IAR assembler. Example 7.1 provides specific guidelines. Search lines marked with <-(1) to see the format.

**2. Vector Entry Setup** As discussed earlier, the CPU uses a vector table to determine the addresses of each interrupt service routine to be executed upon each interrupt type. Explicit code must be included in any program to initialize the vector entries of each active interrupt with its corresponding ISR address. Examples 7.1 and 7.2 illustrate how this is done in IAR. Check for lines marked with <–(2) within the comments.

**3. Interrupt Service Routine** Each active interrupt must have in place an interrupt service routine (ISR). This is the to be executed when the interrupt is triggered. Like all programs, ISRs must observe good programming practices. However, due to the fact that ISRs, once the interrupts are enabled, can be triggered in any point of the code, specific rules must be strictly observed. These include:

- Make ISRs short and quick: Recall that while an ISR is executed, all other interrupts, by default, are disabled. If any other urgent event were triggered while a long ISR is being executed, that other event might experience a long latency that

---

[2] C-language programs do not require explicit stack allocation as it is performed by the compiler.

will make the system appear slow or even compromise the system integrity. In programming terms, short and quick are not synonymous. Avoid lengthly computations and loops or function calls. If there is a need for lengthly computation, relay them to the main program. Search for lines marked with $<$–(3) to see example ISRs in Examples 7.1 and 7.2 below.

- Make ISRs register transparent. This rule takes special meaning when programming in assembly language.[3] Programmers must explicitly push any register used within an ISR prior to its first use and pop it before exiting the function.
- No parameter passing or value return: As ISR invocations are unpredictable within a program, there is no way you can safely pass or return parameters via registers or the stack. In most instances, global variables are used. A lot of care is required if code will be reentrant.
- End your ISR with an interrupt return (RETI). Although this might look an obvious requirement, it is common to find novice assembly programmers[4] attempting to exit an ISR with a conventional return. This error will not be detected by the assembler, and will make your code loose track of the stack, status, and prevent further interrupts in the system. Also try to avoid multiple exit points from an ISR as it might complicate debugging.

**4. Interrupt Enable** For interrupts to take place, all required levels of enabling for an event must be enabled. This requires *at least* two levels of enabling: first at the processor-level by setting the CPU GIE flag and second at the device-level itself by enabling the device to issue interrupts. The interrupt flag in the devices interfaces is sometimes automatically cleared when accessed after they have placed an interrupt request, and some do not, but all need to disable the request signal when served. Make sure their interrupt enable ability is restored before exiting the ISR, or otherwise the device will not issue any further interrupt to the CPU. Search for lines marked with $<$—(4) in the examples below to see the local and global interrupts enabled.

## 7.3.2 Examples of Interrupt-Based Programs

The following examples illustrate the usage of interrupts from both, assembly- and C-language standpoints. The comments in the code allow for following through the stages of coding in the interrupt usage.

**Example 7.1 (Using Interrupts in Assembly Language)** *Consider an MSP430-F2231 with a red LED connected to P1.2 and a hardware debounced push-button connected to P1.3. Write an interrupt-enabled assembly program to toggle the LED every time the push-button is depressed.*

---

[3] C compilers take care of this detail, transparently to the programmer.

[4] In C-language programs the compiler takes care of that.

*Solution:* This problem can be solved by setting P1.3 to trigger an interrupt every time a high-to-low transition is detected in the pin. The ISR just needs to toggle P1.3, which can be done x-oring the pin bit itself in the output port register.

A program implementing this solution is listed below. Within the comments, we have used numbers in parentheses and arrows to denote where each of the four requirements described in Sect. 7.3.1 are satisfied. This solution also invokes a low-power mode in the CPU after the interfaces have been configured. The occurrence of an interrupt wakes-up the CPU, toggles the I/O pin, and returns to sleep mode. Low-power modes are discussed ahead, in Sect. 7.3.6.

```
;==========================================================================
; Code assumes push-button in P1.3 is hardware debounced and wired to
; produce a high-to-low transition when depressed.
;--------------------------------------------------------------------------
#include   "msp430g2231.h"
;--------------------------------------------------------------------------
            RSEG CSTACK                      ; Stack declaration <------(1)
            RSEG CODE                        ; Executable code begins
;--------------------------------------------------------------------------
Init        MOV.W   #SFE(CSTACK),SP          ; Initialize stack pointer <--(1)
            MOV.W   #WDTPW+WDTHOLD,&WDTCTL ; Stop the WDT
;
;------------------------- Port1 Setup  -----------------------------------
            BIS.B   #0F7h,&P1DIR             ; All P1 pins but P1.3 as output
            BIS.B   #BIT3,&P1REN             ; P1.3 Resistor enabled
            BIS.B   #BIT3,&P1OUT             ; Set P1.3 resistor as pull-up
            BIS.B   #BIT3,&P1IES             ; Edge sensitivity now H->L
            BIC.B   #BIT3,&P1IFG             ; Clears any P1.3 pending IRQ
Port1IE     BIS.B   #BIT3,&P1IE              ; Enable P1.3 interrupt <--(4)

Main        BIS.W   #CPUOFF+GIE,SR           ; CPU off and set GIE <---(4)
            NOP                              ; Debugger breakpoint
;--------------------------------------------------------------------------
PORT1_ISR   ; Begin ISR <--------------(3)
;--------------------------------------------------------------------------
            BIC.C   #BIT3,&P1IFG             ; Reset P1.3 Interrupt Flag
            XOR.B   #BIT2,&P1OUT             ; Toggle LED in P1.2
            RETI                             ; Return from interrupt <---(3)
;
;--------------------------------------------------------------------------
;           Reset and Interrupt Vector Allocation
;--------------------------------------------------------------------------
            ORG     RESET_VECTOR             ; MSP430 Reset Vector
            DW      Init                     ;
            ORG     PORT1_VECTOR             ; Port.1 Interrupt Vector
            DW      PORT1_ISR                ; <-------(2)
            END
;==========================================================================
```

Interrupts are even easier to use from a C-language program, as illustrated in the Example 7.2.

**Example 7.2 (Using Interrupts in C-language)** *Consider the MSP430 configuration described above in Example 7.1. Provide an interrupt-enabled, C-language solution to the problem.*

***Solution:*** *The code for this solution is quite simple and straightforward. Observe the there is no need to declare a stack or terminate the ISR with IRET, or program the reset vector, as those details are taken care of by the C compiler. The program fundamentally has the main, the ISR vector programming (#*`pragma`* command in line 19), and the ISR.*

```
//===========================================================================
#include <msp430g2231.h>
//---------------------------------------------------------------------------
void main(void)
{
  WDTCTL = WDTPW + WDTHOLD;          // Stop watchdog timer
  P1DIR  |= 0xF7;                    // All P1 pins as out but P1.3
  P1REN  |= 0x08;                    // P1.3 Resistor enabled
  P1OUT  |= 0x08;                    // P1.3 Resistor as pull-up
  P1IES  |= 0x08;                    // P1.3 Hi->Lo edge selected
  P1IFG  &= 0x08;                    // P1.3 Clear any pending P1.3 IRQ
  P1IE   |= 0x08;                    // P1.3 interrupt enabled
//
  _ _bis_SR_register(LPM4_bits + GIE); // Enter LPM4 w/GIE enabled <---(4)
}
//
//---------------------------------------------------------------------------
// Port 1 interrupt service routine
#pragma vector = PORT1_VECTOR         // Port 1 vector configured <---(2)
_ _interrupt void Port_1(void)         // The ISR code <----(3)
{
  P1OUT ^= 0x04;                      // Toggle LED in P1.2
  P1IFG &= ~0x08;                     // Clear P1.3 IFG
}
//===========================================================================
```

### 7.3.3 Multi-Source Interrupt Handlers

In the interrupt management infrastructure of microprocessors and microcontrollers it is common to find peripherals that generate multiple interrupt events, but have allocated a single vector. These are commonly referred to as multi-source interrupts. The MSP430 is no exception to this case. A quick look at Table 7.1 reveals several multiple source interrupts, like those of I/O ports 1 and 2, each with eight sources, Timers A and B, and the serial interfaces (USC Tx and Rx). When dealing with the ISRs of such sources, there are two rules that must be observed:

A first rule in this case is that unlike single source interrupts, the peripheral flag triggering a multiple source interrupt request is not automatically cleared by the CPU when the peripheral is served. Therefore, this flag must be software cleared within the ISR itself or otherwise there will be an interrupt request for the same event just served upon returning from the ISR. In Example 7.1, this is the purpose of the first instruction in the ISR.

A second rule to observe is that when a multiple sourced interrupt is triggered, the trigger could be from any of its sources. Therefore, the first step that needs to be performed within the ISR is to identify what was the triggering source. This is similar to the process described in Sect. 7.1.4 for non-vectored interrupts.

The identification can be done via polling, by interrogating each of the interrupt flags associated with the peripheral or by performing a calculated branch.

**Identification Via Polling IRQ Flags**

When the identification is made by polling, as in the case of a non-vectored interrupt systems, the code just needs to sequentially test the IRQ flags of the shared sources. The polling order will establish the service priorities.

Figure 7.7 shows an example of an ISR design with three events, Event1, Event2, and Event3. Each event marks it interrupt requests by setting its respective flags Flag1, Flag2, or Flag3. Upon entering the shared ISR, the flags are polled to determine which one triggered the interrupt request, and when determined, the request flag is cleared (Rule 1 above) and the corresponding service rendered.

This example also allows to see that the polling order assigns FLAG1 the highest priority, followed by FLAG2, and with the least priority FLAG3. Note that when in the polling sequence, if Flags 1 and 2 are checked and found clear we still check

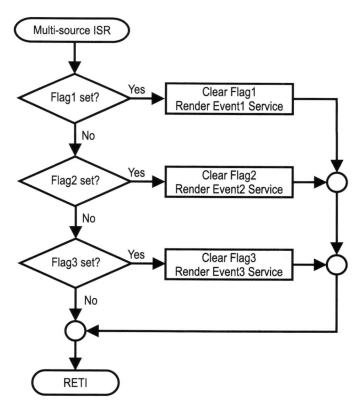

**Fig. 7.7** Flowchart of a service routine for a multi-source interrupt vector

for Flag 3. Although a novice programmer might be tempted to go right away and assume Flag3 was the cause of the interrupt, it is actually recommended to check Flag3 as well. Doing so helps to reduce the chances of responding to a false interrupt trigger.

**Using Calculated Branching**

An even faster technique for identifying the interrupt source in a shared vector is by using *Calculated Branching*. This technique is also known as a *Jump Table* and is commonly used for coding CASE statements.

In this approach, instead of executing a list of successive test and jump instructions, the destination address for the service function is obtained from a branch table (look-up table) holding the addresses of the pieces of code serving each event. The index to enter the look-up table is obtained from the contents of the register holding the service requests of the events shared in the vector. This way, multiple comparisons are avoided, resulting in a much faster and predictable way of branching to the service instructions of each event based on the contents of the shared interrupt request flags. An assembly pseudo-code illustrating the technique for a source shared by three events is listed below:

```
;========================================================================
#include "headers.h"                    ; Header file for target MCU
;------------------------------------------------------------------------
                ...                     ; Preliminary declarations
                RSEG CODE
JumpTable       DW Exit,Serv1,Serv2,Serv2,
                Serv3,Serv3,Serv3,Serv3 ; Branch table with addresses of
                                        ; the 1st line of each service code
Init            ...                     ; First executable instruction
;------------------------------------------------------------------------
; Multi source Interrupt Service routine
; Assumes FlagReg contains individual IRQ flags
;------------------------------------------------------------------------
MultiSrcISR PUSH R15                    ; ISR is R15 transparent
            CLR.W R15                   ; Calculation of jump address begins
            MOV.B FlagReg,R15           ; R15 low loaded with IRQ flags
            AND,B #07,R15               ; Only the three lsb assumed active
            ROL R15                     ; Multiply R15 by 2 to get even address
            MOV JumpTable(R15),R15      ; Load address of selected service code
            JMP @R15                    ; Jump to selected service code
            ...
Serv1       CLR Flag1                   ; Clear Event1 IRQ flag
            Event1 service code         ; Service instructions for Event1
            ...
            JMP Exit
Serv2       CLR Flag2                   ; Clear Event2 IRQ flag
            Event2 service code         ; Service instructions for Event2
            ...
            JMP Exit
Serv3       CLR Flag3                   ; Clear Event3 IRQ flag
            Event3 service code         ; Service instructions for Event3
            ...
Exit        POP R15                     ; Restore R15
            RETI                        ; Exit IRS
;------------------------------------------------------------------------
COMMON INTVEC                           ; Interrupt vector area
```

```
; ------------------------------------------------------------------------
                ORG  RESET_VECTOR
                DW   Init
                ORG  MULTI_SOURCE_VECTOR ; Whatever shared vector
                DW   MultiSrcISR
; ========================================================================
```

This approach takes the same amount of time to get to any of the functions serving each event, which helps to minimize the interrupt latency. In this particular case, the code is written to assign priorities in decreasing order from the most significant bit of the request register. This priority order is established by the organization of the labels in the branch table.

Note that having three request flags can result in eight different request conditions going from 000 when no request is placed to 111 if all sources simultaneously request service. If for example, events one and three place simultaneous requests (101), Event3 will be served first.

An even more efficient implementation can be achieved when the service request flags are encoded. Several shared sources in MSP430 provide prioritized, encoded flags such as GPIO or serial device. In such cases the destination branch can be determined by just adding to the program counter the contents of the flag request service. The code fragments below illustrate hot to perform such operations in assembly language.

```
; ========================================================================
; Multi source Interrupt Service routine for four events
; Assumes FlagReg contains prioritized, encoded IRQ flags organized as:
; 0000 - No IRQ          0004 - Event2          0008 - Event4
; 0002 - Event 1         0006 - Event3
; ------------------------------------------------------------------------
#include "headers.h"          ; Header file for target MCU
...                           ; Preliminary declarations and code
MultiSrcISR
                ADD  &FlagReq,PC    ; Add offset to jump table
                JMP  Exit           ; Vector = 0: No interrupt
                JMP  Event1         ; Vector = 2: Event1
                JMP  Event2         ; Vector = 4: Event2
                JMP  Event3         ; Vector = 6: Event3
Event4          Task 4 starts here  ; Vector = 8: Event4
                ...
                JMP  Exit           ; Return
Event1          Task 1 starts here  ; Vector 2
                ...                 ; Task starts here
                JMP  Exit           ; Return
Event2          Task 2 starts here  ; Vector 4
                ...                 ; Task starts here
                JMP  Exit           ; Return
Event2          Task 3 starts here  ; Vector 4
                ...                 ; Task starts here
Exit            JMP  Exit           ; Return
```

The code fragment below illustrates a branch table for the multi-source case of four events in C-language.

```
//===========================================================================
// Multi source Interrupt Service routine for four events
// Assumes FlagReg contains prioritized, encoded IRQ flags organized as:
// 0000 - No IRQ          0004 - Event2          0008 - Event4
// 0002 - Event 1         0006 - Event3
//---------------------------------------------------------------------------
#include <headers.h>                ; Header file for target MCU
...                                 ; Preliminary declarations and code
// Multi-source ISR
#pragma vector = MultiSrc_VECTOR _ _interrupt void MultiSrc_ISR(void) {
    switch(_ _even_in_range(FlagReg,8)) {
        case 0x00: // Vector 0: No interrupts
            break;
        case 0x02: ... // Vector 2: Event1
            break;
        case 0x04: ... // Vector 4: Event2
            break;
        case 0x06: ... // Vector 6: Event3
            break;
        case 0x08: ... // Vector 8: Event4
            break;
        default: break;
    }
}
```

## 7.3.4 Dealing with False Triggers

In the proceeding discussions we mainly focused in providing vectors and ISRs for active interrupts, meaning by active those that have a hardware or event source to trigger them. Not every system will use every interrupt available in the microcontroller. Those remaining unused are the ones referred to as inactive.

Sometimes, power glitches, electromagnetic interference, electrostatic discharges, or some other form of noise might get coupled into the CPU and corrupt interrupt request lines, registers, or other sensitive system components. As a result, false interrupt triggers might end up occurring in the system. This situation could result in unpredictable consequences for the system integrity, particularly if some of the unused interrupts get to be triggered.

A measure for mitigating this situation can be providing service routines for all sources of interrupts, particularly those that share a common vector. Those that are actually used get the necessary code to be served with provisions to detect false triggering, and those unused can be given a dummy ISRs. A dummy ISR can have just one instruction: RETI. In some instances, as a false trigger might also corrupt other portions of the system, it might be advisable to configure all unused interrupts to point to a common error handler ISR, or a function causing a system reset. For systems working in harsh environments, false triggers might even render interrupt support unusable, resulting more reliable a polled operation [42].

## 7.3.5  Interrupt Latency and Nesting

Interrupt latency refers to the amount of time a peripheral device or system event has to wait from the instant it requests interrupt service until the first instruction of its ISR is fetched. This time is determinant to the perceived and factual speed of a system, and could become a reliability issue in real-time events and time critical processes. The simple rule that must prime in every interrupt managed system is that every event requesting CPU attention shall be served with low interrupt latency.

Interrupt latency in embedded systems is affected by both hardware and software factors.

In the hardware side, the supporting structure in place for handling the signalization involved in serving an interrupt is the dominant component. Examples of factors associated to the interrupt hardware infrastructure include the propagation delay in the path traversed by the interrupt request signal in its way to the CPU, the mechanism used to identify the requesting source, the way priorities are resolved, and the delay in the interrupt acknowledgement path if there were such a signal involved in the scheme.

Software factors are dominated by the scheduling approach programmed into the application to handle interrupt requests and the ISR coding style. These factors are responsible for the largest latency times in interrupt handling, sometimes hundreds of times longer than any hardware induced latency. A fast hardware infrastructure can be significantly slowed down by a poor priority handling scheme or by the style used to program the interrupt handlers (ISRs), or both.

### Interrupt Latency Reduction Methods

Once the interrupt handling structure of a system has been laid, there is little chance to reduce the hardware factors causing interrupt latency. If the designer does have a choice in deciding the hardware components for interrupt management, measures like reducing the number of stages through which request and acknowledgment signals propagate, choosing vectored or auto-vectored approaches over non-vector, and implementing in-service tracking in hardware, are some of the most effective ways to reduce the lag. Most of these provisions are already in place when a programmable interrupt controller is used. Hardware induced latency actually represents the lower bound for the interrupt latency in an embedded system design.

Most opportunities for reducing interrupt latency are tied to good software design practices. Keeping ISRs short and quick, avoiding instructions with long latency, and properly handling interrupt priorities top the list of recommendations. One major source of latency is the disablement of interrupts in the system, that automatically done when an ISR is entered into. In this last aspect, two particular mitigating strategies are commonly used: allowing interrupt nesting and establishing prioritization schemes.

### Interrupt Nesting

Interrupt nesting is achieved by re-enabling the GIE after the processor context has been saved to the stack. This can be done in combination with a prioritization scheme where only interrupts with a higher priority than the currently served are enable. If re-entrancy[5] were desired, equal priority requests had to be enabled as well. These programming practices require careful program control to avoid system havock. Strict software control of the nesting level should be observed to avoid stack overflow. Moreover, if reentrant coding were to be allowed, additional provisions need to be taken in the ISR and the rest of the code to avoid situations that would mess-up the memory system. General recommendations include the following basic rules:

- Avoid Static Variables: A reentrant function shall not hold static (non-constant) data due to the risk of self modification. Although a reentrant function can still work with global data, it is advised to use them with atomic instructions.[6] The best alternative is to have function allocated variables (dynamic allocation), so that every time the function is invoked it allocates its own set of variables. The atomicity observation also applies to the usage of hardware: do not allow a hardware transaction to be interrupted.
- Do not use self-modifying code: A reentrant function cannot contain under any circumstance instructions that modify code memory.
- Do not invoke non-re-entrant functions: All function calls from within a re-entrant function must be reentrant as well.

Additional rules pertinent to ISRs in general also apply, like not passing parameters through the stack or registers and not returning values.

These programming techniques, due to their need of keeping a tight in-service track and centralized management of the interrupt structure, are best suited to be implemented when the hardware infrastructure provided by a programmable interrupt controller. Small microcontrollers usually do not provide such facilities. In such instances, the user software needs to take care of all the details. When this is done, the software usually becomes complex, and in most cases ends-up worsening the interrupt latency itself.

## 7.3.6 Interrupts and Low-Power Modes

Many embedded applications, particularly those in real-time reactive systems have the MCU most of the time waiting for an event to occur. This results in a CPU that is used only sporadically, with long idle periods. Upon this situation we could write the application software in one of two ways:

The first alternative would initialize the peripherals and system resources to then send the CPU into an infinite loop implemented through either polling the expected event or waiting for an interrupt to trigger the CPU response. This is probably the most intuitive way of doing it, but not the most power efficient. By continuously

---

[5] A re-entrant ISR is one that safely allows being interrupted by itself.

[6] Instructions able to perform read-modify-write operations without interruption.

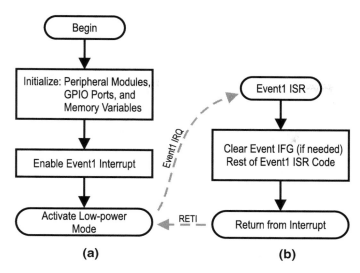

**Fig. 7.8** Flowchart of a main program using a low-power mode and a single event ISR. **a** Main program, **b** Interrupt service routine

polling the event flag or leaving the CPU active in an infinite loop to be broken by an interrupt would waste lots of CPU cycles and energy.

A more energy efficient alternative is to initialize the peripherals, system components, enable the interrupts of the expected events and then send the CPU into a sleep mode. Every time an enabled interrupt is triggered, it will wake-up the CPU to serve the event and go back to sleep. This second alternative will be orders of magnitude more energy efficient as the CPU will consume energy only when waken-up to serve the enabled peripherals and the rest of the time will be in a low-power mode consuming minimal or no energy. Figure 7.8 illustrates a flowchart for a main program and one ISR (Event1 ISR) designed to operate with the CPU in low-power mode. This plan assumes that all the tasks needed in the system can be performed inside the ISR.

If there were additional events to be served in this system, the only additional requirement would be enabling the interrupts of the additional events and providing their corresponding ISRs. Assuming the vectors of all active events are configured, then they all can be served the same way. If you look back at the structure of the programs in Examples 7.1 and 7.2, you will notice that both have the structure illustrated in Fig. 7.8.

Low-power modes are of outmost importance in battery powered applications due to the dramatic reduction in power consumption they induce. But not only battery-powered designs benefit from a low-power consumption. Energy efficiency, as discussed in Chap. 1, must be a primary objective in every embedded design, as it has implications improving system reliability, reducing power supply requirements, size, weight, and overall cost of applications.

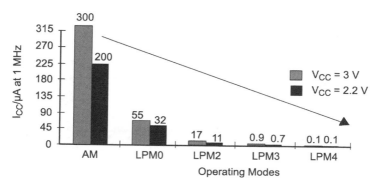

**Fig. 7.9**  Current consumption of MSP430F21x1 devices in their different operating modes (*Courtesy of Texas Instruments, Inc.*)

To have an idea of how dramatic can be the power reduction, consider the case of MSP430F21x1 devices. The activation of their low-power modes can reduce the MCU supply current well below 1 % of the current consumption in active mode. Figure 7.9 shows the current consumption of this MCU family in their different low-power modes.

**Enabling Low-power Modes**

A low power mode in an MCU is activated by writing the processors operating mode bits, typically located in the processor status register, with the bit pattern of the desired operating mode. Most modern microcontrollers feature one or more low-power modes, where the power reduction is achieved by either reducing the clock rate or completely shutting-off the clock generator, and/or selectively powering down unused peripherals, including the CPU itself.

When a microcontroller is sent into a low-power mode, the way to bring it back into activity is through an interrupt. Recall that in the interrupt cycle, one of the steps prior to executing the ISR clearing the status register, deactivating any low-power mode in effect. This highlights the importance of enabling interrupts prior to sending an MCU to a low-power mode. The designer, when planning an application, needs to determine which events will wake-up the CPU, to include in the program the necessary support to have their interrupts functional and enabled prior to sending the CPU into a low-power mode.

At the end of the ISR, if no changes were made to the saved SR, its retrieval by the interrupt return instruction restores the low-power mode in effect prior to the interrupt. That is why the flowchart in Fig. 7.8 shows the program flow returning to the low-power mode in the main upon exiting the ISR.

**Coping with Complex Service Routines**

More than often, the operations to be performed as a result of serving a peripheral module can be complex enough for not being feasible to perform all tasks completely inside an ISR. It should not be forgotten that one of the strongest recommendations in the design of ISRs is to keep them short and quick.

To cope with the requirements of a complex ISRs the solution would be relaying part or all the processing to the main program, while still triggering the code execution through interrupts. This can be easily and efficiently done with the aid of a low-power mode and software defined global flags as outlined below:

1. Declare in the Main program one global variable as flag per each ISR to be partially or completely executed in the Main. Initialize all declared flags cleared. These flags will be used to indicate which interrupt was activated. Also initialize the rest of the system resources as usual.
2. Still in the main program, enable interrupts as usual (GIE and individual events IF).
3. Activate the appropriate low-power mode (if more than one were available). This is where the execution of the main will stop to until an interrupt is triggered.
4. Code in the main, beginning right after the instruction activating the low-power mode, a loop to check the software flags. Each flag detected set will be the indication for executing the code of the corresponding ISR event. Don't forget to clear the flag. After rendering the corresponding service, keep the program in the loop checking additional software flags that might be set until all flags are clear. At this point go back in the main to the instruction where the low-power mode was originally activated.
5. In the ISR of each event to be served, set the corresponding software flag and cancel the low-power mode in the saved status register. This will cause the CPU to restore an Active Mode when the IRET instruction is executed, allowing the main to go executing the loop where the software flags are checked.

At first sight, this approach might seem complex, but it is actually very simple. Figure 7.10 shows a flowchart illustrating this process with a program serving three interrupt events. An advantage of this approach is that it allows reducing the length of time the GIE flag is cleared, since the actual ISR is really short.

Example 7.3 illustrates how simple is to implement this approach in assembly language for an MSP430.

## 7.3.7  Working with MSP430 Low-Power Modes

One of the greatest features of MSP430 microcontrollers is their low-power consumption and low-power modes, where the already low power consumption of the MCU can be reduced to nanoampere range currents, as indicated in Fig. 7.9.

MSP430 devices prior to series x5xx/x6xx feature four low-power modes, denoted LPM0, LPM1, LP2, and LPM3. The default operating mode of the CPU, the Active Mode with the CPU and all clocks active is how the chip operates just after a reset or when entering an interrupt service routine. Changing from the active mode to any of the low-power modes under software control can be achieved by configuring the status register (SR) bits CPUOFF, SCG1, SCG0 and OSCOFF, as indicated below, in Table 7.2.

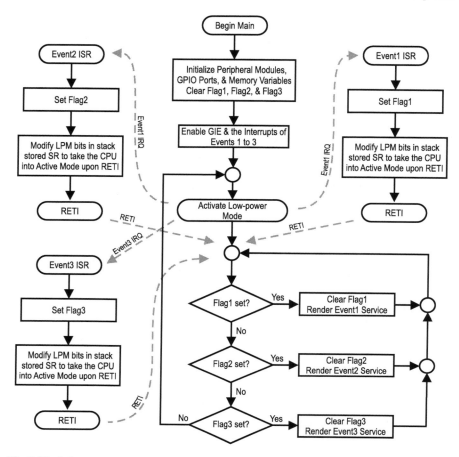

**Fig. 7.10**  Software strategy to execute ISR code in the main

Series x5xx/x6xx feature two additional operating modes denominated LPM3.5 and LPM4.5, which, in addition to disabling the modules listed for LPM3 and LPM4 in the table above, also turn off power to the RAM and register areas of the MCU, enabling an even lower power consumption. These modes result useful in applications where loss data retention capabilities in the MCU registers and RAM can be tolerated. Additional details of these modes can be found in the corresponding device data sheet.

Each operating mode, either Active or LPM$n$, achieves different power consumption rates which vary depending on the particular device being configured.

The selection of a particular low-power mode over another is done based on the modules that must stay active in the chip during the sleep mode to allow its interrupt to wake the CPU. For example if a particular application has a timer triggering an interrupt to wake the CPU, and that timer is fed by ACLK, then the chip can be sent into LPM3, which keeps ACLK on.

**Table 7.2** MSP430 Operating Modes (*Courtesy of Texas Instruments, Inc.*)

| SCG1 | SCG0 | OSC-OFF | CPU-OFF | Mode | CPU & Clocks status |
|---|---|---|---|---|---|
| 0 | 0 | 0 | 0 | Active | CPU and all enabled clocks are active |
| | | | | | Default mode upon power-up |
| 0 | 0 | 0 | 1 | LPM0 | CPU & MCLK disabled, |
| | | | | | SMCLK & ACLK active |
| 0 | 1 | 0 | 1 | LPM1 | CPU & MCLK disabled. DCO & DC |
| | | | | | generator off if DCO is not used |
| | | | | | for SMCLK. ACLK active |
| 1 | 0 | 0 | 1 | LPM2 | CPU, MCLK, SMCLK, & DCO disabled |
| | | | | | DC generator enabled & ACLK active |
| 1 | 1 | 0 | 1 | LPM3 | CPU, MCLK, SMCLK, & DCO disabled |
| | | | | | DC generator disabled, ACLK active |
| 1 | 1 | 1 | 1 | LPM4 | CPU and all clocks disabled |

The example below illustrates a case of low-power mode operation with the servicing code executed in the main program.

**Example 7.3** (**Relying service code to the Main**) *Re-write the push-button/LED program of Example 7.1 to have the servicing instructions executed in the main.*

**Solution:** *In this particular case, there is only one event to be served, thus, there is no need to use global software flags or to poll after resuming the main. See how simple is to modify the status register while still in the stack. Recall that just after entering into the ISR, the top-of-stack (TOS) is at the last push, which corresponds to the SR.*

```
#include  "msp430g2231.h"
;=====================================================================
            RSEG CSTACK                      ; Stack Segment
            RSEG CODE                        ; Executable code begins
;---------------------------------------------------------------------
Init        MOV.W   #SFE(CSTACK),SP           ; Initialize stack pointer
            MOV.W   #WDTPW+WDTHOLD,&WDTCTL    ; Stop the WDT

;           Port1 Setup
            BIS.B   #0F7h,&P1DIR              ; All P1 pins but P1.3 as output
            BIS.B   #BIT3,&P1REN              ; P1.3 Resistor enabled
            BIS.B   #BIT3,&P1OUT              ; Set P1.3 resistor as a pull-up
            BIC.B   #BIT3,&P1IFG              ; Clears any P1.3 pending IRQ
            BIS.B   #BIT3,&P1IE               ; Enable P1.3 interrupt

Main        BIS.W   #CPUOFF+GIE,SR            ; CPU off and set GIE
            NOP                               ; Assembler Breakpoint
            XOR.B   #0x04,&P1OUT              ; Toggle LED
            JMP     Main                      ; Go back to reactivate LPM
;---------------------------------------------------------------------
PORT1_ISR  ; Begin ISR
;---------------------------------------------------------------------
            BIC.C   #BIT3,&P1IF               ; Reset Interrupt Flag
```

```
        bic      #LPM4+GIE,0(SP)          ; CPU in active mode
        reti                              ; Return from interrupt

;-----------------------------------------------------------------------
;          Interrupt Vectors
;-----------------------------------------------------------------------
        ORG      RESET_VECTOR             ; MSP430 RESET Vector
        DW       Init
        ORG      PORT1_VECTOR             ; Port1 Interrupt Vector
        DW       PORT1_ISR
        END
;=======================================================================
```

## 7.4 Timers and Event Counters

This section discusses one of the most useful resources of a microcontroller: timer units. Timers are important because they are used to implement several signature applications. Among them, the implementation of watchdog timers, interval timers, event counters, real-time clocks, pulse width modulation, and baud rate generation. We will explain some of these in Sect. 7.4.4.

Microcontrollers may include one or more timers among their peripheral units. Timers are generally configurable to greatly enhance their basic functionality. They can generate a square signal that can be accommodated to our needs, by adjusting its frequency, duty cycle, or both. Timers also have interrupt capabilities, or that of capturing the specific time at which some external event occurs. By controlling these and other timer features an embedded system designer may relieve the CPU from a long list of time related tasks in many applications.

### 7.4.1 Base Timer Structure

In its most basic form, a timer is a counter driven by a periodic clock signal. Its operation can be summarized as follows: Starting from an arbitrary value, typically zero, the counter is incremented every time the clock signal makes a transition. Each clock transition is called a "clock event". The counter keeps track of the number of clock events. Notice that only a single type of transition, either low-to-high or high-to-low (not both), will create an event. If the clock were a periodic signal of known frequency $f$, then the number of events $k$ in the counter would indicate the time $kT$ elapsed between event 0 and the current event, being $T = 1/f$ the period of the clock signal. The name "timer" stems from this characteristic.

To enhance the capabilities of a timer, further blocks are typically added to the basic counter. The conglomerate of the counter plus the additional blocks enhancing the structure form the architecture of a particular timer.

Figure 7.11 illustrates the basic structure of a timer unit. Its fundamental component include the following:

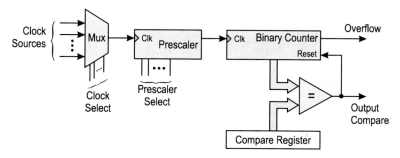

**Fig. 7.11** Components of a base timer

- Some form of a clock selector (Mux) allowing for choosing one from multiple clock sources.
- A prescaler, that allows for pre-dividing the input clock frequency reaching the counter.
- An $n$-bit binary counter, providing the basic counting ability.
- An $n$-bit compare register, that allows for limiting the maximum value the counter can reach.
- A hardware comparator that allows for knowing when the binary counter reaches the value stored in the compare register. Note how a match of these values resets the binary counter.

Below we provide additional information about these components.

**Clocks and Pre-scaler (Divider)**

In general, the clock signal for a timer may come from one or several different sources that include the external or internal clock sources, or even asynchronous events.

The accuracy of the clock signal is particularly important in time sensitive applications. Clocks and their selection criteria were discussed in Chap. 6. In particular Sect. 6.3.2 provides a discussion of factors that need to be considered when choosing a clock source.

In most MCUs and many microprocessor-based systems the clock signal for timers is derived from the system clock. In many applications such a frequency value might result too high for practical purposes. This situation calls for a means of reducing the frequency of the clock signal reaching the binary counter. This can be done with the aid of a timer pre-scaler or input divider. The scaling factor dividing the input frequency is generally software selectable. The output of the divider drives the binary counter.

**Counter and Compare units**

The counter register is probably the most important component of a timer. It is usually of the same size as the rest of the registers in the microcontroller. In most systems, it is also a read/write register.

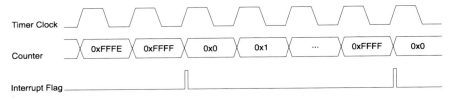

**Fig. 7.12**  Example of signals commonly obtained from a timer. The interrupt signal can be obtained from the compare output

The counter changes its state with each pulse of the driving signal that comes from the pre-scaler. In this way, it "counts" the number of pulses that have elapsed.

With $n$ bits, the counter will count from 0 to $2^n - 1$ and then go back again to 0. When it reaches the highest value, it generates an *overflow signal*. The overflow can be optionally configured to trigger an interrupt to the CPU. Figure 7.12 illustrates the behavior of an overflow signal in a 16-bit counter. The illustration also includes the behavior of an optional interrupt signal triggered by the overflow condition. The unit will count from 0x0000 to 0xFFFF, or 65535, firing the overflow signal when the counting sequence passes from 0xFFFF to 0x0000.

In many applications it results convenient limiting the maximum value that can be reached by the binary counter. In such cases, the availably of a compare register and a hardware comparator comes in handy. The combination of these two components allows for triggering an *output compare signal* when the binary counter reaches the value stored in the compare register. The process is similar to that illustrated by Fig. 7.12 with the appropriate changes in the signal names and the firing instant. For example, assuming the interrupt signal is configured to trigger upon an output compare condition, we might make it fire every 1000 cycles (instead of 65,533) by writing 0x03F8 to the compare register.

**Example 7.4**  *Assume that we want to generate a signal at an output port that divides the counter's input clock by 6.*

*This would mean that one period of our output signal has to cover six periods of the clock signal driving the timer. This in turn would require toggling the I/O pin once every three clock periods. Hence, we load the compare register with the value 3 and use the compare output signal to toggle the output port. This is illustrated in Fig. 7.13.*

*Notice that, in general, to divide the frequency by n, we load the compare register with a value equal to n/2.*

### 7.4.2  Fundamental Operation: Interval Timer vs. Event Counter

Most applications of timers involve one of two basic interpretations to its operation as an event counter, or as an interval timer. As an event counter, a timer counts the

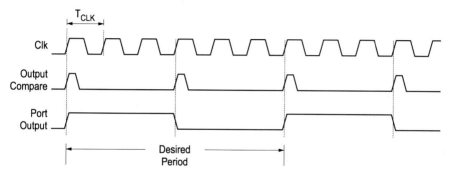

**Fig. 7.13** Signal obtained when loading the compare register with a value of three (3)

occurrence of external events. As an interval timer, it measures the time elapsed after $k$ clock cycles. In the case of event counter operation, the clock signal is derived from the event being counted. For timer operation, the clock input is driven by a square signal of known frequency.

### Event Counter Operation

When operated as an event counter, a timer simply counts the number of events it detects in its clock input. Note that in this case the clock signal needs to be derived from the event being counted. This would yield a clock signal does not necessarily have a periodic behavior, as illustrated in Fig. 7.14b.

Let's illustrate the operation of an event counter with a practical example.

**Example 7.5** *Consider an application where we need to count the number of people passing through a door. Describe how a timer configured as event counter could be used for providing a solution.*

*Solution: In this case we could place in the door frame a digital detector (like an optocoupler) to generate a pulse every time someone passes through. The optocoupler output would then be connected to the timer clock input. With this arrangement, once the timer is enabled, every time someone crosses the door the timer would increment by one.*

Note that in the door example above, the prescaler is assumed to be configured in the divide by one mode. This is an important detail, because otherwise the timer would increase by one after $p$ events have occurred, where $p$ is the prescaler factor.

In the event that we expect to count a number of events larger than the timer capacity, it is easy to extend the timer range. A simple solution would be via a software variable that counts the occurrence of overflows or the number of compare register pulses. Hardware solutions could include configuring the prescaler to a value other than divide by one, or by cascading multiple timers as shown in Fig. 7.15.

In the case of resorting to the prescaler, note that the timer will no longer count single events. Instead it will count sets of $p$ events, where $p$ is the value configured in the prescaler. Example 7.6 sheds some light into this situation.

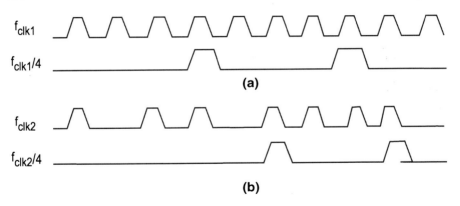

**Fig. 7.14** Dividing by 4 in pre-scaler: **a** periodic signal; **b** non periodic signal

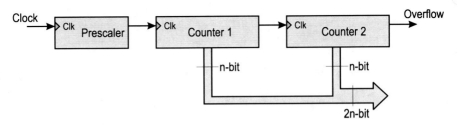

**Fig. 7.15** Cascading of counters

**Example 7.6** *The pre-scaler in a timer has been set for a factor of 4. The counter size is 16 bit. (a) What is the maximum number of events that can be counted? (b) What number should be loaded in the compare register to count 75000 events and fire the compare output signal? (c) Is it possible to count 83253 events? Justify your answer.*

*Solution: (a) Since each counter step covers four events, the maximum number of events that can be registered is*

$$(65,535) \times 4 = 262,140 \; events$$

*(b) The counter must reset when it reaches the value representing 75000 events. Since the count is scaled by a factor of 4, the value to be written on the compare register is*

$$75000 \div 4 = 18750$$

*(c) This event counter cannot register 83253 because the counter steps represent a group of 4 input events, and the desired value is not a multiple of 4.*

**Interval Timer Operation**

When the input clock signal is periodic of frequency $f$, the timer can be used to measure time intervals between two events. If the frequency is $f$ Hz, then the period will be

$$\text{Period } T = \frac{1}{f} \text{ seconds}$$

Therefore, when the counter shows $k$ pulses, it has registered a duration of $kT = k/f$ seconds. **An interval timer**

**Example 7.7** *A* 38 KHz *crystal oscillator is used as a clock source, and passed through the pre-scaler with a factor of* 16. *(a) If the timer's counter is reset at a certain moment and at the occurrence of a particular event the counter is reflecting* 0x8A39, *what is the time interval length between the reset instant and the event occurrence? (b) How can we have the counter trigger a signal every* 50 ms? *What are the absolute and relative errors in the triggering time?*

*Solution: The frequency actually driving the counter is* 38 KHz/16 = 2.375 KHz *or a period of* (1/2.375) ms = 421 μs. *This is the value which we should use in our problem.*

*(a) Since the number of periods that the counter has registered is* 0x8A39 = 35385, *we compute the elapsed time as* 35385 × 421μs = 14.9s.

*(b) Since we need to fire the flag every* 50 ms, *we must count* (50 ms)× (2.375 kHz) = 118.75 *periods. Rounding to the nearest integer, the compare register is loaded with* 119. *The actual interval, disregarding any delay introduced by hardware, would be* 119 × 421μs = 50.105 ms. *This yields an absolute error in excess of* 105 μs, *or* 0.21 %. *Based on the application, the designer will determine if this error is acceptable or not.*

**Example 7.8** *Consider the problem of counting the number of people passing through a door in Example* 7.5. *Assume that we'd like to add to this application the capability of measuring the average rate at which people is passing through the door. Outline a possible solution.*

*Solution: A simple solution would be providing a running average of the time between individual passes. Let's say we want to take the average of the last eight passes and still be able to keep track of the number of people passing through the door.*

*One possibility would be using a second timer driven by the system clock or other clock source of known frequency. Using the optocoupler output to also trigger an interrupt that reads the value of the second timer in every consecutive optocoupler event. By subtracting each new timer lecture from the previous yields the number of clock cycles between passes. Storing the difference in a circular buffer of eight positions and computing the buffer average upon every interrupt would do it.*

*An even simpler solution using a single tilmer could be enabled if the timer unit had input capture capability. An input capture uses a free-running counter, which is read upon each external event, in this case the optocoupler output. The input capture ISR holds a count variable for the number of events detected and the input*

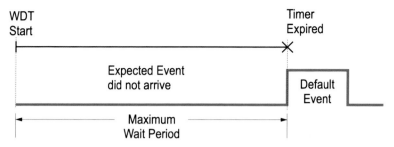

**Fig. 7.16**  Watchdog timer expiration

*capture register automatically holds the timer count between reset and the current event or between two consecutive events. This value can be fed to the running average buffer described above.*

### 7.4.3 Signature Timer Applications

#### 7.4.3.1  Watchdog Timers

In this section we discuss a selection of the most common uses of timers in embedded systems.

A rather common application of timers in embedded systems is that of a watchdog timer (WDT). This is a special case of an interval timer used to perform a certain default event or action, like issuing an interrupt or a reset, if within a predetermined period of time an expected event does not occur. Figure 7.16 illustrates how this works. A maximum allowable period of time is preset by software; a default event will be executed when the preset time period expires unless a certain expected event occurs first.

To prevent a watchdog timer from triggering its default event, the interrupt service routine associated to the expected event must cancel the timer before the preset time. Figure 7.17 illustrates this action.

To illustrate the operation of a watchdog timer, consider the case of a security door operated with an electronic card. To enter through the door a user must swipe a card through a reader. This action unlocks the door to allow the user to open it. But if after a certain time, say 30 seconds, the door has not been pushed open, the lock is activated again. Here, the WDT start occurs when the card is swiped. The default event is locking back the door. The expected event is opening the door.

A watchdog is basically a safety device. Under a normal conditions, it is assumed that the timer will be serviced (WDT cancelled) before its period elapses. Otherwise, something must be wrong and the system must execute the established default action. In embedded systems, watchdog timers can be used for diverse purposes, such as

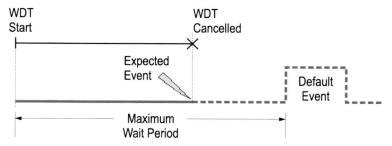

**Fig. 7.17** Watchdog timer cancellation

for cancelling runaway programs, exiting infinite loops, or detecting anomalous bus transactions, among others.

All MSP430 devices activate a WDT upon reset, configured to prevent runaway programs. The wait period in this case is 32,768 clock cycles, the default event is a system reset, and the expected event is the user software explicitly cancelling the WDT. This is the reason why all MSP430 programs begin by cancelling the WDT. Section 7.4.4 provides a detailed explanation of the MSP430 WDT.

### Real-time Clocks

One of the most common applications for timers in embedded systems is that of a real-time clock. A real-time clock is simply a timer configured to measure seconds, minutes, hours, etc. Often, the resolution of real-time clocks might be needed to go down to the level of fractions of a second, becoming chronometers. Other times, functionality can be taken to resolve days of the week, months, and years, becoming a real-time clock calendar (RTCC). In either case, the fundamental requirement to make a timer a real-time clock is the ability to resolve a period of a second with minimum or no error.

Some microcontrollers have specialized timers that can handle all the functions of a real-time clock. In such cases, the timer includes registers for handling seconds, minutes, hours, days of the week, months, and years. Some can even handle leap year calculations. Although the functionality of an RTCC can be fully implemented in software, using a dedicated timer for such purposes saves a lot of CPU cycles as the timer, once configured and enabled, operates with minimal or no necessity of CPU intervention.

Many MSP430 devices feature real-time clocks among its timers. Figure 7.18 shows a simplified block diagram of the real-time clock in MSP430 4xx series devices.

The accuracy of a real-time clock will greatly depend on the frequency and accuracy of the crystal used.

A typical clock source in RTC applications is a 32.768 kHz crystal. Since this base frequency can be successively divided by two until reaching exactly 1Hz $(32,768 = 2^{15}$, it is a preferred value for RTC applications. Care must be taken in providing the appropriate load capacitance when such a crystal is used. The resonant

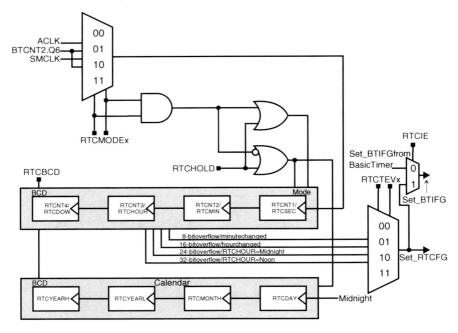

**Fig. 7.18**  MSP430x4xx real time clock simplified diagrams

frequency of a crystal can be affected by the total capacitance loading the oscillator.
Section 6.3 provides a detailed discussion of factors affecting crystal oscillators.

**Baud Rate Generation**

One of the most common ways to produce the periodic signal required by serial
interfaces to produce a desired data transfer rate is through a dedicated timer. In
such applications, the timer's clock frequency is divided by the desired baud rate to
determine the number of counts between successive bits or *bit time*. The resulting
value can then be used as the timer's counter value.

Although baud rate generation using this method is possible, it deserves to mention
that most contemporary microcontrollers include serial communication modules with
dedicated baud rate generators. A detailed explanation of how to implement baud
rate generation is presented in Sect. 9.5.

**Pulse Width Modulation**

Pulse-width modulation (PWM) is another widely used timer application in
embedded systems. A PWM module produces a periodic signal whose duty cycle is
controlled by the MCU. Actually, since the signal frequency is also controlled by the
MCU, we can say the both, signal period and duty-cycle, can be controlled in this
application.

Figure 7.19a illustrates the structure of a PMW module. It fundamentally contains an *n*-bit timer (with clock selector and prescaler) whose count is compared in

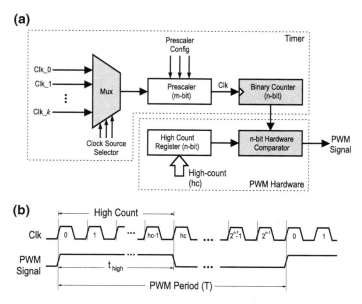

**Fig. 7.19**   Pulse-widt modulation module: (**a**) hardware structure, (**b**) signal output

hardware to the contents of a "high-count register" ($hc$). While the timer has a value less than $hc$ the PWM signal is high. Otherwise the PWM signal is low.

The signal relations can be observed in Fig. 7.19b. Note that the the number of bits ($n$) in the timer determine the PWM resolution, while the value stored in $hc$ controls the signal pulse width (duty cycle = $t_{high}/T$. The signal frequency can be controlled through the prescaler and/or the selected input frequency in the multiplexer (Mux).

Pulse width modulation is widely used in energy control systems as the amount of energy in the PWM is a function of the signal duty cycle. Applications like DC motor control speed, heater's temperature, light intensity in LEDs, and even musical tones can be implemented with PWM. This widespread of applications make PWM modules a useful addition to the list of MCU peripherals.

**Example 7.9**   *Consider an application where we wish to control the brightness of an illumination LED using PWM. In LEDs, brightness is a function of the average current intensity passing through the LED junction.*

*The LED in this application, an SR-05-WC120, produces a maximum brightness of 240 lumens (lm) when applied its maximum current of 720mA. Assume we want to make an MCU controlled dimmer for this LED able to produce 8 different intensity levels: 0lm, 30lm, 60lm, 90lm, 120lm, 150lm, 180lm, and 210lm. Assuming a 3.3V compatible MCU and a 5V power supply for the LED, the interface would require a buffer to drive the LED from an I/O port. The MCU is assumed to provide a maximum of 20mA per pin (See section 8.8 for analysis). Figure 7.20 shows the recommended interface and required duty cycle values.*

Note that each brightness level corresponds to a 1/8 increase in duty cycle. As we are specifying only eight levels, including 0% (LED off), the maximum specifiable brightness value would be 7/8 (why?).

Assuming an 8-bit timer and a selected clock frequency of 32.768KHz, the values to be loaded in the "count-high" register for the corresponding levels of brightness would be in increments of 1/8 of $2^8$, this is: 0, 32, 64, 96, 128, 160, 192, or 224. Note that the value required for maximum LED brightness, 256, would not be possible to be specified under this scheme.

A few observations:

- The default frequency of 32.768KHz was used, meaning the prescaler was ser to divide-by-one.
- The LED blinking rate for the above frequency rating would be 32768Hz/256 = 128Hz. A rate fast enough to avoid visible flickering. (What would be the possible maximum prescaler value if the blinking rate were to be reduced the closest to the barely minimum 0f 24Hz to avoid visible flickering?).
- Although the number of brightness levels in this example was specified at eight, it would be possible to have up to 256 levels. This is the maximum resolution that can be obtained with an 8-bit counter.
- We leave as an exercise to the reader writing a program to implement this solution in the microcontroller of your preference.

### 7.4.4 MSP430 Timer Support

All MSP430 microcontrollers include a Watchdog Timer and at least a timer called Timer_A. For this timer, some models have just a basic configuration while others

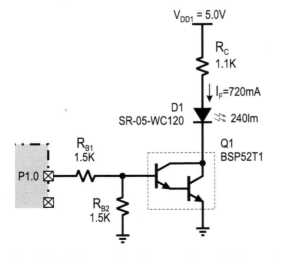

| Brightness (lm) | Duty Cycle |
|---|---|
| 0 | 0% |
| 30 | 12.5% |
| 60 | 25.0% |
| 90 | 37.5% |
| 120 | 50.0% |
| 150 | 62.5% |
| 180 | 75.0% |
| 210 | 87.5% |
| 240 | 100.0% |

**Fig. 7.20** LED connection diagram and required duty cycle values

include a second Timer_B, and the most recent models include yet a third timer called Timer_D. Depending on the model and generation, some models include from one to three Real Time Clocks. In the sections below we discuss the basic features in the timer support offered by MSP430 microcontrollers. For fine details about each type of timer unit, the reader is referred to the corresponding device User's Manuals and data sheets.

## MSP430 Watchdog Timer

The main purpose of the Watchdog Timer (WDT), as explained in Sect. 7.4.3, is basically to reset the system when a software problem happens. The WDT may be put in watchdog mode, regular interval timer mode, or stopped. Since the watchdog mode in the WDT is automatically configured after a PUC, the user must setup, and/or take care to stop the WDT prior to the expiration of the reset interval.

Configuration of the WDT is performed via a 16-bit read/write watchdog timer control register WDTCTL illustrated in Fig. 7.21. This register is password protected. When read, WDTCTL will read 069h in the upper byte. When written, the write password 05Ah must be included in the upper byte, otherwise a security key violation will occur and a power-up clear (PUC) reset will be triggered. Bits 7 to 0 work as follows:

**WDTHOLD (Bit 7)** Watchdog timer hold. If 0, WDT is not stopped; if 1, WDT is stopped and conserves power.

**WDTNMIES (Bit 6)** Watchdog timer NMI edge select. When WDTNMI = 1 it has the following effect:

- 0: NMI happens on rising edge
- 1: NMI happens on falling edge

**WDTNMI (Bit 5)** This bit selects the function for the $\overline{\text{RST}}$/NMI pin: When 0, it has a Reset function; when 1, the pin has an NMI function

**WDTTMSEL (Bit 4)** Watchdog timer mode select: 0 for Watchdog mode; 1 for Interval timer mode

**WDTCNTCL (Bit 3)** Watchdog timer counter clear. Setting WDTCNTCL = 1 – by software – clears the count value of the counter, WDTCNT, to 0000h. This bit is automatically reset.

**WDTSSEL (Bit 2)** Watchdog timer clock source select: 0 for the SMCLK; 1 for the ACLK

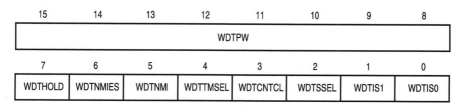

**Fig. 7.21** MSP430 Watchdog Timer Control Register (WDTCTL)

**WDTISx (Bits 1-0)** Watchdog timer interval select. These bits select the watchdog timer interval to set the WDTIFG flag and/or generate a PUC. The alternatives are

- WDTIS0 for 00: Watchdog clock source / 32768 (Default)
- WDTIS1 for 01: Watchdog clock source / 8192
- WDTIS2 for 10: Watchdog clock source / 512
- WDTIS3 for 11: Watchdog clock source / 64

Some useful examples of instructions associated to the WDT configuration are the following.

```
mov #WDTPW+WDTHOLD,&WDTCTL ; To stop WDT

mov #WDTPW+WDTCNTCL,&WDTCTL ;Reset WDT

mov #WDTPW+WDTCNTCL+WDTSSEL,&WDTCTL ; select ACLK clock

;WDT interval timer mode with ACLK, and interval clock/512
mov #WDTPW+WDTCNTCL+WDTMSEL+WDTIS2,&WDTCTL
```

The watchdog timer includes a 16-bit watchdog timer+ (WDT+) counter WDTCNT not accessible to the user, whose operation and source are controlled through the WDTCTL register. As illustrated in the above code lines, WDTCNTCL should be included in any instruction affecting the counter to avoid unwanted results such as an accidental PUC.

Table 7.3 shows different features of the implementation of the WDT across the different MSP430 families. Notice that the number of available software selectable time intervals varies with the MSP430 version from four in the x4xx, x1xx, and x2xx versions to eight in the x3xx and x5xx/x6xx versions. Observe also that the RST/NMI pin function control is absent in the x5xx/x6xx versions but available in all the other versions.

**Table 7.3** Watchdog timer implementation features

| WATCHDOG TIMER | x3xx | x4xx | x1xx | x2xx | x5xx/x6xx |
|---|---|---|---|---|---|
| Software selectable time intervals | 8 | 4 | 4 | 4 | 8 |
| Watchdog mode | Yes | Yes | Yes | Yes | Yes |
| Interval mode | Yes | Yes | Yes | Yes | Yes |
| Password protected write | Yes | Yes | Yes | Yes | Yes |
| Selectable clock source | Yes | Yes | Yes | Yes | Yes |
| Power conservation | Yes | Yes | Yes | Yes | Yes |
| Clock fail-safe feature | No | Yes | Yes | Yes | Yes |
| RST/NMI pin function control | Yes | Yes | Yes | Yes | No |

The clock failsafe protection identified as WDT+ was added starting with the x4xx version of MSP430 microcontrollers. Failsafe means that the clock to the WDT+ cannot be disabled if the watchdog function is enabled. This is important because, if for example ACLK is the clock source for the WDT+, then the WDT+ will prevent ACLK from being disabled and thus LPM4 will not be available. If SMCLK is the clock source for the WDT+ and the user chooses LPM3 the WDT+ will not allow SMCLK to be disabled, increasing power consumption in LPM3.

If either ACLK or SMCLK fail while watchdog mode is enabled, the clock source is automatically switched to MCLK. If this happens with MCLK using a crystal as its source and the crystal fails, then the DCO will be activated by the failsafe to source MCLK. There is no failsafe protection for when the WDT is in interval timer mode. This mode is selected with bit 4, WDTTMSEL, in the WDTCTL.

Two interrupts are associated to the WDT operation: the NMI interrupt and the WDT interrupt. They are signaled by flags MMIIF and WDTIFG respectively, which may be cleared or set by software too. These flags are located in the special function register IF1. Enabling and disabling are handled with bits NMIIE and WDTIE in the special function register IE1 using the instructions

```
bis.b #NMIIE,&IE1; enable NMI interrupt
bis.b #WDTIE,&IE1; enable WDT interrupt
bic.b #NMIIE,&IE1; disable NMI interrupt
bic.b #WDTIE,&IE1; disable WDT interrupt
```

WDTIFG is set at the expiration of the selected time interval. In WDT mode, it generates a PUC. In this case, the WDTIFG flag can be used by the reset vector interrupt service routine to determine the cause of the reset. If WDTIFG is set, then the reset condition was initiated by either the expiration of the time interval or by a security key violation. On the other hand, if WDTIFG is cleared then the reset was caused by some other source.

In interval timer mode, the WDTIFG flag is set after the time interval expires but will trigger an interrupt only if both the WDTIE and GIE bits are set. The WDTIFG is reset automatically in interval mode when the interrupt is serviced. It is important to note that the interval timer interrupt vector is different from the reset vector used in watchdog mode.

Let us close this discussion with an example. The following code was written by D. Dang from Texas Instruments [80].[7] The standard constant WDT_MDLY_32 is defined in the header by the declaration:

#define WDT_MDLY_32   (WDTPW+WDTTMSEL+ WDTCNTCL).

**Example 7.10** *Toggle P1.0 using software timed by the WDT ISR. Toggle rate is approximately 30 ms based on default DCO/SMCLK clock source used in this example for the WDT.* $ACLK = n/a, MCLK = SMCLK = default DCO$

```
;=============================================================================
;MSP430G2xx1 Demo - WDT, Toggle P1.0, Interval Overflow ISR, DCO SMCLK
;By D. Dang - October 2010
```

---

[7] See Appendix E.1 for terms of use.

```
;Copyright (c) Texas Instruments, Inc.
;------------------------------------------------------------------------
#include <msp430.h>
;------------------------------------------------------------------------
            ORG      0F800h                  ; Program Reset
;------------------------------------------------------------------------
RESET       mov.w    #0280h,SP               ; Initialize stackpointer
SetupWDT    mov.w    #WDT_MDLY_32,&WDTCTL     ; WDT ~30ms interval timer
            bis.b    #WDTIE,&IE1             ; Enable WDT interrupt
SetupP1     bis.b    #001h,&P1DIR            ; P1.0 output
                                             ;
Mainloop    bis.w    #CPUOFF+GIE,SR          ; CPU off, enable interrupts
            nop                              ; Required only for debugger
                                             ;
;------------------------------------------------------------------------
WDT_ISR;    Toggle P1.0
;------------------------------------------------------------------------
            xor.b    #001h,&P1OUT            ; Toggle P1.0
            reti                             ;
                                             ;
;------------------------------------------------------------------------
;           Interrupt Vectors
;------------------------------------------------------------------------
            ORG      0FFFEh                  ; MSP430 RESET Vector
            DW       RESET                   ;
            ORG      0FFF4h                  ; WDT Vector
            DW       WDT_ISR                 ;
            END
```

## MSP430 Timer_A: Structure and Counter Characteristics

The MSP430 family supports three timers. Timer_A, present in all models; Timer_B included in all but the legacy 3xx series; and Timer_D, appearing in the 5xx/6xx series. Any timer can be used for applications such as real-time clock, pulse width modulation, or baud rate generation, among others. We focus our explanation on Timer_A in this section since it is the timer included in all MSP430 chips. Timer_B and Timer_D features are only summarized later on. The reader can consult device User's Guide for more details on details of a particular device timer.

A simplified block diagram for Timer_A is depicted in Fig. 7.22. Timer_A is a 16-bit timer/counter (TAR), with at least two capture/compare registers TACCR0 and TACCR1, configurable PWM outputs, and interval timing. The timer also has ample interrupt capabilities with interrupts that may be generated on overflow conditions and from the capture/compare registers. These interrupts are decoded using an interrupt vector register TAIV that encompasses all Timer_A interrupts.

Table 7.4 summarizes the implementation features of Timer_A. Except for the now legacy x3xx series which does not include PWM capability, asynchronous input and output latching, or an interrupt vector register, the implementation is quite uniform across the different series.

The timer's operation is software configurable with the Timer_A control register TACTL, shown in Fig. 7.23. A summarized description of the control bits in TACTL is as follows:

• Bit TAIFG is the counter's interrupt flag, and bit TAIE enables the timer interrupt.

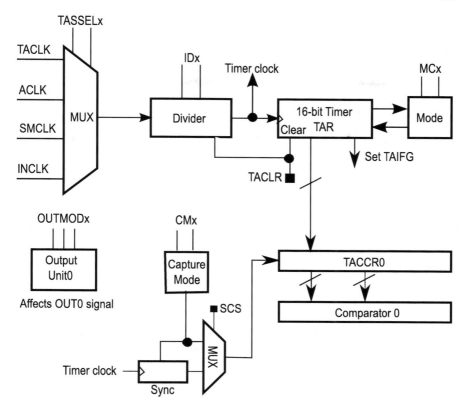

**Fig. 7.22** Simplified MSP430 Timer_A block diagram

**Table 7.4** Timer_A implementation features

| TIMER_A | x3xx | x4xx | x1xx | x2xx | x5xx/x6xx |
|---|---|---|---|---|---|
| 16-bit timer/counter w/4 operating modes | Yes | Yes | Yes | Yes | Yes |
| Asynchronous | No | Yes | Yes | Yes | Yes |
| Selectable and configurable clock source | Yes | Yes | Yes | Yes | Yes |
| Independently configurable CCRs | 5 | 3 or 5 | 3 | 2 or 3 | up to 7 |
| PWM output capability | No | Yes | Yes | Yes | Yes |
| Asynchronous I/O latching | No | Yes | Yes | Yes | Yes |
| Interrupt vector register | No | Yes | Yes | Yes | Yes |
| Second Timer_A | No | Yes | No | No | No |

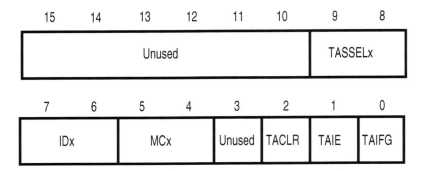

**Fig. 7.23**  MSP430 Timer_A Control Register (TACTL)

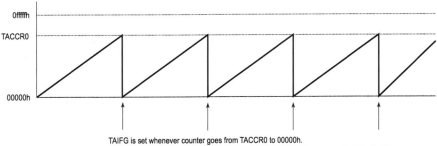

TAIFG is set whenever counter goes from TACCR0 to 00000h.
TACCR0 CCIFG is set whenever the timer counts to TACCR0, i.e. one count prior to TAIFG.

**Fig. 7.24**  Timer operating in up mode

- Setting TACLR clears the counter, the clock divider and the count direction when the timer is in the up/down mode.
- TASSELx bits select the clock source: TACLK (00), ACLK (01), SMCLK (10), or INCLK (11)
- IDx bits determine the frequency division factor in the prescaler: 1 (00), 2 (01), 4 (10), and 8 (11)
- MCx bits set the operation mode: Halt (00), up mode (01), continuous mode (10), and up/down mode (11).

Except when modifying the interrupt enable TAIE and interrupt flag TAIFG, the timer should always be stopped before modifying its settings. Any modification will take effect immediately. Also, to avoid unpredictable results, stop the timer before reading the counter.

The three non-stop modes are illustrated in Figs. 7.24, 7.25, and Fig. 7.26, respectively.

In the up and up/down modes, the timer can be stopped writing a 0 to TACCR0 and started from zero by writing a nonzero value to TACCR0. This register sets the upper count limit, which must be different from 0FFFFh. In both cases the TACCR0

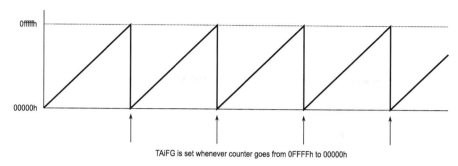

**Fig. 7.25**   Timer operating in continuous mode

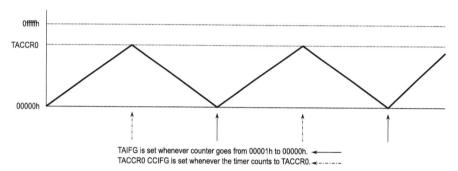

**Fig. 7.26**   Timer operating in up/down mode

CCIFG is set by the (TACCR0 - 1) to TACCR0 transition. Interrupt generation by TAIFG differs depending on the mode:

- In the up mode the TAIFG flag is set by the TACCR0-to-zero transition, and
- in the up/down mode the TAIFG is set when count goes from 1h to 0h.

In the continuous mode the counter always counts up to 0FFFFh and sets the TAIFG interrupt flag in the 0FFFFh to 0h transition. This mode is particularly useful for generating independent time intervals controlled by hardware and output frequencies with an interrupt generated each time a time interval is completed without impact from interrupt latency. Each of the output/compare registers can be used independently to generate different time intervals or output frequencies.

If the contents of TACCR0 were changed while the timer is counting up, and the new period were greater than the current count, it would continue up to the new value. Otherwise, the new period would not take place until the count reaches 0.

We end this discussion on timer_A with an example. The following code was written by D. Dang from Texas Instruments [80].[8]

---

[8] See Appendix E.1 for terms of use.

**Example 7.11** *Toggle P1.0 using software and TA_0 ISR. Toggles every 50000 SMCLK cycles. SMCLK provides clock source for TACLK During the TA_0 ISR, P1.0 is toggled and 50000 clock cycles are added to CCR0. TA_0 ISR is triggered every 50000 cycles. CPU is normally off and used only during TA_ISR.* $ACLK = n/a, MCLK = SMCLK = TACLK = default\ DCO$

```
;=============================================================================
;MSP430G2xx1 Demo - Timer_A, Toggle P1.0, CCR0 Cont. Mode ISR, DCO SMCLK
;By D. Dang - October 2010
;Copyright (c) Texas Instruments, Inc.
;-----------------------------------------------------------------------------
#include <msp430.h>
;-----------------------------------------------------------------------------
            ORG     0F800h                  ; Program Reset
;-----------------------------------------------------------------------------
RESET       mov.w   #0280h,SP               ; Initialize stackpointer
StopWDT     mov.w   #WDTPW+WDTHOLD,&WDTCTL   ; Stop WDT
SetupP1     bis.b   #001h,&P1DIR            ; P1.0 output
SetupC0     mov.w   #CCIE,&CCTL0            ; CCR0 interrupt enabled
            mov.w   #50000,&CCR0            ;
SetupTA     mov.w   #TASSEL_2+MC_2,&TACTL   ; SMCLK, contmode
                                            ;
Mainloop    bis.w   #CPUOFF+GIE,SR          ; CPU off, interrupts enabled
            nop                             ; Required only for debugger
                                            ;
;-----------------------------------------------------------------------------
TA0_ISR;    Toggle P1.0
;-----------------------------------------------------------------------------
            xor.b   #001h,&P1OUT            ; Toggle P1.0
            add.w   #50000,&CCR0            ; Add Offset to CCR0
            reti                            ;
                                            ;
;-----------------------------------------------------------------------------
;           Interrupt Vectors
;-----------------------------------------------------------------------------
            ORG     0FFFEh                  ; MSP430 RESET Vector
            DW      RESET                   ;
            ORG     0FFF2h                  ; Timer_A0 Vector
            DW      TA0_ISR                 ;
            END
```

## MSP430 Timer_A Capture/Compare Units

A brief introduction to the capture/compare registers is provided here. More details can be found in the user guides or data sheets.

We can operate the capture/compare blocks in either the capture mode or the compare mode. The first mode is useful to capture timer data, whereas the compare mode is used to generate PWM output signals or interrupts at specific time intervals. We will explain the capture/compare features of the MSP430 timers using Timer_A, although these blocks are available in other timers as well.

The Timer_A capture/compare registers TACCRx, $x = 0, 1, 2$, are configured with the capture/compare control register TACCTLx, illustrated in Fig. 7.27. The bits and bit groups of this register will be explained as needed.

**Capture mode** To operate in capture mode we must have CAP $= 1$.

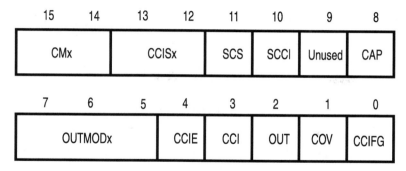

**Fig. 7.27** Timer_A Capture/Compare Control Register TACCTLx

When a capture occurs, the TAR timer data is copied into the corresponding TACCRx register and the TACCRx CCIFG interrupt flag is set. In this way the user can record time computations and use them, for example, to measure time, calculate the speed of an event, or any other applications. Capture is done using either external or internal signals connected to capture inputs which are selected with the CCISx bits. It is also possible to initiate capture by software.

Complementing the capture input selection, the CMx bits allow us to select the capture input signal edge: rising, falling, or both. If a second capture were performed before the previous one was read, then the capture overflow bit or COV, would be set to indicate this condition.

A race condition may occur when the capture input signal is asynchronous to the clock. In order to avoid this situation, it is recommended to synchronize both signals by setting the synchronize capture source bit, SCS.

**Compare mode** To operate in the compare mode we need CAP = 0. As mentioned earlier, this mode is used to generate PWM output signals or interrupts at specific time intervals.

When in capture mode, the capture/compare interrupt flag will be set, i.e. TACCRx CCIFG = 1 when counter TAR counts to the value stored in register TACCRx. This also sets an internal signal EQUx, i.e. EQUx = 1, which triggers an output according to a selected output mode. Furthermore, the capture compare input signal CCI will be latched into the synchronized capture/compare input SCCI.

The output modes available for the capture/compare blocks are discussed next.

### Timer_A Output Units

As it can be seen in the Timer_A block diagram, each capture/compare block has an output unit to generate output signals such as pulse width modulation or periodic signals. Bits 5, 6, and 7, i.e. the OUTMODx bits, in register TACCTLx define the output mode. These modes are described in Table 7.5 [32].

Let us now close this discussion on Timer_A with two examples.

**Example 7.12** *The following piece of code configures Timer_A to select the ACLK clock source and the up mode. After Low Power Mode 0 is entered, the CPU is only*

**Table 7.5** MSP430 output modes. *Courtesy of Texas Instruments, Inc.*

| MODx | Mode | Description |
|------|------|-------------|
| 000 | Output | The output signal OUTx is defined by the OUTx bit. The OUTx signal updates immediately when OUTx is updated |
| 001 | Set | The output is set when the timer counts to the TACCRx value. It remains set until a reset of the timer, or until another output mode is selected and affects the output |
| 010 | Toggle/Reset | The output is toggled when the timer counts to the TACCRx value. It is reset when the timer counts to the TACCR0 value |
| 011 | Set/Reset | The output is set when the timer counts to the TACCRx value. It is reset when the timer counts to the TACCR0 value |
| 100 | Toggle | The output is toggled when the timer counts to the TACCRx value. The output period is double the timer period |
| 101 | Reset | The output is reset when the timer counts to the TACCRx value. It remains reset until another output mode is selected and affects the output |
| 110 | Toggle/Set | The output is toggled when the timer counts to the TACCRx value. It is set when the timer counts to the TACCR0 value |
| 111 | Reset/Set | The output is reset when the timer counts to the TACCRx value. It is set when the timer counts to the TACCR0 value |

*awaken to handle the ISR after an interrupt caused by the capture/compare interrupt flag CCIFG. A LED at pin P1.0 is then toggled.*

```
#include    <msp430.h>
;-----------------------------------------------------------------------
            ORG     0F800h  ; Program Start
;-----------------------------------------------------------------------
RESET   mov   #0280h,SP                     ; initialize SP
        mov   #WDTPW+WDTHOLD,&WDTCTL        ; stop WDT
        bis.b #BIT0,&P1DIR                  ; output pin P1.0
        bic.b #BIT0,&P1OUT                  ; red LED off
        mov   #CCIE,&TACCTL0                ; CCR0 interrupt
        mov   #10000,&TACCR0                ; Load upper bound
        mov   #MC_1+TASSEL_1,&TACTL         ; up mode, ACLK
        bis   #LPM0+GIE,SR                  ; low power mode 0, interrupts
        nop                                 ; only to sync debugger
;-----------------------------------------------------------------------
;   TACCR0 ISR    ; Interrupt Service Routine
;-----------------------------------------------------------------------
TACCR0_ISR   xor.b #BIT0,&P1OUT             ; toggle LED
                                            ; CCIFG automatically reset
```

```
                                        ; when TACCR0 ISR is serviced
             reti                       ; return from ISR
;----------------------------------------------------------------------
;            Interrupt Vectors
;----------------------------------------------------------------------
             ORG     0FFFEh             ; Reset vector address
             DW      RESET              ; RESET label address
             ORG     0FFF2h             ; TACCR0 vector address
             DW      TACCR0_ISR         ; TACCR0_ISR address
             END
```

The following code illustrates the use of Timer_A to generate PWM. It was written by D. Dang from Texas Instruments [80].[9]

**Example 7.13** *This program generates one PWM output on P1.2 using Timer_A configured for up mode. The value in CCR0, 512-1, defines the PWM period and the value in CCR1 the PWM duty cycles. A 75% duty cycle is on P1.2. $ACLK = n/a$, $SMCLK = MCLK = TACLK = default DCO$*

```
;=======================================================================
;MSP430G2xx1 Demo - Timer_A, PWM TA1, Up Mode, DCO SMCLK
;By D. Dang - October 2010
;Copyright (c) Texas Instruments, Inc.
;----------------------------------------------------------------------
#include <msp430.h>
;----------------------------------------------------------------------
             ORG     0F800h      ; Program Reset
;----------------------------------------------------------------------
RESET        mov.w   #0280h,SP    ; Initialize stackpointer
StopWDT      mov.w   #WDTPW+WDTHOLD,&WDTCTL   ; Stop WDT
SetupP1      bis.b   #00Ch,&P1DIR           ; P1.2 and P1.3 output
             bis.b   #00Ch,&P1SEL           ; P1.2 and P1.3 TA1/2 otions
SetupC0      mov.w   #512-1,&CCR0           ; PWM Period
SetupC1      mov.w   #OUTMOD_7,&CCTL1       ; CCR1 reset/set
             mov.w   #384,&CCR1             ; CCR1 PWM Duty Cycle
SetupTA      mov.w   #TASSEL_2+MC_1,&TACTL  ; SMCLK, upmode
                                           ;
Mainloop     bis.w   #CPUOFF,SR            ; CPU off
             nop                           ; Required only for debugger
                                           ;
;----------------------------------------------------------------------
;            Interrupt Vectors
;----------------------------------------------------------------------
             ORG     0FFFEh               ; MSP430 RESET Vector
             DW      RESET                ;
             END
```

## Timer_B and Timer_D

These two timers are briefly discussed below. A more complete treatment of this topic can be found in the user guides.

**Timer_B** While Timer_B is practically identical to Timer_A, there are a few notable differences which are summarized in Table 7.6. In Timer_B, TBCLx rather than TACCRx, is used to determine interrupts, and TBCL0 takes on the role of TACCR0

---

[9] See Appendix E.1 for terms of use.

in count modes. Also, the capture/compare registers can be grouped so that multiple TBCCRs can be loaded together into TBCL. Notice that there is no synchronized capture/compare input SCCI bit function.

**Timer_D** introduces several important features with respect to Timer_A. The most important include:

- Providing for internal and external clear signals.
- Allowing for routing internal signals between Timer_D instances, and external clear signals.
- Introducing interrupt vector generation of external fault and clear signals.
- Generating feedback signals to the Timer capture/compare channels to affect the timer outputs.
- Supporting high resolution mode.
- Allowing to combine two adjacent TDCCRx registers in one capture/compare channel.
- Supporting dual capture event mode.
- Allowing for synchronizing with a second timer.

### Basic Timer and Real-Time Clock

The Basic Timer (BT) was included only in the first two generations of MSP430 devices, i.e. the x3xx and the x4xx. Figure 7.28 shows a simplified block diagram, in which we can appreciate the signals used to control its operation. Probably one of the most important contribution of this timer was the LCD control signal generator which provides the timing for common and segment lines.

The basic features of the Basic Timer are summarized next.

- Two independent 8-bit timers/counters, BTCNT1 and BTCNT2, that can be cascaded to produce a 16-bit counter.
- Selectable clock source for BTCNT2 among ACLK, ACLK/256, and SMCLK. BTCNT1 is driven with the ACLK.
- Interrupt capability.

**Table 7.6**  Timer_B distinctive features

| Feature | x4xx, x1xx, and x2xx | x5xx/x6xx |
|---|---|---|
| Programmable length (8, 10, 12, or 16 bits) | Yes | Yes |
| Double-buffered compare latches w/synchronized loading | Yes | Yes |
| Three-state Output | Yes | Yes |
| SCCI | No | No |
| Double-buffered TBCCRx registers | Yes | Yes |
| Configurable CCRs[1] | 3 or 7 | up to 7 |

Note 1: Timer_A has up to 3 CCRs

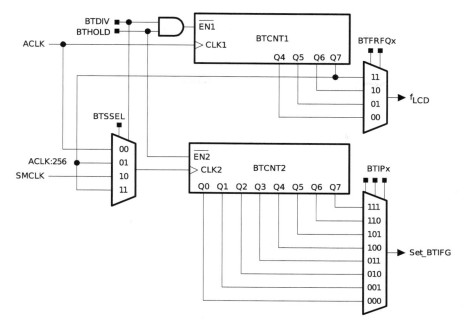

**Fig. 7.28** MSP430 Basic Timer_1 block diagram

As with any timer, the BT could be used as an interval timer or an event counter, or to implementing a real-time clock. Although RTC is not present in series x1xx or x2xx, the x5xx/x6xx series include three separate real time clocks: RTC_A, RTC_B, and RTC_C.

The Real Time Clock in the x4xx series has the following features:

- 32-bit counter module with calendar function.
- Calendar and clock mode.
- Automatic counting of seconds, minutes, hours, day of week, day of month, month, and year in calendar mode.
- Selectable clock source.
- Interrupt capability.
- Selectable BCD format.

Table 7.7 summarizes the features included of the three different real time clocks implemented in the x5xx/x6xx MSP430 devices.

**Final Remarks on Timer_A Interrupts**

Timer_A interrupts are maskable, which means that the GIE bit should be set in the Status Register to call an ISR. The module has two interrupt vectors associated with it: the TACCR0 interrupt vector for TACCR0 CCIFG, and the TAIV interrupt vector associated with all other CCIFG flags and with TAIFG.

**Table 7.7**  MSP430x5xx/x6xx RTC_A, B, and C implementations

| x5xx/x6xx<br>REAL TIME CLOCK | RTC_A | RTC_B | RTC_C |
|---|---|---|---|
| Configurable for calendar<br>  function or counter mode | Yes | No | Yes |
| Provides second, minutes,<br>  hours, day of week, day of<br>  month, month, and year<br>  in calendar mode | Yes | Yes | Yes |
| Leap year correction | No | Yes | Yes |
| Interrupt capability | Yes | Yes | Yes |
| Selectable BCD or binary format | Yes | Yes | Yes |
| Calibration logic for offset<br>  correction in RTC mode | Yes | Yes | See below |
| Crystal offset error and<br>  temperature drift | No | No | Yes |
| Programmable Alarms | No | Yes | Yes |
| Operation in LPMx5 | No | Yes | LPM3.5 |
| Protection for RTC<br>  registers | No | No | Yes |
| Operation from a<br>  separate voltage supply<br>  with programmable charger<br>  (device-dependent) | No | No | Yes |

By hardware, the CCIFG flag is set in capture mode when a value is captured in register TACCRx or in compare mode when TAR counts to the value in TACCRx. The CCIFG action is enabled with the CCIE bit.

When an interrupt request is serviced, i.e. when RETI is executed, the CCIFG flag will be reset. On the other hand, we could also set or clear the CCIFG flag using software. Whenever the corresponding CCIE and GIE bits are set, the CCIFG flags request an interrupt. Timer_A interrupt flag, TAIFG, is bit 0 of the Timer_A Control Register, TACTL whereas CCIE and CCIFG correspond to bit 4 and bit 0 from the Capture/Compare Control Register (TACCTLx).

The TACCR0 CCIFG flag has the highest priority from among the Timer_A interrupt sources. This means that, if the interrupt flags corresponding to TACCR0 and any of the other TACCRx, say TACCR1 are set, then the TACCR0 interrupt will be serviced first, as long as there is not a higher priority interrupt pending. The rest of Timer_A interrupt sources share a common interrupt vector. This means that TACCR1 CCIFG, TACCR2 CCIFG, and TAIFG are sourced by a single interrupt vector. In this case, register TAIV is used to determine which flag has requested an interrupt, and also as an arbiter for multiple Timer_A interrupts. Figure 7.29 illustrates the TAIV register.

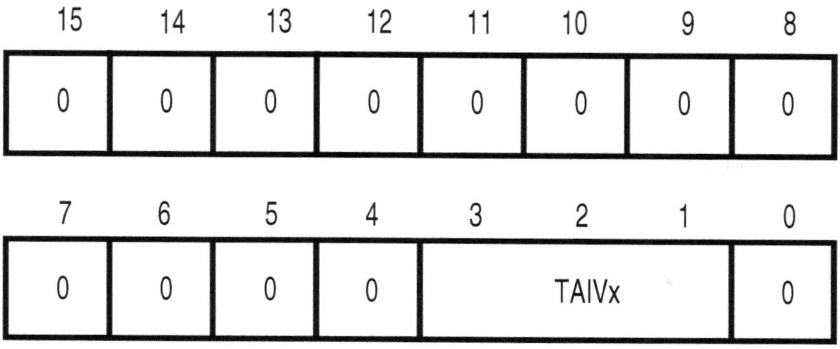

| 15 | 14 | 13 | 12 | 11 | 10 | 9 | 8 |
|----|----|----|----|----|----|----|----|
| 0 | 0 | 0 | 0 | 0 | 0 | 0 | 0 |

| 7 | 6 | 5 | 4 | 3 | 2 | 1 | 0 |
|----|----|----|----|----|----|----|----|
| 0 | 0 | 0 | 0 | | TAIVx | | 0 |

**Fig. 7.29** Timer_A Interrupt Vector Register TAIV

Only bits 3, 2, or 1 in TAIV can have a nonzero value. This means that the only values available are even from 00h to 0Eh. From this set, the following cases are highlighted:

- 00h: No interrupt pending
- 02h: Interrupt source is capture/compare 1 (TACCR1 CCIFG), with highest priority.
- 04h: Interrupt source is capture/compare 2 (TACCR2 CCIFG)
- 0Ah: Interrupt source is Timer Overflow (TAIF), with lowest priority.

All other values are reserved. It should be noted that reading from or writing to the TAIV register will automatically reset the highest pending interrupt flag.

The following example from [32] shows an skeleton code used as an illustration on how to handle the different Timer_A interrupt sources.

**Example 7.14** *This example from [32]*[10] *is used to show the recommended use of TAIV. Note how the TAIV value is added to the PC register in order to reference the appropriate piece of code from the interrupt handler or interrupt service routine. Courtesy of Texas Instruments, Inc.*

```
;   Interrupt handler for TACCR0 CCIFG
CCIFG_0_HND
;   ... ; Start of handler
reti
;   Interrupt handler for TAIFG, TACCR2, and TACCR1
TA_HND
...
add &TAIV,PC; Add offset to Jump table
reti    ; Vector 0: No interrupt
jmp CCIFG_1_HND ; Vector 2: TACCR1
jmp CCIFG_2_HND ; Vector 4: TACCR2
reti    ;  Vector 6: Reserved
reti    ;  Vector 8: Reserved
TAIFG_HND   ;  Vector 10: TAIFG Flag
```

[10] See Appendix E.1 for terms of use.

```
... ;   Task starts here
reti
CCIFG_2_HND        ;   Vector 4: TACCR2
... ;   Task starts here
reti  ;  Back to main program
CCIFG_1_HND           ;   Vector 2: TACCR1
... ;   Task starts here
reti     ;   Back to main program
```

## 7.5  Embedded Memory Technologies

Memory technologies in embedded systems provide the fundamental functions of storing programs and data, like in any other computing system. In embedded applications, however, the requirements of storage have traditionally being different in the sense of size and speed. Most small and distributed embedded applications require small amounts of data memory to hold process variables. These values typically require frequent modification so loosing them when the system is shut-down has no lasting effect. Most applications use volatile data memory and rarely require more than a few hundred bytes of space.

Program memory requirements are also moderate. As embedded applications are single functioned, in most cases, a few kilobytes of program memory results sufficient to accommodate the programs needed to complete an embedded application. Unlike data memory, program memory is required to hold its contents when the system is powered-down. This way the embedded system can regain its functionality when powered-up anew. Therefore program memory is provided in the form of non-volatile memory, and it is quite common to find that most embedded systems require less than 64KB of storage for programs.

Of course, if we move into the arena of high-performance systems, the memory requirements might dramatically increase. Depending on the number of tasks to be performed by the system and the amount of data it needs to process to perform its function, will be the requirements of program and data memory. However, when considering the universe of embedded applications, these cases requiring large amounts of memory are the least frequent.

### 7.5.1  A Classification of Memory Technologies

Embedded memory can be classified according to two main criteria: *storage permanence* and *write ability*. The first criterion refers to the ability of a programmed memory to hold its contents, while write ability, refers to how easily the memory contents can be modified.

Looking at all the memory technologies that have reached mainstream (ROM, PROM, EPROM, EEPROM, FLASH, and RAM), masked, read-only memory (ROM) earns the mark as the least writable and most permanent of all types. The zeros and ones defining a ROM contents are hardcoded in the chip through the metallization layer in their fabrication process. This characteristic also makes the memory non-volatile. Recall that volatility in memory refers to the loss of contents when the chip is de-energized.

Programmable ROM (PROM) closely follows ROM in terms of write ability and permanence. PROM contents is written by selectively blowing internal metal fuses in each cell via a dedicated programming circuit. This writing method makes cells non-volatile, but only allows writing their contents once as blown fuses cannot be brought back to an un-blown state. PROMs are also called One-time Programmable non-volatile memories or OTP NVM. They are considered more writable than ROM because un-programmed, blank chips can be programmed in the field.

Erasable PROM (EPROM for short), marked a breakthrough in memory technology as it provided a non-volatile cell that could be erased and reprogrammed many times. This was enabled by the introduction of floating-gate transistors, a MOSFET arrangement able to trap charges in its gate by magnifying the effect of "hot carriers". Charges trapped in the double gate structure increase the transistor threshold voltage. Thus the minimum voltage needed to turn on a non-charged transistor will not turn on a charged one, making possible to use the non-stored versus stored charge states as ones and zeros, respectively, in the memory cells. Moreover, the charge retention in the floating gate is retained even when the cell is de-energized, making it non-volatile. Conveniently, bombarding the gate with high energy particles, like UV light photons, allows for removing the charge accumulation, thus erasing the cell.

Erasing an UV EPROM required extended exposure (up to 15 min) to an UV light source. EPROMS were fabricated with a convenient transparent window facilitating the erasure procedure. Writing the cell required application of relatively high voltages (12–18 V) provided by an external circuit for relatively extended periods (milliseconds) compared to the typical read access time of the cell (nanoseconds). Due to these limitations, although EPROMs could be re-written, they needed to be physically removed from the application to be erased and then re-programmed to modify their contents. Nevertheless, with its ability to reprogram its contents, EPROMs moved up in the write-ability scale with respect to PROM. However, their susceptibility to loose contents due to radiation and gradual charge loss gave it a lower mark in storage permanence than ROM and PROM technologies. Under normal conditions an EPROM can retain its contents for over a decade and tolerate erase-write cycles up to thousands of times (the double gate charge trapping ability is lost with time).

The form of modifying EPROM cells brought-up the issue of how the write-ability was provided in a memory chip. In this respect, it is common to distinguish between off- and in-system write ability. The "in-system" denomination is given to the type of memory whose contents can be modified without physically removing it from the system or board where it is installed. Off-system writable memories require moving the chip to specialized hardware to erase them and then re-writing new contents in order to be modified.

EPROM technology further evolved to give floating transistors the ability to release charge by the application of an electric field. This improvement led to the technology of Electrically Erasable-programmable Read-only Memory (EEPROM). This new cell technology shared the characteristics of EPROMs in terms of non-volatility and storage permanence, but without the necessity of UV to erase the cells. Moreover, the provision of on-chip circuitry to erase and reprogram the cells, enabled this technology to eventually become in-system re-writable, moving up its score in terms of write ability. Further development in floating-gate-based cells and their interface lead to the FLASH technology, discussed in more detail in Sect. 7.5.2.

While non-volatile memory technologies evolved, volatile memory has also seen improvement in different aspects. Non-volatile technologies moved from magnetic core-based to semiconductor-based technologies. Two types of volatile memory are predominant now-a-day: static and dynamic cells.

A static cell uses a flip-flop as storage element. The flip-flop retains its contents as long as the cell is energized, reason for the static denomination. A dynamic cell uses a capacitor as storage element. Due to unavoidable charge leak, the capacitor contents needs to be periodically refreshed to avoid information loss. This refresh requirement is what gives the dynamic denomination to this type of memory. The capacitor contents is also lost when the cell is de-energized, making the technology volatile.

Semiconductor-based volatile memories score the highest mark in terms of write ability due to easiness of being quickly read and written. At the same time, its volatility makes it score the lowest mark in terms of storage permanence. Because of historical reasons, volatile, read-write memory has been called RAM.[11] Two important parameters associated to RAM technologies are their reading and writing speeds, quantified as the memory access time. Static RAM (SRAM) is the fastest, but dynamic RAM (DRAM) is preferred for large memory arrays because of its density. Newer technologies like Ferroelectric RAM (FRAM) are begining to show up in the market. This technology uses ferroelectric materials to implement a micro-magnetic capacitor as storage element, inheriting the non-volatility of early core memory cells with their inherent read-write ability. Current FRAMs are not as dense as the other types of semiconductor memories, but because of its non-volatility advantages result an attractive alternative for embedded applications.

Contemporary embedded systems use fundamentally static RAM for data memory. The relatively small requirements of RAM can be easily accommodated on-chip. Moreover, the modest program memory requirements, coupled with their non-volatility characteristic are accommodated in most contemporary systems using FLASH. The storage permanence, density, and in-system programmability characteristics of embedded FLASH in contemporary microcontrollers make necessary to delve a bit deeper into this technology, as discussed in the next section.

---

[11] In its origin, Random Access Memory (RAM) designated the type of memory whose contents could be randomly addressed, unlike Sequential Access Memory (SAM) where contents had to be sequentially accessed.

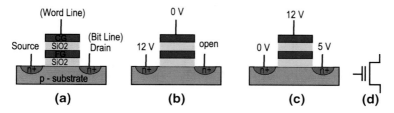

**Fig. 7.30** Floating Gate Mosfet (FGMOS) (**a**) Basic structure; (**b**) erasing process; (**c**) Programming; (**d**) Symbol

## 7.5.2 Flash Memory: General Principles

Flash is a type of EEPROM differentiated by its own name because it may be written and read in blocks. Other EEPROM's need to be completely erased byte by byte, or by very small blocks in some cases, making the process slow.[12]

Memory chips based on floating gate transistors were introduced in the early 1970s. The first flash memory was built by Toshiba in 1984, but it was not until late 1980s that it became reliable enough for mass production. Nowadays, flash memory is the dominant memory type wherever a significant amount of non-volatile, solid state storage is needed.

When flash memory operates in *read mode*, it works like any ROM. Flash memory offers fast read access times, as fast as dynamic RAM, although not as fast as static RAM. The other modes of operation are *erase* and *program*. In the first case, the cells are taken to an "erased state", which may be a logic 0 or logic 1, depending on the hardware configuration. In the programming mode, the cells are taken to (or left in) a logic 1 or logic 0, depending on the programming needs.

Another important and distinctive feature of flash is its capability to write in one block while reading another one, as far as the two blocks are not the same one. It is not possible to program and read simultaneously in the same block.

The basic element used in flash memory is a floating gate MOS transistor (FGMOS), whose structure is illustrated in Fig. 7.30a. This is a MOSFET transistor with two gates, one floating gate (FG) electrically isolated from the rest of the structure, and a control gate (CG) connected to the word line (address). The drain is connected to the bit line, that is, the data. Charge put on floating gate affects the threshold voltage. Since the FG is surrounded by an insulator, the trapped charge remains practically forever, unless modified by the user. When enough electrons are present on this gate, no current flows through the transistor. When electrons are removed from the floating gate, the transistor starts conducting. If a pull down p-mosfet is used, in the first case we have a logic 0, and in the second case a logic 1. The opposite happens if there is no pull-up transistor.

---

[12] The term "flash memory" was coined because of the speed, like a flash.

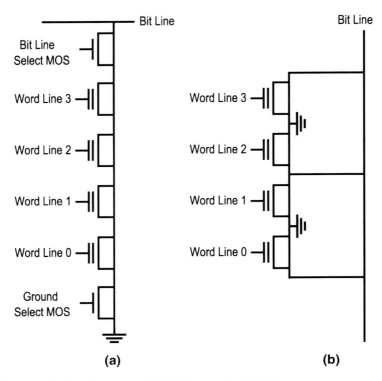

**Fig. 7.31**  Basic flash architectures: **a** NAND Flash, and **b** NOR Flash

Charging and discharging the FG is done with a relatively high voltage, say 12 V, applied to the CG. Ground or a lower voltage is applied to either the source or drain, as illustrated in Fig. 7.30b, c.

There are two main types of flash memory, NAND and NOR logic gate, so named because of their structures, illustrated in Fig. 7.31. NAND type flash memory may be written and read in blocks (or pages) which are generally much smaller than the entire device. Compared to the NOR type, they are faster to write but slower to read. The NOR type has better endurance and it allows a single machine word (byte) to be written or read independently. Both types are used in several applications such as personal computers, digital audio players, digital cameras, and so on. However, NOR type is preferred for code storage and NAND type for data storage. The reader may consult for more detailed hardware descriptions and working principles elsewhere.

Flash memory is not forever, of course. It can withstand only a limited number of program-erase cycles. The NOR flash endures more cycles than the NAND type. However, this depends not only on the architecture, but also on how the cycles are executed. If a particular memory block is programmed and erased repeatedly without writing to any other blocks, it will wear out before all the other blocks and the whole storage device will become useless. Also, different cells can be or become worn out or not functioned correctly, causing malfunction of the memory.

To take care of these issues, and to manage the program-erase cycles in flash memory devices, or in systems with this type of memory, engineers include a flash memory controller.

**Flash Memory Controller**

A flash memory controller, or flash controller, manages the data stored on flash memory and communicates with the CPU, computer or whatever electronic device the memory is serving.

Initially, at fabrication, the flash controller starts by formatting the flash memory and to ensure that the device is operating properly. It maps out bad flash memory cells, and allocates spare cells to be substituted for future failed cells. Part of the spare cells is also used to hold the firmware which operates the controller and other special features for the particular storage device. Finally, the controller creates a directory structure to allow conversion of requests for logical sectors into the physical locations on the actual flash memory chips.

The system communicates with the controller whenever it needs to read data from or write data to the memory. The controller uses a technique called *wear leveling* to distribute writes as evenly as possible across the flash blocks to prevent unfair wearing of blocks. Ideally, wear leveling technique enables every block to be written to its maximum life.

### 7.5.3 MSP430 Flash Memory and Controller

Except for the original 'x3xx series, now obsolete, all other series have models with flash memory. MSP430 flash memory is partitioned as follows:

- Information segments A, B, C and D.
- Bootstrap loader (BSL) segments A, B, C and D
- Main flash memory, partitioned in banks

  – Banks are partitioned into segments 0, 1, 2, ....

Only series '5xx/'6xx have BSL segments. Small models and some series may contain only one or two information segments. Also, the main memory may consist of only one bank, depending on the size. Flash memory sizes may be as small as 0.5 kB (model 'G2001) or up to 512 kB as found in some models of the 'x6xx series.

All flash-based MSP430 devices incorporate a flash controller that allows for the execution of code from the same flash module in which the software is concurrently modifying the data or re-programming code segments. The flash module of MSP430 consists of three blocks:

- Control logic: State machine and the timing-generator control of flash erase/ program

**Table 7.8**  Flash control registers in MSP430 models

| Series | FCCTL1 | FCCTL2 | FCCTL3 | FCCTL4 |
|--------|--------|--------|--------|--------|
| x1xx   | Yes    | Yes    | Yes    | No     |
| x2xx   | Yes    | Yes    | Yes    | Yes1   |
| x4x    | Yes    | Yes    | Yes    | Yes1, 2 |
| x5/xx  | Yes    | No     | Yes    | Yes    |
| x6/xx  | Yes    | No     | Yes    | Yes    |

1: Not available in all devices. Must consult the specific device data sheet 2: The higher byte is 00h on reset for this series

- Flash protection logic: Protection against inadvertent erase/program operations
- Programming voltage generator: An integrated charge pump that provides all voltages required for flash erase/program

The module controller consists of four 16-bit Flash Controller Control registers, FCCTLx, $x = 1, \ldots 4$, not all present in all models, as shown in Table 7.8. All registers are password protected. The high byte is 96h at reset or after a writing execution, and must be A5h when writing.[13] The low bytes are illustrated in Fig. 7.32.

In addition to these registers, the Interrupt Enable 1 register, which contains information for several peripheral modules, houses as bit 5 the *Access Violation Interrupt Enable* bit, ACCVIE, which is used to enable Flash Memory Access Violation Interrupts (ACCVIFG).

Bits FSSELx from control register FCCTL2 select the clock source from ACLK, SMCLK, or MCLK for the flash timing generator. The frequency of the generator must comply with some specifications and often require the clock source frequency to be divided. This is accomplished with the FNx bits.

Memory size writing is controlled by the bits BLKWRT and WRT; erasing is controlled with MERAS and ERASE. The size of writing/erasing is as shown in Tables 7.9 and 7.10, respectively.

Information segment A is locked separately from all other segments with the LOCKA bit. When LOCKA = 1, Segment A cannot be written or erased, and all information memory is protected from erasure during a mass erase or production programming. When LOCKA = 0, it is treated as any other flash memory segment.

Below is a code fragment, courtesy of Texas Instruments, Inc. for writing a block of data to the flash memory of an MSP430 of the x4xx family. User guides provide an extensive detailed treatment of flash memory and also several code examples as well as recommended management strategies.

```
; Write one block starting at 0F000h.
; Must be executed from RAM, Assumes Flash is already erased.
; 514 kHz < SMCLK < 952 kHz
; Assumes ACCVIE = NMIIE = OFIE = 0.
```

---

[13] This is a case where standard constant naming is not as standard as one should expect. In series 5/6, the name for the reading and writing passwords are, respectively, "FRPW" and "FWPW". In all other cases, the constants are named "FRKEY" and "FWKEY", respectively. The user should consult the guides or the header files if necessary.

# FCCTL1

| bit7 | bit6 | bit5 | bit4 | bit3 | bit2 | bit1 | bit 0 |
|------|------|------|------|------|------|------|-------|
| BLKWRT | WRT | Reserved* | EEIEX** | EEI** | MERAS | ERASE | Reserved |

\* : SWR in Series '5xx/'6xx
\*\* : Reserved in series '1xx/'5xx/'6xx

# FCCTL2

| bit7 | bit6 | bit5 | bit4 | bit3 | bit2 | bit1 | bit 0 |
|------|------|------|------|------|------|------|-------|
| FSSELx | | FNx | | | | | |

# FCCTL3

| bit7 | bit6 | bit5 | bit4 | bit3 | bit2 | bit1 | bit 0 |
|------|------|------|------|------|------|------|-------|
| FAIL** | LOCKA* | EMEX** | LOCK | WAIT | ACCVIFG | KEYV | BUSY |

\* : SWR in Series '1xx
\*\* : Reserved in series '1xx/'5xx/'6xx

# FCCTL4

| bit7 | bit6 | bit5 | bit4 | bit3 | bit2 | bit1 | bit 0 |
|------|------|------|------|------|------|------|-------|
| LOCKINFO* | | MRG1 | MRG0 | | | | VPE* |

\* :  Series '5xx/'6xx only

**Fig. 7.32**  MSP430 low bytes of flash memory control registers

**Table 7.9**  Flash memory writing mode

| BLKWRT | WRT | Action |
|--------|-----|--------|
| 0 | 0 | No write |
| 0 | 1 | byte/word write |
| 1 | 0 | long word (32 bits) write[1] |
| 1 | 1 | Mass erase ( |

1: Series '5xx/'6xx only

```
        MOV #32,R5                      ; Use as write counter
        MOV #0F000h,R6                  ; Write pointer
        MOV #WDTPW+WDTHOLD,&WDTCTL      ; Disable WDT
L1      BIT #BUSY,&FCTL3                ; Test BUSY
        JNZ L1                          ; Loop while busy
        MOV #FWKEY+FSSEL1+FN0,&FCTL2 ; SMCLK/2
        MOV #FWKEY,&FCTL3               ; Clear LOCK
        MOV #FWKEY+BLKWRT+WRT,&FCTL1 ; Enable block write
```

**Table 7.10** Flash memory erasing mode[1]

| MERAS | ERASE | 'x1xx, 'x2xx, and 'x4xx | 'x5xx/'x6xx |
|---|---|---|---|
| 0 | 0 | No erase | No erase |
| 0 | 1 | Individual segment | Individual Segment |
| 1 | 0 | Main memory segments | One bank erase |
| 1 | 1 | Erase all flash memory[2] | Mass erase all banks |

1: MSP430FG461x have another bit, GMERAS, for more options. 2: Information Segment A is not erased when locked with LOCKA = 1.

```
L2      MOV Write_Value,0(R6)           ; Write location
L3      BIT #WAIT,&FCTL3                ; Test WAIT
JZ      L3                             ; Loop while WAIT = 0
        INCD R6                        ; Point to next word
        DEC R5                         ; Decrement write counter
        JNZ L2                         ; End of block?
        MOV #FWKEY,&FCTL1              ; Clear WRT,BLKWRT
L4      BIT #BUSY,&FCTL3               ; Test BUSY
        JNZ L4                         ; Loop while busy
        MOV #FWKEY+LOCK,&FCTL3         ; Set LOCK
        ...                            ; Re-enable WDT if needed
```

## 7.6 Bus Arbitration and DMA Transfers

Direct Memory Access (DMA) techniques offer a convenient way to accelerate data transfers from peripherals or memory to memory or viceversa. In embedded systems with such a capability a DMA also offers enhanced opportunities for implementing low-power solutions for data transfer intensive applications.

DMA controllers are however a special kind of peripherals with bus master capabilities, as they are able to substitute the CPU in controlling the bus activity. As such, in order for a DMA to gain control of the buses, a bus arbitration procedure has to mediate. In this section we introduce bus arbitration as a preamble to the discussion of DMA controllers. The next section is taylored to provide a better understanding of the fundamental concepts on the subject of bus arbitration.

### 7.6.1 Fundamental Concepts in Bus Arbitration

Bus arbitration is the process through which devices with bus mastering capabilities can request to a default bus master the control of a set of shared buses (address, data, and control) and become temporary bus masters.

In a computer system, a bus master capable device is one that has the resources to command the system buses. This is, a device capable of issuing addresses, managing control signals, and regulating the activity on the data bus.

In single processor systems and small embedded applications, the CPU is typically the only device with bus master capabilities, and thus the bus master by default. In such a system, every transaction occurring in the system buses is regulated by the CPU. An example of such an structure is the classical computer system organization depicted back in Chap. 3, when the general architecture of a microcomputer system was introduced and explained with Fig. 3.1.

When a system contains devices with bus mastering capabilities other than the CPU, such as math co-processors, graphical processing units, direct memory access (DMA) controllers, or even other CPUs like occurs in multi-core or multi-processor systems, the access to a shared system bus becomes a source of contention. At any given moment, multiple devices might need to use the buses, but only one master at a time can have control of them. This situation creates the necessity of establishing a mechanism in which each potential bus master can request control of the buses and, in an orderly fashion, control can be granted to requesting devices without creating conflicts. This is the scenario where bus arbitration becomes a necessity.

When a system has multiple bus master capable devices, the device that is most likely to request the buses or the one that controls the buses when no other potential master is requesting them is designated as the default bus master. Usually, the default bus master is given the task of booting the system upon a power-up or soft reset condition. Therefore, bus arbitration transactions place requests to the default bus master.

**Basic Bus Arbitration Protocol**

Consider a simple scenario where a CPU and one additional bus master capable device coexist, as illustrated in Fig. 7.33. For a bus arbitration transaction taking place between these two devices, both of them must have been designed to support bus arbitration. This requires an ability in the involved devices for accepting requests from external potential masters ($\overline{BRQ}$ CPU input), some way for the CPU notifying

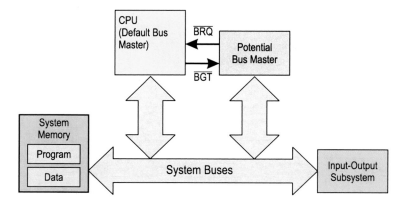

**Fig. 7.33** Simple scenario for a bus arbitration transaction

the potential master when a request has been granted ($\overline{BGT}$ CPU output), and some mechanism for the CPU learning that buses are free anew.

Assuming the CPU is the default master, the bus arbitration transaction begins when the potential master asserts its bus request signal $\overline{BRQ}$ to the default master. This situation could result either because the potential master has its own tasks to perform, or because the CPU delegated some task to it. If the CPU were in the middle of an instruction or in some other state where it could not honor the bus request, it would just complete the current instruction or continue its current task until reaching a state where the buses could be granted.

When the CPU becomes able to accept the bus request (assuming $\overline{BRQ}$ is still asserted) it would proceed to disable its internal bus drivers, isolating itself from the system buses, and then would assert the bus grant signal $\overline{BGT}$. At this point if the CPU had internal or local resources that allow for it to continue processing without accessing the buses, it would continue to do so. Otherwise it would enter a suspend mode waiting for the release of the buses.

The assertion of the $\overline{BGT}$ line would be the indication to the requesting master that it can take control over the buses. Thus, it would now enable its own internal bus drivers, taking control of the buses to perform whatever task for which it requested to be master.

When the potential master completes its task, it would disable its internal bus drivers, isolating itself from the bus and then would unassert the $\overline{BRQ}$ line. This would be the indication to the CPU that the buses are free. The CPU would then proceed by unasserting the $\overline{BGT}$ signal and enabling its bus drivers, becoming again the bus master. Figure 7.34 shows the timing in signals $\overline{BGT}$ and $\overline{BGT}$ during this process.

There are different ways in which the request-grant-release process can be indicated. Intel x86 processors use a single bidirectional line that is first pulsed by the potential master to make a request, then pulsed by the default master to grant the buses and last pulsed by the potential master to release the buses. Other processors like the ARM use a 3-line scheme where a request and grant signals operate as described above but granting the bus for a single clock cycle. A third line, BLOK (Bus Lock) prevents the potential master from releasing the buses from one cycle to another. In general, handshaking protocols change from one processor to another, but the transaction is remains the same.

**Bus Arbitration Schemes**

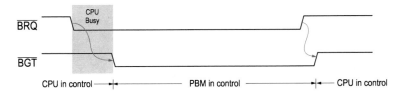

**Fig. 7.34** Timing of a bus request and granting process

When a system has multiple potential masters, the arbitration solution requires a few additional considerations than those illustrated in the previous section as it would become necessary to introduce prioritization schemes to handle simultaneous request from multiple potential masters.

In a sense, this problem is not different to that of interrupt priority discussed earlier in Sect. 7.1.5. The solutions to the priority arbitration problem use the same approaches as for interrupt prioritization: serial daisy chains or central arbitration.

In a daisy-chain solution, simple hardware links like those for the interrupt approach are used, with the same simplicity advantages and hard-wiredness disadvantages that were discusses for the interrupt case.

A central arbiter is usually a more flexible solution, as like in the case of interrupts it provides rotative priorities, programmability, and masking levels that centralize the solution.

### 7.6.2 Direct Memory Access Controllers

Conventional data transfers from peripheral devices to or from memory or from memory to memory in microprocessor-based systems can occur either via polling or via interrupts. In either case, the CPU plays a central role in two aspects.

First, every transfer is the result of the executing of some data transfer instruction that needs to be fetched and decoded before it can be executed. The second aspect is that as transfers require specifying source and destination addresses, these two actions cannot occur in a single bus cycle. So, the CPU needs to perform them with two separate instructions, the first reading the source data into a CPU register and the second transferring the character from the CPU register into the destination address. Figure 7.35 graphically illustrates this process for an I/O to memory transfer.

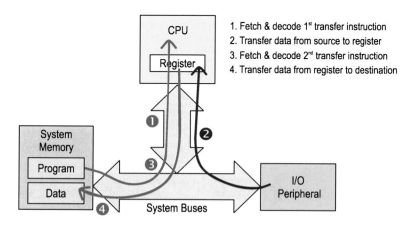

**Fig. 7.35**  Sequence of events in conventional data transfer

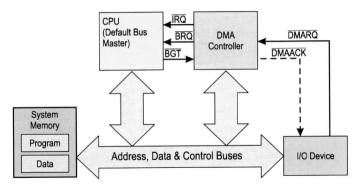

**Fig. 7.36** Connecting a DMA controller to an MPU system and I/O device

By considering also the additional overhead introduced by either the polling or interrupt-based scheme (checking a flag or branching to the corresponding ISR and returning), it is possible to realize the considerable number of clock cycles it takes to make a transfer under the traditional scheme.

Although many embedded applications will not be affected by this overhead, high-speed systems might face severe limitations in their throughput by this scheme. Consider for example the embedded application retrieving bits and bytes from a terabyte hard drive spinning at 7200RPM through a magnetic pick-up head, or the embedded controller sending streaming video data to a digital TV screen. None of these applications would tolerate the overhead of a conventional transfer. A solution for such high speed applications would be using a DMA controller (DMAC).

**DMA Operation**

A DMA controller can operate as an I/O device when is configured by the CPU or as a bus master device. When operating as a bus master, the DMA replaces the CPU in regulating data transfers in the system buses in such a way that data can be directly accessed to and from memory without executing CPU instructions or passing the data through the CPU. This is what gives this controller the "Direct Memory Access" denomination. Figure 7.36 illustrates the way a DMA controller is inserted in a microprocessor-based system to support memory-to-memory or memory to/from I/O transfers.

A DMAC receives transfer requests from its associated I/O devices and can optionally respond with an acknowledgment signal. As a result of a DMA request, the DMA initiates a bus arbitration transaction with the CPU to gain control of the system buses. When the buses are granted, the DMA, as bus master, can generate the address signals to point to memory locations and/or select the I/O device such that data transfers can directly occur between memory and the device without the intervention of the CPU. The DMA also generates the required control signals to establish the transfer direction.

When operated as an I/O device, a DMAC needs to be configured by the CPU to specify the initial addresses where transfers will take place, the number of words that

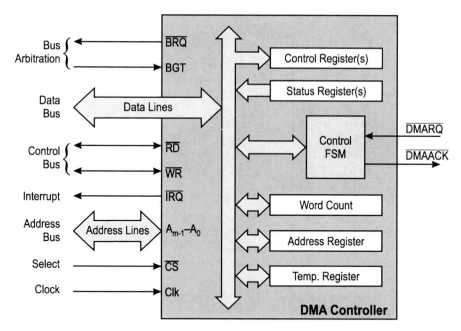

**Fig. 7.37**   Minimal structure of a one-channel DMA controller

it needs to transfer, and the modality and desired type of transfer to perform. After configuration, then DMAC needs to be enabled, allowing it to initiate bus arbitration transactions when its associated peripheral needs to perform a transfer. Transfer requests can also be initiated by the CPU. In order to support such operations, the minimal structure of a DMA controller is expected to be similar to that illustrated in Fig. 7.37.

An address register is needed to specify the initial address of a transfer. A second address register and a temporary data register are also required if memory-to-memory transfers are to be supported (discussed below as two-cycle transfers).

A word count register is required to specify how many words the DMAC needs to transfer in a session.

Finally, one or several control and status registers will be required for configuring and indicating the modes and status conditions developed during operation.

The CPU side interface of the DMAC includes bidirectional connections to address, control, and data buses to allow the dual operation of the DMA as I/O peripheral when accessed by the CPU or as bus master when coordinating transfers. In the device side, the minimum requirement calls for a DMA request input to accept transfer requests from a peripheral I/O device. Optionally a DMA acknowledgement signal might be also provided.

Most DMA controllers are able to support more than one I/O peripheral by featuring multiple *DMA Channels*. One DMA channel is just an arrangement of Word

Count, Address, and Temp. Register per each supported I/O device. Each supported I/O device has a dedicated DMARQ line, and internally the DMAC takes care of prioritizing requests received from the supported peripherals.

**Burst Mode Versus Cycle Stealing**

A DMA controller can request the buses to transfer data in one of three modalities: *burst*, *cycle stealing*, or *transparent*.

In a burst mode transfer, the DMAC retains command of the buses until performing the transfer of an entire block of data. This mode is preferred when performing memory-to-memory transfers, transfers from fast peripherals, or when devices feature a multi-word buffers that need to be retrieved or filled in a single session.

In cycle stealing mode, the DMA controller performs a bus arbitration transaction for every word transferred. This mode is convenient for slower devices whose data throughput could generate idle time in the bus while the DMA waits for new data or when interleaved operation with the CPU is desired. After each word transfer, the buses are released and requested anew when the peripheral device places the next DMA transfer request. This operation can also be controlled via a timer-counter for establishing uniform sampling rates in data acquisition applications.

The transparent mode requires additional signalization between the DMAC and the CPU for determining periods where the CPU is not using the buses. This way the DMA can become bus master without even performing a bus arbitration transaction, operating transparently from the CPU point of view. As this mode requires a DMAC capable of sensing the processor bus activity, it is supported in only a reduced number of controllers.

Regardless of the transfer modality, DMA controllers are typically configured to perform transfers of blocks of data containing hundreds or thousands of words. As this process might take some time to complete, an interrupt signal can be optionally generated to the CPU to indicate the end of a transfer, allowing the CPU to react to the event.

**One-cycle Versus Two-cycle DMA Transfer**

Depending on the path through which data traverses and the number of cycles necessary to complete a transfer, a DMAC can perform one- or two-cycle transfers.

**Two-cycle Transfers:** In two-cycle transfers the DMA controller uses two bus cycles to complete a transfer. In the first cycle, the DMAC performs a read access from the source location and stores the datum into a temporary register inside the DMAC. In the second cycle the DMAC performs a write access to the destination location transferring there the contents of the temporary register. This process is illustrated in Fig. 7.38 with an a memory to I/O transfer.

The sequence of steps taking place for performing the transfer illustrated in the figure can be outlined as follows:

Step 1. A $\overline{\text{DMARQ}}$ signal is issued by the I/O device. This could be initiated by the CPU through the I/O device interface to send data to its associated peripheral or by the peripheral itself.

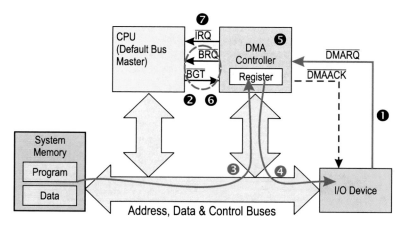

**Fig. 7.38** Sequence of events in a two-cycle DMA transfer

Step 2.  The DMA controller initiates a bus arbitration transaction to gain control of the buses. The CPU becomes isolated from the buses and grants the buses to the DMA controller. At this point the DMA becomes the bus master.

Step 3.  This is the first cycle of DMA transfer. The DMAC sends the memory address of the source data, activates the $\overline{RD}$ signal, and the datum is transferred from the source location into the DMA temporary register via the data bus.

Step 4.  This is the second cycle in the DMA transfer. The DMA controller places in the address bus the destination address, which corresponds to the I/O device interface, activates the $\overline{WR}$ signal, and the datum is transferred from the temporary register into the I/O device interface.

Step 5.  Internally, de DMAC updates the source address and transfer counter. If the transfer were in burst mode, go back to Step 3 until the entire block of data is transferred (transfer count = 0). In cycle stealing mode, continue to Step 6.

Step 6.  The DMAC releases the buses and the CPU becomes bus master.

Step 7.  If the transfer counter is zero, optionally, the DMAC, as an I/O peripheral, issues an interrupt to the CPU.

Observe that as two different bus cycles are used, one for reading the source and other for writing the destination, this modality allows performing memory-to-memory transfers since two different addresses can be issued, one in each cycle. This is the most commonly used transfer mode in DMA controllers.

**One-cycle Transfers:** In one-cycle transfers the DMA controller performs a complete a transfer within a single bus cycle. Two conditions are necessary to achieve such a task. First, a DMAACK signal from the DMAC to the I/O device interface becomes necessary, and second the CPU must be able to issue separate signals to strobe memory and I/O transfers.

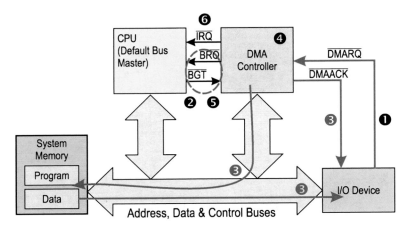

**Fig. 7.39** Sequence of events in a one-cycle DMA transfer

The first condition allows using the DMAACK signal as a select strobe for the I/O device peripheral such that it can be selected without having to issue its explicit address. The second condition implies having separate read and write signals for memory and I/O locations, an strategy called *separate memory and I/O spaces*.[14]

Assuming both conditions are satisfied, the way the transfer is completed is simple. When the DMAC becomes master, within a single bus cycle it issues the address of the memory location involved in the transfer along with the memory access strobe ($\overline{\text{MemRD}}$ or $\overline{\text{MemWR}}$) AND asserts the $\overline{\text{DMAACK}}$ with the corresponding I/O strobe ($\overline{\text{IOWR}}$ or $\overline{\text{IORD}}$) allowing data to flow directly from source to destination within a single bus cycle. This process is illustrated in Fig. 7.39 with a memory to I/O transfer.

The sequence of steps taking place for performing the transfer can be outlined as follows:

Step 1.  A $\overline{\text{DMARQ}}$ signal is issued by the I/O device.
Step 2.  The DMA controller initiates a bus arbitration transaction, becoming the bus master.
Step 3.  The DMAC sends the memory address of the source data, activates the $\overline{\text{MemRD}}$ signal selecting the source memory location, and simultaneously activates $\overline{\text{DMAACK}}$ signal selecting the I/O device. While still in the same cycle, the DMAC also activates the $\overline{\text{IOWR}}$ strobe making possible for the word coming out from memory in the data bus to be strobed into the I/O device. This way the whole transfer is completed in just once bus cycle.
Step 4.  Internally, de DMAC updates the source address and transfer counter. If the transfer were in burst mode, go back to Step 3 until the entire block of data is transferred (transfer count = 0). In cycle stealing mode, continue to Step 5.
Step 5.  The DMAC releases the buses and the CPU becomes bus master.

---

[14] This modality uses independent address spaces for memory and I/O devices.

Step 6. If the transfer counter is zero, optionally, the DMAC, as an I/O peripheral, issues an interrupt to the CPU.

One-cycle DMA transfers can achieve higher transfer rates than two-cycle transfers, but the conditions imposed on the system hardware to make them possible make this a restrictive transfer type. First, additional decoding logic is needed in the I/O device interface to accommodate the DMAACK signal as a second device select line. Second, separate memory and I/O spaces is rarely supported in contemporary systems. Most microprocessors now-a-day use memory-mapped I/O[15] For these reasons only a limited number of systems are able to support this type of transaction.

### 7.6.2.1 Programming Fundamentals

In a direct memory access process using a data transfer controller, the CPU must provide information to the controller about the process. Although each data transfer controller has its own features that may distinguish it from others, there are some general guidelines to be considered.

First, the user must decide what device(s) will trigger the DMA. Alternatively, transfers might be initiated by a software trigger.

Second, for each DMA channel active the user must provide starting addresses for source and destination, as well as the number of words to be transferred. If the controller has the option, the increasing/decreasing direction of addresses in transfers involving multiple words. Note that this is independent of wether the controller would work in burst mode or cycle stealing.

Third, the user must take the application into consideration when planning the transfer mode. If all data is needed before proceeding further, or the system bus will not be used while transfer is taking place, then the burst or block transfer mode is the natural choice. For example, to take the average of a data set, we must have all data available first. If this is the system's task, to acquire data and find average, there is nothing to do while collecting data, so the CPU may be put into a low power mode.

If, on the other hand, buses are needed for other tasks, cycle-stealing mode is preferred.

Most controllers work on a two-cycle method, mainly because most contemporary architectures use memory-mapped I/O. If one-cycle transfers are required, the user should look for an appropriate controller, preferably with the two options and ensure the used processor will be able to support it.

---

[15] Memory-mapped I/O places memory and I/O peripherals in the same address space.

## 7.6.3 MSP430 DMA Support

Apart from the data transfer controller that may be included in the ADC10, several MSP430 models have a DMA controller peripheral. There are several channels available in the different models. Its features include:

- Several independent transfer channels, depending on MSP model
- Configurable DMA channel priorities
- Supports only two-cycle transfers, taking two MCLK clock periods per transfer
- Ability to perform byte-to-byte, word-to-word, and mixed byte/word transfers
- The maximum block sizes are up to 65,535 bytes or words
- Configurable transfer trigger selections
- The ability to select edge or level-triggered DMA transfer requests
- Support for single, block, or burst-block transfer modes

The DMA is in general configured with several control registers, and each channel has four associated registers for operation. The entire DMA controller module features a single, multi-sourced interrupt vector shared by all supported channels. This makes necessary to use either flag poll or calculated branching in the DMA ISR.

The main differences among the different MSP430 models supporting DMACs are summarized in Table 7.11

Below we describe the registers used the MSP430x2xx family as an example. The reader is referred to the specific User's Guide and dataasheet of the device used for more information.

### 7.6.3.1 MSP430 DMA Registers for 'x2xx Series

The DMA has three general registers and in addition each channel has four associated registers. The general registers include:

DMA control 0 (DMACTL0)    This register defines what event triggers each channel. The event may be a software request or an interrupt request from one peripheral. The selection is device dependent, so the user must consult the respective data sheet.

**Table 7.11** DMA Registers and channels in MSP430 families

| Series | Number of Control Reg. | Number of Channels | Interrupt vector Register |
|--------|------------------------|--------------------|---------------------------|
| 'x1xx | 2 | 3 | No |
| 'x2xx | 2 | 3 | Yes |
| 'x4xx | 2 | 3 | Device specific |
| 'x5xx&'x6xx | 4 | Up to 8 | Yes |

DMA control 1 (DMACTL1)    It works with three bits:

- **DMAONFETCH** (Bit 2): 0 if the transfer occurs when triggered; 1 if transfer occurs after next instruction;
- **ROUNDROBIN** (Bit 1): 0 for DMA channel priority DMA0 - DMA1 - DMA2; 1 for DMA channel priority changes with each transfer; and
- **ENNMI** (Bit 0): 0 if transfer cannot be interrupted by a non maskable interrupt; 1 if it can. In this case, the DMA finishes current transfer and then stops.

DMA interrupt vector (DMAIV)

Each channel has the following registers (x=0, 1, or 2):

DMA channel x control (DMAxCTL):    This register determines

- the transfer mode,
- independent up/down directions for destination and source addresses in block transfers,
- independent source and destination byte/word sizes,
- triggering of the process.
- turns off or on the DMA channel,
- includes the interrupt flag and interrupt enabling, and
- includes a flag to signal interrupted transfer.

DMA channel x source address (DMAxSA)    It contains the source address for single transfers or the first source address for block transfers. The register remains unchanged during block and burst-block transfers.

DMA channel x destination address (DMAzDA)    It contains the destination address for single transfers or the first destination address for block transfers. The register remains unchanged during block and burst-block transfers.

DMA channel x transfer size (DMA0SZ)    it defines the number of byte/word data per block transfer, up to 65,535. DMAxSZ register decrements with each word or byte transfer. When DMAxSZ decrements to 0, it is immediately and automatically reloaded with its previously initialized value.

### 7.6.3.2  DMA Transfer Modes

DMA transfers occur when they are triggered by an event, either requested by software of by the peripheral. The transfer mode is defined by the DMADTx bits 14-12 of the DMAxCTL register, and activated with the DMAEN (DMA enable) bit. The transfer modes are:

Single line transfer:    Each transfer requires a trigger, and DMAEN is disabled after transfer

Block transfer:    A complete block transfer occurs when triggered, and DMAEN is disabled after the complete transfer is achieved.

Burst-block transfer:     Is a block transfer with interleaved CPU activity. DMAEN
    is disabled at the end of the block transfer.
Repeated single line transfer:     Each transfer requires a trigger, but DMAEN remains
    enabled.
Repeated block transfer:     Each block transfer requires a separate trigger, but
    DMAEN remains enabled.
Repeated burst-block transfer:     Each burst-block transfer requires a separate trig-
    ger, but DMAEN remains enabled.

### 7.6.3.3 Triggering and Stopping DMA

DMA channels are independently configured for the trigger source with the DMAxT-
SELx bits with the DMAEN bit = 0. The operation may be started by software of
by peripherals such as timers, ADC, DMA channel, etc. The set of peripherals and
the selection is device specific.

The trigger can be edge-sensitive or level sensitive. When DMALEVEL = 0, the
DMA is triggered when the signal goes from low to high; when DMALEVEL = 1,
the DMA operates while the triggering signal is high. Pay attention to the fact that
the only level-sensitive trigger is the external DMA signal.

Maskable interrupts do not affect DMA operation. On the other hand, DMA can
stop an interrupt service execution. It is therefore necessary to disable the DMA prior
to ISR execution if interruption is unacceptable.

Any type of transfer may be interrupted by an NMI only if the ENNMI is set.
Block and burst-block transfers can also be halted by clearing DMAEN.

Finally, the DMAONFETCH bit controls when the CPU is halted for a DMA
transfer. The CPU may halted immediately when a trigger is received and the transfer
begins when DMAONFETCH = 0 . When DMAONFETCH = 1, the CPU finishes
the currently executing instruction before the DMA controller halts the CPU and the
transfer begins. The DMAONFETCH bit must be set when the DMA writes to flash
memory to prevent unpredictable results.

### 7.6.3.4 Application Examples

To configure the DMA operation, consider at least the following actions:

1. Configure trigger source with appropriate control register (DMACTL0 for families '1xx, '2xx and 4'xx)
2. Define source and destination addresses, as well as the number of transfers
3. Define the addressing mode for transfer
4. Define the direction (unchanged, up or down) for source and destination addresses
   – default is unchanged.
5. Enable DMA and turn on the DMAONFETCH bit if necessary.

Let us work some configuration examples for the DMA in '2xx series:

*Example 1* We want to reverse the upward storage sequence of N bytes in a RAM segment starting at address `SourceAddr`.

**Solution:** For that reason, we store it in a reverse order at a RAM segment whose highest address is `Dest_up_Addr`. Transfer is done in one block transfer by software request. The following sequence of instructions do the work.

(A) Assembly instructions:

```
Setup_DMA:   mov.w   #SourceAddr,&DMA0SA        ; Source Block Address
             mov.w   #Dest_up_Addr,&DMA0DA      ; Destination Block Address
             mov.w   #N,&DMA                    ; Block size
             mov.w   #DMADT_2+DMADSTINCR_2+DMASRCINCR_3,&DMA0CTL
                                                ; Block Trans., decrement
                                                ; Dest. Addr, and Inc.Src. addr.
             bis.w   #DMASRCBYTE|DMADSTBYTE,&DMA0CTL    ; Byte transfers
             bis.w   #DAEN,&DMA0CTL             ; Enable DMA
                                                ;
                                                ;
Init_trans   bis.w   #DMAREQ,&DMA0CTL           ; Software trigger for DMA
```

(B) C instructions:

```
DMA0SA = SourceAddr;                              // Source block address
DMA0DA = Dest_up_Addr;                            // Destination single address
DMA0SZ = N;                                       // Block size
DMA0CTL = DMADT_2+DMADSTINCR_2+DMASRCINCR_3;      // Block Trans., decrement
                                                  // Dest. Addr, and Inc.Src. addr.
DMA0CTL |= DMASRCBYTE|DMADSTBYTE        ;          //Byte transfers
DMA0CTL |= DAEN,&DMA0CTL                ;          //Enable DMA
                                                  //
                                                  //
DMA0CTL |= DMAREQ;                                //  trigger transfer
```

*Example 2* (*Courtesy of Texas Instruments*: MSP430x26x Demo - DMA0, single transfer Mode UART1 9600, ACLK, by B. Nisarga, Texas Instruments, 2007.)[16]

The string "Hello World" is transferred as a block to U1TXBUF using DMA0. UTXIFG1 WILL trigger DMA0. The rates is 9600 baud on UART1. The Watchdog atimer triggers block transfer every 1000ms. Level sensitive trigger used for UTX-IFG1 to prevent loss of initial edge sensitive triggers—UTXIFG1 which is set at POR.

ACLK = UCLK 32768Hz, MCLK = SMCLK = default DCO 1048576Hz

Baud rate divider with 32768hz XTAL @9600 = 32768Hz/9600 = 3.41 (0003h)

(A) Assembly code

```
#include "msp430x26x.h"
;-------------------------------------------------------------------------
LF          EQU     0ah                      ; ASCII Line Feed
CR          EQU     0dh                      ; ASCII Carriage Return
;-------------------------------------------------------------------------
```

---

[16] See Appendix E.1 for terms of use.

```
            RSEG    CSTACK              ; Define stack segment
;----------------------------------------------------------------------
            RSEG    CODE                ; Assemble to Flash memory
;----------------------------------------------------------------------
RESET       mov.w   #SFE(CSTACK),SP     ; Initialize stackpointer
StopWDT     mov.w   #WDT_ADLY_1000,&WDTCTL  ; WDT 1000ms, ACLK,
                                        ; interval timer
            bis.b   #WDTIE,&IE1         ; Enable WDT interrupt
SetupP3     bis.b   #BIT6+BIT7,&P3SEL   ; P3.6,7 = USART1 TXD/RXD
SetupUSCI1  bis.b   #UCSSEL_1,&UCA1CTL1 ; ACLK
            mov.b   #3,&UCA1BR0         ; 32768Hz 9600 32k/9600=3.41
            mov.b   #0,&UCA1BR1         ; 32768Hz 9600
            mov.b   #UCBRS_3,&UCA1MCTL  ; Modulation UCBRSx = 3
            bic.b   #UCSWRST,&UCA1CTL1  ; **Initialize
                                        ; USCI state machine**
SetupDMA0   mov.w   #DMA0TSEL_10,&DMACTL0  ; UTXIFG1 trigger
            movx.a  #String1,&DMA0SA    ; Source block address
            movx.a  #UCA1TXBUF,&DMA0DA  ; Destination single address
            mov.w   #0013,&DMA0SZ       ; Block size
            mov.w   #DMASRCINCR_3+DMASBDB+DMALEVEL,&DMA0CTL; Repeat,
                                        ; inc src
Mainloop    bis.w   #LPM3+GIE,SR        ; Enter LPM3 w/ interrupts
            nop                         ; Required only for debugger
                                        ;
;-------------------------------------------------------------
WDT_ISR;    Trigger DMA block transfer
;-------------------------------------------------------------
            bis.w   #DMAEN,&DMA0CTL     ; Enable
            reti                        ;
;-------------------------------------------------------------
String1     DB      CR,LF, 'Hello World'
;-------------------------------------------------------------
            COMMON  INTVEC              ; Interrupt Vectors
;-------------------------------------------------------------
            ORG     WDT_VECTOR          ; Watchdog Timer
            DW      WDT_ISR
            ORG     RESET_VECTOR        ; POR, ext. Reset
            DW      RESET
            END
```

## B) C Code:

```c
#include "msp430x26x.h"

const char String1[13] = "\nHello World";

void main(void)
{
  WDTCTL = WDT_ADLY_1000;               // WDT 1000ms, ACLK, interval timer
  IE1 |= WDTIE;                         // Enable WDT interrupt
  P3SEL |= BIT6 + BIT7;                 // P3.6,7 = USART1 TXD/RXD
  //Configure USCIA1, UART
  UCA1CTL1 = UCSSEL_1;                  // ACLK
  /* baud rate = 9600 */
  UCA1BR0 = 03;                         // 32768Hz 9600 32k/9600=3.41
  UCA1BR1 = 0x0;
  UCA1MCTL = UCBRS_3;                   // Modulation UCBRSx = 3
```

```
/* Initialize USCI state machine */
UCA1CTL1 &= ~UCSWRST;

// Configure DMA0
DMACTL0 = DMA0TSEL_10;              // UTXIFG1 trigger
DMA0SA = (int)String1;              // Source block address
DMA0DA = (int)&UCA1TXBUF;           // Destination single address
DMA0SZ = sizeof String1-1;          // Block size
DMA0CTL = DMASRCINCR_3 + DMASBDB + DMALEVEL;
                                    // Repeat, inc src, byte transfer
   __bis_SR_register(LPM3_bits + GIE); // Enter LPM3 w/ interrupts
   _NOP();                          // Reqd  for debugger
}

#pragma vector = WDT_VECTOR         // Trigger DMA block transfer
__interrupt void WDT_ISR(void)
{
   DMA0CTL |= DMAEN;                // Enable
}
```

## 7.7 Chapter Summary

Embedded peripherals give MCUs the versatility that allow for configuring them into single-chip computers. Essential to embedded peripherals is an interrupt system that allows for timely servicing devices needing CPU attention. This chapter focused in the support structures allowing MCUs to use interrupts, timers, flash, and DMA.

An interrupt is a signal or event that changes the normal course of execution a program to transfer control to a special code called an interrupt service routine or ISR. When the special instruction RETI (return from interrupt) is executed, program control is transferred back to the main program.

For interrupts to work in a system, four fundamental requirements must be met:

- There must be a properly allocated stack.
- The interrupt vectors must be configured.
- An ISR must be in place.
- Interrupts must be enabled at both, the GIE flag level and at the local peripheral level.

There are maskable, and non-maskable interrupts, identifying whether the interrupt can or cannot be disabled by the GIE flag.

The discussion of timers revealed that a timer is just a binary counter fitted with an input frequency divider (prescaler) and one or more output compare registers that facilitate either counting time or external events. Key applications of timers were discussed including watchdog timers, real-time clocks, and pulse-width modulation.

The section on memory provided an overview of memory technologies considering their ability to be modified and the permanence of its contents. Special attention was placed on FLASH memories, as this is the mainstream non-volatile technology in contemporary embedded controllers.

380 7 Embedded Peripherals

The last section of the chapter was devoted to bus arbitration and direct memory access controllers. This particular subject brought up the discussion of how to attain high transfer throughput and enabling further opportunities for low-power design in embedded systems.

## 7.8 Problems

7.1 Can you explain why it is necessary for the MSP430 to finish the instruction it is currently executing before attempting to service an interrupt?

7.2 Consider an embedded application of TV remote control. The microcontroller function in this application is to react to a key press, debounce the key, assign a numeric code to the depressed key, and wrap it according to the communication protocol defined between remote and TV, and transmit the encoded value via the infrared (IR) transmitter.

Assume the remote control user depresses, in average, one key every minute. This would allow for a reasonable key operation rate considering an average usage cycle where the user is either flipping channels, watching a show with sporadic volume adjustments, or not using the control at all (TV off periods). In all of them the TV remote is on and, with the processor running at 1 MHz, it takes 300 μs to process each keystroke. Let's also assume the MCU consumes 1 mA in active mode and 100 μA in sleep mode, and runs on batteries with a capacity of 1,200 mAh. Assume all active chores are completed with the MCU active power budget, except the IR transmitter, that takes 50 of the 300 μs to send each key code and consumes 20 mA.

(a) Devise an algorithm to operate the remote control via polling. Provide a flowchart for a modular plan depicting how to structure a main program and all necessary functions to operate the remote control.

(b) For the data provided, estimate how long would the battery last under the polling scheme devised in part (a).

(c) Repeat part (a) but instead of using polling, provide an interrupt-based approach, where the main program initializes all required peripherals, including interrupts, and then sends the CPU into sleep mode.

(d) Repeat part (b), now based on the system behavior induced by the software plan in part (c).

(e) Estimate the improvement in energy usage induced by the low-power algorithm devised in part (c) with respect to the polling approach used in part (a).

7.3 List three advantages of using the low power modes in the MSP430. Briefly justify each.

7.4 Is there any advantage in having a flag to deactivate the (Non) maskable interrupts in the MSP430? Would you use them in a critical application or a real-time application? Justify your answers.

7.5 Does it make any difference recognizing an interrupt during a rising or falling edge of the trigger difference? Briefly explain your answer.

7.6 Show the sequence of instructions in assembly language to

(a) Enable Timer_A Capture Compare Register 1 interrupt.
(b) Load Capture Compare Register 1 with 32768.
(c) Select the SMCLK clock.
(d) Configure the timer for continuous mode of operation.

7.7 An Engineering student needs to generate a square waveform to drive a device. The student decides to use the MSP430's Timer_A with a duty cycle of 60% at 32.768 khz. Show the sequence of instructions in assembly language that the student might have used to configure Timer_A to accomplish the requirements.

7.8 What is the race condition that could develop if the capture input signal is asynchronous to the clock signal in an MSP430 capture/compare unit?

7.9 Setup the MSP430 LaunchPad to use Timer_A to toggle the red LED in Port 1 each time the timer counts to 50,000. You should be able to do this using:

(a) A monolithic assembly language program.
(b) An assembly language program that calls a subroutine to configure Timer_A to toggle the LED every 50,000 cycles.
(c) An assembly language program that sets up an interrupt from Timer_A every 50,000 cycles.
(d) Repeat the previous parts (a) through (c) using C language instead of assembly.

7.10 A periodic signal with a period equals to 10 times the MCLK clock in the MSP430 is needed. A designer wants to use one of the compare units in Timer_A to accomplish this. What are the values that need to be written in the Timer_A Control register TACTL and in the Timer_A Capture Compare register (TACCRx) in order to accomplish this? Show the corresponding instructions in both assembly language and C.

7.11 It was stated in this chapter that the watchdog timer could be configured as an interval counter. Configure the MSP430 Watchdog Timer to toggle the red LED in the MSP430 LaunchPad every three (3) seconds. To conserve energy the MSP430 needs to enter a low power mode after your configuration is complete.

7.12 An embedded application requires a signal with a duty cycle of 30% to drive four LEDs. It is desired to obtain such signal using an MSP430 device. How would you configure the output unit to obtain the necessary signal assuming everything else is configured correctly? What changes would you have to make to the output unit if a different clock source is chosen?

7.13 Write a small program to test the WDT+ failsafe feature of the MSP430. Have the MSP430 perform some function like toggling one of the LaunchPad LEDs and then attempt to disable all clocks.

7.14 The frequency at which an event is occurring needs to be determined. Show how to configure the MSP430 watchdog timer, with the SMCLK clock source, as an interval timer in order to accomplish this task. What formula would you use to determine the frequency of the even?

7.15 On page 673 (need a marker on the page) a piece of assembly code for writing a block of data to the flash memory of the MSP430x4xx device was shown. Write the corresponding code in the C language. As explained in the comments, the code shown must be executed from RAM, what is the reason for this? Could we have executed the program from Flash rather than from RAM?

7.16 Explain the need for wear leveling. What would be the effect of not using it?

7.17 It was explained hat the IE1 register houses the ACCVIE bit used to enable the Flash memory access violation interrupts. How could this bit be used while the system is executing programs from Flash?

7.18 Answer the following questions:

(a) Explain what is a "bus arbitration" transaction.
(b) What methods can be used for dealing with multiple simultaneous bus requests in a bus arbitration transaction.
(c) Explain the concept of "Direct Memory Access"
(d) Describe the three basic modes for supporting direct memory access transfers.
(e) Identify two limitations for implementing one-cycle DMA transfers in an MSP430.

7.19 Describe the difference between one-cycle and two-cycle DMA transfers. Emphasize the weakness and strength of each method with respect to the other.

7.20 Describe important registers to be found in almost any DMA controller. Why are these registers so important?

7.21 What are the basic setup steps that should be considered when programming DMA transfers. Provide pieces of code for an MSP430 device where those steps are illustrated.

7.22 How do you configure a DMA in the MSP430 family for a repeated burst-block transfer of 4 KB from Loc_A to Loc_B?

7.23 A set of three devices DEV0, DEV1 and DEV2 are to be connected in a daisy chain configuration with priority scheme DEV0-DEV1-DEV2. Each device can generate a bus-request signal, and starts transmission upon a bus-acknowledge signal received. Once a device has been granted bus control, it generates a busy signal that prevents the other devices to request bus control. The device deactivates the busy signal after transfer is completed. Design a daisy chain logic system for these devices.

# Chapter 8
# External World Interfaces

Once the basic interface of a microcontroller-based system is established, the next natural step is to use general purpose I/O (GPIO) pins to connect the MCU to the external world. This Chapter focuses on the tasks related to interfacing external devices to microcontrollers using general purpose I/O lines.

The first section provides a thorough discussion of GPIOs covering their structure, capabilities, configuration, and electrical characteristics. The section is crowned by a detailed coverage of GPIO characteristics in MSP430 generations.

The rest of the chapter deals with the interface of real world devices via GPIO. A thorough discussion of how to interface switches, displays, large DC and AC loads, and motors to microcontrollers is included, highlighting MSP430 supporting functions and capabilities, enhanced by hardware and software examples that illustrate their applicability.

## 8.1 General Purpose Input-Output Ports

One of the most valuable resources in microcontrollers are its general purpose Input/Output (GPIO) ports. GPIO ports or just I/Os provide MCUs with the ability of communicating with the external world via digital lines that can be configured to operate as either inputs or outputs. A GPIO port consists of a group of I/O lines that can be, independently or as a group, configured as inputs or as outputs. Most MCUs associate eight I/O lines to a port, although it is possible to find MCUs with ports of only four, six or other number of I/O lines in a port. GPIO lines become available to the external world through either dedicated or shared package pins. Figure 8.1 shows the package outline of an MSP430G2452IN20, denoting the pins where I/O lines become available. This particular MCU provides 16 I/O lines available as two 8-bit ports: Port 1 (P1) and Port 2 (P2). The individual lines in a port are denoted P$i.j$, where $i$ denotes the port number and $j$ represents the line number in the port,

M. Jiménez et al., *Introduction to Embedded Systems*,
DOI: 10.1007/978-1-4614-3143-5_8,
© Springer Science+Business Media New York 2014

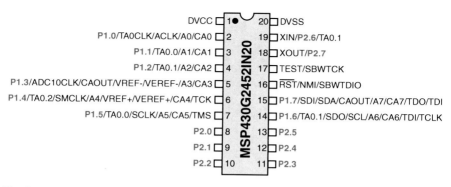

**Fig. 8.1**   Package outline of an MSP430g2452 denoting I/O pins

labeled from 0 to 7. This package also show how the I/O lines in P1 are all shared with other functions, while those in P2 have dedicated lines.

An I/O line configured as input allows the MCU to accept incoming information about the binary status of an external device connected to it. Whenever we need to learn about the status of an external event, device status, or any binary signal whose value can be represented through either a logic high or logic low level, input ports are the hardware resource of choice. Examples of such conditions include determining if a device is on or off, open or closed, up or down, left or right, set or clear, etc.

When configured as output, an I/O line provides the MCU with the ability of controlling the binary status of external devices or signals. Actions such as turning on or off an LED indicator, a buzzer, or anything that can be managed by setting an external voltage level to either a logic high or logic low can be done with an output I/O line.

### 8.1.1 Structure of an Input-Output Pin

To better understand how to configure an I/O port, it deserves to understand the structure of an I/O interface. Figure 8.2 shows a generic structure of an I/O pin driver.

To communicate the CPU with the external world using a digital I/O line, it is necessary an interface that enables connecting a physical I/O pin to the processor buses, with the ability of operating as either input or output. To perform an Input operation (In), the minimum requirement calls for an input buffer. Such a requirement is provided in Fig. 8.2 by the buffer labeled "*In*". The input buffer is typically a standard logic buffer with tri-state control. The tri-state capability allows for configuring the pin as either input or output while avoiding signal contention. Note the a logic "0" in the port direction latch will enable the *In* buffer, while disabling the "*Out*" buffer, making the pin operate as an input. Under this mode, the input data flows through the input buffer to the $P_{i,j}\_In$ line accessing the data bus.

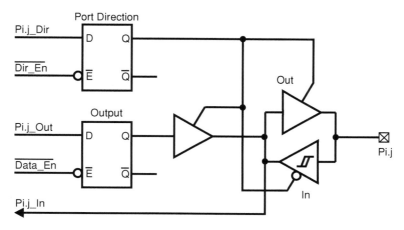

**Fig. 8.2** Basic structure of an Input/Output pin driver

Sometimes, the signal being fed to an input port might have a slow changing behavior or noisy contents around the nominal logic levels. Either of these conditions will cause an undesired behavior in the digital side of the circuit. In such cases, a Schmitt-trigger input buffer becomes convenient. Section 8.2.2 on page 395 discusses the advantages of Schmitt-triger inputs. At this point it is enough to point out that this type of input buffer has the ability of speeding-up slow transitions and eliminating most noise from a digital input. The *In* buffer in Fig. 8.2 is of this type. The hysteresis loop inside the buffer symbol indicates it is a Schmitt-trigger capable input pin.

Contemporary microcontrollers such as the MSP430 typically provide Schmitt-trigger capable GPIO inputs in their ports, saving the designer the burden of externally providing them when working with slow or noisy signals.

An output transaction (Out) in the I/O pin would require the value written by the CPU through the $P_{i,j}\_Out$ data line to stay on the line until a later CPU write access changes the pin to a different value. This calls for a latched output. The latch labeled *Output* performs this function in Fig. 8.2. The output latch is double buffered to allow for being configured as either input or output. In particular, the I/O pin ability to source or sink current will be determined by the characteristics of the "Out" buffer.

The direction in which the pin is used, either as input or as output, is achieved by using tri-state buffers for the input and output functions, and connecting them to form a transceiver. The transceiver control is exerted through line $P_{i,j}\_Dir$, via a latch to allow for retaining the desired In or Out functionality. Additional logic not shown in Fig. 8.2 takes care of interfacing lines $P_{i,j}\_In$, $P_{i,j}\_Out$, and $P_{i,j}\_Dir$ to a single bidirectional data bus line. This logic would also provide for decoding the address, timing, and control lines to drive signals $\overline{Dir\_En}$ and $\overline{Data\_En}$ allowing for a physical connection to the processor buses. The configuration of such an interface logic will depend on the signal architecture of the CPU buses. Figure 8.3 shows a possible configuration assuming a Motorola style bus timing.

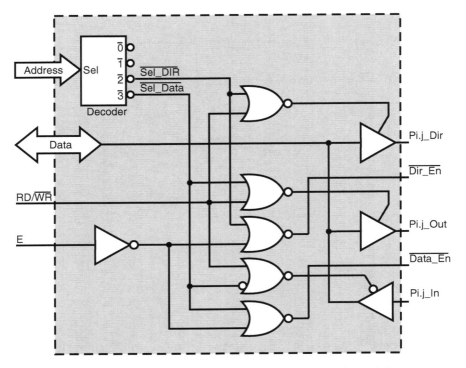

**Fig. 8.3**  Interface logic for an Input/Output pin assuming a Motorola style bus timing

The three tri-state buffers in the pin interface side allow accessing a bidirectional data bus line. The NOR gates decode control signals $RD/\overline{WR}$, E, and the decoder outputs provide unique access for the corresponding data lines. The decoder assigns unique addresses to the direction and data latches. Note that $P_{i,j}\_In$ and $P_{i,j}\_Out$ share the same address, the first as input to the CPU and the other as output, while $P_{i,j}\_Dir$ is a write-only location.

### 8.1.2  Configuring GPIO Ports

The insight gained into the structure of a bidirectional I/O pin in the previous section denotes that for its usage, we simply need to configure the pin direction (as either In or Out), and then access (read or write) its data contents. As MCUs group I/O pins into ports, all I/O lines in a given port are accessed at the same address location. The same rule applies for all direction control bits for the port, making it possible to use simple instructions to configure and access a port. Figure 8.4 illustrates a simple port model (Port.x) containing one data and one direction register.

Let's use an example to illustrate how to read and write I/O pins in a GPIO port.

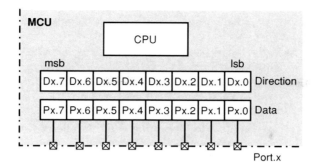

**Fig. 8.4**  Programmer's model for a simple GPIO port containing only data and direction registers

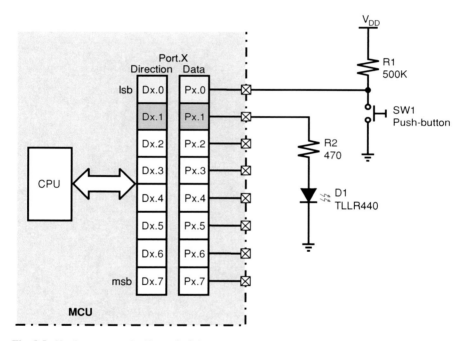

**Fig. 8.5**  Hardware setup for Example 8.1

**Example 8.1  (Push-button/LED Interface)** *Consider an 8-bit GPIO port, denoted as Port X (Px), with a momentary push-button SW1 connected to pin 0 (Px.0) and a light-emitting diode (LED) D1 connected to pin 1 (Px.1). A hardware setup for this example is illustrated in Fig. 8.5. Assume Px pins are by default GPIOs with direction and data registers at addresses PxDir and PxDat, respectively. Also assume that a direction bit set to "1" makes its corresponding port bit work as output, otherwise it will work as input. Provide a code fragment to properly configure the port and to toggle the LED whenever SW1 is depressed.*

*Solution: In this setup,* Px.0 *needs to be configured as input and* Px.1 *as output. SW1 has been wired to provide a logic low level when SW1 is depressed, and a "1" otherwise. D1 is wired to turn-on when* Px.1 *is set to high. The software solution needs to configure pin* Px.0 *as in and* Px.1 *as output. It will then enter an infinite loop where we poll the switch until detected pressed.*

*When SW1 is depressed we'll toggle* Px.1 *and proceed to read again the switch status. To avoid LED flickering due to the time SW1 remains depressed with respect to the MCU speed, we'll also detect when the switch is released before returning to poll it for low. The code assumes SW1 is debounced and a header file provides declarations for port registers* PxDir, PxDat. *The pseudo-assembly code fragment below shows a solution for this problem.*

```
; ========================================================================
; Addresses PxDir and PxDat are assumed declared in header file
; Assumes Dir = 1 for In
; ------------------------------------------------------------------------
#include "headers.h"
; ------------------------------------------------------------------------
              . . .
              AND #0FEh,&PxDir    ; Set Px.0 bit 0 as input (FEh=11111110b)
              OR #002h,&PxDir     ; Set Px.1 as output (02h=00000010b)
              AND #0FDh,&PxDat    ; Begin with D1 off
              . . .
loop_low      TEST #001h,&PxDat   ; Check the value of Px.0
              JNZ loop_low        ; If set, continue to poll Px.0
              XOR #002,&PxDat     ; Otherwise, toggle Px.1
loop_high     TEST #001,&PxDat    ; Check if SW1 was released
              JZ loop_high
              JMP loop_low        ; When released go back to poll depress
              END
; ========================================================================
```

### 8.1.3 GPIO Ports in MSP430 Devices

MSP430 microcontrollers provide a rich assortment of GPIO pins in its devices. Depending on the particular family member, up to eleven GPIO ports can be found in an MSP430 MCU, enumerated from P1 through P11. The first generation of MSP430 MCUs, series x3xx featured a port P0, but none of the succeeding generations have featured it. Series x5xx/x6xx devices, being the largest devices in pin count per package, feature ports P1 through P11.

Each MSP430 GPIO port has up to eight pins. Every I/O pin can be individually configured to operate as input or output, and each I/O line can be individually read or written to. When configured as input, each I/O line has individually configurable, internal, pull-up or pull-down resistors.

Ports P1 and P2 have interrupt capability, each with its own interrupt vector. The service routine of each port must poll the corresponding interrupt flag register (PxIFG) to determine which pin triggered the interrupt. Interrupts from individual

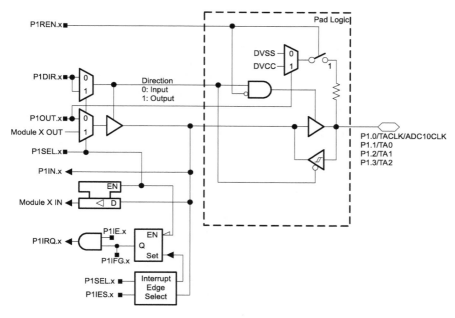

**Fig. 8.6** Structure of pin P1.0 in an MSP430F2274 MCU (*Courtesy of Texas Instruments, Inc.*)

I/O lines of each port can be independently enabled/disabled or configured to be triggered with the rising or falling edge of an input signal. Figure 8.6 shows the structure of P1.0 in an MSP430F2274. The particular topology of an I/O pin in the MSP430 will depend on the device itself and the functions shared in the pin. Despite the specific details, it is possible to identify the I/O logic in the pad, the pull-up/pull-down resistor, and the interrupt logic shared by all P1 and P2 pins.

A set of up to ten registers, depending on the device series, are used to configure and operate each I/O port in the MSP430. These registers include:

**PxSEL** —(Port x Function select) This register selects the functionality to be adopted by a particular pin. When cleared, this bit makes the corresponding pin a GPIO.
**PxDIR** —(Port x data direction) If a pin is set as a GPIO pin with PxSEL, the corresponding bit in this register sets the pin direction: '0' = Input or '1' = Output.
**PxOUT** —(Port x Out) If a pin is set as output, writing PxOUT will cause the corresponding port pin to be driven to the logic level written to the pin. PxOUT is a R/W register and when read will return that last value written to the pin. In MSP430x5xx/x6xx devices with optional pull-up/pull-down resistors in their input mode, the bits of PxOUT are used to select the pulling action of the embedded resistors (when enabled). Setting a PxOUT bit will select the corresponding pin resistor as pull-up, otherwise, when cleared, will set it as pull-down.
**PxIN** —(Port x Input) If a pin is set as input with PxDIR, this register will reflect the logic level present at the corresponding GPIO pin. PxIn is a read-only register.

**PxIFG** —(Port x interrupt flag) This flag is set when the corresponding pin experiences the appropriate signal edge transition(either high to low or low to high).

**PxIES** —(Port x Interrupt edge selection) Establishes the type of transition setting PxIFG. When clear, a low-to-high transition in the corresponding input pin will set PxIFG. Otherwise, a high-to-low transition will set PxIFG.

**PxIE** —(Port x Interrupt enable) When this bit is set, an event setting the corresponding PXIFG will generate an interrupt.

**PxREN** —(Port x resistor enable) This register is present only in x2xx and x5xx/x6xx devices. When present, setting this bit will enable the corresponding pull-up/pull-down resistor in the input port.

**PxDS** —(Port x Drive strength) When present, setting this bit makes the corresponding pin a high-current driver, allowing up to 30 mA while keeping the output voltage within the nominal $V_{IL}/V_{IH}$ value. This increased strength is gained at the expense of a higher power dissipation and increased electromagnetic interference (EMI). This feature is exclusive of x5xx/x6xx devices.

**PxIV** —(Port x interrupt vector) Provides a priority encoding of the interrupts within a port. When the $n$ lowest interrupts of a port are set, PxIV reads $2n + 2$ allowing this number to be used as a displacement to jump to the corresponding ISR segment where the pin is served. This feature is exclusive to x5xx/x6xx MCUs.

Some MSP430 MCUs have a built-in pin oscillator function in some of their output pins that can be activated for facilitating external sensor interfacing. This feature is particularly useful to interface capacitive touch sensors, as no external components other than a conductive pad is needed in the interface. The conductive pad forms a capacitor that is part of the pin oscillator itself, making the resulting oscillation frequency a function of the loading state of the external pad. Devices featuring such a function, like the MSP430G2553, use internal timer channels in conjunction with pin oscillator inputs for determining the loading state of the external pad. The pin oscillator configuration is device dependent, thus the data sheet of the particular device being used needs to be consulted for specific information. Application note "MSP430 Low-cost PinOsc Capacitive Touch overview" provides a detailed explanation of how to use this feature [45].

Additional details related to the configuration and operation of particular I/O pins in MSP430 devices can be found in the corresponding Family user's manual and in the particular device data sheets. Since topologies and features change from one family to another and within a family from one particular device to another, it becomes necessary to have both pieces of documentation when interfacing or programming a particular chip.

## 8.2 Interfacing External Devices to GPIO Ports

When interfacing embedded systems to the real world, there will be many instances when the devices to be driven by the MCU GPIO pins, or the signals we need to feed into an I/O will not meet the electrical requirements of the port. When that happens, it becomes necessary to provide a feasible interface to make possible connecting the MCU to the real world.

To determine whether or not an interface will be needed between the MCU I/Os and a particular external device, the first step we need to perform is to revise and understand the electrical characteristics of both, the MCU GPIO's and the external device's. This will unavoidably lead us to the device's data sheets.

To understand the capabilities and limitations of an input/output port, the first thing we need to realize is that like any electronic circuit, GPIOs have voltage and current limits that we need to adhere to when interfacing to them. There are both, electrical and switching characteristics associated to an I/O pin that we need to become familiar with.

In this section we review the fundamental electrical characteristics of GPIOs to help understand their strengths and limitations, and then provide several typical cases when interfaces become necessary for their connection.

### 8.2.1 Electrical Characteristics in I/O Pins

Like any digital circuit, the electrical characteristics of an I/O pin are dominated by the specifications of the currents and voltages that can be safely managed by the pin. In addition, these also include the chip power supply requirements and thermal limits of the chip itself.

The voltage specifications of an I/O pin are described by parameters of the voltage-transfer characteristics (VTC) of its input and output buffers. Specifically, an input GPIO pin will have the thresholds $V_{IL}$ and $V_{IH}$ of its input buffer. An output GPIO pin will exhibit the voltage limits $V_{OL}$ and $V_{OH}$ of its output driver. These four voltage parameters are described as follows:

$V_{IL}$ is the pin input-low voltage. It establishes the maximum voltage level that will be interpreted as a low by the pin input buffer. The value of $V_{IL}$ is lower bounded by the reference level ($V_{SS}$) used by the chip, typically ground level or 0 V.

$V_{IH}$ is the pin input-high voltage. It represents the minimum voltage level that will be interpreted as a high by the I/O pin. The value of $V_{IH}$ is upper bounded by the supply level in the MCU ($V_{DD}$).

$V_{OH}$ is the maximum voltage level exhibited by a digital output pin. The nominal value of $V_{OH}$ is specified for a no-load condition in the pin. In the practice, the value of $V_{OH}$ decreases as the current driven by the pin in its high level increases.

$V_{OL}$ Represents the minimum level observed in a digital output pin. This voltage is also specified for a no-load condition, but in the practice its level increases with the current driven by the pin when in its low state.

A fifth voltage parameter, rarely specified for GPIOs is the pin threshold voltage $V_M$. By definition, this is a voltage level that when fed to a buffer input will produce exactly the same voltage level in its output. In symmetric CMOS logic $V_M = V_{DD}/2$.

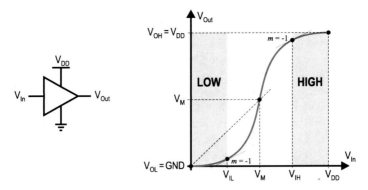

**Fig. 8.7**  Symbol and voltage transfer characteristic of a conventional buffer

Figure 8.7 shows the voltage transfer characteristic of a conventional non-inverting buffer, denoting all its voltage parameters, including $V_M$.

The nominal regions of low- and high-level input voltages are clearly marked in the figure. They are lower-bounded by $V_{SS} = 0V$ and upper bounded by $V_{DD}$ (or $V_{CC}$), meaning that no input shall be applied outside this range. The portion of the input ranging between $V_{IL}$ and $V_{IH}$ is denominated the *transition region* of the pin. Valid digital voltages are expected to avoid this region as they might lead to degraded voltage levels within the input port logic.

An input level falling in the transition region of a pin might actually be internally regenerated to a valid logic level after it passes through a few internal gates. The regenerated level will end up being low or high depending on wether the applied input level is correspondingly below or above $V_M$. This is why regular logic is said to have a single threshold level $V_M$.

Despite the regeneration possibility, a good interface is expected to provide valid logic levels, with logic high levels greater than or equal to $V_{IH}$ and logic low levels lower than or equal to $V_{IL}$ to GPIO inputs. If at some point in a design, the voltage fed from a peripheral to a GPIO or viceversa does not meet the input threshold requirements, some form of level-shifting interface becomes necessary. This situation frequently arises when the supply levels of the MCU and one or more of its external peripherals are different. We need to ensure voltage compatibility in the input side. The basic rule here is: if the output levels of the driver ($V_{OL}$ or $V_{OH}$) fall outside the valid regions of low or high for the load, then a voltage level-shifter must be provided between both sides regardless of who is the driver and who is the load. The rule can be summarized as established in Eqs. 8.1 and 8.2.

$$\text{If } V_{OH_{driver}} \begin{Bmatrix} < V_{IH_{load}} \\ \text{or} \\ > V_{DD_{load}} \end{Bmatrix} \Rightarrow \text{lever-shifter needed} \tag{8.1}$$

$$\text{If } V_{OL_{\text{driver}}} \begin{cases} > V_{IL_{\text{load}}} \\ \text{or} \\ < V_{SS_{\text{load}}} \end{cases} \Rightarrow \text{lever-shifter needed} \qquad (8.2)$$

Nowadays many MCUs and peripherals are designed to operate from a 3.3 V supply, but a considerable number of circuits still work at 5.0 V, a voltage level inherited from the TTL era. When both technologies encounter each other in the same application, it becomes necessary to verify compatibility and in most cases taking some measures to make signals compatible.

The solution to the compatibility problem will depend on the direction of the signals. When the driver works at 3.3 V and the load at 5.0 V, in most cases the connection can be directly made. A standard TTL gate with $V_{IHmin} = 2.0$ V and $V_{ILmax} = 0.8$ V will generally accept a 3.3 V CMOS driver with $V_{OH} = 2.7$ V and $V_{OL} = 0.6$ V. However, when the connection is in the opposite direction, a level-shifter becomes unavoidable. The simplest solution is usually obtained with a resistor-based voltage divider. This solution works in situations where a high-impedance constant load is to be driven. Low impedance or high-speed loads should not be interfaced with resistor-based dividers as the integrity of signals and edge speed could be compromised. In such cases, active shifters and dedicated buffers offer a better solution. Example 8.2 illustrates a typical situation and a feasible solution.

**Example 8.2 (Mixed $V_{DD}$ MCU Interface)** *Consider the electrical interface of a 5 V, HD44780-based LCD module to an MSP430F169 GPIO fed from a 3.3 V supply. Provide an interface such that the LCD memory can be read back from the MCU.*

*Solution: In most cases, LCD modules can be connected in write-only mode by hardwiring their $R/\overline{W}$ input to GND. Under this condition the MCU acts as driver ($V_{OH} \simeq 3.3$ V, $V_{OL} \simeq 0$ V) and the LCD as load ($V_{ILmax} = 0.6$ V and $V_{IHmin} = 2.2$ V). In such a case we can get away with a direct connection.*

*However, the case here calls for an ability to read back the LCD memory. Such a condition makes the LCD act as a driver with $V_{OH}$ anywhere between 2.4 and 5.0 V, and $V_{OL}$ between 0 and 0.4 V. The MSP430F169 GPIO becomes a load with requirements of $V_{IL} \leq 1.09$ V and $V_{IH} \geq 1.65$ V, upper bounded by $V_{DD} = 3.3$ V. These specs present no problem at all for the low-level voltages. However, the LCD output-high level voltage would result harmful to the MCU as it exceeds the absolute maximum rated input voltage of any MSP430F169 input, specified at $V_{CC} + 0.3$ V = 3.6 V. In this case a level shifter becomes mandatory.*

*A feasible solution in this case could be provided with a voltage divider [46]. Note that the MSP input leakage in any GPIO pin is only 50 nA, making them indeed, high impedance loads. Moreover, signal speed are limited to around 2 MHz by the 250 nS cycle width spec in the LCD driver. The interface for this case is illustrated below, in Fig. 8.8.*

*The values of $R_1$ and $R_2$ must be chosen to produce satisfactory low- and high-level voltages in the load even under the worst case voltage conditions. Moreover, they must be appropriately sized to let the gate leakage current ($I_Z$) circulate without*

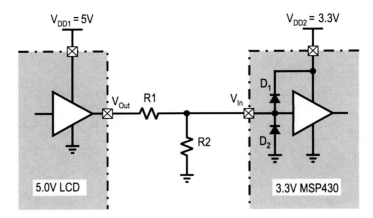

**Fig. 8.8**   A practical 5.0–3.3 V level shifter

**Fig. 8.9**   Worst case voltages
and resistances in divider
network

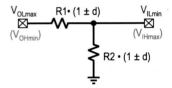

*perturbing the load input logic levels. These requirements are satisfied under the
following three conditions:*

1. *A worst case, low-level, driver output $V_{OLmax}$ shall produce a voltage equal or less
   than the nominal low level input in the load. As $V_{IL}$ has maximum and minimum
   limits, the worst case must satisfy the minimum low-level input $V_{ILmin}$*
2. *A worst case, high-level, driver output $V_{OHmin}$ shall produce a voltage equal or
   higher than the worse, high-level input in the load $V_{IHmax}$.*
3. *The voltage drop caused by the GPIO input leakage ($I_Z$) on the equivalent voltage
   divider impedance shall be negligible when compared to the voltage levels applied
   to the load input.*

*We also need to consider the deviations caused by the resistors' tolerances from
their nominal value. Assuming both resistor have the same tolerance $\pm d$ %, then
their nominal values are affected by a factor $(1 \pm d)$, where $d \ll 1$. The voltage and
resistor conditions in the divider are illustrated in Fig. 8.9.*

*Considering the worse case conditions of $R_1$, $R_2$, and the driver-load voltages
limits as discussed above, a simple circuit analysis allows for obtaining limiting
expressions for the resistor values. Arriving to Eqs. 8.3 through 8.5 requires introducing approximations $(1 + d)/(1 - d) \simeq (1 + 2d)$ and $(1 - d)/(1 + d) \simeq (1 - 2d)$,
as $d \ll 1$.*

$$\frac{R_1}{R_2} \geq \frac{(V_{OHmin} - V_{IHmax})}{V_{IHmax} \cdot (1 - 2d)} \tag{8.3}$$

$$\frac{R_1}{R_2} \leq \frac{(V_{OLmax} - V_{ILmin})}{V_{ILmin} \cdot (1 + 2d)} \tag{8.4}$$

$$R_1 \parallel R_2 \ll \frac{V_{DD2}}{I_Z} \tag{8.5}$$

*To finally compute the resistances we plug-in the voltage values in 8.3 through 8.5. The LCD data sheet provides values* $V_{OHmin} = 2.4\,V$, $V_{OLmax} = 1.5\,V$. *The MSP430169 values were obtained by extrapolating those in the data sheet as specific input thresholds for* $V_{DD} = 3.3\,V$ *are not listed. A quick linear extrapolation on the Schmitt-trigger thresholds yields* $V_{IHmax} = V_{Tmax}^+ = 2.16\,V$, *and* $V_{ILmin} = V_{Tmin}^- = 1.09\,V$, *and assuming resistors with 5% tolerance, Eqs. 8.3 to 8.5 result* $\frac{R_1}{R_2} \leq 1.378$, $\frac{R_1}{R_2} \geq -0.575$, *and* $R_1 \parallel R_2 \ll 66\,M\Omega$.

*The negative value in the lower bound for* $R_1/R_2$ *only means that the low-level output voltage is already below the minimum required input, which makes it compliant regardless of the resistor ratio. So we can assign* $R_1/R_2 = 1$, *which would satisfy both conditions. The third condition is also relaxed so we could assign both* $R_1 = R_2 = 100\,K\Omega$, *which satisfies all conditions.*

*Note that in this particular case the LCD can still be written back by the MCU through this level shifter. The LCD input leakage* $I_{Zlcd}$ *is specified at* $1\,\mu A$ *for its entire voltage range. The voltage drop caused by the level shifter when driven by the MCU would be* $V_{drop} = I_{Zlcd} \cdot R_1 = 1\,\mu A \cdot 100\,K\Omega = 0.1\,V$, *thus the LCD input would still have an input voltage high enough to qualify as a logic-high level. The MCU would not see a significant load:* $I_{MCUout} = I_{Zlcd} + I_{R2} = 1\,\mu A + 3.3\,V/100\,K\Omega = 34\,\mu A$, *which is indeed a light load for the GPIO pins.*

*One last observation about the solution in Example 8.2: the assignment of* $R_1$ *and* $R_2$ *has tradeoffs. The higher their value, the lower the power dissipation as the current through the divider will be lower. However, the time constant resulting from the equivalent resistance and the load input capacitance will go up making the interconnect slower.*

*A better solution in terms of switching speed is offered by a pass-transistor operating as a bidirectional level shifter like that illustrated in Fig. 8.10. This solution effectively shifts voltages in both directions and the switching times are improved as the on-resistance of a MOSFET is typically less than* $200\,\Omega$. *The circuit analysis is left as an exercise to the reader.*

## 8.2.2 Schmitt-Trigger Inputs

GPIO input ports with Schmitt-trigger capability behave differently from standard logic inputs. For such inputs, instead of $V_{IL}$ and $V_{IH}$, the parameters of interest are the positive- and negative-going input thresholds $V_T^+$ and $V_T^-$. A Schmitt-trigger capable

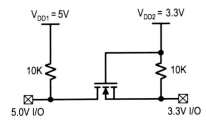

**Fig. 8.10**  A bidirectional level-shifter

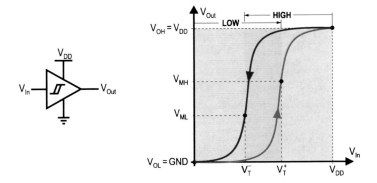

**Fig. 8.11**  Symbol and VTC of a Schmitt-trigger buffer

input is necessary when interfacing slow rising or noisy signals to a digital circuit.. It is common in contemporary MCUs to feature Schmitt-trigger capable inputs in part or all its GPIO inputs.

A Schmitt-trigger buffer adds hysteresis to its behavior by establishing two different threshold voltages, $V_T^+$ and $V_T^-$. A rising input signal would need to reach $V_T^+$ to make the output change from low-to-high. Similarly, a falling input signal would need to drop at least to $V_T^-$ to cause a high-to-low output transition. This behavior can be observed in Fig. 8.11, where the symbol and voltage transfer characteristic of a typical Schmitt-trigger buffer are illustrated.

The slope in the transition region in a Schmitt-trigger buffer VTC is much steeper than that of a conventional buffer. This is an effect of the internal positive feedback used in the buffer to have hysteresis. This instability makes the buffer exhibit abrupt output transitions when the input voltage reaches the low- or high-going thresholds. This property is exploited to sharpen the edges of slow changing input signals. This behavior is illustrated in Fig. 8.12a where a slow rising input is illustrated feeding Schmitt-trigger buffer. Also, when compared to a conventional, single threshold ($V_M$) buffer, a Schmitt-trigger, when fed in its input a noisy signal, provides a cleaner output than that of a conventional buffer. Figure 8.12b compares the outputs from a both types of buffers in front of a noisy input. Note how the undesirable output

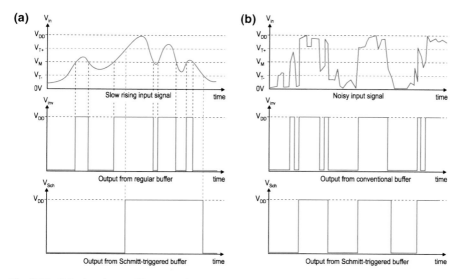

**Fig. 8.12** Schmitt-trigger effect on noisy and slow rising inputs. **a** Response to slowly changing input, **b** Response to noisy input

transitions caused by noise in the conventional buffer output are eliminated from the Schmitt-triggered output.

### 8.2.3 Output Limits in I/O Pins

The practical values of the output levels $V_{OL}$ and $V_{OH}$ in a GPIO change with the current sank or sourced by the pin. This bring us to the definitions of the current parameters of an I/O pin.

In standard logic, five different current values might be specified associated to a GPIO pin. These include the high- and low-level input currents $I_{IH}$ and $I_{IL}$, the high- and low-level output currents $I_{OH}$ and $I_{OL}$, and for tri-state buffers the fifth specified values is the high-impedance leakage current $I_Z$, which quantifies the current leaking through the pin when in a high-impedance state.

These current values in well designed devices reduce to three: the output strength in either low- or high-level $I_{OL}$ and $I_{OH}$, which are dependent on the load, and the input leakage $I_Z$. The input leakage describes in a single value the former $I_{IL}$, $I_{IH}$, and $I_Z$ when inputs have been designed to have a high impedance input regardless of their state.

By convention, manufacturer's data sheets assign to each of these currents positive signs when they flow into a digital pin and negative otherwise. Despite their sign the current magnitude is what determines the internal voltage drop caused in a driver,

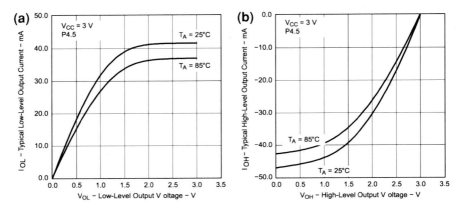

**Fig. 8.13** Driving strength as a function of the output voltage in an MSP430F2274 (*Courtesy of Texas Instruments, Inc.*) **a** $I_{OL} = f(V_{OL})$, **b** $I_{OH} = f(V_{OH})$

defining the net output voltage seen at its output. This explains why we have varying specifications for $V_{OL}$ and $V_{OH}$ in function of the load current.

To exemplify this dependence, let's consider the case of an MSP430F2274. Its leakage current $I_Z$ is specified at $\pm 50\,nA$, while its driving capability is given as a function of the pin current for each logic level. Figure 8.13 reproduces the manufacturer's specification in GPIO P4.5 when only one output is loaded at a time and the supply voltage is 3.0 V.

Both, Fig. 8.13a, b show how the output voltages change with the load current in both the low and high states. For an output voltage $V_{OH} = 80\,\% V_{DD}$ we see that the maximum level-high current source capability would be around 25 mA. Similarly for a nominal $V_{OL} = 0.6\,V$, the maximum $I_{OL}$ results approximately 22 mA. Both values were estimated at 25 °C. The curves denote that the current capability at any given voltage is lower at a higher temperature. Although not shown in the figure, the output current capability also depends on the supply voltage level. According to the chip data sheets, if the supply voltage were reduced to 2.2 V, the maximum currents for $V_{OH} = 80\,\% V_{DD}$ and $V_{OL} = 0.6\,V$ would have resulted in approximately 12 and 17 mA, respectively [48].

The limiting current values of an output pin must be observed when it is being loaded. A design shall not cause an I/O pin to degrade its output voltage at any logic level, as this might take the voltage levels to the point where the high and logic levels cannot be differentiated. Even worse, making the pin current reach or exceed the device's absolute maximum current ratings would cause irreversible damage on the chip.

As a general rule, a designer needs to ensure that the total load current connected to an output pin never exceeds the pin rated current for the nominal load voltage. A simple KCL (Kirchhoff's Current Law) verification shall suffice to determine the loading conditions. Given an output $k$ driving $n$ loads, where each load $i, i = 1, 2, \ldots, n$, requires current $I_{IL(i)}$ when driven low and $I_{IH(i)}$ when driven high, as illustrated in Fig. 8.14, we must make certain that the rules in (8.6) and (8.7) are both

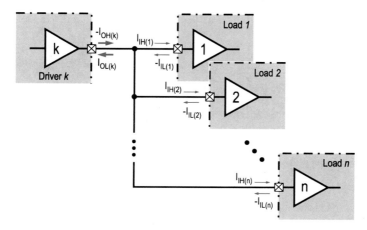

**Fig. 8.14** Currents in a loaded driver

satisfied.

$$\sum_{i=1}^{n} |I_{IL(i)}| \leq |I_{OL(k)}| \tag{8.6}$$

$$\sum_{i=1}^{n} |I_{IH(i)}| \leq |I_{OH(k)}| \tag{8.7}$$

If we need to drive a total load that exceeds the current limit of the pin in either state, a buffer providing sufficient current driving capability is needed. Section 8.8 discusses several buffering options when driving large loads.

### 8.2.4 Maximum Power and Thermal Dissipation

Besides the total current per pin, the total power dissipation of the chip must also be kept within bounds. An MCU might have all its I/O pins driving currents below their individual limits, but that does not mean the chip is being safely operated. There is a limit in the maximum power dissipation that a chip can safely withstand. This limit is determined by the maximum temperature the die inside the chip case can withstand. Some manufacturers explicitly provide in their data sheets a limit for the total power dissipation of their chips, but this specification is more commonly given in the form of the thermal limits of the packaged device, due to its dependence in the ambient temperature.

A packaged MCU, when operated at a certain supply voltage and clock frequency, will generate heat from its die. The total temperature accumulated in a die

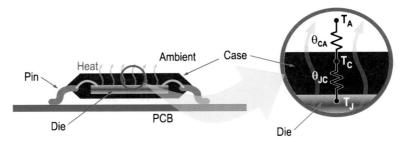

**Fig. 8.15** MCU heat dissipation denoting thermal resistances $\theta_{JC}$ and $\theta_{CA}$

(junction temperature $Tj$ will result from the addition of the heat caused by the chip power dissipation and the ambient temperature ($T_A$). The chip ability to release this heat accumulation will depend on the thermal conductance of the die surrounding materials. The lower the heat conductivity, the higher the thermal resistance of a material. The thermal resistance is measured in centigrade degrees per watt (°C/W). Figure 8.15 shows a sectional view of a packaged IC. There we can observe that heat would need to pass from the die to the case (junction to case thermal resistance $\theta_{JC}$) and from the case to the surrounding air (case to ambient thermal resistance $\theta_{CA}$. The values for these thermal resistances for a chip in a given case are provided in the device data sheets or in the package specifications.

The total thermal resistance from junction (die) to ambient when no heat dissipation mechanism has been added to the chip can be written as:

$$\theta_{JA} = \theta_{JC} + \theta_{CA} \tag{8.8}$$

The die temperature can be estimated as the sum of the external temperature and that generated by the power dissipation as:

$$T_J = T_A + \theta_{JA} \cdot P_{\text{diss}}, \tag{8.9}$$

where $P_{\text{diss}}$ is the total chip power dissipation. $P_{\text{diss}}$ can be obtained from the supply voltage $V_{DD}$ and the sum of all I/O currents handled by chip the as:

$$P_{\text{diss}} = V_{DD} \cdot \left( I_{DD(avg)} + \sum_{allpins} |I_{IO(avg)}| \right) \tag{8.10}$$

Note that making a precise estimate of the total power dissipation of an embedded design is a complex problem. Power in digital circuits is a function of the operating voltage, driven load, clock frequency, and processor activity. From this list of parameters, perhaps the most difficult to obtain is the processor activity since it depends on the information being processed by the CPU. The expression in Eq. 8.10 is a rough estimate assuming a constant supply voltage and the average currents have been adjusted to account for low power modes in use, use duty cycle, supply voltage, and clock frequency. For a deeper insight into the problem of estimating power in embedded systems, the reader is referred to [49].

If we were interested in the chip maximum power dissipation, it can be obtained from (8.9) using the ambient temperature $T_A$, and the manufacturer's provided maximum junction temperature $T_{J(max)}$ and the junction-to-ambient thermal resistance $\theta_{JA}$ as:

$$P_{diss(max)} = \frac{(T_{J(max)} - T_A)}{\theta_{JA}} \tag{8.11}$$

If the load on a chip makes the IC to exceed its maximum power dissipation, its junction temperature will reach values where early thermal failure may occur. To reduce the chip junction temperature, possible alternatives include the following approaches (or their combination):

1. Reduce the chip power dissipation. This can be achieved by placing external buffers to reduce the current demand on the IC or by lowering the chip supply voltage level, if that were an option. Modern MCUs, including the MSP430, provide a valid range of supply voltage options. Also, improving the power consumed by the application by exploiting low power modes and good hardware and software design practices can be very effective in reducing the MCU power consumption.
2. Reduce the total thermal resistance of the chip. A reduction on $\theta_{JA}$ can be obtained by two means: attaching a heat sink to the IC package and/or forcing the air circulation around the chip with a fan. The thermal resistance reduction induced by a heat sink will be a function of the heat sink area. When a heat sink is attached to an IC, the junction to ambient resistance formula is modified to replace $\theta_{CA}$ with the heat sink thermal resistances from case to sink $\theta_{CS}$ and form sink to ambient $\theta_{SA}$. In this case the total thermal resistance is:

$$\theta_{JA} = \theta_{JC} + \theta_{CS} + \theta_{SA} \tag{8.12}$$

If a fan were included, the air volume per unit time in cubic-feet per minute (CFM) moved by the fan would determine an adjustment factor $\gamma < 1$ that multiples $\theta_{SA}$, lowering the net thermal resistance.
3. Reduce the ambient temperature $T_A$. As the net junction temperature is a direct function of the ambient temperature, a reduction of the surrounding chip temperature will help.

Figure 8.16 shows a solution combining heat sink and fan on a microprocessor to keep the die junction temperature cool.

**Fig. 8.16** IC cooling
approach combining *heat
sink* and *fan*

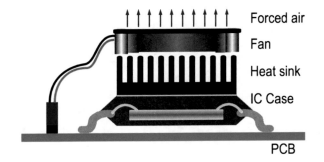

## 8.2.5  Switching Characteristics of I/O Pins

The second important set of specifications in GPIO regards their timing specifications. These define how fast a GPIO pin can be switched.

The maximum switching frequency of an I/O pin is generally defined by the voltage supply level, load capacitance, and the main clock frequency. Input pins have a minimum pulse width they can detect. For example, an MSP430F5438 specifies the minimum detectable pulse width in a GPIO input enabled to trigger interrupts at 20 nS.

Output pins would generally be limited by the maximum system clock and the external capacitance being switched. In the MSP430F5438 the maximum output frequency in an 18 MHz part is 25 MHz when the load capacitance does not exceed 20 pF. As the load capacitance or output resistance increase, the load time constant increases, softening signal edges and making transitions slower. This is manifested by an increase in the rising and falling times of the signals being propagated. Moreover, I/O pins associated to specific internal peripherals have their own timing constraints, the data sheets specific to a given part number or their particular peripheral specification need to be consulted to learn about the specific limitations that each case might have.

## 8.3  Interfacing Switches and Switch Arrays

One of the components most commonly interfaced to MCUs GPIO ports are mechanical contact switches. Switches provide for intuitive user interfaces and practical detection mechanisms for a vast number of applications. The states of a switch, either open or closed, perfectly match with the binary detection ability of an input GPIO line.

Depending on the application, switches might be used individually, like the case illustrated in Example 8.1. Other applications use arrays of switches, such as the

**Fig. 8.17** Symbols of four common types of switches

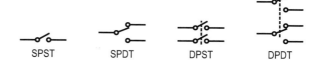

**Fig. 8.18** Common types of momentary push-buttons

case of conventional keypads or keyboards. Depending on the case, particular switch interfaces can be used to interface them to GPIO ports.

## 8.3.1 Switch Types

Many variants of switches can be found in electronic circuit applications. A common classification is based on the number of contacts (poles) operated an the positions (throws) adopted by the contacts. Under such a classification, a *single-pole, single-throw* (SPST) switch has only one contact that is either open or closed, while a double-pole, single-throw (DPST) has two contacts that are either open or closed and moved by a single throw mechanism. Figure 8.17 illustrates a few typical switch configurations.

One particular type of switch widely used in MCU interfaces is the momentary push-button (MPB). A normally-open (NO) momentary push-button is a SPST switch that remains open at all times, except when it is being depressed. A normally closed (NC) MPB will remain closed at all times except when actuated. Figure 8.18 shows three types of push-button switches.

## 8.3.2 Working with Switches and Push-Buttons

The interface of a single switch to a GPIO input pin is quite simple. It only needs to provide a means of ensuring that the two possible switch states, OPEN or CLOSED, are represented with appropriate logic levels. Figure 8.19 shows two simple connections for a momentary push-button (SW). In the first case, node $(x)$ provides a logic low level when the switch is closed, while in the second, it provides a logic high for the same switch condition. It results straightforward to see why node $x$ in Fig. 8.19a goes low when SW is closed: it gets directly connected to GND. When SW is open, we need to ensure that node $x$ goes high. To this end we attach pull-up resistor $R_{pull-up}$ to the node. Note that without this resistor, when SW opens, node

**Fig. 8.19** Common interfaces for momentary push-buttons. **a** MPB with pull-up resistor, **b** MPB with pull-down resistor

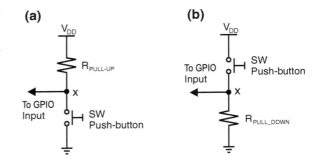

$x$ would simply float to a high impedance state, which would not necessarily be interpreted as a valid high level.

The case illustrated in Fig. 8.19b is similar to the above in terms of providing a reliable logic level when the switch is open. Except that in this last case the OPEN state is encoded a logic low. The pull-down resistor $R_{pull-down}$ does the job in this case.

In either case the value of the resistor is chosen to satisfy two criteria: first, to allow for having a valid logic level when SW is open and second for limiting the current circulating to ground when SW is closed. In the case of a pull-up resistor, the voltage at $x$ is expected to be within the valid range of a logic high for the GPIO, i.e. $V_x \geq V_{IH}$. Thus, if the amount of current leaking into the GPIO pin in high were $I_{IH}$ then the value of $R_{pull-up}$ would need to satisfy (8.13).

$$V_{DD} - R_{pull-up} \cdot I_{IH} \geq V_{IH} \tag{8.13}$$

Thus, an upper bound for the pull-up resistor would be:

$$R_{pull-up} \leq \frac{V_{DD} - V_{IH}}{I_{IH}} \tag{8.14}$$

If we establish a limit to the current to ground when the switch is closed, to be for example $I_{SWmax} = 10 I_{IH}$, then we can also set a lower bound to the resistor. This however might be unnecessary since in most cases the limit obtained from $V_{IH}$ would yield a satisfactory value. In this case we would say

$$R_{pull-up} \geq \frac{V_{DD}}{I_{SWmax}} \tag{8.15}$$

If a $V_{IH}$ specification were not readily available we can safely assume $V_{IH} = 0.9 V_{DD}$. Some MCUs have a very small input current in either state, reason for which this current is listed as an input leakage, denoted by $I_Z$.

**Example 8.3** *Consider an MCU operating with $V_{DD} = 5.0\,V$, with an input current of $1\,\mu A$ in its GPIO pins for either low or high levels and $V_{IHmin} = 0.7V_{DD}$. Find a suitable value for a pull-up resistor to interface a NO-MPB to one of its input pins.*

*Solution:In this case we will assume $V_{IH} = 0.9V_{DD}$ which satisfies the minimum $V_{IH}$. Applying Eq.(8.14) we'd obtain*

$$R_{pull-up} \leq \frac{0.1V_{DD}}{1\,\mu A} = 500\,K\Omega \tag{8.16}$$

*If we assign $I_{SWmax} = 20I_{IH} = 20\,\mu A$ then an upper bound for the resistor would be*

$$R_{pull-up} \geq \frac{V_{DD}}{20I_{IH}} = \frac{5.0\,V}{20\,\mu A} = 250\,K\Omega \tag{8.17}$$

*Thus, any value in the range $250\,K\Omega \leq R_{pull-up} \leq 500\,K\Omega$ would satisfy these specifications. For the sake of low-power, choosing the largest value would yield the lowest current when the switch is closed.*

When using a pull-down resistor, as illustrated in Fig. 8.19b the reasoning would be similar to the previous case, except that we'd need to look into providing $V_x \leq V_{IL}$ when the switch is open. Thus we'd use the specification for $I_{IL}$ to find the upper limit of $R_{pull-down}$, and, if necessary, a $I_{SWmax}$ criteria for a lower bound.

### 8.3.3 Dealing with Real Switches

So far we have assumed that switches, when operated will cleanly close or open such that their interfaces produce one sharp high-to-low or low-to-high when opened or closed. That would be true only under ideal conditions. Real switches have multiple non-ideal characteristics that need to be considered when used in the real world. Among such non-idealities we can list the following:

- Non-zero "ON" Resistance: although small, the resistance between switch terminals, when closed is never zero.
- Limited contact current capability: there is a maximum current the switch can safely handle when closed. Exceeding this value accelerates contact wear and leads to early failure.
- Limited Dielectric Strength: This is the maximum voltage the open contacts can withstand before arcing.
- Limited Mechanical Life: Number of times the switch is rated to be operated.
- Finite Contact Bounce Time: Time required to contacts to settle after operated.

From this list, a non-ideality of concern in embedded systems design is contact bounce. When two mechanical contacts are operated, they do not open or close cleanly. Instead, they rebound a certain number of times, sometimes over one hundred times, before steadying at their final state. The maximum time taken by the contacts

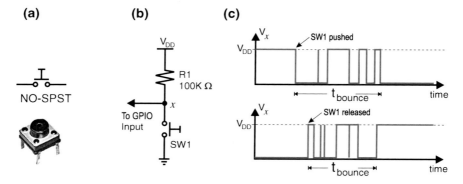

**Fig. 8.20** Bouncy behavior of mechanical NO-MPB. **a** Actual Switch, **b** Interface Circuit, **c** Output waveform. The *top* and *bottom* waveforms correspond to the switch closing and opening, respectively

in a switch to reach steady state after being operated is called the *switch bounce time*. Figure 8.20 shows a common type of push-button used in many embedded applications, its interface, and the manifestation of contact bounce in the interface output when operated.

### 8.3.4 Characteristics of Switch Bounce

The waveforms in Fig. 8.20c illustrate the effect of bouncing in the switch output, denoting that the phenomena occurs in both, switch closures and openings. The top plot shows the closing bounce, while the bottom waveform shows the behavior upon switch opening. Typical bouncing times changes from one switch to another. An informal study carried by Ganssle shows that although most switches bounce for 10 ms or less, it is possible to find some switches with bouncy behavior extending for over 150 ms [52]. Quality switches typically have less than 10 ms bounce time. If we are uncertain about the bouncing time of a particular switch, a simple experiment with a digital storage oscilloscope (DSO) can help.

### 8.3.5 Problems Associated to Contact Bounce

Bouncing causes that a single switch throw be interpreted as multiple operations, causing in many cases incorrect system operation. Consider for example an embedded system implementing a TV remote control. If the channel flipping button were not properly debounced, depressing the channel up key would actually make the unit jump several channels at once. Other scenarios can be devised.

Switch bouncing is particularly troublesome when the switch action is set to trigger an interrupt or other edge triggered action. Every bounce will account for a trigger, potentially unleashing a chaotic software behavior. Polling might also be affected by bouncing, particularly if the polling interval is shorter than the bounce time. To avoid such situations, the solution is to implement some sort of switch debouncing approach in every design using mechanical switches. Note that this recommendation is valid not only for manually operated switches, but also for any application where system operation were driven from the opening or closure of mechanical contacts.

## 8.3.6 Switch Debouncing Techniques

The problem of bouncing can be approached from either a hardware or software standpoint. A hardware approach will insert circuit components such as a filter or some form of digital logic to suppress at the digital circuit input the transient pulses caused by the switch bounce. Software-based approaches allow the bouncy signal to enter the MCU, and then deals with the transient in the program, preventing the interpretation of the transitions associated to the switch bounce. Below we present some of the most widely used hardware and software bounce techniques.

## 8.3.7 Hardware Debouncing Techniques

Hardware debouncing techniques attempt to filter the switch transient behavior, feeding to the digital input pin one clean transition each time the switch is operated. There are different ways to debounce a switch by hardware. Below we discuss representative hardware debouncing circuits.

**SR Debouncing Circuit:** One of the most effective ways of debouncing a switch is by using a Set-Reset (SR) latch between the switch and the digital input, as illustrated in Fig. 8.21.

The SR latch circuit will debounce SW1 regardless of its bounce time. R1 and R2 in this circuit act as simple pull-up resistors. When SW1 is in the up position, the latch will remain set, yielding a logic high at the output. When the switch is flipped, the latch will be reset and the output will transition from high to low. When the contact bounces, the latch will just remain reset regardless of the number bounces. A similar situation will happen when SW1 is toggled to the set position. If multiple switches were to be debounced using this method, a 74HC279 with four SR cells in a single package, or a 74HC373, which has eight, might become handy. One disadvantage of this circuit is SW1. It needs to be a SPDT switch, which is more expensive and bulkier than a regular NO momentary pushbutton.

**RC Debouncing Circuit:** A cost-effective solution to debounce the contacts of a switch can be achieved by using an RC network as debouncer, as illustrated in Fig. 8.22.

**Fig. 8.21** SR latch-based
switch debouncing circuit

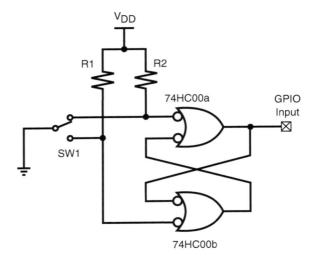

**Fig. 8.22** RC network-based
switch debouncing circuit

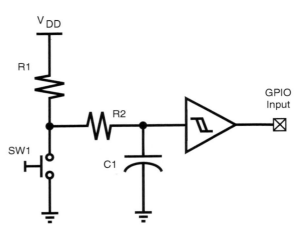

An RC debouncer uses a resistor-capacitor pair to implement a delay in the switch line. To understand how it works, let's refer to Fig. 8.22. Assume $SW_1$ has been open long enough, to have C1 charged to $V_{DD}$. Note that while $SW_1$ is open, $C_1$ charges through $R_1 + R_2$. When SW1 is closed, C1 begins to discharge through $R_2$. If the time constant $R_2C_1$ is large enough, when $SW_1$ contacts bounce and open, voltage $V_x$ at the Schmitt trigger input would not have reached the gates's input-low threshold $V_T^-$, preventing the circuit output from making a transition. To determine the values of $R_2$ and $C_1$, we just need an estimate of the bounce time $t_{bounce}$.

The formula describing the voltage at node $x$ would be that of discharging $C_1$ through $R_2$:

$$V_x = V_{DD} \cdot \exp^{-\left(\frac{t_{bounce}}{R_2 C_1}\right)} \tag{8.18}$$

Since (8.18) has two unknowns, $R_2$ and $C_1$, we need to estimate one of them. It is usually easier assigning a value to $C_1$ and solving for $R_2$ since the latter can be more easily accommodated with a standard resistor value. Thus, considering that we want to have $V_x \geq V_T^-$ at the end of $t_{bounce}$, we obtain the value for $R_2$ as:

$$R_2 \geq \frac{t_{bounce}}{C_1 \cdot \ln\left(\frac{V_{DD}}{V_T^-}\right)} \tag{8.19}$$

To calculate the circuit requirements for debouncing the switch aperture, let's assume that before the switch is released, the contacts have been closed long enough to completely discharge $C_1$, making $V_x(0) = 0V$. Opening $SW_1$ will cause $C_1$ to begin loading through $R_1 + R_2$. Again, we want the time constant $(R_1 + R_2)C_1$ to be long enough such that $V_x$ remains below $V_T^+$ before the bouncing finishes. So, with $C_1$ and $R_2$ already assigned, all we need to do is to compute the value of $R_1$. In this case the value of $R_1$ can be obtained via the charging equation for $C_1$, that describes $V_x$ as:

$$V_x = V_{DD} \left[ 1 - \exp^{-\left(\frac{t_{bounce}}{(R_1+R_2)C_1}\right)} \right] \tag{8.20}$$

Considering that we want $V_x \leq V_T^+$, the sum of $R_1$ and $R_2$ needs to satisfy:

$$(R_1 + R_2) \geq R_{min}, \tag{8.21}$$

where

$$R_{min} = \frac{t_{bounce}}{C_1 \cdot \ln\left(\frac{V_{DD}}{V_{DD}-V_T^+}\right)} \tag{8.22}$$

Thus, resulting in $R_1 \geq (R_{min} - R_2)$.

This circuit solution shall work in most cases. Some designers might recommend replacing $R_2$ by a shortcircuit. This alternative in addition to risking not providing enough delay for $SW1$ closure, will also cause an accelerated wear of $SW_1$ contacts due to the current peak resulting from the shorting of $C_1$. Depending on the threshold values of the Schmitt trigger buffer, there is a possibility of ending up with a value of $R_{min} > R_1$, which makes the expression for $R_1$ an absurd. In such a case, a modified version of the circuit in Fig. 8.22 would be recommended, inserting a diode in parallel to R2, with its anode connected to node $x$ [52]. This would make the circuit to charge $C_1$ through $R_1$ and the diode, while the discharge would occur exclusively through $R_2$. It is left as an exercise to the reader obtaining the expressions for the circuit components.

**IC Debouncer Circuit:** Sometimes, instead of using discrete components to implement a hardware debouncer, it might be more convenient an integrated solution. This might be particularly convenient if a debouncing solution for multiple switches must include externally to the MCU all resistors, capacitors, and Schmitt trigger buffers. In such cases, commercial off-the-shelf alternatives such as the ELM410

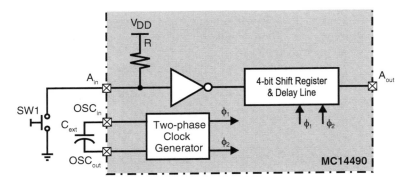

**Fig. 8.23**  IC-based switch debouncing circuit using the MC14490

from Elm Electronics or the MC14490 from ON-Semiconductors might result appropriate. As an example, let's consider On Semiconductor's MC14490 Hex Contact Bounce Eliminator. The chip contains six independent debouncing circuits for equal number of switches. It provides on-chip pull-up resistors, buffers, and a reloadable shift registers to provide independent rising and falling edge debouncing for all six switches. A single on-chip clock generator provides the internal frequency $\phi$ driving the shift registers. Figure 8.23 shows a block diagram of the debouncer structure.

The debouncing strategy in this IC assumes the switch contacts have settled when no bouncing transitions are detected for four consecutive cycles of internal clock $\phi$. The frequency of $\phi$ is determined with external capacitor $C_{ext}$ (one per chip). Some of this IC solutions might result handy and compact, but they are also expensive and parts that might not be offered by all chip distributors.

## 8.3.8 Software Debouncing Techniques

Software debouncers have the attractiveness of not requiring external components beyond the switch and pull-up resistor illustrated in Fig. 8.19. However, they require CPU cycles to remove the bouncy portion from the switch signal. There are a few general rules that should be followed to design better software debouncing routines:

- A software debounced switch shall not drive an interrupt input. Interrupt inputs usually drive the clock lines of input flip-flops, which might behave erratically with the very short pulses generated by the switch bounce.
- Avoid wasting CPU cycles by sending the CPU to do-nothing NOP cycles or executing useless instructions while you wait for the switch contacts to settle. This will waste not only CPU cycles, but also energy. Activate some low-power mode while you wait or get the CPU to do some other useful task.
- Keep response time the shortest possible. Humans can detect response delays as small as 75 ms, which make the system feel "slow".

- Keep solutions simple. A debouncing routine is not expected to be complex. Afterwards, it is just a switch what is being interfaced.

A few representative software approaches are described below.

**A Polling Debouncer:** This simple alternative polls the switch port with a constant polling period longer than the expected switch bouncing time. If for example the mean bounce time of a switch were 15 ms, then polling the switch about every 20 or 25 ms would suffice to avoid the switch bounce. To avoid wasting cycles while waiting for $t_{bounce}$ to expire, one could use a timer to set the poll time and send the CPU to a low-power mode. This alternative, although simple, might fail to debounce a switch with bounce time that exceeds the assumed value of $t_{bounce}$.

**A Counting Debouncer:** A better approach is to assume the contacts have settled if they have not bounced for a certain number of samples $n$ [52]. If $n$ consecutive settled samples are collected, then a valid edge indicator is returned. Otherwise, if a change in the switch status were detected before collecting the $n$ samples, then the count would be reset and counting restarted. A feasible implementation could use a timer to periodically sample the switch. Typical values of inter sample time could be between 1 and 10 ms, while setting $n$ to around a dozen samples. The pseudo assembly fragment below would provide a suitable implementation of the timer ISR.

```
;=========================================================================
; Timer service routine for debouncing a switch closure (Closed = Low)
; Global variables "S_Count" and "EdgeOK" initialized to FFFFh and FALSE
;-------------------------------------------------------------------------
Timer_ISR    BIT.B #SW1,&PxDat      ; Load switch status into carry flag (C)
             JC Bouncing            ; If high then contact was found open
             RLC.W &S_Count         ; Valid contact state detected & counted
             CMP #F000h,&S_Count    ; Is it the 12th steady sample?
             JNZ Exit               ; Valid closure but not there yet
             MOV.B #TRUE,&EdgeOK     ; Count top reached
Bouncing     MOV #FFFFh,&S_Count    ; Restart counter
Exit         RETI
;=========================================================================
```

The assembly function above is compact, simple, effective, and portable. However it only debounces the switch closure and would delay the notification of the received edge. Example 8.4 illustrates an improved solution that debounces both the switch closure and opening while also preventing edge repetition due to long switch closures and providing early notification of the switch state change.

**Example 8.4 (SW Debounced MPB & LED)** *Redesign the software solution in Example 8.1 to provide for a software debounced switch interface for SW1. Solution is to maintain the toggle functionality without repetition due to the switch being held pressed.*

*Solution: In this case we adopt the counting approach coupled with an edge detection capability that allows detecting either a closure or opening of the switch. A complete solution is provided where the main program initializes all structures: I/O port, Timer, and variables, and then sends the MCU to a low-power mode. Timer A0 wakes the MCU approximately every 2.2 ms to sample SW1. The whole debouncing*

**Fig. 8.24** Flowchart for
software debouncing timer
ISR in Example 8.4

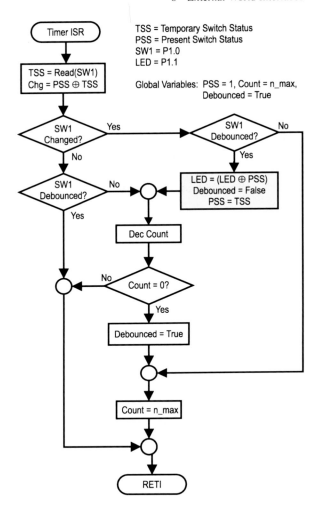

An assembly implementation of this solution is provided below. The implementation targets an MSP430x2xx, like that in TI's LaunchPad. The solution makes use of Timer A channel 0 to space samples in time. Although the discussion of timers is ahead, in Chap. 7, the provided comments should guide the reader through the solution steps.

```
;===================================================================
; Symbolic constants and addresses are declared in MSP430 header file
; Assumes fclk = 32.768KHz from LF XTAL1
;-------------------------------------------------------------------
#include "msp430g2231.h"
```

```
;-------------------------------------------------------------------
Top_Cnt EQU 12 ; Number stable SW samples to be debounced

RSEG DATA16_N
Count DS 1        ; Sampling counter
PrsSwStat DS 1 ; Present switch status  R7
Debounced DS 1 ; Debounced state  R8

RSEG CSTACK     ; Stack segment

RSEG CODE
Reset MOV.W #SFE(CSTACK),SP         ; Initializes the stack area
      MOV.W #WDTPW+WDTHOLD,&WDTCTL ; Stops the MSP430 WDT

;----->Port 1 Setup
      BIC.B #BIT1,&P1DIR ; Set Px.0 bit 0 as input
      BIS.B #BIT0,&P1DIR ; Set Px.1 as output
      BIC.B #BIT0,&P1OUT ; Begin with D1 off

;----->Timer A0 Setup
      MOV.W #TASSEL_2+MC_2,&TACTL ; Source from ACLK, continuous up
      MOV.W #72,&TACCR0            ; Sampling set to about 2.2ms
      MOV.W #CCIE,&TACCTL0         ; TACCR0 interrupt enabled

Main  BIS.B #BIT0,&PrsSwStat ; Present SW state is OPEN
      BIS.B #BIT0,&Debounced ; Present status is debounced
      MOV.B #Top_Cnt,&Count  ; Initialize Count
      BIS.W #LPM0+GIE,SR     ; Enable interrupts, LPM0
      NOP


;-------------------------------------------------------------------
; Timer A0 service routine for debouncing SW1 and toggling LED1
;-------------------------------------------------------------------
TimerA0_ISR PUSH R5 ; Used for temporary switch state (TSS)
            PUSH R6
            MOV.B &P1IN,R5 ; Load current switch status into R5
            AND.B #BIT4,R5 ; Screen-out SW1 bit
            RRA.B R5       ; Align LED and SW1 bits
            MOV R5,R6      ; Save TSS into R6
            XOR.B &PrsSwStat,R5; Detect if switch state changed
            JZ NoChg ; No change wrt previous state

            BIT.B #BIT0,&Debounced  ; Check if already debounced
            JZ ChgNDbc              ; Changed but not debounced
            XOR.B &PrsSwStat,P1OUT  ; Debounced & changed => Toggle LED
            CLR.B &Debounced        ; Now SW1 is undebounced
            MOV R6,&PrsSwStat       ; Change the present switch state

NChgNDbc    DEC.B &Count            ; Update count of stable samples
            JNZ Exit                ; If less than top count then exit,
            BIS.B #BIT0, &Debounced ; else, set debounced as true

ChgNDbc     CLR.B &Count            ; Reset the sample counter
            JMP Exit

NoChg       BIT.B #BIT0,&Debounced  ; Check if already debounced
```

```
            JZ NChgNDbc                 ; No Change & not debounced
Exit        POP R6                      ; Restore registers
            POP R5
            RETI

; -----------------------------------------------------------------
COMMON INTVEC ; Interrupt Vectors
; -----------------------------------------------------------------
            ORG TIMERA0_VECTOR
            DW TimerA0_ISR
            ORG RESET_VECTOR
            DW Reset
END
; =================================================================
```

## 8.3.9  Switch Arrays

Many applications require multiple switches to be interfaced to the MCU. Under such situations, switch arrays come in handy to facilitate the interface. Two general types of switch arrays can be identified: Linear arrays, also called DIP Switches, and matrix type, also referred to as keypads or keyboards.

## 8.3.10  Linear Switch Arrays

A linear switch array contains multiple independent SPST slide switches, typically in a dual-in-line package, which lends them the denomination of DIP Switches. DIP Switches come in groups from two to about twelve, being the DIP-8 one of the most commonly used in MCU systems. Figure 8.25 shows the symbol and actual package of a DIP-8 switch array.

DIP switches are frequently used to provide hardwired codes or configuring hardware components in a easily modifiable way. Each switch in a linear array needs to be individually interfaced, and if they are meant to be directly read by the MCU, each will require its own I/O pin. Moreover, if the switches are meant to be read by the CPU while they are operated, debouncing techniques will also be required.

If a linear switch array were to be hardware debounced, the same techniques available to individual switches would apply. One must consider however the impact on the design recurrent costs if this modality were to be applied.

Software debouncing techniques usually offer a more cost effective alternative when dealing with switch arrays (shall CPU load permit). In this case, instead of individually debouncing each switch, it is more convenient to debounce them all simultaneously. This is particularly simple if all the switches are connected to contiguous pins in the I/O port(s). An approach similar to that of the counting debouncer

**Fig. 8.25** A typical DIP switch: **a** DIP-8 symbol, **b** actual DIP-8 switch

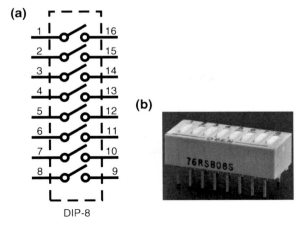

**Fig. 8.26** A 3-by-4 keypad showing the internal arrangement and appearance. **a** Switch arrangement, **b** Actual keypad

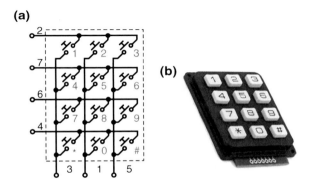

in the previous section could be followed, except that instead of acquiring and checking a single switch line, the entire switch port(s) is treated at once.

The algorithm depicted in Fig. 8.24 could easily be adapted to debounce a switch array by changing the instances where the switch is read and its changes detected by accesses to the entire switch I/O port. The adaptation of the function is left as an exercise to the reader.

## 8.3.11 Interfacing Keypads

The second type of switch arrays correspond to keypads and keyboards. Keypads have switches arranged in a matrix form where individual switch status can be decoded via a scan algorithm applied to rows and columns. Figure 8.26 shows a schematic diagram and an actual picture of a twelve-key keypad arranged as a three-by-four matrix.

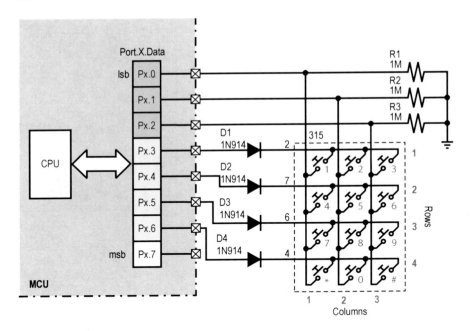

**Fig. 8.27**   Interfacing a 3-by-4 keypad to a GPIO port

The main advantage of arranging the switches in a matrix form is the reduction in the number of I/O lines required to read the entire switch array. Consider for example, a 64-key keyboard. Arranging the keys to form an 8-by-8 matrix allows using only sixteen IO lines to interface the array, instead of the 64 lines it would have required should we had chosen to interface each key individually. In general, an array of $N$ keys can be interfaced using $2\sqrt{N}$ I/O lines when arranged in matrix form.

Although a matrix approach simplifies the interface, decoding de array to detect individual key closures requires a dedicated keypad scanning algorithm. Two possible types of solutions can be chosen: devote CPU cycles and I/O lines to scan the keypad, or using a dedicated interface chip.

**CPU-based Keypad Scanning:** An interface for having the CPU scanning a keypad requires allocating I/O lines to connect row and column lines to the MCU. Row lines can be connected as inputs and columns as outputs (or viceversa). This arrangement allows for issuing a scan code through the output lines and reading the return sequence through the input lines. The designation of either rows or columns as inputs can be arbitrarily done if the key array has an equal number of columns and rows. When the number of rows and columns is different, it is generally preferred to designate the group with the lesser number of lines as inputs. This is because the set of lines designed as inputs need to contain key interfaces to the I/O port. Figure 8.27 shows a schematic diagram of an interface for a 3x4 keypad to an MCU I/O port.

The connection illustrated in Fig. 8.27 presumes input lines, in this case P1.0 to P1.2, are bounce-free. If a software debouncing algorithm were to be implemented, an

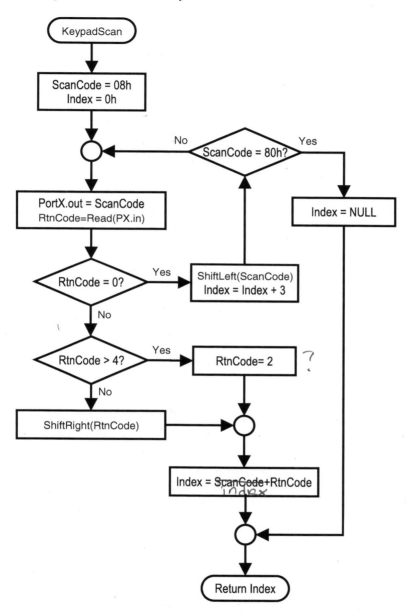

**Fig. 8.28** A scanning algorithm for the 3x4 keypad in Fig. 8.28

approach serving all input lines at once, like that described in Sect. 8.3.10 for a linear array, could be applied. If a hardware debouncing solution were preferred instead, each input line would require the addition of hardware debouncing components.

To understand the basic scan algorithm let's assume the keypad rows are connected to port outputs and the columns to inputs using pull-down resistors as illustrated in Fig. 8.27. As in the case of single switches, the pull-down resistors will cause the corresponding switch to be read as a low-level when open. If a high-level voltage were applied to only one of the keypad rows by setting the corresponding output pin to high, reading the keypad columns via input pins would unequivocally give us the state of all the keys in the row that was driven to high. For example, setting row 2 (PortX.4) to high while key "8" is depressed, would return code "010" through lines PortX.0 through PortX.2.

Thus, by performing a sequence that sends scan codes "0001", "0010", "0100", and "1000" would activate each row one at time. If after sending each individual scan code we read the column lines (return code), the combination of each scan code with the corresponding return code would unequivocally allow us to identify any key depressed in the entire keypad. A keypad scan algorithm would just need to perform this sequence in a loop. Thus, if while this algorithm were running keys 1, 5, and 8 were depressed, one at a time, the corresponding seven-bit codes resulting from their scan-return codes would be "0001001", "0010010", and "1000010", respectively. Note that in these bit sequences, the upper four bits represent the scan code and the lower three bits contain the corresponding return code.

The trick that makes this basic algorithm work without missing the detection of any single keystroke is the microcontroller's ability to perform the scan sequence much faster than a human can type.

A few observations can be made in regards to this algorithm and its supporting hardware. First, if multiple keys in the same row were simultaneously depressed, the row return code would allow their identification within a single scan. However, if multiple keys of the same column were depressed, lines from different rows would become joined, with the potential to create a short circuit between the output port lines. Some keypads have embedded diodes in their keys, which prevent this situation. Inexpensive keypads rarely have this provision. In such cases it is advisable to include diodes in the scan code lines, as illustrated in Fig. 8.27 with diodes D1 through D4. Once the scan/return code combination has been obtained, all that is left is just assigning a meaning to each of the collected key codes.

Figure 8.28 shows a flowchart of a simple procedure to assign sequential numeric codes to the scan of a 4x3 keypad connected as illustrated in Fig. 8.27. The returned "Index" would assign a binary between 00h and 0Bh to the depressed key in the order "123456789*0#". If no key were depressed a null character is returned. If multiple keys were depressed, only the one with the highest return code will be returned.

**Dedicated Keypad Interfaces:** In some applications, devoting CPU cycles for scanning, debouncing, and encoding a keypad might not be feasible. In such instances it might be convenient to consider the usage of a dedicated keypad interface.

A dedicated keypad interface provides the convenience of accommodating in a single chip the necessary logic to scan, debounce, and encode the keys of matrix or non-matrix type keypads and keyboards. Such interfaces relieve the MCU from allocating I/O pins and cycles, reducing code complexity in keypad interfaces at the expense of increasing the count number of board components and increasing the recurrent system costs.

Specific features of dedicated keypad interfaces vary from one vendor to another. Some commonly found features include:

- Debouncing Capabilities
- Key repetition
- Interrupt capabilities
- Serial output

A representative example of such type of interfaces is the EDE1144. This chip incorporates the ability of handling matrix type keypads up to 4x4, providing serial output and even the ability of tactile feedback via an external buzzer. Another example of dedicated keypad interface is the 74c922. This chip provides a simpler interface including the ability of handling keypads up to 5x4 and supporting key debouncing and encoding.

Despite their convenience, these types of chips tend to be expensive and might not be available with every vendor. For these reasons, when a dedicated keypad interface becomes necessary in an application, some designers prefer to devote a low cost MCU solely devoted to interfacing keypads or key arrays. For example, a small MCU like the MSP430G2231 could be programmed to serve a keypad up to 5x5, allowing to define by software all the desired features in the keypad performance. An application note for the MSP430F123 published by Texas Instruments under the number SLAA139 describe an ultra low-power keypad interface that provides a good starting point to such an application.

## 8.4  Interfacing Display Devices to Microcontrollers

The second most common type of interfaces in embedded systems are display devices. Coupled with switches and keypads, visual display components form the most common type of user interface that allows humans and other species interact with embedded systems.

Nowadays there is a broad selection of display devices that can be attached to microcontrollers to provide visual information. These range from discrete light emitting diodes (LEDs) to indicate binary conditions to intricate colorful plasma, LED, and Liquid-crystal Displays (LCD) flat panels capable of displaying real life still images and videos.

A broad classification of displays can be made, based on the amount of visual information they are able to provide. These include discrete indicators, numeric displays, alphanumeric displays, and graphic panels.

- **Discrete Indicators** refer to the usage of discrete light emitting diodes (LEDs) or some other display mechanism to provide single dot indicators. Traditional red, green, blue, and yellow LEDs provide for simple monochromatic, binary indicators able to provide an On/OFF or Yes/No indications. Polychromatic single dot indicators are also possible when using RGB LEDs. An RGB LED is just a combination of three monochromatic LEDs conforming the three basic colors: red, green, and blue. RGB LEDs, with their ability to produce colors, can be used to provide more than simple binary indications. Eight different indications ($2^3 = 8$) are possible by the binary combinations of its three basic LEDs. An even larger gamma of indications can be achieved by regulating the intensity of each individual basic color.

- **Numeric Displays**: Typically seven-segment displays that allow only displaying numeric information with decimal digits 0 through 9. Both, LED- and LCD-based displays of this type can be found.
- **Alphanumeric Displays**: These displays are able to produce both numbers and alphabetic characters using either denser segment arrays, like 14- and 16-segment displays, or the most commonly used, dot matrix displays. In this last category we predominantly find LCD- or LED-based displays, although there is a niche of household and audio applications where gas-discharge displays are predominant. A class of alphanumeric, dot-matrix-based displays deserves special mention: the dot-matrix type LCD display, widely used in many small embedded applications.
- **Graphic Panels**: These displays have the characteristic of being pixel addressable making possible to reproduce graphic information. A wide gamma of devices fall in this category. In the low end we find small monochromatic graphic LCDs with 128x64 dots or less, organic LED displays (OLED), and small LED matrix displays like those used in banner signs. In the high-end we find large LED screens like those in electronic billboards, and dense LCDs and plasma displays like those used in television receiver sets and computer monitors. Other graphic technologies include electronic ink (e-ink) displays like those used in early electronic readers.

In this section we analyze the most common types of displays used in small embedded applications. Their principles of operation are the same as those in larger display technologies, forming a basis for understanding how they operate in a larger scale.

## 8.5  Interfacing LEDs and LED-Based Displays to MCUs

Light Emitting Diodes (LEDs) provide the most basic display element used nowadays in embedded systems. An LED is simply a PN junction in which the photon emission phenomena due to conduction is enhanced. When electrons from the N-side in a forward biased PN junction cross to the P side and recombine with holes, the energy lost by the carriers due to the recombination process causes photons to be released. This is an intrinsic phenomena in all PN junctions. In an LED, wide bandgap materials and their dopant concentrations are manipulated to enhance the emission of photons and control their wavelength to produce light of different colors. A lens on the top of the LED package concentrates and focuses the emitted light to direct it in a particular

**Fig. 8.29** Structure of a typical THT LED and LED symbol. **a** Structure, **b** Symbol

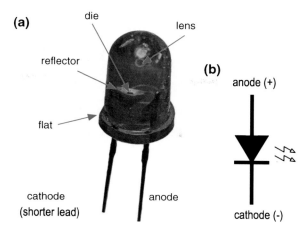

**Fig. 8.30** Basic interconnect of a discrete LED

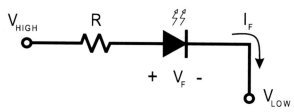

direction or just difusses it. Figure 8.29 shows the structure of an LED, along with its symbol.

Like any other type of diode, LEDs conduct and therefore, emit light, only when they are forward biased. LED specifications include their forward bias voltage ($V_F$), typically 1.2 V or larger, their forward current ($I_F$), and their brightness (B) for a given current intensity, measured in mcd (millicandela). Roughly speaking, one candela is about the amount of light delivered by a candle lit in a dark room.

### 8.5.1 Interfacing Discrete LEDs

The simplest LED interface requires a voltage source and a resistor $R$ in series with the LED to limit the forward current circulating through the LED. Figure 8.30 shows the basic circuit of an LED interconnect.

The value of R is chosen to allow $I_F$ circulate through the LED. Considering generic voltages $V_{LOW}$ and $V_{HIGH}$ applied to the anode and cathode of the diode as illustrated in Fig. 8.30, the value of $R$ can be obtained with (8.23).

$$R = \frac{V_{HIGH} - V_F - V_{LOW}}{I_F} \tag{8.23}$$

**Fig. 8.31**  Driving LEDs with
I/O pins

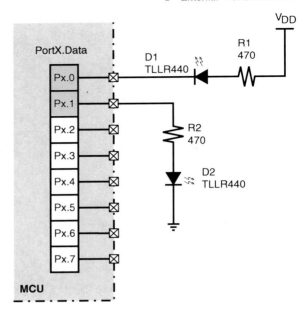

In most cases, an individual LED can be directly driven by an MCU output pin. The main criteria to determine if a direct connection is possible are based on whether or not the output pin has enough driving capability to handle the LED forward current, and if the port output voltage is large enough to produce across the LED the required $V_F$. Most LEDs used for signalization require $V_F$ around 1.8–2.2 V with current between 5 and 20 mA, which are manageable by most GPIO ports. LED connections requiring larger currents or voltages than those directly provided by a GPIO pin directly need buffering and/or voltage level shifting interfaces like those discussed in Sect. 8.8.

Depending on how the LED is connected, two different configurations are possible: LED turned-on with a logic high or turning-on with a logic low. Figure 8.31 shows both of these configurations. $D_1$ is connected to turn-on when Px.0 is low, while $D_2$ would be on with Px.1 is high. The values of $V_{OL}$ and $V_{OH}$ in each case provide the respective levels of $V_{LOW}$ and $V_{HIGH}$. In each case we must ensure the I/O pin currents $I_{OL}$ for $D_1$ and $I_{OH}$ for $D_2$ are large enough to sink and source $I_F$ in each case.

### 8.5.2 Interfacing LED Arrays

One of the earliest applications of LED-based displays used seven-segment numeric LED arrays. A seven-segment LED display features an arrangement of seven bar-shaped LEDs in a pattern that allows to form the shape of any digit between 0 and 9

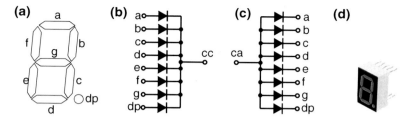

**Fig. 8.32**  Ordering of individual LEDS in a 7-segment display. **a** Segment arrangement, **b** Common cathode, **c** Common anode, **d** Actual Component

**Fig. 8.33**  Directly driving
a 7-segment display module
from a GPIO port

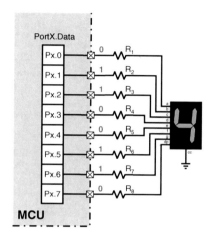

and other characters by selectively turning on or off specific LED segments. Many seven-segment display units also include an eighth LED to provide a decimal point to the right of the digit. Figure 8.32 shows the arrangement given to the seven LEDs forming a 7-segment LED display plus decimal point.

The segments in an LED-based seven-segment display are internally connected to either have a common anode or a common cathode terminal, as illustrated in Fig. 8.32b, c. This eases the interconnection of the modules in typical applications. The codes to illuminate a particular digit will depend on the numeral being represented. For example, to obtain the digit "4" on the display, assuming the least significant bit of the port is connected to segment "a", "b" to the next significant bit, and so on, we'd need to send the binary code "01100110", or 66h, assuming a common cathode display. This case is illustrated in Fig. 8.33 with a single seven-segment display being directly driven by a GPIO port.

From an electrical point of view, driving the segments of a single 7-segment display is not different from driving set of discrete LEDs. For each segment we'd need to provide $I_F$ at the specified $V_F$ to have it lit. The values of resistors $R_1$ through $R_8$ are chosen the same way as in the case of individual LEDs. This assumes that the I/O lines have enough driving capability to source the $I_F$ of each segment. In the event that $I_F$ exceeds $I_{OH}$, a buffer would be required. Although individual transistors could still be used, dedicated seven-segment drivers offer a more compact solution. For example, the 74HC47, a BCD-to-seven segment could be used. This IC, besides providing enough current for each segment, also features inputs accepting BCD digits, which reduces the input pin requirement.

Interfacing multi-digit displays is usually done by time-multiplexing individual seven-segment modules. This technique, when compared to using a dedicated GPIO port per digitt, saves I/O pins at the expense of a slightly more elaborated software component.

A multiplexed, multi-digit seven-segment display shares the segment lines among all modules, while the common cathodes (or anodes in the case of CA modules) are used to individually turn on one module at a time. To show all digits in an $n$-digit number, the seven-segment codes for each digit are sent one at a time, each followed by a control code that activates only the seven-segment module where the digit is to appear, and leaving the other digit drivers deactivated.

If this operation were performed for all the $n$ digits in the display, and the entire sequence repeated in a loop such that all the digits were refreshed at least 24 times per second, a human eye would see all the digits lit simultaneously, each with their corresponding digit displayed. The reason why this trick works is due to the persistence of vision phenomena in the human eye.

To achieve a suitable refresh frequency for the entire display, a timer channel could be devoted to set the digit refresh rate. Note that if we desire a refresh rate of $f_{refresh}$ in an $n$-digit display, each individual digit will need to be refreshed at a frequency equals to $(n \times f_{refresh})$. At the same time, in each refresh cycle, for the entire display, each digit will be lit $1/n$th of the cycle. This will reduce the brightness of each digit in the same proportion, which could lead to dim displays if no counter measures were taken.

One alternative to increase brightness in such a situation would be by increasing the current through each segment in the same proportion the duty cycle is reduced, an objective easily reached by reducing the limiting resistors in each segment. When this strategy is used, care must be taken not to exceed the maximum current allowed per segment at a given duty cycle or per I/O port.

Another important observation is that the CA or CC terminal will circulate the sum of the currents of all ON segments. This sum could be as high as $8I_F$ if all segments are on. For most GPIOs this amount of current would exceed the port driving capability, requiring an external buffer. Example 8.5 illustrates a multiplexed LED design that considers these conditions.

**Example 8.5  (Multiplexed multi-digit seven-segment display)** *An interface for a three-digit multiplexed seven-segment CC LED display is desired to operate with a refresh frequency of 33.3 Hz. Provide a suitable hardware setup and a timer-based ISR to set the desired refresh rate.*

*Solution: In this case we will provide a solution using eleven I/O lines: eight for driving the segments in Port1, and three in Port2, bits 0 to 2, for controlling the common cathodes. Assume the MCU is an MSP430F1222 operating at 3.3 V, with*

**Fig. 8.34** Multi-digit seven-segment display arrangement

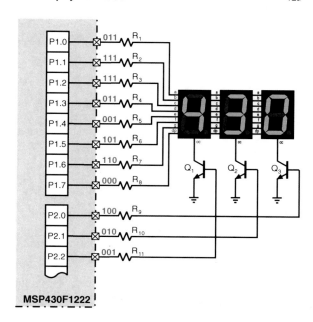

$f_{clk} = 32.768 \, KHz$. *Figure 8.34 shows a diagram illustrating a connection of the three modules to form a three-digit display.*

*Note that lines "a" through "f", driven by Port1, are shared by all three modules. The common cathode lines are controlled with lines P2.0 through P2.2, with the most significant digit driven by PY.2. Transistor Q1 through Q3 are necessary to drive the entire module current, which can be up to $8 \cdot I_F$, when all segments are on. Section 8.8 discusses how to design transistor interfaces.*

*Let's use commercial off-the-shelf, three-digit, common cathode, seven-segment modules similar to Kingbright's BC56-12SRWA [54]. These modules specify a typical brightness of 34 mcd at $I_F = 10 \, mA$. This is a rather high brightness level for a seven-segment display. We would set it to provide an average of 5.67 mcd, which would allow enough brightness even on daylight. As the luminous intensity is proportional to $I_F$, the continuous current needed would be 1.67 mA. However, as we have three digits, the duty cycle would be 33%. So we will set $I_F = 5 \, mA$ to obtain the desired average brightness.*

*At 5 mA forward current, the display module specifies a forward voltage $V_F$ of approximately 1.75 V. Note that the common anode could, in the worst case, manage 40 mA per digit, justifying the usage of transistor buffers Q1 through Q3 in the seven-segment CC terminals. For such a current, a Fairchild's 2n3904 NPN general purpose transistor would suffice. The 3904's data sheets specify at $I_{Csat} = 40 \, mA$ a current gain factor $\beta_{sat} = 10$, $V_{CEsat} = 0.075 \, V$, and $V_{BEsat} = 0.82 \, V$ at room temperature of 25 °C. Thus, the values of the limiting resistors $R_1$ through $R_8$, and base resistors $R_9$ through $R_{11}$ would result:*

$$R_1 = \frac{V_{OH} - V_F - V_{CEsat}}{I_F} = \frac{3.1\,V - 1.75\,V - 0.075\,V}{5\,mA} = 255\,\Omega$$

$$R_9 = \frac{V_{OH} - V_{BEsat}}{I_{Bsat}} = \frac{3.1\,V - 0.82\,V}{4\,mA} = 570\,\Omega$$

*The circuit setup shows next to the port outputs the binary codes to be sent if we wanted to display the value "430" on the display. Note there are three columns of binary numbers there. Each column represents an instant in time. The upper 8 bits represent the seven-segment (plus decimal point) code, while the lower three bits represent the corresponding control code.*

*For setting a display refresh frequency of 33.3 Hz with three digits we would need at least a 100 Hz digit refresh rate. This implies refreshing each digit every 10 ms. If we configure a timer channel to trigger an interrupt every 10 ms, that would yield the desired refresh rate. Figure 8.35 shows a flowchart for the timer ISR. The implementations details are left as an exercise to the reader.*

Some applications might either not have enough I/O pins or the CPU might be too busy to devote cycles to directly manage the refreshing of a multiplexed display. In such cases, using an external LED display driver might result appropriate. Several off-the-shelf alternatives can be identified, including the Motorola MC14489B, Intersil's ICM7218, and Maxim's MAX6951. They all provide the ability of driving up to eight seven-segment displays, providing serial interface to the MCU side and taking care of driving and multiplexing functions.

## 8.6 Interfacing Liquid Crystal Displays

Liquid Crystal Displays (LCD) have become the dominant display technology in household and portable devices. Its low-power consumption and compactness have been the major drives to this tendency. The once dominant cathodic ray tube technology has now become obsolete, being displaced by LCD flat screens in desktop and portable computers. In the arena of portable electronics, there has never been any technology with more dominance than LCDs.

### 8.6.1 Principles of Operation of LCDs

All LCD technologies are based on the property of crystalline solid materials of twisting the polarization angle of coherent light. Liquid crystalline materials, also called nematic state materials, were first discovered in 1888, and their thin-layer properties studied for many years. In 1970, eighty-two years after their discovery, liquid crystals found their way to mass consumed electronics in the form of numeric displays for

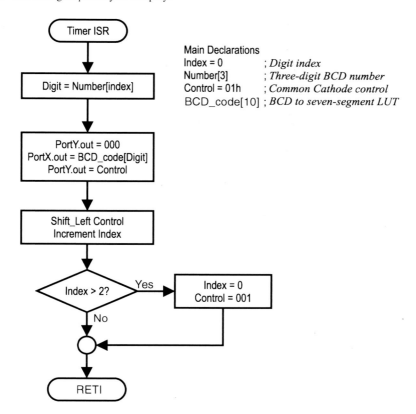

**Fig. 8.35**  Timer service routine for multi-digit display

wristwatches. Nowadays, liquid crystals are the base for the most common type of display technology employed in embedded applications.

Liquid crystal materials do not emit light. Instead, they manipulate the light coming from some other source when it passes through the crystal body. Typical light sources in LCD applications include ambient light or some form of backlighting illumination. The manipulation of light in the liquid crystal can be made to produce patterns of shadows in front of the light source, allowing the formation of images.

What makes possible the manipulation of light inside a liquid crystalline material resides in the properties of its molecules. Nematic state materials are composed by rod-like crystal molecules that in their natural state are organized in a helix form similar to that of a DNA strand. These helixes work as light wave guides, twisting the angle of polarization of a light wave passing through them. However, when the material is subjected to an electric field, its molecules change their helix alignment, loosing their light twisting properties.

Ambient light is composed of a virtually infinite number of traveling photon wavefronts with a random variety of transversal orientations. Each of these wavefront orientations, perpendicular to the traveling light direction, corresponds to a polarization direction. When ambient light passes through an infinitesimally narrow slit, only those wavefronts aligned with the slit will pass, resulting in polarized light. A surface made of parallel slits oriented in a particular direction will form a linear polarizing filter. Polarized light is also called coherent light.

When two polarizing filters, the first with an orthogonal orientation with respect to the second, are placed one in front of the other, no light would pass through them. The first filter would allow the passing of light with only one polarization direction. When such a wavefront reaches the second filter, their orientations will not match and therefore no light would pass.

If a thin layer of a liquid crystalline material, such as 4-n-pentyl-4-cyanobiphenyl ($C_{18}H_{21}N$), were placed between the two orthogonal polarizing filters, assuming the layer thickness has been tuned to twist just 90°, the crystal rods in the material would twist the polarization angle of the light coming from the first filter in the same amount, allowing it to pass through the second filter.

By placing transparent patterned electrodes made from Indium-Tin Oxide (ITO) above and below the liquid crystal material layer, it is then possible to control what areas of the liquid crystal will and will not twist light, allowing the creation of the dark and light features we see on a conventional LCD image. Figure 8.36 shows the fundamental structure of a twisted nematic LCD screen.

### 8.6.2 Control Signals for LCDs

Turning on dark features on an LCD screen is not as simple as just applying a DC voltage to the features' electrodes. When ITO electrodes are subjected to an electrical

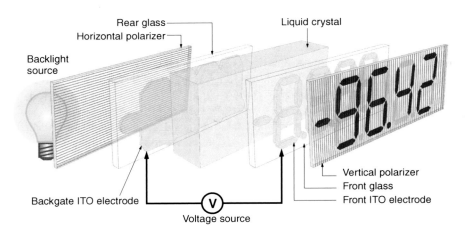

**Fig. 8.36** Structure of a twisted nematic LCD

**(a)**  **(b)**

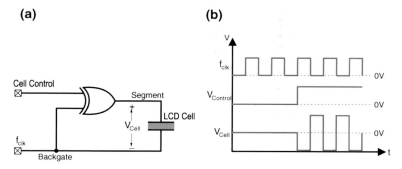

**Fig. 8.37** LCD AC drive circuit and waveforms. **a** AC driver circuit, **b** Timing waveforms

field for extended periods of time their chemical characteristics change, loosing their transparency. Moreover, the chemical characteristics of the liquid crystal material itself changes when subjected to permanent DC voltages. A solution to overcome these difficulties is driving the LCD electrodes with an alternating voltage signal with average DC value of zero.

A commonly used technique to produce such an AC signal from a square wave input is illustrated in Fig. 8.37.

Assuming a digital clock signal with 50% duty cycle, and a CMOS XOR gate, the cell control signal makes the XOR a selective inverter: Control = 0 the gate is a non-inverting buffer, while Control = 1 makes it an inverter. Thus with control in low, the same voltage is applied to both the backplane and cell, yielding a net voltage o zero on the cell. When Control is high, the segment and backplane signals are 180° out of phase, resulting in the cell a $2V_{DD}$ peak-to-peak AC waveform. This way the average DC voltage applied to the cell is zero, avoiding a destructive condition on the cell.

It deserves to note that unlike LEDs and other display elements, liquid crystal cells do not respond to the instantaneous peak value of the applied signal. Instead, they respond to the RMS value of the applied signal. The RMS value of the AC signal controls the contrast obtained in the feature. To have a feature appearing dark requires its cell to be driven with a signal whose RMS value exceeds the LCD threshold for darkness $V_{Ton}$. A feature will appear clear when the RMS value of the activation voltage is below $V_{Toff}$. Between $V_{Ton}$ and $V_{Toff}$ there is a transition zone where the feature appears grey.

In the process of driving a cell, it is observed that the AC signal will be actually turning the cell on and off. The human eye will not see the flickering if it happens faster than 24 Hz. For this reason the refresh frequency is specified at 30 HZ or above.

Power consumption in an LCD cell is mainly due to the driver circuit, as the cell itself behaves fundamentally as a capacitor. Thus, the higher the clock frequency the higher the power consumed by the display controller. Considering this fact, to maintain a low-power consumption the refresh frequency is usually set to a value between 30 and 60 Hz.

The complexity of the signalization necessary for driving an LCD screen grows rapidly as the number of segments on the screen increases. Note that besides the control signal, each segment needs to be fed an AC signal with average DC value of zero.

Three major techniques can be identified for driving LCDs, each resulting feasible depending on the total number of segments being controlled on the screen. These include:

- Direct LCD Driving
- Multiplexed Driving
- Active Matrix Driving

### 8.6.3 Direct LCD Drivers

As its name implies, a direct driver will directly drive each LCD segment with one dedicated line. This approach results cost effective for simple displays with a relatively small number of segments. LCD screens like those used in digital watches and meters typically have less than 36 segments, which make then good candidates for direct driving.

LCD screens designed to be directly driven have a single backplane electrode for all its segments, and a number of control lines equal to the number of segments in the display. Figure 8.38 shows a typical 3.5-digit LCD meter module and its internal connections. The dark area represents the backplane connection, while the light features denote the twenty-eight electrodes required for directly driving the module.

A controller for this module shall provide a 28-bit latched output for equal number of control signals. These, when XORed and referenced to a clock line for the refresh ratio would produce the required AC drive for the module. Implementation alternatives include using GPIO pins and a timer channel. Other alternative is using a dedicated, external direct-drive controller, like National Semiconductor's MM5452. This chip provides a serial connection to a host MCU, requiring only two GPIO lines. It features an internal oscillator for generating the refresh clock, plus 32 undecoded,

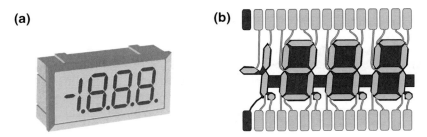

**(a)**                                              **(b)**

**Fig. 8.38** 3.5-digit meter LCD module with direct drive electrodes. **a** Module, **b** Electrode arrangement

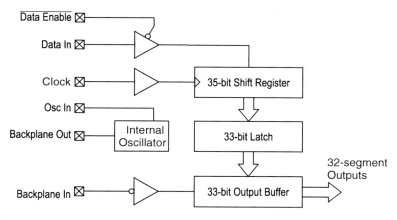

**Fig. 8.39** Block diagram of MM5452 LCD direct driver

latched segment outputs that produce an equal number of direct-drive AC segment lines. The chip does not contain a character generator. It can be serially cascaded to allow for driving LCDs with more than 32 segments. Figure 8.39 shows a block diagram of the MM5452.

Despite their disadvantage in the number of segment pins, direct drive LCDs offer the best contrast ratio and viewing angle of all LCD driving techniques.

### 8.6.4 Multiplexed LCD Drivers

LCD screens featuring a large number of segments, like those used in calculators or in dot-matrix displays, might have hundreds or even thousands of segments. For such displays, the direct drive method results impractical due to the large number of required lines. In such cases, some form of multiplexing approach becomes necessary.

In a multiplexed approach, electrodes are arranged in a matrix form. The backplane electrode is partitioned to form $n$ rows while the segment electrodes are grouped to form $m$ columns. To operate a particular segment located at the crossing of row $i$ and column $j$, the row and column voltages are set to values $V_{BPi}$ and $V_{Sj}$ respectively. If the segment is to appear dark (active), $V_{BPi}$ and $V_{Sj}$ are assigned values such that the—$|V_{BPi} - V_{Sj}| \geq |V_{Ton}|$, where the vertical bars denote absolute value. For making the segment appear clear, it is necessary to make—$|V_{BPi} - V_{Sj}| \leq |V_{Toff}|$.

Covering a frame requires a time-division multiplexing scheme of $n$ time slots. In each time slot $i$ backplane row $BPi$ is applied voltage $V_{BPi}$ and all columns simultaneously driven with their corresponding voltages segment voltages. The activation of each row in a time slot actually requires two voltage pulses, one positive and one negative, to comply with the requirement of a resultant AC waveform with zero net DC value. Recall that the LCD cell will respond to the RMS value of the AC signal.

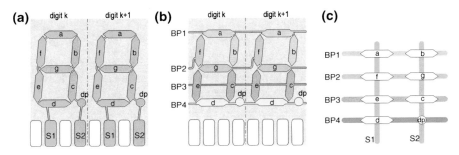

**Fig. 8.40** Electrode arrangement for a 4-mux scheme in a 7-segment LCD. **a** Segment electrodes, **b** Backplane electrodes, **c** Digit electrode matrix

Multiplexing schemes are frequently designated as *n-mux*, to denote the number of backplane rows running on the LCD. The most commonly used approaches in seven-segment type LCDs are 2-mux, 3-mux, and 4-mux. Figure 8.40 shows a 4-mux electrode arrangement for a seven-segment plus decimal point LCD. The figure illustrates two consecutive digits. Observe the reduction in the number of used pins: two pins per digit plus four for an entire row. A 3.5-digit LCD under this scheme would require only eleven pins, compared to 28 when using a direct driver.

The signal sequence driving the rows and columns of a multiplexed LCD depends on the multiplexing ratio, the electrode arrangement, and the order chosen for the signal phases, among other factors. In a 4-mux mode for a seven-segment plus digital point LCD each refresh cycle will require $2 \times 4$ time steps to cover one screen frame, four positive and four negative. To visualize how the scheme works, let's assume we want to display character "3" in the LCD structure illustrated in Fig. 8.40. Assume the use of four voltage levels, $V1$ through $V4$, to excite the LCD and let's assign the signal phases all positive first and then all negative. Figure 8.41 shows this particular case. Sub-figure (a) shows the character to be displayed, (b) its activation matrix, and (c) the signal sequence applied to rows and columns in each time slot. In this last sub-figure the dashes denote when analog voltage values other than the signal maximum or minimum are applied. The timing diagrams in Fig. 8.41d show the signal waveforms sent to the electrodes. In particular, the last four diagrams show the net voltages seen by the LCD cells corresponding to segments a, f, e, and d.

Signals $BP1$ through $BP4$ activate the backplane one row at a time. Each frame takes eight steps, and then repeats. Column signals S1 and S2 are encoded from the segment activation patterns to determine which segments would come dark and which are light. Notice how the LCD cells see an AC waveform with net DC value of zero. At each time step the voltage in the cell is obtained as the difference between $V_{BPi}$ and $V_{Sj}$. See how segments a and d have large impulses exceeding $V_{Ton}$, which will make them dark, while segments f and e never see such large impulses, remaining clear.

In general, the assignment of the number of analog voltage levels and their values for implementing a multiplexing LCD controller needs to consider the particular

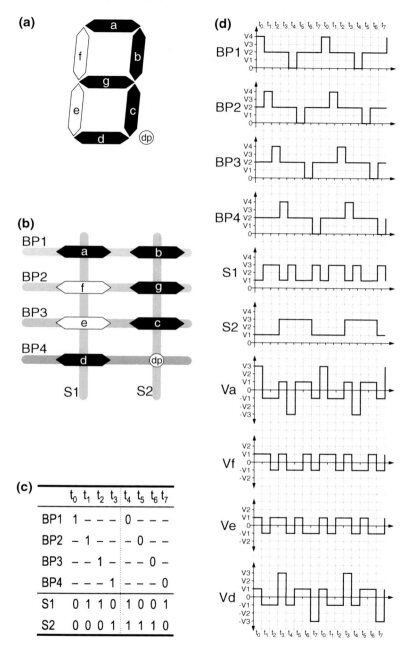

**Fig. 8.41** Displaying one 7-segment character in 4-mux format. **a** Displaying character "3", **b** Activation matrix, **c** Signal sequence, **d** Timing waveforms in column S1

LCD characteristics and be optimized for the peak contrast between dark and clear features. Usually, a resistor ladder with as many stages as intermediate voltage steps is connected via an analog multiplexer to supply these voltages. Resistor values are assigned according to the optimal voltage ratios. Moreover, when extended temperature ranges are to be supported, a thermal compensation network is required.

Higher order multiplexing approaches further reduce the pin count in the interface, at the expense of requiring additional time slots to refresh the display and a more intricate signal pattern. As the number of time slots increases, the duty cycle per cell is reduced, resulting in a lower RMS value of the signal applied to each cell. This translates into lower contrast and reduced viewing angle in the display. To worsen contrast, multiplexed LCDs also suffer from cell crosstalk, which is the effect of one cell's activation voltage into its neighbors. In dense dot matrix arrays, crosstalk causes partial activation in the surrounding cells, diminishing contrast.

Another limiting factor taking relevance when the multiplexing ratio increases, is the response time of the LCD cell. Changing the cell state from dark to light or viceversa requires moving the crystal molecules inside the viscous liquid crystal material. This action is by nature slow, taking tens of milliseconds at room temperature, and worsening as temperature lowers. This slow response limits the minimum refresh time per cell and therefore the maximum number of segments that can be driven with via multiplexing.

Improved LCD technologies have encountered limited success in alleviating these problems. For example, super twisted nematic (STN) LCDs, which twist light nearly 270° instead of the 90° obtained in TN devices, have been found to alleviate the loss of contrast in higher-order multiplexing schemes. This technology has stretched multiplexing ratios, reaching up to 480 backplane lines. Despite such an improvement, multiplexed LCDs fall short in brightness and contrast in applications requiring higher line count and faster response time, like in displays used as graphic monitors or in color displays like those used in computer monitors and television sets.

### 8.6.5  Active Matrix LCDs

One of the most important improvements in LCD technology has been the introduction of the active matrix (AM) cell. An AM LCD incorporates a switching element within each cell, allowing for directly driving charge into the equivalent capacitor of the LCD cell. In today's LCDs, thin-film transistors (TFT) are the most widely used switching elements embedded in the cells. Figure 8.42 illustrates an array of cells, highlighting the cell detail and its electric circuit. A typical TFT in an LCD cell occupies less than 2 % of the cell area, which makes it minimally non-obtrusive for the cell visibility.

The inclusion of switching elements on the screen is achieved by etching the transistors directly on the glass substrate. The circuit formed with the transistor and the liquid crystal (LC) cell acts like a memory cell (notice the similarity with a DRAM cell), making possible to break the dependence of the cell activation duty

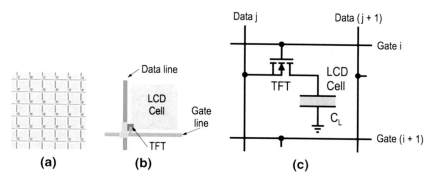

**Fig. 8.42** Structure of an active matrix LCD cell. **a** Cell array, **b** Single cell, **c** Circuit

cycle from the size of the array. This cell design allows for using fast electronics to access the transistors while each transistor holds the charge level in the cell for as long as needed by the LC cell to obtain an optimal contrast ratio. Large LCD screens might include an additional capacitor per cell, in parallel to the LC capacitor, to improve charge retention.

The transistors can be operated in their linear region, allowing for precisely controlling the level of voltage applied to each cell. This permits controlling different amounts of light passing through the cell (cell light transmittance), making possible to obtain multiple precise levels of grey on the screen.

Although the construction of a TFT LCD is more complex than that of a multiplexed LCD, the signalization for an active matrix display is simpler. The scan sequence for a frame has two phases. In the first phase all transistors in one row (TFT gate) are simultaneously activated by applying them a positive voltage pulse. At the same time each column is presented with a positive voltage level that establishes the level of light transmittance of each LC cell in that row. At this point, the activation voltage is stored in the cell capacitor. The row voltage is then driven to the low level (a negative value), holding the column voltage in the cell. The same sequence is repeated for the next row. When all rows are scanned, the second phase begins. Here the sequence is repeated with the column voltages reversed with respect to the preceding pass. This way each individual cell is applied an AC signal with zero DC value. Figure 8.43 illustrates the basic signals driving the row and column lines of a particular cell.

The operation of a color TFT LCD is based on the same principle: a matrix of TFT-driven LC cells driven by a row-column scan circuit. The difference in a color screen is that each color pixel is composed of three individually addressable cells, each with one color filter on top, one for each of the three basic colors: red (R), green (G), and blue (B). By modulating the light transmittance of each cell and combining the three basic colors it is possible to produce images with millions of colors on the screen. Figure 8.44 shows the structure of a color LCD screen. Pixels are only a few tens microns in size, so the naked eye cannot see them individually.

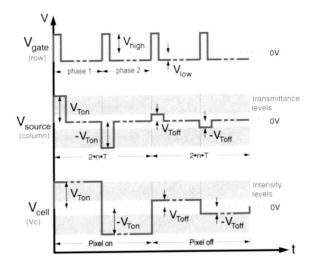

**Fig. 8.43** Signal timing in a single active matrix LCD cell

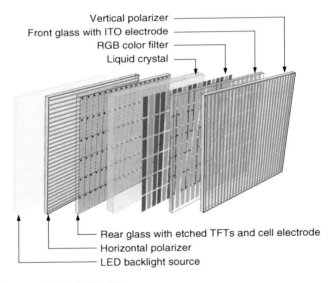

**Fig. 8.44** Structure of color TFT LCD screen

## 8.6.6 Commercial LCD Modules

Despite the complexity in generating the signals to drive the rows and columns of multiplexed and active matrix LCDs, most commercial display modules feature integrated controllers that hide all the hassle associated with driving the display. The

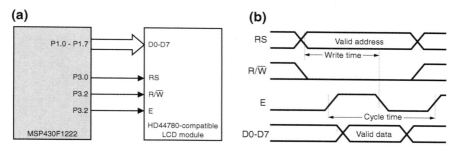

**Fig. 8.45** Connecting an alphanumeric LCD module through GPIO lines. **a** GPIO connection, **b** signal timing

controllers' communication protocols provide for direct attachment through either parallel or serial interfaces.

A common characteristic of most LCD controllers is that after a power-up event, the controller must be initialized before it can process display commands or accept display data. The particular initialization sequence and commands used to control the display depends on the specific controller embedded in the module. Below we discuss a couple of representative examples.

**Parallel Interface LCD Controllers**

These controllers allow connection with 8-bit/4-bit microprocessor buses following either a Motorola 6800 or Intel 8085 signal style. Despite the MPU timing style, it is always possible to generate the control signals from I/O lines, making possible their interface through GPIO lines. Some of these devices allow to select the signal style and/or data width interface while being hardwired or during the command initialization procedure. Examples of these type of controllers include Hitachi HD44780U and compatibles, used for alphanumeric dot-matrix multiplexed LCDs and Samsung's Electronics KS0108B and compatibles, used for 64-mux, bare LCDs. From these, the HD44780-based modules are most widely used due to their embedded character generator and simple interface. KS0108B modules are more oriented to bit-mapped graphic displays, although the controller itself does not provide character generation or graphic support functions. These are typically provided by a host processor.

To illustrate a typical LCD module in more details, consider a Hitachi HD44780-compatible LCD module. The HD44780 is designed to drive dot-matrix type, multiplexed LCDs in alphanumeric mode. The controller features an internal 80-character display RAM, a character generator supporting both western and Japanese kana symbols, and an internal $16 \times 40$ multiplexed LCD driver. It provides a 6800-type bus interface customizable to operate in 4- or 8-bit mode, but can be connected through an MCU GPIO lines as illustrated in Fig. 8.45a. Line $R/\overline{W}$ could be hardwired to ground as the module can be operated as a write-only device.

**Table 8.1**  Signals in an HD44780-compatible alphanumeric LCD module

| Signal | Type | Description |
|--------|------|-------------|
| $RS$ | Control | Command/Data: '0' = Command, '1' = Data |
| $R/\overline{W}$ | Control | Read/Write line: '0' = Write, '1' = Read |
| $E$ | Control | Enable strobe |
| $D0 - D7$ | Data | Bidirectional data lines |

**Fig. 8.46**  A common 2 × 16 alphanumeric LCD module

The HD44780 control signals are shown in the timing diagram in Fig. 8.45b. A description of each signal is provided in Table 8.1. The timing signals can be readily generated via software in the MCU. Two important considerations to be made in such a connection are: (i) ensuring voltage compatibility of the voltage levels, as some modules might work at a voltage different from the MCU, (ii) verifying that minimum *Cycle* and *Write* times are satisfied. Once the interface is completed it only remains initializing the controller to make it operational. The initialization sequence in the HD44780 requires fundamentally six steps:

1. A warm-up delay of 40 ms or more after power-up.
2. Specify whether an 8-bit (default) or 4-bit protocol will be used.
3. Another delay, this time of at least 4.1 ms.
4. Resend the command in line 2 two more times.
5. A function set command specifying the number of lines and character set.
6. Three commands in sequence, issuing a display off, a clear, and set entry mode.

Figure 8.46 shows a picture of a fairly common 2 × 16 alphanumeric LCD module. The module data sheet provides detailed information on commands and other module features.

## Serial Interface LCD Controllers

A growing number of LCD controllers with serial interfaces can nowadays be found in the market. The main attractiveness of these modules is their reduced number of I/O lines to connect with the host MCU, and the advantage of some of these devices in providing embedded support for character generation and some even including basic graphic functions.

Many of the devices are fundamentally adaptations onto multiplexed dot-matrix displays or TFT displays developed from joining a bare LCD module and a custom-designed MCU-based interface that besides driving the LCD provides an external serial interface.

Serial interface options concentrate in fundamentally three types of interfaces: Universal Asynchronous Receiver-Transmitter (UART) compatible, three- or four-wire Serial Peripheral Interface (SPI), or Inter-Integrated Circuit (I2C). Some stand-alone modules using an Universal Serial Bus (USB) can also be found. All these protocols are discussed in Chap. 7 in detail. Aside from the standard serial protocol assumed in these interfaces, their particular command sets allowing to initialize or operate the modules changes from one implementation to another. Some initiatives have begun to emerge providing standardized graphic languages for LCD modules, like 4DGL from the Australian company 4D Systems. These approaches still have limited market coverage and support a limited number of display alternatives. Below we provide a summarized overview of some common serial interfaces LCD modules.

- **UART-based LCD Modules** provide interfaces through the standard RxD and TxD serial lines (plus ground), typically at a default baud rate of 9,600BPS. Command-based interface supports changing the default baud rate and interacting with the pixels on the screen.
- **SPI LCD Modules** with three- or four-line interfaces provide standard synchronous communication using SDI and SDO lines for serial data-in and data-out and SCL for serial clock. Four-wire SPI interfaces also include a chip select (CS) line for module selection. Some modules also provide custom lines for indicating the type of information being fed into the module (command or data) and a dedicated reset line.
- **I2C Modules** adhere to the standard specification of using bidirectional clock (SCL) and data (SDA) lines. Since I2C provides addressing and multi-drop capabilities, multiple modules can share the serial bus. Internal logic differentiates command from data transmissions.

## 8.7 LCD Support in MSP430 Devices

Some MSP430 devices include dedicated hardware to facilitate interfacing liquid crystal displays to the MCU. In particular the MSP430x4xx and MSP430x6xx include on-chip support for direct drive and multiplexed type LCDs. The on-chip support provides the AC voltage signals required for handling both numeric and dot-matrix type multiplexed LCDs.

The features provided by both MCU series include display memory, signal generation, configurable frame frequency, blinking capability, and support for multiple

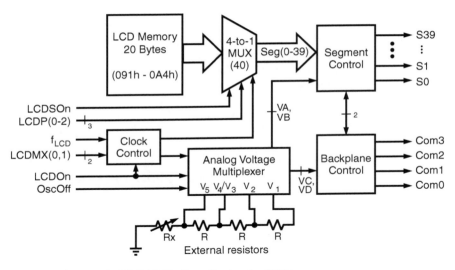

**Fig. 8.47** Structure of MSP430 LCD Controller in x4xx MCUs

multiplexing modes. The sections below summarize the salient characteristics in both MCU families.

### 8.7.1 MSP430x4xx LCD Controllers

MSP430x4xx devices introduced the fundamental structure for supporting liquid crystal displays in MSP430 microcontrollers. The *LCD Controller* in this MSP430 generation was aimed fundamentally at managing seven-segment type LCDs. The controller structure, illustrated in Fig. 8.47, incorporates memory, digital and analog multiplexers, and segment control logic to drive either directly or in multiplexed mode up to 160 segments, depending on the particular x4xx device chosen.

A total of twenty bytes in the address range from 091h to 0A4h are allocated for the LCD memory. The memory contents is mapped assigning one bit per segment for a total of 160 segments. Segment *S*0 is aligned to the least significant bit of byte position 091h.

The multiplexer block features 40 4-to-1 units, used as either 1-to-1, 2-to-1, 3-to-1, or 4-to-1 depending on the multiplexing mode chosen. The segment control block routes the analog voltages applied to the segment electrodes through lines *S*0 through *S*39. Lines *Com*0 through *Com*3 allow application of backplane voltages. *Com*0 is used for static (direct) mode, *Com*0 and *Com*1 in 2-mux mode, and so on.

The LCD controller requires an external resistor ladder network to generate the LCD voltages. The network will contain from one to four 680 K$\Omega$ resistors, depending on the selected mux mode. The LCD clock frequency ($F_{LCD}$) is obtained from

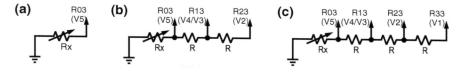

**Fig. 8.48** Resistor networks for different mux modes in x4xx devices **a** Static (direct) **b** 2-mux mode **c** 3- and 4-mux modes

**Table 8.2** Salient characteristics of x4xx multiplexing mode

| MUX mode | LCDMX$x$ bits | Max. no. segments | Backplane signals | Memory alignment |
|---|---|---|---|---|
| Static | 00 | 40 | Com(0) | The lsb of each nibble |
| 2-MUX | 01 | 80 | Com(0,1) | Two lsb's of each nibble |
| 3-MUX | 10 | 120 | Com(0-2) | Three lsb's of each nibble |
| 4-MUX | 11 | 160 | Com(0-3) | All bits of each nibble |

Timer1. Figure 8.48 shows the network configuration for each mode. Placing a potentiometer at Rx allows for manual contrast control on the LCD.

Operating the LCD requires control bits LCDON = 1 and OscOff = 0 to enable the module. Toggling LCDSON (LCD Segment On) allows blinking all segments simultaneously. Control bits LCDMX(0,1) set the multiplexing mode for the controller as well as voltage levels $V_A$, $V_B$, $V_C$, and $V_D$, driving segment and backplane electrodes.

Part of the LCD output pins are shared with GPIO pins. The functionality of shared pins is assigned via the I/O port register PxSELx. All LCD controller functions are configured via a single LCD control register (LCDCTL). Salient characteristics of each multiplexing mode are listed in Table 8.2.

## MSP430x4xx LCD_A Controller

The latest device entries in the x4xx generation introduced a second on-chip LCD controller named *LCD_A*. LCD_A, whose block diagram is illustrated in Fig. 8.49, is a revised version of the basic LCD controller. It retains the basic functionality and structure of the original LCD Controller while improving in several aspects that include:

- The LCD frequency $f_{LCD}$ is now directly obtained from ACLK, making no longer necessary to use a timer channel. Control bits LCDFreq(0-2) allow choosing a divider for ACLK, setting the value for $f_{LCD}$.
- An internal LCD Bias Generator including a resistor ladder and a charge pump circuit is used for obtaining voltages $V_1$ through $V_5$. The external resistor ladder network required in the previous generation controller is no longer necessary,

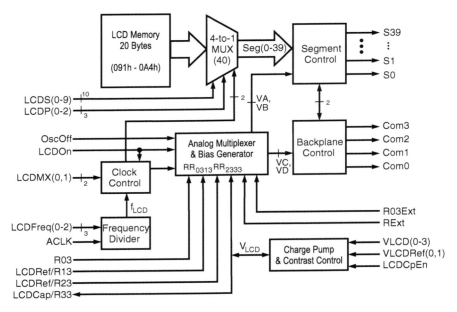

**Fig. 8.49** Structure of MSP430x4xx LCD_A Controller

although if desired, it can still be used. The charge pump circuit requires an external 4.7 μF capacitor connected to R33 pin.

- An on-chip contrast control circuit now makes possible to adjust the LCD contrast ratio via software. Seven different contrast levels are possible, plus an eight level where segments are off.
- An enhanced, more flexible segment enable control logic is now included, allowing activation of segments in increments of four, starting at S0.

Due to the additional functionality embedded into the controller, configuring LCD_A takes now five registers: one LCD_A control register (LCDACTL), two LCD_A port control registers (LCDAPCTL(0,1)), and two LCD_A voltage control registers (LCDAVCTL(0,1)). The reader is referred to the MSP430x4xx Family User's Guide (SLAU056J), available at TI's MSP430 MCU home page for programming details.

### 8.7.2 MSP430x6xx LCD Controllers

The x6xx generation of MSP controllers feature two different LCD controllers: LCD_B and LCD_C. Both, like previous generations, continue aimed at servicing multiplexed LCDs, while introducing several new enhancements.

## MSP430x6xx LCD_B Controller

The LCD_B Controller inherits most characteristics of LCD_A controller, supporting static, 2-mux, 3-mux, and 4-mux modes, and up to 160-segment LCDs. Like its predecessor, it also provides on-chip LCD bias generation, software controlled contrast ratio, and an internal clock divider that does not require a dedicated timer channel. In addition, LCD_B features several improvements that include:

- An expanded frequency divider for generating $f_{LCD}$ that now allows using VLO-CLK as clock source, in addition to ACLK.
- Support for a segment blink capability that allows blinking individual segments. This is achieved by adding a second 20-byte memory block used for indicating the blinking status. Each bit in the blinking memory indicates the blinking status of its homologous bit in the segment memory. The blink frequency is provided by a dedicated blinking frequency divider fed from the same source as the LCD frequency divider.
- The blinking memory can double as a display memory if not used for blinking purposes. When used as a secondary display memory, it allows to switch under software control between the two buffers by solely changing the LCDDISP bit in the memory control register (LCDBMEMCTL).
- Four interrupt sources (LCDFRMIFG, LCDBLKOFFIFG, LCDBLKONIFG, and LCDNOCAPIFG) served by a single interrupt vector. Each source has an independent enable and status flag that allows to determine the triggering source. LCDFRMIFG can be configured to trigger when a frame boundary is reached. LCDBLKOFFIFG and LCDBLKONIFG can be individually set to be triggered by the blink clock (BLKCLK) when a frame boundary that either turns off or on the segments is reached. The fourth source, LCDNOCAPIFG can be enabled as a warning service to detect when the LCD biasing charge pump has been enabled without an external capacitor attached to the LCDCAP pin.

Configuring LCD_B is achieved through a set of eleven 16-bit registers that include two for general control (LCDBCTL(0,1)), one for blinking control (LCDB-BLKCTL), one for memory control (LCDBMEMCTL), one for voltage control (LCDBVCTL), four for port control (LCDBPCTL(0-3)), one for controlling the charge pump (LCDBCPCTL), and one for interrupt control (LCDBIV). All control registers are reset by a PUC event.

## MSP430x6xx LCD_B Controller

The LCD_C Controller is, so far, the most powerful LCD controller embedded into MSP430 devices. LCD_C supports all features exhibited by its predecessor, LCD_B, and expands to support up to 8-mux mode LCDs. This allows the enhanced controller to drive display modules with up to 320 segments. This expanded segment driving capability makes LCD_C able to drive small dot-matrix displays with as many as eight $5 \times 7$ characters plus a nonblocking cursor. In terms of standard seven-segment characters, the controller can drive up to forty digits with decimal points.

Such an expanded multiplexing capability imposes limitations to the contrast attainable on the screen. In the case of 8-mux mode, the best voltage contrast ratio achievable is 1.446, a reduction of 16.5 % with respect to the best 4-mux mode contrast. This is still a viewable contrast for most LCD modules, but keep in mind that the higher the multiplexing ratio the lower the maximum attainable contrast.

Additional configuration and operation details for LCD_B and LCD_C controllers can be found in the MSP430x5xx/MSP430x6xx Family User's Guide available from Texas instruments in its MCU web page.

## 8.8  Managing Large DC Loads and Voltages

Managing direct current (DC) loads rated at voltages or currents beyond those directly manageable by an I/O pin require level-shifting and/or current buffering interfaces. These interfacing requirements can be met either using discrete components of packaged driver solutions. Here we discuss representative approaches.

### 8.8.1  Using Transistor Drivers

Perhaps one of the most common interfaces for managing DC loads is based on a transistor buffer. A transistor buffer not only allows to boost the amount of current that can be controlled by an I/O pin, but also allows for driving a load at a voltage different than that feeding the MCU. Figure 8.50 shows generalized topologies with bipolar (BJT) and field-effect (MOSFET) transistors used as buffer-driver.

In the cases illustrated, a generic DC load is being driven in subfigure (a) by an NPN BJT and in (b) by an N-channel MOSFET. In either case the transistors are turned-on with a logic-high voltage $V_{OH}$ in the I/O pin. Diode $D_C$ is optional for resistive loads, however it must be included when inductive loads are switched to protect the transistor from overvoltages due to the transient when the inductor is turned off.

The buffer topology can also be implemented using PNP or PMOS transistors to drive the load. The main consideration for such a change is that voltage polarities and current directions are reversed with respect to those in N-drivers, thus requiring a logic-low level voltage to turn the transistor on.

The MCU and load supply voltages could be the same ($V_{DD1} = V_{DD2}$) or different. We could, for example, drive a 12VDC load with a 3.3 V MCU. As long as we can provide enough base or gate voltage to turn on and off the transistor, the interface would work.

When a transistor is used to interface a binary load (ON/OFF), as is the case here, we want the transistor to operate as a switch, regardless of its use as a current

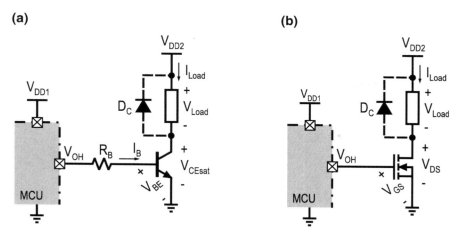

**Fig. 8.50**  Generalized topology of a transistor buffer using N-devices. **a** NPN buffer, **b** NMOS buffer

buffer, level-shifter, or both. If the interface is based on a BJT, we need to ensure the base resistor allows enough current into the transistor base to saturate it with $I_{C(sat)} = I_{Load}$. When a BJT saturates it exhibits a low $V_{CEsat}$, typically around 0.1 or 0.2 V, allowing virtually all the load supply voltage to drop across the load. To keep the base current at the minimum required, it is advised to keep the BJT at the verge of saturation.

In the case of a MOSFET driver the criteria is very similar to the above. The MCU side of the interface does not need a gate resistor, as seen in Fig. 8.50b, as MOSFETs are activated by voltage, not current. The gate voltage $V_{GS} = V_{OH}$ shall allow for placing the NMOS in the linear region with $I_D \geq I_{Load}$. This way, its voltage drop $V_{DS}$ would be small, allowing for most of $V_{DD2}$ to drop across the load. In a MOSFET, the maximum current in linear mode is obtained near pinch-off, but that would require $V_{DS} \approx (V_{GS} - V_{th})$, which might be too large a $V_{DS}$. Typically a mid-linear region value provides a better $V_{DS}$ value.

The main criteria for choosing the BJT (MOSFET) to be used for a particular driver/buffer are:

**Current**  The transistor must be able to handle the load current. In the case of a BJT (MOSFET) the maximum collector (drain) current $I_{Cmax}$ ($I_{Dmax}$) must be larger, at least 15 %, than the load current.

**Voltage**  The maximum $V_{CE}$ ($V_{DS}$) would develop when the transistor is off. Thus $V_{CEmax}$ ($V_{DSmax}$) must be large enough to accommodate $V_{DD2}$.

**Power**  The transistor will dissipate power when conducting. Therefore, the device maximum power dissipation must satisfy $P_{Cmax} \geq V_{CEsat} \times I_{Load}$ ($P_{Dmax} \geq V_{DS} \times I_{Load}$). A safe margin of at least 15 % above it shall be allowed.

BJT drivers are usually faster than equivalent MOSFET drivers and require smaller control voltage ($V_{BE}$) to be switched, but as they are activated by current $I_B$, and in

**(a)**                                         **(b)**

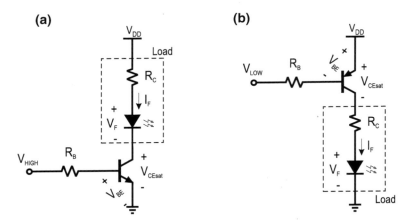

**Fig. 8.51**  Using bipolar transistors to buffer high-current LEDs. **a** NPN buffer, **b** PNP buffer

saturation $\beta$ is small, large $I_C$ values might lead to $I_B$ values exceeding the GPIO capability. In such cases, multi-level drivers, like a Darlington pair could become necessary. Example 8.6 on the facing page illustrates one such case. The advantage of a Darlington pair is its large current gain $\beta_{eq} = \beta_1 \times \beta_2$, where $\beta_1$ and $\beta_2$ are the current gains of its two transistors. The pair, however, requires a larger control voltage ($V_{gs(eq)} = 2 \cdot V_{BE}$) and higher $V_{CE(eq)}$ as the output driver does not actually saturate.

A situation exemplifying the use of a transistor driver is the case of a light emitting diode whose $I_F$ exceeds the port driving capability. A transistor buffer would offer a simple solution. For such a case, the transistor is chosen to handle the LED current. Figures 8.51 and 8.52 show a few possibilities. The topologies in Fig. 8.51 show how to solve the problem using a bipolar transistor.

Using the transistor data, we first choose $R_C$ to limit the current passing through the diode and to fix its forward drop $V_F$. Taking as reference the NPN-based buffer, the value of $R_C$ can be obtained as:

$$R_C = \frac{V_{DD} - V_F - V_{CEsat}}{I_F} \qquad (8.24)$$

Assuming the BJT is at the verge of saturation, we use the minimum value of $\beta$ as the saturation current gain $\beta_{sat}$. Some manufacturers provide explicit values of $\beta_{sat}$ in their data sheets. This allows obtaining $I_{Bsat} = I_F/\beta_{sat}$ which then leads to calculating $R_B$ as:

$$R_B = \frac{V_{HIGH} - V_{BEsat}}{I_{Bsat}} \qquad (8.25)$$

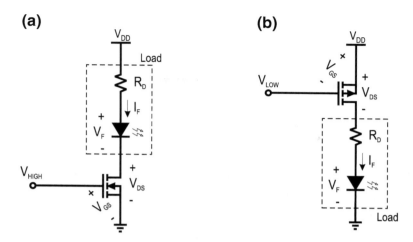

**Fig. 8.52**  Using MOSFETs to buffer high-current LEDs: **a** an NMOS buffer, **b** a PMOs buffer

The process for obtaining the values of $R_C$ and $R_B$ for the PNP buffer in Fig. 8.51b is fundamentally the same, except that now voltages and currents are reversed with respect to the NPN case.

A final note about the transistor buffers is that we typically connect the load such that the conduction biasing is not affected by voltage fluctuations in the load. In the case of BJTs the load is connected to the collector. Connecting the load to the emitter would introduce feedback in the $V_{BE}$ voltage that could take the transistor out from saturation. Due to similar reasons, in MOSFETs, the preferred terminal is the drain.

The value of $R_D$ is obtained considering the MOSFET biased in the linear region close to pinch-off for a small $V_{DS}$. Obtaining the equations for $R_D$ is left as an exercise.

**Example 8.6**  *A LED-based flash for a smart phone camera is to be driven by an MSP430 MCU. The flash is designed to produce a 1/60 s burst of light when the camera shutter is depressed, if the subject is too dark. The LED, a Luxeon SR-05-WN100 part, requires a forward voltage of 3.2 V and 700 mA to produce a 180 Lumens output. Assume a boost converter elsewhere in the camera provides a 5VDC/1.5A supply for the flash, and a light sensor, already embedded in the camera provides the indication of darkness. Design a suitable interface to trigger the flash from an I/O pin of the MSP430F1222.*

*Solution:*  *For this application, the load current requirement of 700 mA cannot be directly provided by the MCU through an I/O pin. The MSP430F1222 data sheet indicates that at room temperature, to maintain $V_{OH}$ compliant with $V_{IHmin} = 80\,\%V_{CC}$, $I_{OH}$ shall not exceed 20 mA. A typical solution with a single stage driver results in a $\beta_{sat} \simeq 10$, yielding $I_{Bsat} \gg 20\,mA$. For this reason, the recommended buffer in this case shall be a darlington pair. A suitable driver in this case would be an ON Semiconductor's BSP52T1. Rated at $I_{Cmax} = 1.0A$ in a convenient surface-mount*

**Fig. 8.53** Darlington-based
LED buffer

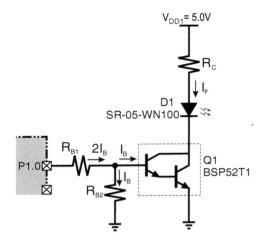

*package and $V_{CEmax} = 80\,V$, this driver comes out as a convenient choice. It's data sheet specifies that at $I_{Csat} = 700\,mA$ it features a $V_{CEsat} = 1.0\,V$ and $\beta_{sat} \simeq 1000$. Figure 8.53 shows a schematic of the LED connection.*

*Note that in this case, feeding the LED with $V_{DD} = 3.3\,V$ will not allow for having a $V_F = 3.2\,V$ in the LED due to the required $V_{CEsat} = 1\,V$ of $Q_1$.*

*Applying Eq. 8.24, the value of $R_C$ would result in approximately $1.2\,\Omega$ at $0.5\,W$. For $\beta_{sat} = 1000$ the base current would be $0.7\,mA$. At the darlington's base it results convenient the addition of resistor $R_{B2}$ to accelerate the transistor turn-off time. Otherwise it would turn-off via the I/O port leakage, which might result too slow if pulsed operation were desired. In this case we would want to have similar turn-on and turn-off times, so we will set the current through $R_{B2}$ to be the same as $I_B$, making the current through $R_{B1} = 2I_B = 1.4\,mA$. According to the BSP52T1 data sheet, for $I_{Csat} = 700\,mA$ the turn-on voltage $V_{BEsat}$ would be $0.95\,V$, allowing to obtain the values of $R_{B1}$ and $R_{B2}$ as:*

$$R_{B1} = \frac{V_{OH} - V_{BEsat}}{2I_B} = \frac{3.1\,V - 0.95\,V}{1.4\,mA} \simeq 1.5\,K$$

$$R_{B2} = \frac{V_{BEsat}}{I_B} = \frac{0.95\,V}{0.7\,mA} \simeq 1.4\,K$$

## 8.8.2  Using Driver ICs

Some applications might need more than just a few buffering interfaces. When many lines need to be interfaced for either boosting current or level-shifting, using discrete transistors might become cumbersome. For such cases, IC drivers become an attractive alternative.

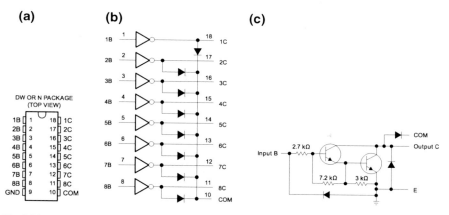

**Fig. 8.54**  TI's ULN2803A Darlington transistor array (*Courtesy of Texas Instruments, Inc.*) **a** Package, **b** Logic diagram **c** Internal circuit

IC drivers can accommodate several individual buffers in a single package, simplifying interfaces that require handling multiple lines. Examples of such ICs include Texas Instrument's ULN2803A and similar devices offered by other semiconductor manufacturers. TI's chip offers eight independent Darlington transistor arrays in a convenient 20-pin package. Figure 8.54 reproduces a package top view, internal logic diagram, and driver schematic of the chip.

Some of the features of this IC include a driving capability of up to 500 mA per driver at voltages as high as 50 V, internal clamping diode to manage inductive loads, and fast response time. The reader is referred to the device's data sheets for a complete list of features and specifications.

## 8.9  Managing AC Loads

Unlike DC loads, alternating current (AC) loads cannot be directly driven by I/O pins. In this case an isolating interface is recommended. The simplest form of controlling AC loads from a GPIO is through a relay.

### 8.9.1 Mechanical Relays

A relay, in its most classical form is an electromechanical device where an electromagnet is used to change the status of a mechanical switch. This allows controlling the mechanical status of the switch contacts via an electrical signal. Relays can be considered the precursors of transistors in the sense of acting as an electrically controlled switch. Figure 8.55 shows the structure of an electromechanical relay and its schematic symbol.

**(a)**

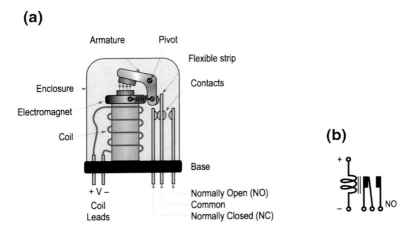

**(b)**

**Fig. 8.55**  Electromechanical relay with SPST NC/NO contacts, **a** Structure, **b** Symbol

A voltage applied to the coil terminals energizes the electromagnet, which mechanically moves the armature, changing the contacts state: NO contacts close while NC open. The relay remains in this state until the coil is de-energized. Contacts are enclosed to reduce ambient induced deterioration. A relay is considered an isolating interface as the contacts and coil are physically separated.

The specifications of a relay minimally provide the contact and coil ratings:

**Coil Voltage**:  Nominal voltage necessary in the coil in order to cause the armature to flip. This can be either an AC or DC voltage.

**Coil Current**:  Amount of current in the coil circuit when the relay is activated. Some manufacturers provide instead the coil impedance, which could readily yield the current once the coil voltage is known.

**Contact Current Rating**:  Maximum amount of current the contacts can withstand in a permanent regime. Exceeding this value will accelerate contact deterioration. This must be selected to safely handle the load maximum current.

**Contact Voltage Rating**:  Magnitude of the maximum voltage an open contact can withstand. The contact voltage can be either AC or DC, regardless of the specified coil voltage. This specification must be large enough to cover and safely exceed the load voltage.

Other relay specifications include their response time (time to change contact state upon a valid coil input), and operating lifetime (expected number of contact operations before failure).

Mechanical relays require a certain amount of current in their coils to be activated. Small DC relays could be directly connected to GPIO pins if their coil current and voltage specs do not exceed those of the pin. Note that relay coils are inductive loads and therefore any control interface must include a clamping diode to protect the pin buffer. The response time of a mechanical relay is large, typically dozens of milliseconds.

A preferred method for driving relay coils is through a buffering transistor, in a topology similar to those illustrated in Fig. 8.50a, b, including the clamping diode $D_C$.

## 8.9.2 Solid-State Relays

Solid-state relays (SSR) replace the electromagnet and mechanical contacts of traditional relays with some form of silicon controlled rectifier (SCR), typically a triac or an arrangement of power MOSFETs, eliminating all mechanical parts in their operation. The coil circuit in an SSR is also modified, being replaced by an optical switch that triggers the SCR gate. These modifications make solid-state relays more durable and faster than their electromechanical counterparts.

SSR are designed to be replacement for electromechanical relays, therefore both types have similar interfacing and specification requirements. In the load side, like their counterparts, SSR can switch large AC and DC loads. Given the electronic nature of their control circuits, SSRs can be controlled with lower currents and voltages in the MCU side than mechanical relays, therefore becoming more appropriate for embedded systems interfaces.

Figure 8.56 shows a simple way to interface an SSR to an MCU for controlling an AC load. The major internal SSR components are also denoted to better understand the interface. The SSR input fundamentally connects to the SSR internal opto-switch. The load is driven by a triac, serving as switching element. An internal zero-crossing detector is used to keep the triac on while the SSR input is active. The relay in the figure corresponds to a *zero-switching* SSR as it would turn-on or -off the load at the next point the AC signal crosses zero after the input has been commanded.

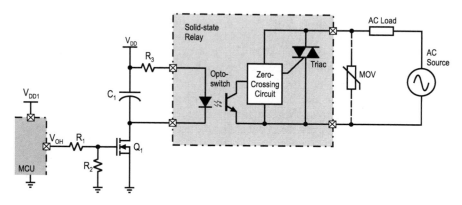

**Fig. 8.56** Solid-state relay: internal structure and MCU interface

The SSR input circuit is driven by an NMOS buffer that allows for driving the current and voltage requirements of the SSR input. Divisor $R_1/R_2$ allows setting the desired $V_{GS}$ at $Q_1$. $R_2$ also helps accelerating turning-off $Q_1$. In the case $V_{OH}$ satisfies the SSR input voltage requirement and the pin current were able to drive the opto-switch, $R_1$ and $Q_1$ could be omitted. Resistor $R_3$ is used to limit the current through the opto-switch LED. Capacitor $C_1$ serves as a filter to reduce the amount of power supply noise reaching the SSR input.

In the output, the SSR terminals are in series with the AC source and load. The optional Metal Oxide Varistor (MOV) in parallel to the SSR output protects against surges in the AC line. The values of all external components will be dependent on the particular SSR being used and the MCU controlling the circuit.

## 8.9.3 Opto-Detectors and Opto-Isolators

Opto-detectors and opto-isolators are based on the same principle: the ability of semi-conductors to generate an electrical current when bombarded by energetic particles like photons or other forms of radiation. When the depletion region of a *pn* junction is exposed to sufficiently energetic photons, electrons are liberated from the material, producing an electrical current. This is the inverse phenomenon that generates light in an LED. A *pn* junction exposed to light becomes a photodiode.

Photodiodes are typically used in reverse bias. When light hits a reverse biased *pn* junction, a reverse current directly proportional to the intensity of the incident light is generated. This principle is used to build light detectors and other light operated devices. Solar cells offer an example of a large area photodiode where the current generated by the photoelectric emission is collected and used as energy.

A phototransistor is a BJT with the collector-base junction exposed to light. The electrons liberated by the photoelectric emission in the collector enter the base region, where they are amplified by the transistor current gain factor ($\beta_F$).

In microcontroller applications, photosensitive devices like photodiodes and photo-transistors are frequently paired with light-emitting diodes to form optical switches. An optical switch operates by detecting the current changes caused in a photodiode or phototransistor by the blockage or exposure to the LED light.

Photodiodes can be tuned to respond to light of specific wavelengths. Their pairing with LEDs of the same wavelength, allows for developing photo-switches able to discriminate light from undesired sources. For example, a TV remote control uses infrared LEDs and photodiodes in the transmitter and receiver, respectively, to only respond to the excitation produced by the transmitter and not to ambient light. Moreover, pulsing the LED on the transmitter allows serially encoding bit patterns into streams of light/darkness impulses that are decoded as zeros and ones at the detector side. Figure 8.57 shows an LED-phototransistor pair forming an opto-coupler and a possible form to interface it to an MCU.

Resistor $R_1$ limits the LED current, while $R_2$ establishes the maximum current through the phototransistor. Setting low the output port driving the LED makes it

**Fig. 8.57** Interfacing an
optocoupler to an MCU

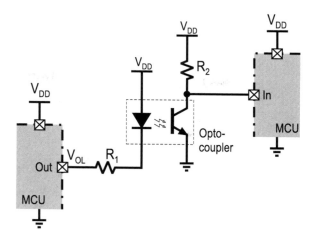

conduct, shining light on the phototransistor. When the phototransistor is exposed
to light, it begins to conduct, bringing to low the voltage at node $x$. This condition
can be read by an input GPIO pin. The values of $R_1$ and $R_2$ are obtained using actual
application conditions.

## 8.10  MCU Motor Interfacing

A number of applications in embedded systems require precise mechanical movement. A few common examples include plotters, inkjet printers, and CNC (Computer
Numerical Control) machines where precise movement under computer control is
needed. The list of alternatives to produce computer controlled movement include DC
motors, servo-mechanisms, and stepper motors. All three can be defined as electromechanical devices capable of transforming electric power into mechanical power
in the form of a rotating shaft. However, there are fundamental differences among
them in terms on how the electro-mechanical conversion is made and particularly on
how they can be controlled.

### 8.10.1  Working with DC Motors

The shaft of a DC motor continuously spins when applied energy. Assuming a
constant mechanical load, the output mechanical power of a DC motor will be proportional to the input electrical power, up to its rated capacity. Thus at constant load,
the motor speed would be a function of the applied voltage.

**Fig. 8.58** Open-loop DC
motor speed control via PWM

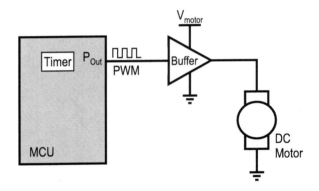

Applications where only the spinning speed is important can be readily served
with DC motors. A variable speed electric fan is a representative example of this type
of application. The spinning speed of a DC motor can be readily controlled from a
microcontroller using pulse-width modulation (PWM). A pulse-width modulated
signal is a square wave whose duty cycle can be controlled by the MCU. The amount
of power carried by a PWM signal is proportional to the waveform duty cycle.

Generating a PWM signal using an MCU results straightforward with an internal
timer and Output pin. Thus, controlling the speed of a constant load DC motor from
an MCU is a simple task. Figure 8.58 shows a diagram of an open-loop DC motor
speed control using PWM that can achieve such an objective. The buffer in the
figure can be implemented as discussed in Sect. 8.8. Chapter 7, in Sect. 7.4.3 offers a
discussion on how to produce a PWM signal with an MCU.

If a reversing control were desired in the DC motor interface, an *H-Bridge* driver
could be employed. An H-bridge contains two pairs of complementary transistor that
allow for reversing the voltage applied to a load under the control of an MCU output
pins. Figure 8.59 shows a basic topology for a reversible DC motor H-bridge driver
implemented with bipolar transistors.

The circuit is controlled with complementary inputs *Fwd* and *Rev*. Setting
*Fwd* = 1 and *Rev* = 0, transistors $Q_1 - Q_3$ are simultaneously turned on, allowing current flow from left to right in the motor, as indicated by the dash-dot line in
Fig. 8.59. Making *Fwd* = 0 and *Rev* = 1, will activate transistors $Q_2 - Q_4$ instead,
reversing the current direction and therefore the polarity of the voltage applied to the
load. Diodes $D_1$ through $D_4$ protect their respective transistors from the inductive
turn-off transient. Inputs *Fwd* or *Rev* could be driven by a PWM signal to also allow
for speed control. The circuit is also frequently implemented with MOSFETs instead
of BJTs, allowing similar functionality.

Although the basic functionality of an H-bridge can be obtained using discrete
devices, IC-level solutions offer a better alternative. Consider for example, TI's
DRV8833, a dual CMOS H-bridge motor driver IC. This chip encapsulates two
full H-bridges able to handle a continuous 3 A load at up to 10.8 V while providing
short circuit protection, thermal shut-down, and supporting low-power modes.

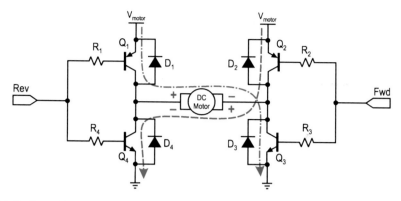

**Fig. 8.59** Reversible speed H-bridge DC motor driver

**Fig. 8.60** Pictorial view of internals of a small servomotor

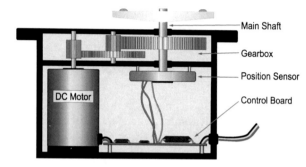

Using a basic DC motor with an open-loop PWM speed control results fairly appropriate in applications where speed control is not tight and loading conditions can be considered constant. If instead, a more precise control of the rotational speed were needed or if accurate positional control were needed, the DC motor control would require to include feedback and some form of control strategy.

## 8.10.2 Servo Motor Interfacing

A servo-motor is a DC motor with feedback control that allows for precise position control. A servo-motor includes four basic components: a DC motor that provides the basic electromechanical conversion, a control board housing the feedback electronics, a set of gears that slow-down the DC motor rotational speed, and a position sensor, typically a potentiometer. Figure 8.60 shows a pictorial diagram of the components in a servomotor.

The external shaft of a servo motor does not continuously spin like that in a DC motor. Although its internal motor may make multiple turns at once, the gearbox reduces the rotation ratio resulting in a restricted travel angle of about 200° or less (typically 180°). The positional sensor is attached to the main shaft axis, providing feedback positional information to the control circuit, which implements a PID feedback controller, allowing precise shaft angular position control.

The external interface of a servo motor has only three wires: $V_{DD}$, $GND$, and *Control*. Power is applied trough the $V_{DD} - GND$ terminals while the desired position is specified through the *Control* pin with a PWM signal. The period of the PWM signal is typically 20 ms ($f = 50$ Hz), and the width of the ON portion of the signal specifies the desired angular position. Specs might slightly change depending on the shaft travel angle, but typical specs define 1000 μS to 2000 μS for the extreme positions ($-90°$ to $+90°$), with 1500 μS for center position. Interfacing to an MCU requires a single PWM output, and the control pin in most cases can be directly driven from an I/O pin or simply a level shifter.

Servo motors come in a wide variety of sizes, from small units used in radio-controlled models, to industrial sizes, in DC or AC versions moving heavy equipment.

### 8.10.3 Stepper Motor Interfaces

A stepper motor shaft moves in discrete increments in response to digital pulse sequences applied from a controller. The structure of a stepper motor is different from that of a DC motor, as it incorporates multiple windings to make possible the stepping behavior. Depending on how the rotor and stator are designed, stepper motors are classified into three types: variable reluctance, permanent magnet, and hybrid.

### 8.10.4 Variable Reluctance Stepper Motors

A variable reluctance (VR) stepper motor has a soft iron, non-magnetized, multi-toothed rotor and a wounded stator with three to five windings. The motor is designed with a larger number of poles in the stator than teeth in the rotor. Figure 8.61 shows a sectional view of a VR stepper with four windings, A, B, C, D in the stator (eight poles) and six teeth in the rotor.

The stator windings in the motor are unipolar, meaning that they can be polarized in only one direction. Torque is developed as poles and teeth seek to minimize the length of the magnetic flux path between the stator poles and rotor teeth. The sequence and combination of pole activation determines the rotation direction and step size. Two different step sizes can be produced: full step, and half step. In full step mode the motor will advance in angular increments equal to its nominal resolution. In half-step mode the resolution is doubled as the motor advances only half the nominal angular resolution.

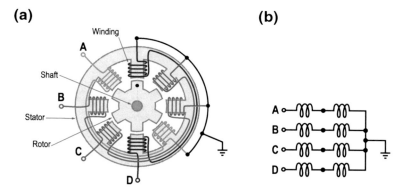

**Fig. 8.61** Sectional view of a variable reluctance stepper motor and winding arrangement. **a** Motor cross section, **b** Winding arrangement

Full-step resolution can be achieved two ways: using a single phase sequence (also called wave sequence) or using a two-phase sequence (routinely called full-step mode).

The simplest form of driving a stepper motor is in the single-phase (wave) mode. In this mode the stator windings are activated one at a time in an orderly sequence. For a VR motor like that in Fig. 8.61 a single-phase operation can be obtained with an activation sequence A-B-C-D repeated over and over. This sequence would cause the rotor move in a clockwise direction with a resolution of 15° per step, for a total of 24 steps per revolution. If the number of stator phases $p_s$ and rotor teeth $t_r$ are known, the step size $\theta_s$ can be obtained as

$$\theta_s = \frac{360°}{t_r \times p_s}. \tag{8.26}$$

Figure 8.62 shows the positions assumed by the rotor with a single phase activation sequence. The dot in the rotor allows to track its position. Assuming the rotor was initially at position 0°, as indicated in Fig. 8.61, the sequence (A-B-C-D) would cause a total rotor displacement of four steps or 60°.

A two-phase sequence can be obtained by activating two consecutive windings simultaneously. One such a sequence would be AB-BC-CD-DA. This sequence would also cause the rotor to move in steps of 15° clockwise, however, the alignment would be 7.5° out of phase with respect to a single-pole activation sequence as the rotor teeth position that minimizes the air-gap flux falls in-between the two excited poles. This operation mode produces the highest torque in the motor and also consumes the most power. Figure 8.63 shows the rotor positions for consecutive DA and AB activations.

A half-step operation can be obtained by combining the single and dual phase mode sequences in a way that entries of each activation sequence are alternated. This would produce a sequence DA-A-AB-B-BC-C-CD-D. Under this sequence the

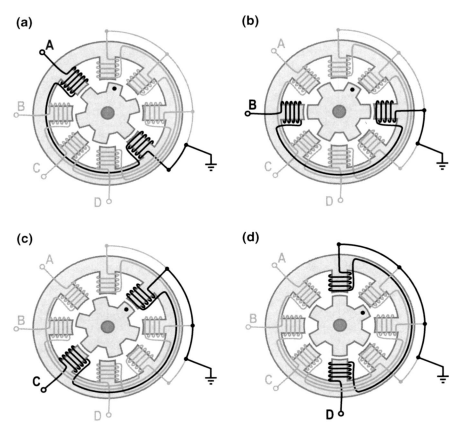

**Fig. 8.62** Full-step wave activation sequence A-B-C-D in a 4-phase VR stepper motor. **a** Phase A excited: angle 15°, **b** Phase B excited: angle 30°, **c** Phase C excited: angle 45°, **d** Phase D excited: angle 60°

rotor teeth would minimize the flux path when aligned to stator poles with single phase excited and in-between the next two excited phases, resulting in an angular travel only half the step size. For the VR motor used in the illustration, this sequence would yield a resolution of 7.5°, duplicating the number of steps per revolution. The increased resolution causes a smoother motor operation but comes at the expense of a reduction in the motor torque.

From an MCU standpoint, either of these modes can be software controlled. An interface would require four I/O pins, each with unipolar drivers. Figure 8.64 shows a block diagram of an interface. In this case the buffers can be just a power PMOS providing both current buffering and level-shifting. The clamping diodes in the diagram protect the drivers against the motor inductive transient.

The functional mode, full- or half-step would depend on the programmed sequence. The speed of rotation can be controlled through the frequency with which the

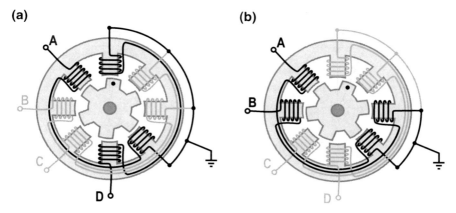

**Fig. 8.63** Two-phase activation sequence DA-AB in a 4-phase VR stepper motor. **a** Phases D and A excited: angle 7.5°, **b** Phases A and B excited: angle 22.5°

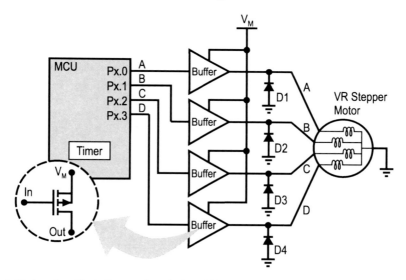

**Fig. 8.64** Interface block diagram for a 4-phase VR stepper motor

winding activation sequence is fed to the stator. Continuous rotation is possible by sending out the activation sequence in a loop. Tables 8.3 and 8.4 show the command sequences to be sent through the designated I/O pins to advance a four-phase VR stepper like that shown in Fig. 8.61 when using single-phase full-step and half-step modes, respectively. Each row in the tables correspond to the code sent in each time slot of operation. Note that the last column in both tables refer to the angular position of the shaft. Every command will rotate the shaft by the corresponding step amount (15° or 7.5°) with respect to the previous position. Sending out the sequence in the reverse order would cause the motor to rotate in counterclockwise direction.

**Table 8.3**  Command sequence for full-step, single-phase operation of 4-phase VR stepper motor

| Px.0 (A) | Px.1 (B) | Px.2 (C) | Px.3 (D) | Position |
|----------|----------|----------|----------|----------|
| 1        | 0        | 0        | 0        | 15°      |
| 0        | 1        | 0        | 0        | 30°      |
| 0        | 0        | 1        | 0        | 45°      |
| 0        | 0        | 0        | 1        | 60°      |

**Table 8.4**  Half-step control sequence for a 4-phase VR stepper motor

| Px.0 | Px.1 | Px.2 | Px.3 | Phase | Position |
|------|------|------|------|-------|----------|
| 1    | 0    | 0    | 1    | DA    | 7.5°     |
| 1    | 0    | 0    | 0    | A     | 15.0°    |
| 1    | 1    | 0    | 0    | AB    | 22.5°    |
| 0    | 1    | 0    | 0    | B     | 30.0°    |
| 0    | 1    | 1    | 0    | BC    | 37.5°    |
| 0    | 0    | 1    | 0    | C     | 45.0°    |
| 0    | 0    | 1    | 1    | CD    | 52.5°    |
| 0    | 0    | 0    | 1    | D     | 60.0°    |

## 8.10.5  Permanent Magnet Stepper Motors

Permanent magnet (PM) stepper motors are constructed differently than VR motors. Unlike VR's, the rotor of a PM motor is built using permanent magnets and has no teeth. PM motor stators have multiple windings, which can be configured to operate as either unipolar, bipolar, or bifilial windings. Torque in a PM motor occurs as a result of the excited stator poles attracting opposite poles and repulsing similar poles among the magnets in the rotor.

### Unipolar Windings

Unipolar PM steppers typically have two stator windings, each split between two opposite poles, resulting in a total of four stator poles. The center taps of both winding is made externally available. Center taps can be independently brought outside the stator carcass providing a total of six terminals: four for the stator poles and two for the center taps. Some motors might have the center taps internally interconnected and brought out as a single terminal, providing five external wires.

Figure 8.65a shows a sectional view of a permanent magnet stepper motor with two unipolar stator winding and six magnetic poles in the rotor. Sub-figure (b) shows the arrangement of its windings. The winding terminals are labeled $A - \overline{A}$ for the first winding, with center tap $A1$ and $B - \overline{B} - B1$ for the second winding.

Operation of a PM stepper can be achieved by applying a fixed DC source $V_M$ to the stator center taps and selectively grounding the winding ends. This way, the current

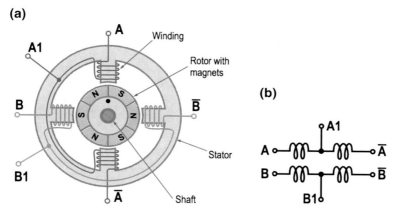

**Fig. 8.65** Sectional view of a permanent magnet stepper motor and windings connections. **a** Motor cross section, **b** Winding arrangement

circulating through each winding determines the magnetization in the corresponding stator pole and thus determining which rotor poles will be attracted or repelled. Note that under this scheme only one half of the winding is used at a time, which limits the maximum torque the motor can generate. Operation from a microcontroller can be accomplished in a straightforward manner by using controlled switches which when closed, will ground the stator poles selected for being excited.

For the motor illustrated in Fig. 8.65, by having terminals $A1$ and $B1$ permanently connected to a power source ($V_M$), and grounding the winding terminals in a sequence $A - \overline{B} - \overline{A} - B$ repeatedly will cause the rotor to move in clockwise direction, incrementing its angular position by $30°$ in each step. Note that in this case, the number of steps per revolution is equal to the number of poles in the rotor. Figure 8.66 shows the rotor positions for each step in a sequence $A - \overline{B} - \overline{A} - B$, assuming and initial position of $0°$ as illustrated in Fig. 8.65. The rotor dot in the illustration allows tracking the shaft position after each step.

Permanent magnet motors can also be operated in one- and two-phase modes. The sequence illustrated in Fig. 8.66 corresponds to a single-phase or wave mode operation. Activating phases in pairs, in this particular case $BA - A\overline{B} - \overline{BA} - \overline{A}B$, moves the motor in single-step two-phase mode, which produces a higher torque at the expense of consuming twice the power as in the single-phase case. This mode also produces a half-step phase shift in the shaft angular position with respect to the positions reached in single phase mode. Combining, in an interleaved way, the single- and double-phase sequences has the same effect as in a VR motor: doubling the resolution, making the motor operate in half-step mode. Figure 8.67 illustrates the rotor dynamics in this mode for a partial sequence of three steps. In this case the resulting step resolution would be $15°$, raising the number of steps per revolution from 12 to 24.

Interfacing a unipolar PM stepper motor to an MCU is quite similar to the arrangement for a VR motor. It requires only four unipolar driving elements that could be

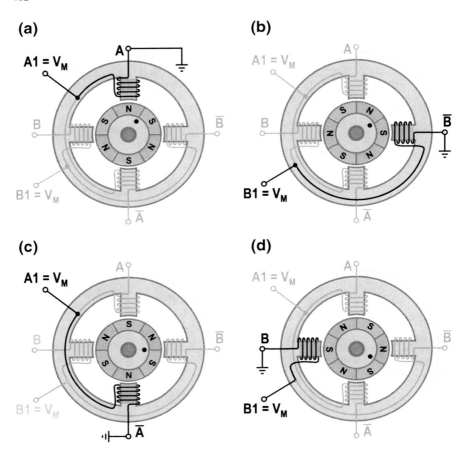

**Fig. 8.66** Full-step wave activation sequence $A - \overline{B} - \overline{A} - B$ in a PM stepper motor. **a** Phase A excited: angle 30°, **b** Phase $\overline{B}$ excited: angle 60°, **c** Phase $\overline{A}$ excited: angle 90°, **d** Phase B excited: angle 120°

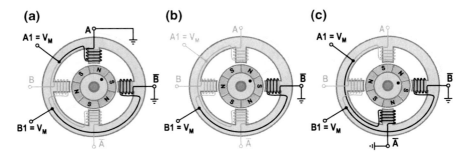

**Fig. 8.67** Half-step sequence $A\overline{B} - \overline{B} - \overline{B}\overline{A}$ in a PM stepper motor with six rotor poles. **a** Phase $A\overline{B}$ excited: angle 45°, **b** Phase $\overline{B}$ excited: angle 60°, **c** Phase $\overline{B}\overline{A}$ excited: angle 75°

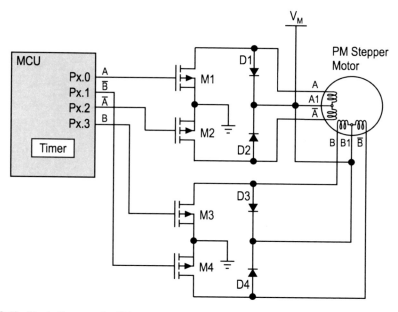

**Fig. 8.68**  Block diagram of a PM stepper motor to MCU interface

**Table 8.5**  Command sequence for full-step, single-phase operation of 4-phase PM stepper motor

| A | $\overline{B}$ | $\overline{A}$ | B | Position |
|---|---|---|---|---|
| 1 | 0 | 0 | 0 | 30° |
| 0 | 1 | 0 | 0 | 60° |
| 0 | 0 | 1 | 0 | 90° |
| 0 | 0 | 0 | 1 | 120° |

accommodated with power MOSFETs. In this case power NMOS devices would be recommended since we want to switch the connection to ground in the controlled windings. Figure 8.68 shows a block diagram of the connection. Note that NMOS transistors $M1$ through $M4$ have their source terminals connected to ground, performing the desired switching action. Diodes $D1$ through $D4$ provide clamping protection for the transistors against the inductive transient of the motor winding.

Table 8.5 shows the code activation sequence for controlling the stepper motor exemplified in this discussion in single-phase, single-step mode from an MCU interface like that of Fig. 8.68. This table assumes the shaft is at position 0°, as illustrated in Fig. 8.65 prior to the first command. Note that the binary sequence is the same as in the case of a VR stepper with unipolar winding, so figuring out the code activation sequence for half-step operation can be readily determined from Table 8.4.

**(a)**

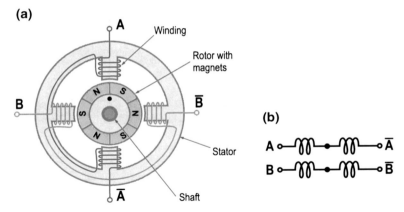

Fig. 8.69 Sectional view of a bipolar permanent magnet stepper motor and windings connections. a Motor cross section, b Bipolar winding arrangement

## Bipolar Windings

A permanent magnet stepper motor with bipolar windings provides no external connections to the center taps, having only four external wire connections. In this configuration the activation of a stator pole uses the entire winding, generating the maximum torque the winding can provide. Control of stator magnetic poles is achieved by changing the direction of the current circulation in the winding. Figure 8.69 shows a sectional view of a PM stepper with bipolar windings, and the internal configuration of its coils.

The necessity of reversing the windings polarity calls for an external circuit more complex than that used in the case of unipolar windings. A polarity reversing driver can be satisfied by using an H-bridge per winding. Figure 8.70 shows a connection of the windings of a bipolar PM stepper motor using two H-bridges, one per winding.

Transistors $M1$ through $M4$ and $M5$ through $M8$ make up the two bridges, the first controlling winding A and the second for winding B. $D1$ through $D8$ are protective clamping diodes. Optional gate resistors (not shown in the figure) might be added to accelerate the MOSFET turn-off (see Fig. 8.56 for reference).

The software for operating a bipolar PM motor is not radically different from that used in the previous motor types analyzed above. Single- and double-phase full step operation is possible by activating one or both windings at a time. The interface circuit will apply $V_M$ to A and $GND$ to $\overline{A}$ when Px.0 is low and Px.1 is high, or viceversa when reversed. A similar situation occurs with $B$ and $\overline{B}$ by controlling Px.2 and Px.3. Setting both Px.0 and Px.1 to high will just deactivate winding A. Similarly will happen to winding B by setting Px.2 and Px.3 to one simultaneously. Therefore, single-phase operation can be achieved by sending-out pins Px.0 through Px.3 the bit patterns in Table 8.6.

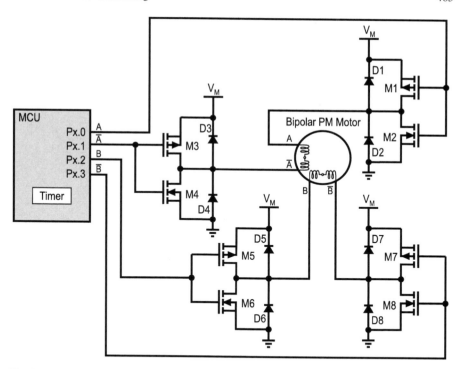

**Fig. 8.70** Block diagram of a bipolar PM stepper motor to MCU interface

**Table 8.6** Control sequence for single-phase, wave operation of a bipolar PM stepper motor

| Px.0 | Px.1 | Px.2 | Px.3 | Phase | Position |
|------|------|------|------|-------|----------|
| 0 | 1 | 1 | 1 | $A$ | 30° |
| 1 | 1 | 0 | 1 | $B$ | 60° |
| 1 | 0 | 1 | 1 | $\overline{A}$ | 90° |
| 1 | 1 | 1 | 0 | $\overline{B}$ | 120° |

Controlling the motor with a double phase activation in a sequence $\overline{B}A - AB - B\overline{A} - \overline{AB}$ produces a higher torque than a single phase excitation, while consuming twice the power. It causes a rotation in full step increments, shifted half a step forward with respect to a single-phase sequence. This shift allows for combining the single- and double-phase sequences to operate the motor in half-step resolution, as has been the case for the previous motor types. Obtaining the command sequence for half-step operation can be readily accomplished from the pattern in Table 8.6 once it is realized that the phase activation sequence would be $\overline{B}A - A - AB - B - B\overline{A} - \overline{A} - \overline{AB} - \overline{B}$. The obtention of the actual command table is left as an excursive to the reader.

(a)                          (b)                          (c)

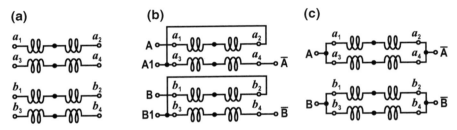

**Fig. 8.71**  Winding arrangement in a bifiliar stepper motor, illustrating their series and parallel connections. **a** Internal winding arrangement, **b** Serial connection as unipolar, **c** Parallel connection as bipolar

**Bifiliar Windings**

Some stepper motors provide the flexibility of being configurable to operate as either unipolar or bipolar by including dual windings in each phase and pole, with their terminals externally available. These are called bifiliar winding stepper motors. A permanent magnet, bifiliar stepper motor provides eight external wires that allow for configuring its windings as desired. Figure 8.71 shows the internal winding arrangement for this type of motors, illustrating how to externally connect them to produce either unipolar or bipolar equivalent windings.

The windings of each pole can be connected in series as illustrated in Fig. 8.71a to produce an equivalent unipolar winding. Leaving the center tap unconnected would produce a series bipolar winding (not illustrated in the figure), which allows for high voltage motor operation. The alternative connection in Fig. 8.71b illustrates how to make a parallel connection to produce equivalent unipolar windings. This alternative allows operating the motor with a lower voltage than the series connection.

Once a series or parallel connection as either unipolar or bipolar mode is established, interfacing and controlling the motor can be made as was discussed in the previous sections for their corresponding equivalent mode.

## *8.10.6  Hybrid Stepper Motors*

Hybrid stepper motors combine features from both variable reluctance and permanent magnet steppers to produce a machine with the ability to produce high torque at low and high speeds with very fine resolution.

A hybrid stepper motor has two multi-toothed, soft iron disks in its rotor with a permanent axial magnet between them. The soft iron disks become extensions of the magnet ends, assuming each the characteristics of their respective end of the magnet. The disks are accommodated to have their teeth interleaved, allowing to accommodate a large number of interleaved magnets. Figure 8.72a shows the arrangement of components in a 16-teeth hybrid rotor. Subfigure 8.72b shows a frontal view of the rotor allowing to see the interleaved magnetic poles.

**(a)**

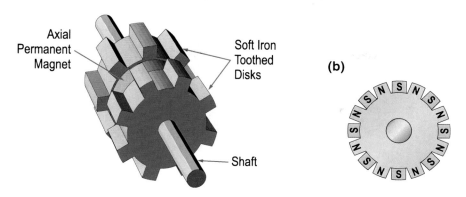

Fig. 8.72 Rotor arrangement in a hybrid stepper motor, illustrating components and frontal view. a Rotor components, b Rotor front view

The stator of a hybrid motor also has teeth in its poles to define preferred points of alignment with the rotor poles. The combination of toothed rotor and stator poles allows for accommodating a dense arrangements of alignment points in the structure, yielding motors with a large number of steps per revolution. Hybrid motors can produce up to 800 steps per revolution in full-step mode and twice that number when driven in half-step mode.

Hybrid stepper motors might have either unipolar, bipolar or bifiliar windings, therefore, operating them from an MCU will require the same types of interfaces and control algorithms that were described earlier for these types of stator windings. Figure 8.73 shows a sectional view of a hybrid stepper motor with a 16 teeth rotor and a two-phase bipolar stator, producing a resolution of $11.25°$ per step or 32 steps per revolution. Driving this motor in half-step mode would result in 64 steps per revolution.

## 8.10.7 Microstepping Control

Besides operating in either half- or full-step modes, modulating the pulses applied to the windings of a stepper motor can further reduce the step size. This driving modality is called *microstepping*. A microstepping driver, instead of just exciting and de-exciting in a binary way the windings of a stepper motor, gradually transitions power from a winding to another. This creates a smoother motor operation, and depending on how many levels are designated for the transition is the number of microsteps resulting from a single step. To better understand the concept, let's consider a two-phase, bipolar stepper motor with windings A and B. In a full-step single phase mode, switching from phase A to B simply turns off phase A and turns on phase B,

**(a)**

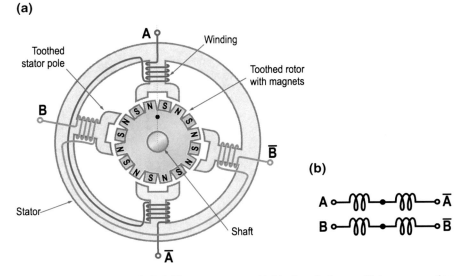

**Fig. 8.73** Sectional view of a hybrid stepper motor with bipolar windings. **a** Motor cross section, **b** Bipolar winding arrangement

alternating polarities. The resulting voltage waveform in windings A and B would be like the continuous staircase waveforms shown in Fig. 8.74.

The ideal waveform for a smooth behaviour would be a sine/cosine waveform like the dotted line in Fig. 8.74. In a practical implementation, a discrete approximation of the sinewave like that in the dot-dash waveform in Fig. 8.74 would be used. Each discrete level would produce a stable position for the rotor, resulting in a microstep. For the case illustrated, the motor would operate with four microsteps. Assuming a sixteen-pole hybrid motor like that illustrated in Fig. 8.73, this technique would increase the step resolution from originally 32 steps per revolution (11.25° per step) in full-step mode to 128 steps per revolution or 2.8° per step. This scheme requires only two bits to encode the four discrete levels used in the sequence. Commercial microstepping drivers are capable of producing up to 32 discrete levels per step (6-bit), significantly increasing the step resolution. The increased resolution comes at the expense of reduced torque. Note that most of the time the windings will be excited below their maximum value. Also, a microstepping driver is more complex than a conventional full/half stepper driver.

### 8.10.8 Software Considerations in Stepper Motor Interfaces

The interface examples shown in the previous sections illustrated ways of controlling a stepper motor directly with I/O ports. In direct I/O control, a software routine is

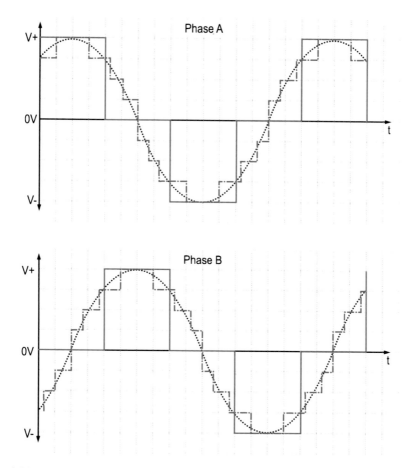

**Fig. 8.74** Waveforms in a microstepping driving approach

required to keep track of the control sequence to make possible the correct shaft rotation. The software requirements for the processing routine are relatively simple:

- A look-up-table (LUT) holding the command sequence. If a PM bipolar motor were used, the LUT would have only four entries: 07h, 0Dh, 0Bh, and 0Eh, assuming byte entries with the least significant four bits actually holding the control code. This would allow easy alignment with the hardware connection illustrated in Fig. 8.70.
- A circular pointer to orderly access the sequence. By circular we mean rolling over the pointer from bottom to top of the LUT or viceversa when the LUT boundaries are reached.
- A timer function to establish the rotation speed. The usage of a timer channel allows for precise periodic updates of the codes in the motor control I/O pins.

**Fig. 8.75** Timing control
waveforms for a PM bipolar
stepper motor

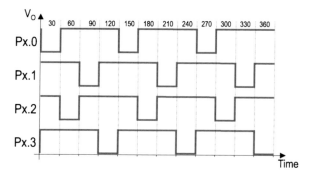

With the pointer initialized to the top of the LUT, every time the timer function is triggered, its ISR updates the motor control I/O pins with the next code entry in the table. Updating the pointer in a circular way allows for continuous control code output. If the pointer update is done by incrementing its value, rotation would be clockwise, while by decrementing would produce counterclockwise rotation. Considering for example the case of a permanent-magnet stepper with two bipolar windings, the resulting waveforms in the I/O pins caused by the described software routine would be similar to that illustrated in Fig. 8.75. Note that this timing diagram is just the same sequence described in Table 8.6.

In applications where devoting CPU cycles to keep track of a stepper motor sequence might become difficult, a dedicated stepper motor controller can become an option. A dedicated stepper motor controller has internal logic to keep track of the motor stepping sequence, providing for forward and reverse operation. Numerous commercial off-the-shelf options from different vendors can be identified. A couple of representative examples are Motorola's MC3479 and Texas Instrument's DRV8811.

Motorola's MC3479 is an example of a simple stepper motor driver, capable of operating a two-phase, bipolar stepper motor in either full-step or half step modes in forward or reverse direction. The chip is capable of directly driving coils with voltages in a range from 7.2 V to 16.5 VA with up to 350 mA per coil. Its digital interface allows specifying the step mode ($\overline{F}/HS$), rotation direction ($\overline{CW}/CCW$), and step clock ($CLK$), among other signals. The analog side allows feeding the motor voltage $V_M$ and connection to the motor windings $L1 - L2$ and $L3 - L4$ for the respective winding phases. Figure 8.76 shows an internal block diagram and typical application of this chip. For more details on this chip, the reader is referred to its data sheets [66].

TI's DRV 8811 provides a more complete solution for stepper motor interface. It supports two-phase bipolar stepper motors up to 2.5 A per winding. Besides the basic indexing functionality and FWD/REV and full/half stepping capabilities, the 8811 supports microstepping motor control with up to eight different current levels. The chip also features thermal shut down, short-circuit, and undervoltage lockout protections. The external interface requires only a few discrete resistors. The chip package features an exposed pad to remove heat through the PCB that allows operation without

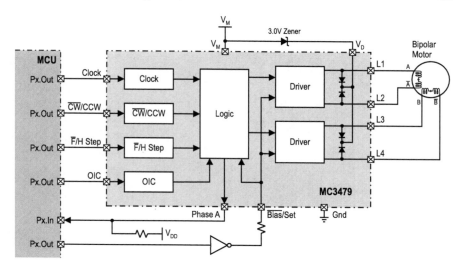

**Fig. 8.76**   Interfacing a bipolar stepper motor using the MC3479

a heatsink in most application conditions. Figure 8.77 shows the internal structure and typical application diagram of a DRV8811. For more details on configuring and operating this chip, the reader is referred to its data sheets [67].

### 8.10.9 Choosing a Motor: DC Versus Servo Versus Stepper

When compared against other types of motors, DC motors are relatively inexpensive and simple. When powered, their open loop rotational speed can be reasonably controlled using PWM or a DAC interface if the motor operates under constant load. Under general loading conditions however, the lack of a feedback control mechanism will cause cumulative deviations from the desired set point, being this a reason why they cannot be used without feedback in applications requiring precise control. By including some form of feedback, like a tachometer or rotary encoder, optional gears, and adding a control circuit their application range can be expanded to a wide number of areas. However this solution involves additional hardware and software components.

DC motors are the preferred solution when the controlled process can be dealt with the motor speed. A number of applications allow open-loop operation, like in the case of variable speed fans or electric toys. One weakness of DC motors is the need of brushes and commutators to pass energy to the rotor winding. This components tend to reduce the motor life due to mechanical wear. Although brushless alternatives for DC motors are available, these need additional electronics to synchronize the stator magnetic field with the rotor permanent magnet poles to produce spin, making them more reliable, but also more expensive.

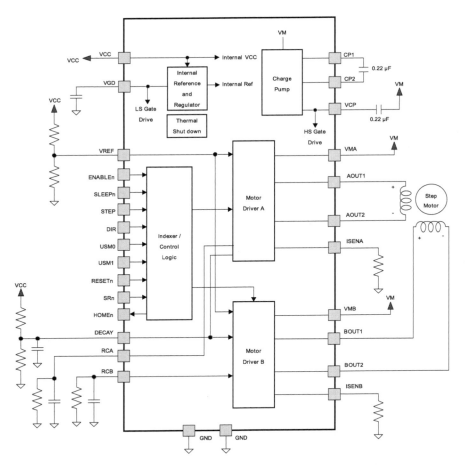

**Fig. 8.77** Interfacing a bipolar stepper motor using TI's DRV8811 (*Courtesy of Texas Instruments, Inc.*)

Servo motors, although based on DC motors, provide an elegant solution for positional control applications as they include an internal feedback loop, gears, and a control circuit that allow for precise angular position control. Servo motors are more expensive than either DC or stepper motors because of the inclusion of gears and the control electronics necessary for their operation. Servos are available with brushed and brushless DC motors as well as with AC motors, offering a wide range of sizes and power. Brushed servo motors suffer from the same weakness as brushed DC motors in the limited life of brushes and commutators. In addition, those using potentiometer-based feedback sensors, the potentiometer usually becomes a cause of failure. Nevertheless, servos are very efficient, lighter than steppers, produce better torque at high speeds, and provide a better reserve of power for momentary torque surges.

Stepper motors offer the most precise open-loop position control because of their inherent stepping nature. For equal power rating, steppers are less expensive than servo motors and allow direct shaft loading. Also, stepper motors offer a higher reliability than any brushed DC motor option as no brushes or commutators are needed for their operation, and in comparison to servos, steppers have less parts prone to failure. Errors due to friction or other factors do not accumulate in stepper motors, and if operated with a load less than or equal to their nominal torque, they rarely miss a step. Steppers however, tend to be noisier and produce more heat than DC or servos counterparts, and are also less tolerant to overload.

## 8.11 Summary

This chapter provided concepts and applications of developing interfaces for microcontroller-based systems using their general purpose input-output capabilities. The Chapter introduced concepts of GPIO structure and elaborated to the point that allowed for integrating user interfaces, large loads, and motors.

General purpose I/Os form the most important means for microcontrollers to interact with the external world. Using their Data-In/Data-Out registers, configuring their direction registers, and enabling their interrupt capabilities were the main subject of Sect. 8.1. Next the fundamental considerations to interface switches and switch arrays, LEDs, LED arrays, numeric alphanumeric, and graphic LCDs, were discussed in detail in Sects. 8.3 and 8.4.

The subject of handling large loads, like heaters, solenoids, motors, lightning fixtures, etc. were discussed in Sects. 8.2 and 8.10. Every section featured how MSP430 devices support each of the base functions discussed, along with hardware and software examples illustrating their applicability.

## 8.12 Problems

8.1 An embedded system incorporates an MSP430F2274 clocked by a 4.096 MHz at 3.3 V, driving two seven-segment displays and two discrete LEDs. The seven-segment display draws 7.5 mA per segment at 1.8 V and the discrete LEDs operate with 10 mA each at 1.6 V. Three push-buttons are needed for completing the user interface. Design a suitable interface to connect all LEDs, seven-segments, and push-buttons. Estimate the circuit power requirements and recommend non-regulated power supply and a suitable regulator.

8.2 Assume the system described in Problem 1 is to be fed from a 4.0 V, 2400 mAH lithium battery. Estimate the expected battery life, assuming LEDs are driven dynamically at 30 % duty cycle. What would be the regulator efficiency? Determine the MCU thermal dissipation for the given loading conditions and verify if it is operating in a safe temperature range.

8.3 Provide an initialization sequence for the MCU in problem 1, and write a program that will make the system operate as a down counting stopwatch. Label the three keys as "b1", "b2", and "b3" and write a short program to display the key name on the seven-segment display while the corresponding key is depressed.

8.4 Devise an algorithm for joining a keypad scan function and a software debouncing function for the keypad such that a seven-segment display shows the character corresponding to the depressed key. Assume only single key depressing will be allowed.

8.5 Write a program to return on R5 the ASCII code corresponding to the depressed key in a 16-key keypad. Show how shall the keypad be interfaced to the MCU.

8.6 Provide a stand-alone program that will make a dedicated 12-key keypad scanner from an MSP430 Launchpad, placing in a predesignated memory location the ASCII code of the depressed key.

8.7 Design a dumb HEX terminal using 16-key keypad and a 4x20 alphanumeric LCD on an MSP430F169. Provide a keypad buffer in the MCU with a capacity of 20 characters with a last-in first-out replacement policy for the data sent to the LCD.

8.8 A 120VAC fan is to be controlled by an MSP430F2273. Assuming the maximum fan current is 2A, provide a solid-state relay interface using a Sharp S108T02. Verify both voltage and current compatibility in the control interface. Add a single push-button interface to the system and a status LED to be lit when the fan is ON. Provide a short program to toggle the fan with the push-button.

8.9 A standard Futaba S3003 servo-motor operated at 5.0 V moves between its two extreme positions 180° apart when fed a 50 Hz signal with 5–10 % duty cycle. Nominally the servo moves to its center position with a 7.5 % duty-cycle signal. Assuming the digital signal requires less than 1mA when pulsed, and the motor consumes anywhere between 8 mA and 130 mA in a load range from idle to full torque. Provide a suitable interface to control the servo from an MSP430 Launchpad I/O pin and provide a software-delay based function that would allow controlling the shaft with a resolution of 0.5°.

8.10 A 12VDC, 1.5A per phase, two-phase PM bipolar stepper motor with eight rotor poles is to be controlled from an MSP430F2274. Provide a discrete component H-bridge interface to control the motor with a GPIO port in the MSP and indicate the actual components to be used. Provide a safe to operate interface and a function accepting as parameters(via registers) the direction in which the stepper will move and the number of steps in each move. Assume the motor is to be operated in half-step mode. Determine the number of steps per revolution to be obtained in this interface.

8.11 Provide an enhanced interface for the motor in problem 10 by using a DRV8811. Upgrade the driving software function to operate the motor in microstepping mode with eight microsteps per full step of the motor. Calculate the new step resolution of the motor.

# Chapter 9
# Principles of Serial Communication

Serial channels are, without doubt, the main form of communications used in digital systems nowadays. Diverse forms of serial communication formats and protocols can be found in applications ranging from short inter- and intra-chip interconnections, to the long range communication with distant spaceships traveling to other planets. Virtually, all forms of communications used nowadays in consumer electronics are supported via serial channels. The Universal Serial Bus (USB), Blue-tooth, Local Area Networks, WiFi, IrDA, the traditional RS-232, FireWire, $I^2C$, SPI, PCIe, and many other protocols and formats are all based on serial links.

## 9.1 Data Communications Fundamental

A serial communication channel, when transmitting an $n$-bit character, instead of using $n$ simultaneous signal links, one per bit, as is done in a parallel communication, uses only one signal link, with the $n$ bits composing the character sequentially transmitted over the channel. Each bit serially transmitted takes a pre-determined amount of time $t_{bit}$, requiring $n \cdot t_{bit}$ seconds to transmit the entire character. Figure 9.1 illustrates the conceptual[1] difference between a parallel and a serial 8-bit channel transmitting character 0A5h.

The transmission rate of a serial channel is determined by the amount of time taken to transmit each bit ($t_{bit}$). Two speed metrics commonly used in serial channels are the *bit rate* and the *baud rate*. The bit rate expresses the number of bits-per-second or *bps* transmitted in the channel. Given $t_{bit}$ we can obtain the bit rate as $bps = 1/t_{bit}$. For example, a serial channel with $t_{bit} = 10$ ns in NZR will have a bit rate of 100 Mbit-per-second or 100 Mbps.

The second metric, the *baud rate*, refers to the number of symbols per second sent through a serial channel. When each symbol represents one bit, as is the case of

---

[1] The actual appearance of bits in the channel will depend on the signal carrier and modulation scheme. Figure 9.1 illustrates a voltage carrier in Non-Zero Return (NZR) format.

M. Jiménez et al., *Introduction to Embedded Systems*,
DOI: 10.1007/978-1-4614-3143-5_9,
© Springer Science+Business Media New York 2014

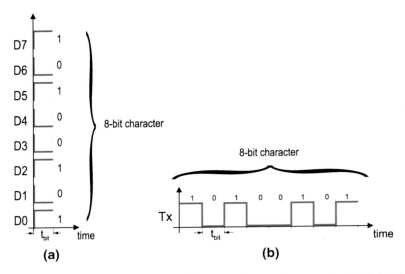

**Fig. 9.1** Signal diagrams for serial and parallel communication channels. **a** Parallel. **b** Serial

NZR, then bit rate and baud rate are the same. By using more complex modulation schemes it is possible to encode more than one bit per signal change. For example, schemes based on phase modulation, like Phase-shift Keying (PSK), can encode two or more bits per phase change. In such cases the bit rate will be higher than the baud rate.

Communication links can be implemented in diverse forms. Wired channels can use either *single-ended* or *differential* links. When a *single-ended* connection is used each link consists of ground referenced wire. An $n$-bit, single-ended parallel channel would require $n + 1$ wires to make the connection: one for each link plus one for the ground reference. For the same type of connection, a serial channel would use only two wires: one for the link and one for the ground reference.

In a *differential channel* each link is represented by the voltage difference between two wires. This type of link is more robust than a single-ended line due to the common mode rejection ratio (CMRR) of the receiving differential pair. A parallel channel using differential signaling would require $2n + 1$ wires to make the connection, versus only three used in a serial connection. Figure 9.2 shows the difference between a single-ended versus a differential link.

A wireless link could be optical like in infrared transceivers, acoustic like in underwater channels, or via a radio-frequency (RF) like in WiFi. In either of these cases a common ground reference between transmitting and receiving ends is not necessary, but still $n$ links would be needed for establishing a parallel connection, versus only one for a serial link.

This simple analysis reveals one of the fundamental advantages of serial channels over parallel: their cost. It is much cheaper to have a single serial link for transmitting

**(a)**                                              **(b)**

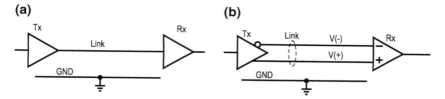

**Fig. 9.2** Wired connection modalities for serial channels. **a** Single ended. **b** Differential

a 16-bit character than sixteen simultaneous links for a parallel transmission. The
cost advantage of serial links escalates as the distance they cover increases.

Assuming equal bit times in serial and parallel channels, in principle, parallel
communications would be $n$-times faster than serial. Up to a certain bandwidth and
distance this comparison holds true. However, as communication speeds and/or distances increase, transmission line effects, cross-talk, and other noise manifestations
begin to take a toll on parallel channels that limit their practical application. This has
been one of the drivers behind the proliferation of high-speed serial links in modern
communication channels.

## 9.2 Types of Serial Channels

Serial channels are said to be *simplex, half duplex, or full-duplex*, depending on their
type of connectivity.

A simplex serial channel transmits permanently in only one direction over a dedicated link. One end of the communication channel features a dedicated transmitter
while the other has a dedicated receiver, as illustrated in Fig. 9.3. Simplex channels
have no means of acknowledging or verifying the correct reception of data. Examples of simplex channels include conventional radio and TV broadcasts channels,
write-only printers and displays, or read-only devices interfaced to a CPU.

A half-duplex serial channel features a single link that allows communication
in either direction, but only in one direction at a time. Both ends of the channel
feature interfaces operating as *serial transceivers*, denoting their ability of working
as either a transmitter or a receiver. Whenever the transmission direction needs to

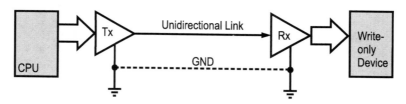

**Fig. 9.3** Simplex serial channel connecting a CPU to a write-only device

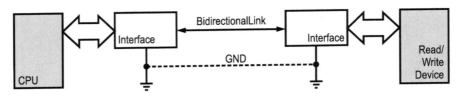

**Fig. 9.4**  Half-duplex serial channel connecting a CPU to a read/write device

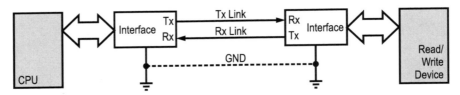

**Fig. 9.5**  A full-duplex serial channel connecting a CPU to a read/write peripheral

be changed, the interfaces in both ends need to switch their modes. Such a change requires rules to avoid having both ends attempting to simultaneously transmit. The set of rules deciding how devices at the ends of a communication channel behave and how they organize and interpret the data in the channel is called a *communications protocol*. Protocols avoid conflicts but also introduce a certain amount of latency in the communications. Figure 9.4 illustrates a topology for a half-duplex channel, denoting its single link. The ground connection is required for wired channels.

Full-duplex serial channels feature two separate links, one dedicated to transmit and another to receive, allowing for simultaneous communication in both directions. The interfaces at each end of the channel have the ability of handling both links simultaneously, requiring no switching, and therefore enabling uninterrupted bidirectional communication. Most serial connections in contemporary embedded systems use full-duplex channels. Figure 9.5 shows the topology of a full-duplex serial channel. Note the Tx→Rx and Rx←Tx connections in their interfaces.

The links used in each of these communication modalities could either be wired or wireless. In the case of wired links, both sides of the channel must have a common ground connection, and therefore, a ground wire must be shared between both ends of the channel. In wireless links such as RF, optical, or other medium, there is no need for a common ground.

The links illustrated in Figs. 9.3, 9.4 and 9.5 correspond to point-to-point topologies, as they contain only two devices, one at each end of the channel. This is the simplest topology in a communication channel and a common one.

Other topologies may incorporate more than two devices in the channel. These are called multi-point or multi-drop links. There are several topologies for multi-point channels, most of them studied as computer networking technologies. In embedded applications, multi-drop serial buses are quite common. Later in this chapter we discuss some of the most common serial bus topologies found around microcontrollers.

## 9.3  Clock Synchronization in Serial Channels

All serial channels require a stable clock signal to establish their transmission and reception rates and to internally synchronize the operation of their interfaces.

Depending on how the clock signal is provided, a channel might be asynchronous or synchronous. Asynchronous channels use independent clock generators at each end, while synchronous transmit the clock along with the data.

Both asynchronous and synchronous serial communication protocols divide a message to be sent through the channel into fundamental units called *data packets* or *datagrams*. Each packet contains three sections: a header, a body, and a footer. The header and footer are leading and trailing handshaking information fields added by the communication protocol to the portion of the message making the body of packet being transmitted.

A packet header includes data indicating the beginning of the packet, optional addressing information necessary in multi-drop channels, and an also optional message length and type fields needed in protocols accepting different types and lengths packets. A packet footer fundamentally contains data delimiting the packet and an optional field with error checking information. Headers and footers in asynchronous and synchronous channels vary in their length format, depending on the protocol. Below we describe distinguishing characteristics of each of them.

### 9.3.1  Asynchronous Serial Channels

An asynchronous communication channel, as indicated earlier, has independent clocks at each end of the channel. To make possible their communication, the interfaces at each channel end are configured to produce the same data rate, as well as having the same parameters defining the exchanged data packets. Figure 9.6 illustrates the topology of an asynchronous, full-duplex serial channel denoting its independent baud rate clock sources.

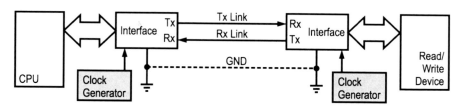

**Fig. 9.6**  An asynchronous serial channel denoting independent clock sources at channel ends

## 9.3.2 *Asynchronous Packet Format*

Asynchronous data packets have a simple structure. Their headers contain only one bit called a *start bit*, which is always a logic zero or space symbol.[2] The packet body contains a five- to eight-bit character arranged from its least- to its most-significant bit. The footer contains an optional *parity bit* that allows for error checking and one or more *stop bits* to indicate the packet end. A stop bit is always a logic one (mark). Figure 9.7 shows the format of an asynchronous packet.

The idle state of an asynchronous channel is in the mark level (logic high). A *start bit* indicates a channel transition from idle to active. A valid start bit has a length of one full bit time. When a one-to-zero transition is detected while in the idle state, this is assumed to be a start bit. To confirm that in fact a start bit is being received, the line is sampled again, half $t_{bit}$ later, at the middle of the bit time, and if it is still found low, then a valid start bit is accepted. Otherwise, the condition will be flagged as a framing error.

After a start bit has been validated, the line is sampled at intervals separated by one bit-time to detect all remaining bits in the packet. In the transmitter side, the length of each bit is determined by the transmission rate, obtained from the clock in the transmitter side. In the receiver end another clock establishes the reception rate. Although transmission and reception clocks are independent, both sides must agree on the same bit rate to make communication possible. The interfaces at each end of the channel use frequency dividers to obtain the correct bit rate from their corresponding local clocks.

Due to the drift that usually develops between independent clocks, the length of asynchronous packets is kept short, and the start bit re-synchronizes them. Also, the sampling rate at the receiver end is always higher than the bit rate of the transmitter by a factor $k$, with typical values of 16, 32, or 64.

The number of bits per character in most contemporary channels is seven or eight, although we might find channels that still maintain compatibility with early teletype machines and thus use five- or six-bit characters. Seven-bit characters support only the base ASCII set, while those with eight bits also support extended ASCII symbols. Both ends of the channel must agree on the number of bits per character to be able to communicate.

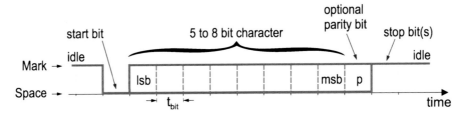

**Fig. 9.7** Format of an asynchronous packet

---

[2] In digital communications jargon, a logic zero is called a *space* and a logic one a *mark*.

### 9.3.3 Error Detection Mechanism

The parity bit field is optional, meaning that communicating devices must agree on wether using it or not. Parity check is an elementary error checking mechanism that allows for detecting single bit flips in a transmitted character. It counts the number of bits set (logic one) in the character an adds an extra bit, called the *parity bit*. The parity bit is set or clear according to the character itself to ensure every transmission will have a pre-determined parity: even or odd.

When the sides of a channel agree on enabling parity check, they must also specify wether parity will be even or odd. In even parity, the parity bit is set or clear to always make even the number of ones in the transmitted character plus parity. In odd parity, the parity bit forces the number of bit in one in the transmitted bitstream to be odd. When a character completes reception, the receiving end strips out the start and stop bits and then calculates the parity of the character plus parity bit. If the check matches the agreed parity, the parity bit is stripped and the character accepted. It the parity does not match, the condition is flagged as a *parity error*.

The last field in the packet, the *stop bit*, marks the end of the packet. Stop bits are always logic high and introduce a forced idle interval between characters. The length of the stop bit might be one or two bit times, depending on the interface, but both sides must agree in the minimum length. Failure to detect the minimum stop bit length is another condition for triggering a framing error.

**Example 9.1 (Asynchronous Bitstreams):** *Consider transmitting the 8-bit character 0B7h with even parity enabled and one stop bit. The bitstream to be transmitted after adding the parity bit would be 111011010 (before attaching the start and stop bits). These leftmost eight bits correspond, from left to right and starting with the least significant bit, to the character to be transmitted; followed by the parity bit in the rightmost position. In this case the parity bit was zero because the character already had even parity. Note that even parity does not mean the character is even. The complete bitstream sent over the channel, including start and stop bits, is illustrated in Fig. 9.8.*

*If instead, character 046h were to be transmitted, the resulting bitstream plus parity would be 011000101 (before start and stop bits). In this case the parity bit is set, forcing the parity of the transmitted character plus parity bit to be even.*

**Fig. 9.8** Transmitting character 0B7h with even parity

Start, parity (if enabled), and stop bits are automatically inserted by the serial interface module in the transmitter side, and also automatically removed by the receiving interface. Similarly, the detection of potential errors like framing of parity are also automatically performed by the adapter. The user software only provides the characters to be sent or processes the received characters. When an error is detected, user software shall decide the proper course of action, if any. An interface module for an asynchronous serial communication channel is called an UART: Universal Asynchronous Receiver and Transmitter. A discussion of UART structure and capabilities is provided in Sect. 9.3.4.

In summary, devices communicating over an asynchronous serial channel must program their UARTs to agree in the baud rate, the number of data bits per character, wether or not parity will be used or not, and the number of stop bits to use when communicating. If parity is used, then the type of parity expected must also be the same in both devices.

### 9.3.4 Asynchronous Communication Interface Modules

In order to have a processor communicating over a serial channel, an interface module is needed. Serial interfaces, or adapters, fundamentally convert data from parallel to serial and viceversa, allowing the CPU to communicate through the serial channel.

In asynchronous serial channels the most widely used interfaces are UARTs. UARTs are available as stand alone peripherals or as embedded modules within MCUs either by themselves or as part of a larger communication peripheral. Common examples of such bundles are USARTS (*Universal Synchronous/Asynchronous Receiver and Transmitter*), which combine in a single module serial interfaces supporting asynchronous and synchronous communications channels.

A USART operates in one mode at a time, either asynchronously or synchronously. In asynchronous mode is just an UART, while their synchronous operation will support a particular protocol, such as SPI, I$^2$C, or other. The next subsections discuss the structure, functionality, and usage of UARTS. A similar discussion for synchronous adapters is included in Sect. 9.4.

### 9.3.5 UART Structure and Functionality

The structure of a UART combines the functionality of a dedicated transmitter and a dedicated receiver into a single module to provide a serial interface with full-duplex capability. Figure 9.9 shows a diagram of a simplified UART module denoting its most important components.

The internal components of a UART are centered around two shift registers: a Parallel-Input Serial Output (PISO) in the transmitter, and a Serial-Input Parallel-Output (SIPO) in the receiver. To perform a transmission, the CPU writes into the

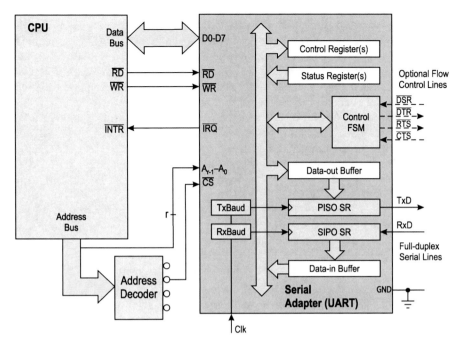

**Fig. 9.9** General organization of a Universal Asynchronous Receiver and Transmitter (UART)

Data-out buffer the characters to be sent. Each character, taken one at time, is then framed with header and footer portions by the UART finite state machine (FSM) to form an outgoing packet and places it into the PISO shift register. In this step, the control FSM calculates and inserts the parity bit (if enabled) as part of the footer when chosen. The contents of the PISO is then transmitted over the serial channel at the rate configured in the transmitter baud rate generator. When all characters in the data-in buffer complete transmission, the condition is flagged by a TxReady flag, denoting that the transmitter portion of the UART is ready to accept a more characters to be sent. Writing the data-out buffer automatically clears TxReady. Simple UARTs have a single character buffer, while more advanced devices can accept multiple characters, reaching even kilobyte capacities.

In the receiving process a valid start bit resets the SIPO and begins loading it with an incoming packet. When all bits have entered the SIPO SR, the FSM strips out the start and stop bits, performs a parity check (if enabled), removes the parity bit (if any), and places the received character in the data-in buffer where the CPU can read it. This event is flagged by a RxReady flag, indicating the receiver is ready with a newly received character. Reading the data-in buffer automatically clears RxReady. If parity or framing errors were detected in the reception process the condition would be flagged.

If during the reception process the data-in buffer were filled faster than the CPU were retrieving the incoming characters, some data could be lost as incoming characters could overwrite previously received data no yet retrieved by the CPU. Such a condition is a reception error known as an *overrun error*. Data-in buffers with multi-character capability help alleviating the situation but the CPU has to promptly remove received characters to avoid the data loss.

The PISO and SIPO shift registers are clocked by the transmitting and receiving clocks, respectively, establishing the transmission and reception data rates. All indicators denoting the transmitter, receiver, and error status are accessible through the designated status registers. The control register(s) allow for configuring all UART parameters defining its operation. The number and functions of control and status registers will depend on the UART capabilities.

Internal control register(s) allow for configuring the diverse options in the module such as character length, enabling/disabling parity, parity type, stop bit length, and baud options. In addition, some of the control bit allow for enabling the transmitter and/or receive portions of the UART and their ability to generate interrupts to the CPU. The specific list of setting options changes depending on the specific unit being configured.

The status register(s) allow for determining specific channel and device status. Typical flags include those indicating error conditions, such as overrun, framing, and parity. The transmitter and receiver statuses are also held here and the pending interrupts. As with control registers, the topology of status registers will vary from one unit to another. Actual examples of such registers are included in the discussion of MSP430 communication interfaces in Sect. 9.5.

### 9.3.6 UART Interface

A UART interface can be seen from three different perspectives: the CPU interface, the clock interface, and the channel interface. The CPU interface becomes relevant when interfacing a UART chip to the CPU address, data, and control buses. The clock interface deals with the way the time base signals used to generate the transmission and reception baud rates are provided. The Channel-side interface connects to the actual serial channel, and becomes relevant in every UART application. The paragraphs below discuss considerations for each of these interfaces.

**CPU-Side Interface**

A UART connects to the CPU buses through a set of bidirectional data lines (typically 8-bit) that transfer data into or out from the adapter, read/write control lines to specify the transfer direction, and selection lines, $A_{r-1}$–A0 and $\overline{CS}$, like any other I/O interface. This model is illustrated in Fig. 9.9. Line $\overline{CS}$ is driven by an address decoder that assigns the module base address. Lines $A_{r-1}$–A0 allow for addressing $2^r$

internal module registers. For example, if the module base address were 0F520h, assuming a byte-wide data bus, the connection of line (A0) would enable internal addresses 0F520h and 0F521h. If lines A0 and A1 were provided, the module would host four consecutive locations starting at its base address.

The CPU side of the UART also includes one or more interrupt request lines that allow for the UART to place service requests to the CPU, enabling interrupt-based servicing of the adapter. Depending on the particular UART used, one or multiple independent $\overline{IRQ}$ lines could be provided. A single $\overline{IRQ}$ would feature a single multi-source vector for Rx, Tx, and possibly error events. Multiple request lines allow for separating individual vectors for different UART events.

When UARTs are embedded within MCUs, the CPU-side interface is hidden inside the chip and the developer only deals with the channel side lines. Clock signals are in most cases internally configured via control registers.

**Clock Interface**

All UART interfaces also include one or more clock input lines that feed the baud rate generators inside the module. These clocks provide the base frequency $f_{clk}$ used by internal frequency dividers TxBaud and RxBaud to establish the transmitter and receiver baud rates. Most MCUs derive this frequency from the system clock generator, although some might allow dedicated external oscillators. The frequency dividers might be integral part of the UART or provided from an on-chip timer. Section 9.3.7 on p. 487 provides additional details about the selection and configuration of the clock source.

**Channel Interface**

The main lines in the channel side of a UART are TxD and RxD, which along with the signal ground (GND) carry the incoming and outgoing serial streams. TxD (Transmitted Data) is the serial output and is driven by the PISO, while RxD (received data) is the serial input, driving the SIPO SR input.

Dedicated UART chips use single-functioned pins for these lines. In most MCUs, however, these lines are shared with GPIO or other functions, making necessary to configure them as UART lines when intended to be used as such.

Some UARTs may also include in their channel side a set of handshaking lines that allow for modem and/or hardware flow control. These include:

DSR Data Set Ready, an input to the UART indicating an external data communication equipment (DCE, originally a modem[3]) is ready to receive data.

---

[3] MODEM = MOdulator-DEModulator: a communication interface that converts digital logic levels into tones and viceversa, allowing for sending them over a telephone line.

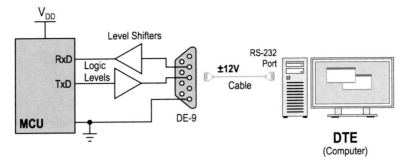

**Fig. 9.10** Level shifting for physical standard compliance: RS-232C

DTR  Data Terminal Ready, an output from the UART (originally designated as a data terminal equipment or DTE) indicating it is ready to operate.

RTS  Request to Send, a UART output indicating to the modem it desires to send data.

CTS  Clear to Send, a UART input from the DCE or modem indicating it will accept new data. CTS is asserted in response to a RTS.

These lines were originally introduced for modem control, and along with the signal ground (GND) and two other lines not pictured in Fig. 9.9 (Ring Indicator (RI) and Data Carrier Detected (DCD)) conformed the signal set designated by the Electronics Industries Association (EIA) for the RS-232C standard. Some dedicated UART chips still in the market, like National Semiconductor's 8250 and 16550 directly support these lines. When provided, these lines can be used for hardware flow control in the channel. MCU embedded UARTs rarely feature dedicated channel handshaking lines. If for some reason these were needed, their functionality can be implemented with GPIO lines under software control.

All signals running into or out from a UART in the channel side must use the same voltage levels as those of the digital logic levels in the UART chip itself, dictated by the supply voltage. For example, a dedicated UART chip operated at 5VDC will send and tolerate TTL compatible signals. A UART embedded in a 3.3VDC MCU, will provide and accept signals in a 0–3.3 V range. Using logic levels directly to transmit and receive serial data will limit the maximum distance signals can travel to only a few inches before being garbled by noise, particularly as baud rates increase. To enable longer transmission distances, different voltage levels and signal formats become necessary.

Physical standards for serial communication use large voltage swings and/or signalization schemes that improve the noise immunity of the signals transmitted over the channel. These levels, are however, not directly compatible with the logic level signals of bare UART chips, requiring level shifters or voltage translators for their interconnection. This situation is exemplified in Fig. 9.10 for the particular case of an RS232 channel. Failing to observe this rule would cause irreversible damage to the UART or MCU chip. Section 9.3.8 on p. 489 discusses the physical requirements of RS-232C and other standards used in serial channels, describing required voltage levels, current limits, and signalization schemes for proper operation.

## 9.3.7  UART Configuration and Operation

Configuring and operating a UART from a software standpoint is a simple task when a systematic approach is followed to configure, enable, and access the UART registers.

### Configuration

Configuring the UART requires the following steps:

- **Choosing the clock source for the baud rate generator(s):** Microcontrollers might have several clock sources to choose from. Although one of them is typically set as default source, it is worth verifying that it will satisfy the channel requirements or else choose one that would result appropriate. This will also establish the input clock frequency $f_{clk}$ to the baud rate generator.

- **Configuring the baud rate generator:** With ($f_{clk}$) known, achieving a given baud rate $BR$ only takes dividing $f_{clk}$ by a factor $N$ determined as:

$$N = \frac{f_{clk}}{BR} \tag{9.1}$$

Simple UARTs frequently use a timer to set the baud rate. In such cases, $N$ would be set through the product of the timer prescaler (if any) and the timer channel terminal count. It is always desirable to choose an $f_{clk}$ value that when divided by $BR$ yields an integer value of $N$. When a non-integer $N$ results, the nearest integer is taken, resulting in an actual baud rate deviated from the desired value. This deviation introduces a *baud rate error* with respect to the other channel end. Small baud rate errors ($\pm 2\%$ to $\pm 4\%$) are usually tolerable, particularly at low baud rates. At higher baud rates, as the bit time becomes shorter, increasing baud rate errors translate into severe reception errors that might render the channel unusable. Baud rate "sweet" $f_{clk}$ values producing integer $N$ with 0% error are listed in Table 6.4 on p. 279.

Modern UARTs include more complex baud rate generators able to handle the fractional part of $N$ while introducing small error. In such units, although the principle for calculating $N$ is the same as above, the particular scheme used to handle the fractional part of $N$ changes with the UART topology.

- **Choosing the correct synchronization mode:** When a UART is contained within a USART, it becomes necessary to configure the corresponding bits in the control register to make the module operate as a UART.

- **Choosing and configuring parity check:** As parity check is optional, both channel sides need to agree on whether or not it will be used. If enabled, both sides must be configured to use the same type of parity (even or odd), and configure the corresponding control bits.

- **Configuring character and stop bit length:** Many UARTs can accommodate different character lengths, as discussed earlier. Both sides must be configured for the same character length. Similarly, UARTs with configurable stop bit lengths should agree on the same length. In some cases this is a soft requirement as even when using more than one stop bit, many UARTs only sample the first stop bit.

### Enabling

Depending on the communication direction and whether operation will be handled via polling or via interrupts, several enabling actions are required. Some UARTs require enabling their transmitter and/or receiver portions individually to be operational, some enable the entire UART or USART. If interrupt servicing were desired, the transmitter and/or receiver interrupt enable bits must be set. For simplex communication, only the corresponding Tx or Rx is enabled, while for duplex operation both must be enabled. Moreover, if error conditions were to trigger interrupts in the UART operation, their enable bit(s) shall also be set. Many UARTs use multi-source shared vectors in their operation, requiring properly designed ISRs capable of identifying the trigger source.

### Operation

The actual operation of the channel is fairly simple. When polled operation is being used (although discouraged), the user software must always poll the readiness of the transmitter and receiver flags before any operation in the UART data buffers. Before writing a new character for transmission in the data-out buffer, the TxReady flag must be polled to determine readiness. Otherwise data loss might occur.

When receiving by polling, the RxReady flag must be polled before retrieving incoming data from the data-in buffer to avoid repeated retrieval of the same data. After each character is received, error flags must be checked to determine if an error was detected in the retrieved character. The user software decides the action to take place in the case of detecting an error.

When interrupt-based operation is enabled, the software is even simpler. Activation of the receiver interrupt means a new character was received and therefore the ISR can directly proceed with its retrieval, polling the error flags if they do not trigger interrupts by themselves. In an analogous way, a transmitter IRQ signals the

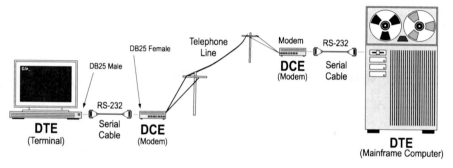

**Fig. 9.11** RS-232 DTE-DCE connections used in early serial data communication channels

readiness of the transmitter to accept new characters to be sent through the channel and the ISR can directly proceed to write into the data-out buffer.

### 9.3.8 Physical Standards for Serial Channels

The physical-level standard for a serial channel establishes the signal definition for carrying bits from one end to another. In wired channels it establishes voltage levels, signalization formats, connectors, and line definitions and notations. In wireless channels, the physical standard defines carriers, frequencies, modulation schemes, and signal strength, among the most important parameters.

#### RS-232

For long time, the most used serial standard in computers and peripherals was the RS-232. The standard was originally developed to specify the signal voltages, timing, handshake protocol, and mechanical connectors between terminals and modems. In those days, computer communication was fundamentally used to connect terminals and peripherals to remotely located mainframe computers via modems over telephone lines, as illustrated in Fig. 9.11. As such, the RS-232 standard assumed connections would mostly happen between a Data Terminal Equipment (DTE) (computer or terminal) and a Data Communication Equipment (DCE) (modems, printers, or other serial devices).

The original RS-232 specification called for a 25-pin male connector (DB25) on the DTE and a female DB25 connector on the DCE with signals at $\pm 25$ V. Due to space and cost savings manufacturers began using a 9-pin connector instead, which omitted the least frequently used signals. By the time of the advent of the RS-232C revision, which reduced the voltage requirement to $\pm 12$ V, the most widely used serial connector had only nine pins. Figure 9.12 shows the organization of signals in

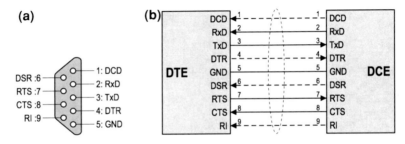

**Fig. 9.12**  Organization of RS-232C signals in a DE9 connector and serial cable. **a** DE-9 Connector **b** Standard serial cable

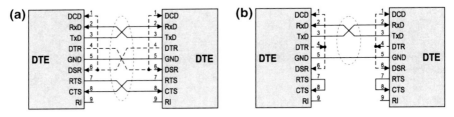

**Fig. 9.13**  Null-modem connection with and without handshaking. **a** Null-modem cable, **b** null-modem without handshaking

a DE9 RS-232C connector and how a standard DCE-DTE modem cable is wired. Observe that in a standard modem connection the inputs on the DTE are outputs on the DCE and viceversa.

This arrangement of signal, although convenient for DTC-DCE connection, becomes awkward for a short DTE to DTE connection, where a modem is not required. A slight modification of the ordering in the lines allowed getting rid of the modem, yielding a null-modem serial connection. Exchanging TxD and RxD, and RTS/CTS allowed for bypassing the modem. Supporting the full handshake protocol included also exchanging DTR and DSR, as illustrated by the dashed lines in Fig. 9.13a. An even simpler connection was enabled by completely bypassing the handshaking protocol, as shown in Fig. 9.13b. As the handshaking protocol became less prevalent, the most common configuration became that including only three wires: TxD, RxR, and GND.

This is the mostly used in embedded applications and it is still the format used in most serial asynchronous interfaces. In such cases, the microcontroller and peripheral both play the role of DTEs and no handshaking signal are provided at all. When a peripheral does require flow control lines (typically RTS/CTS), their functionality in the MCU side can be provided using GPIO lines under user software control.

When a connection is made through a compliant RS232 channel or port as was illustrated in Fig. 9.10 on p. 486, level shifters become mandatory to provide for voltage compatibility. One of the most widely used level shifters for RS232 channels was introduced by Maxim Integrated, the MAX232. This chip features internal

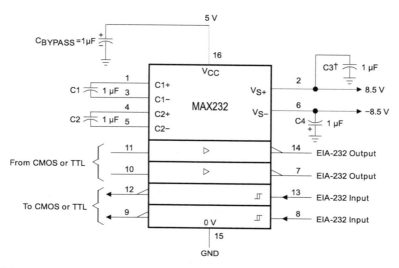

**Fig. 9.14** Typical connection of a MAX232 driver/receiver (*Courtesy of Texas Instruments, Inc.*)

charge pumps that allow for producing RS232C-compatible voltage levels without the usage of a $\pm12$ V bipolar power supply. The original chip used a single 5.0 V TTL-compatible supply, but later versions like the MAX13223E also offer 3.3 V and 3.0 V operation and low-power modes. Several manufacturers nowadays produce MAX232 compatible chips.

The MAX232 provides two pairs of buffers, a first pair supporting TTL to RS232 conversion, and the second pair supporting RS232 to TTL conversion. This arrangement allows for driving and receiving RS232C levels in a full-duplex channel with basic handshaking signals RTS/CTS within a single IC. External component requirement included only five capacitors of typically 1 $\mu$F. Figure 9.14 shows the recommended connection of the chip in a variant produced by Texas Instruments.

Since its introduction in 1962 by the EIA, RS232 has gone through multiple revisions (A through F), being the most prevalent revision C. RS232 dominated the serial interfaces in computers and peripherals for more than three decades, and it is still found in many computer applications. However, its inherent limitations that include support for only point-to-point connections, narrow bandwidth (up to 20 Kbps at 50 ft), and inconsistencies requiring deviation from the standard, gave way to newer and improved specifications. At present, new designs have migrated to standards like FireWire and the Universal Serial Bus (USB), being the later the predominant standard in contemporary serial interfaces. USB is a more robust form of serial communications that includes a full-fledged multi-drop protocol and allows for faster transfer rates than were attainable with the aging RS232 standard. Section 9.3.9 on p. 494 offers additional details on the USB standard.

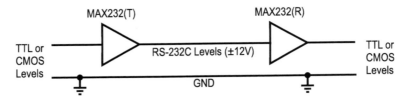

**Fig. 9.15**  Single-ended RS232C line

**RS-422 and RS-485**

Single-ended serial links like those used in RS232 channels, offer a cost effective way of carrying signals over wires, since they require only a single wire per signal plus a common ground. This topology is illustrated in Fig. 9.15. These connections however have a limited ability to fight noise. The main robustness factor they feature is their noise margin. RS232C for example uses up to $\pm15\,$V in the transmitter side, while the receiver needs only $\pm3\,$V to discriminate between marks and spaces (1's and 0's), yielding up to 12 V of noise margin. This large noise margin is however ineffective for rejecting noise coupled into the line by the unavoidable voltage loss caused by the line impedance and electromagnetic interference (EMI) picked-up by wires when passing through magnetic fields.

A differential channel overcomes this weakness by providing two wires per signal and representing signals by the relative voltage of one wire with respect to the other. This signalization scheme allows for cancelling any form of noise coupled in common mode into the lines. Noise caused by EMI, balanced voltage drops, and ground bounce are naturally eliminated in a differential channel due to its high common mode rejection ratio.

These characteristics also allow for using narrower voltage swings, which combined with the improved noise immunity enables higher transmission rates over longer distances. A drawback of differential channels is their cost, space, and routing complexity due to the requirement of using two wires per signal. Despite these factors, differential channels are the top choice for high-speed serial communications channels.

Two widely accepted physical standards based on differential connections are RS-422 and RS-485.

**RS-422:** This is a differential data transmission standard that offers robust mechanisms for communications over noisy environment. Officially designated as ANSI TIA/EIA-422, RS-422 uses a shielded, differential twisted pair with $100\,\Omega$ characteristic impedance and $16\,$pF/ft distributed capacitance. The signal swing can be between $\pm2\,$V and $\pm5\,$V with respect to ground and up to $\pm10\,$V from line to line, with a maximum short-circuit current output in its line drivers of $150\,$mA. Despite this seemingly narrow swing, when compared to RS232, RS422 can achieve data rates up to $10\,$Mbps at $50\,$ft or less, and up to $100\,$Kbps at a distance of $4,000\,$ft, a

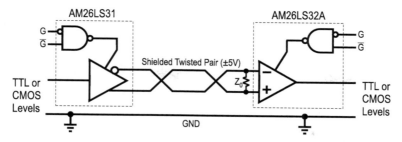

**Fig. 9.16** Differential RS-422 line with AM26LS31/AM26LS32A driver/ receiver pair

whopping improvement with respect to that of RS-232. Figure 9.16 shows a typical RS-422 differential line.

RS-422 is designated as a *simplex multidrop* standard implying that a single RS-422 line can have only one driver but up to ten receivers without the need of repeaters or buffers. Its interconnections are ended by a termination impedance $Z_0$ that matches the characteristics impedance of the line in about $\pm 20\%$ to suppress far-end reflections in the line. Traditional drivers and receivers for RS-422 include the driver/receiver pair AM26LS31/AM26LS32A by Texas Instruments. This matched pair allows for enabling/disabling all chip buffers with a single control signal (G/$\overline{G}$), as illustrated above in Fig. 9.16, providing a more flexible control of the channel. RS-422 buffers can also be used as custom replacements for RS-232 line drivers, allowing to extend speed and distance in the channel. Such a change, although effective, requires additional logic to connect with standard RS-232 ports.

**RS-485:** This standard offers improved characteristics over RS-422 by specifying bidirectional drivers that allow for operating a single differential serial line as a half-duplex bus. This implies that the channel can be written by more than one driver, although only one bus driver can be active at a time. The standard, officially designated TIA/EIA-485, also calls for lines with lower characteristic impedance (54 $\Omega$) and receivers with higher input impedance (120 K$\Omega$) than RS-422, increasing the driver's fan-out. This change allows for accommodating 32 drivers and 32 receivers in a single bus topology and extending to longer line lengths. The remaining electrical and timing characteristics are similar to those of the RS-422 standard.

Figure 9.17 shows an RS-485 bus topology featuring three bidirectional, half-duplex nodes connected via RS-485 compatible transceivers. The shown transceiver is an SN65HVD10 driver/receiver by Texas Instruments. This transceiver features one driver and one receiver per IC, the latter with 1/8 unit load, extending the maximum number of nodes in the channel to 256. Independent, complementary driver and receiver enable lines $\overline{RE}$ and DE allow for individual or complementary direction control.

Since RS-485 lines are bidirectional, termination impedances ($Z_0$) are required at all ends of the line to suppress reflections.

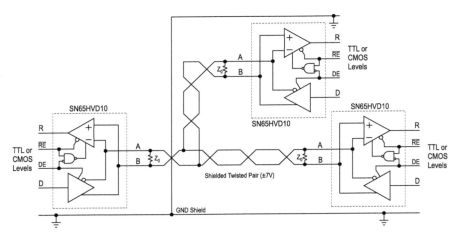

**Fig. 9.17** Differential RS-485 line denoting its multidrop capability

Electrical RS-485 signals are compatible with those of RS-422, allowing the former to be used on RS-422 channels. Although RS-422 and RS-485 offer significant improvements over RS-232, they never reached the widespread usage level of RS-232.

### 9.3.9  The Universal Serial Bus (USB)

The Universal Seral Bus, or USB, was introduced in the mid nineties as a standard to provide a simple, yet consistent, fast, and robust way to communicate peripherals such as printers, scanners, keyboards, media players, digital cameras, and other similar devices to a host computer. USB was also designed to provide power to low-power peripherals, freeing each of them from requiring individual power supplies.

The acceptance gained by the protocol allowed it not only to displace power chargers and earlier computer communication standards like RS232, PS2, and Centronics. It also penetrated to the broad market of embedded applications providing an effective and reliable standard communicating and powering smart phones, industrial equipment, medical, military, and even space applications. USB has been the most successful computer standard in the history of computers.

In its initial 1996 conception, USB 1.0 supported both low- and full-speed data transfer rates of 1.5 Mbps and 12 Mbps, respectively, in half-duplex mode. USB 2.0, released in year 2000, increased the bandwidth with an additional high-speed data rate of up to 480 Mbps. The latest USB generation, USB 3.0 released in 2008, boosts transfer rates with a Super-Speed mode reaching up to 5 Gbps with the ability of supporting full-duplex communication. Every newer version is backward compatible, allowing to support earlier generation devices.

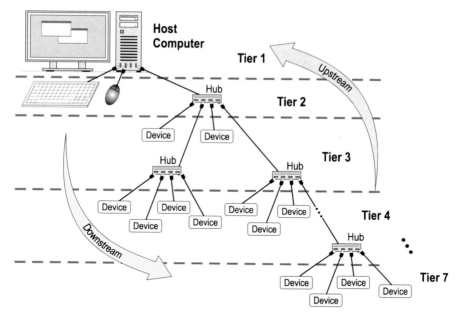

**Fig. 9.18** Tiered structure of a Universal Serial Bus

## USB Architecture

Unlike RS232 and RS485, USB is more than just a standard for cables, voltages, and connectors. The Universal Serial Bus is a complete communication protocol that includes physical, link, session, and application layers enabling host-supported functionality with a low-cost and consistent specification.

The architecture of a USB includes three types of elements: Host, devices, and cables. In its original version a USB topology can feature only a single host computer (master) and up to 127 devices in a tiered star configuration connected by USB cables, as illustrated in Fig. 9.18.

The USB host, although originally thought to be a personal computer, can be any kind of computing platform fitted with a USB host controller. The host controller is responsible for most of the functions that control the communication in the serial bus. It includes hardware and software layers that take care of the functionality supporting hot swapping operations, device enumeration, session and flow control, power management, and communicating with the host CPU application. All communications in the bus can only be initiated by the host, which acts as bus master.

USB devices, identified in the standard as "*functions*", are the particular peripherals connected to the bus. From the point of view of the host, devices behave like slaves. A USB device functionality will depend on the type of peripheral they support, be it a keyboard, a mouse, a digital camera, a flash drive or any other kind.

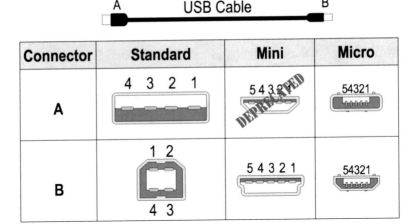

**Fig. 9.19**  Connectors in USB 2.0 cables

One special type of device is a USB hub: A device featuring multiple USB ports that allow for connecting additional devices or hubs to the bus. The host controller includes a hub by itself, creating the first tier in the bus. Each subsequent external hub located downstream in the bus establishes a new tier supporting multiple USB ports. A hub typically provides four ports, although a larger number is possible. A single bus can have a maximum of 127 devices in at most, seven tiers.

**Electrical and Physical Characteristics**

USB cables provide the physical link connecting devices in each tier to either the host or to the corresponding hub. Each USB cable can be up to 5 m (12.7 ft) long, implying the entire bus can extend to a theoretical length of 35 m (90 ft).

The form of a USB cable depends on the standard. A USB 2.0 or lower carries four shielded wires: two for power ($V_{DD}$ (+5 V) and GND) and two forming a differential twisted pair for carrying data(D+ and D−). The cable has two different connectors: connector "A" for the upstream side and connector "B" for the downstream side. The shape and size of the connector includes three sizes: standard, mini, and micro. Figure 9.19 shows the forms of connectors "A" and "B" for the different connector sizes. Table 9.1 lists the signal assignment for each connector type.

The most commonly used USB cables feature a standard connector A for the upstream and a Mini or Micro B connector for the downstream. This arrangement provides for a standard size for the PC/hub side and a small footprint on the device side, highly convenient for miniature devices like smart phones and cameras. This usage pattern has become so prevalent that the Mini-A (also known as Mini-AB) connectors have been removed from the standard.

**Table 9.1** Signal assignment in USB 2.0 connectors

| Pin No. | Standard | Mini | Micro |
|---|---|---|---|
| 1 | $V_{Bus} = 5\,\text{V}$ | $V_{Bus} = 5\,\text{V}$ | $V_{Bus} = 5\,\text{V}$ |
| 2 | D− | D− | D− |
| 3 | D+ | D+ | D+ |
| 4 | GND | NC | NC |
| 5 | N/A | GND | GND |

USB signals are encoded using NRZI (non return to zero inverted) with a unipolar differential signaling driver. A logic one is transmitted by pulling up D+ $\geq$ 2.8 V via a 15 K$\Omega$ resistor, and pulling down D− $\leq$ 0.3 V via a 1.5 K$\Omega$ resistor. A logic zero is obtained by swapping the voltages on D+ and D− with their respective pulling resistors. Connections to drivers and receivers do not require termination resistors.

USB 3.0 includes a second differential link in its cable and connector, allowing for full-duplex communication. The connectors for USB 3.0 are downward compatible with earlier versions, implying that they accept USB 1.1 and 2.0 devices, although a USB 3.0 function will not necessarily work in a USB 2.0 port.

The power bus ($V_{Bus}$–GND) allows powering low-power devices directly from the USB port. It provides a nominal voltage of 5 V $\pm$ 5 %. The current limit per unit load in ports up to USB 2.0 is 100 mA, with a maximum total of five unit loads (500 mA). In USB 3.0 the unit load current is 150 mA and allows a maximum of six unit loads (900 mA).

When a function is plugged into a USB port, it is first recognized as a low-power device and granted a current limit of one unit load (100 mA in USB 2.0). After an enumeration transaction, the device can ask to be granted a high-power status, which if granted, allows it to draw the maximum allowed current by the version (500 mA in USB 2.0 or 900 mA in USB 3.0). Current consumption by functions (devices) in the bus is monitored and if any of them exceeds the granted limit, the infracting function will be disabled.

## USB Interfacing and Programming

Interfacing and programming for USB devices can be seen from two different perspectives: from the perspective of a function developer creating from scratch interfaces that comply with the USB standard, or from the perspective of the application developer that seeks to use COTS parts to enable embedded applications that communicate over an existing USB infrastructure.

For those in the first group, that might need to possibly develop silicon for USB hardware and software compliance, the best resource would be the complete USB specification published by the USB Implementer's Forum [71, 72]. These documents provide all the details necessary for developing from scratch applications that meet the physical, electrical, and functional specifications of the USB standard.

For those in the second group, seeking to develop applications with COTS parts, capable of communicating over an existing USB infrastructure without having to implement the protocol from scratch, there are a number of resources that allow for developing embedded applications able to comply with USB function requirements and communicate with a host. Most of these solutions are based on ICs that implement the USB protocol and provide simple interfaces that can be directly handled by an MCU without the burden of having to develop an entire USB compliant function protocol.

The pool of available options includes ICs able of converting UART, $I^2C$, or SPI channels to USB, enabling easy integration of custom developed applications as USB functions. Specific examples include FTDI Chip's FT201X, FT220X, and FT231X, able to interface USB to $I^2C$, SPI, and UART, respectively. Many of these ICs are available conveniently packaged within cables or drop-in modules that can be directly driven from MCU ports, providing transparent translation between formats.

## 9.4 Synchronous Serial Communication

Synchronous serial channels are characterized by having both, transmitting and receiving devices synchronized with the same clock signal. This is achieved by having both, data and clock transmitted over the channel. Figure 9.20 shows a synchronous full-duplex channel configuration illustrating the concept in a point-to-point connection.

Synchronous serial channels usually operate in a master/slave mode where the master device initiates transfers and provides the clock signal driving the timing and synchronization in the channel. A slave device is controlled by the master to either receive or send information when instructed to do so.

Since both master and slave devices are driven with the same clock signal, datagrams in synchronous channels can be longer than those used asynchronous channels. Also, by using suitable line drivers, many synchronous channels allow multidrop configurations enabling the establishment of a network of devices communicating over the same channel.

Several synchronous protocols have been developed, each one establishing its own set of rules of how data and clock signals are routed through the channel and

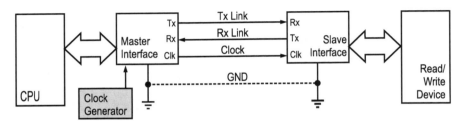

**Fig. 9.20** A synchronous, full-duplex serial channel denoting the transmitted clock

how interconnecting devices behave. Some protocols establish a single master device per channel with the rest behaving as dedicated slaves, while others allow devices to behave as either master or slave. In the latter case, specific protocol rules are introduced to ensure that only one device in the channel behaves as master at a given time.

Examples of synchronous serial protocols include the Serial Peripheral Interface (SPI), the Inter-Integrated Circuit (IIC or $I^2C$), and the Controller Area Network (CAN) among others. Several of these protocols are explained in the sections below.

## 9.4.1 The Serial Peripheral Interface Bus: SPI

The Serial Peripheral Interface (SPI) is a synchronous serial bus standard with full-duplex capability introduced by Motorola to support communications between a master host processor and one or multiple slave peripheral devices.

SPI is one of the simplest synchronous communications protocols ever developed, as it only establishes a basic mechanism to relay packets between a dedicated master and one or more slaves, without specifying any data, session, or higher-level protocol. This simplicity translates into a protocol with little overhead, capable of achieving a high efficiency in the channel usage, particularly in point-to-point connections.

An SPI controller is developed around a single shift register that serves as both, receiver and transmitter, synchronized by the transmission clock. Figure 9.21 illustrates the structure of an SPI master-slave pair implementing a point-to point channel. The CPU side of the interface connects to the data lines D0-D7 and the selection and control lines like any other interface, omitted in the figure for simplicity,

The channel side features four signals whose functions are:

- SCLK: Serial clock, sent by the master and synchronizing both master and slave.

- SDO: Serial data-out, the serial output stream from the device.

- SDI: Serial data-in, the serial input stream into the device.

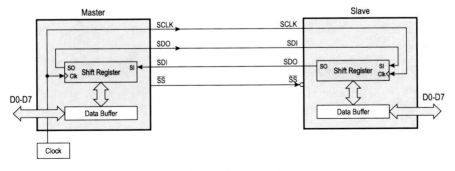

**Fig. 9.21** SPI synchronous bus for a point-to-point connection

**Fig. 9.22** SPI single-master, multi-slave connection

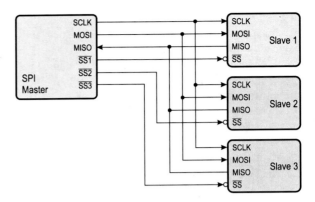

- SS: Slave select, a selection line to enable the slave. The SS line is omitted in point-to-point interconnects, grounding the slave SS input.

A character sent by the CPU is stored in the interface data buffer and copied into the shift register. The master initiates the transfer by activating the slave select signal. In a point-to point, the slave select signal can be hardwired, allowing a 3-wire connection with the master.

Since the master and slave shift registers are tied together and synchronized by the same clock, both interfaces simultaneously transmit and receive. As bits are shifted out from the master they are also shifted into the slave and viceversa. A device doing a transmit-only transfer simply discards the received frame.

Since an SPI does not specify an address field, additional hardware is required for managing multiple slaves. The slave selection lines can be implemented with GPIO lines or an external decoder and user software must activate the corresponding slave. Figure 9.22 illustrates a single master, multi-slave SPI configuration.

The SPI clock signal can be derived from the system clock, from an external clock source, or from a timer channel, depending on the particular implementation. Interrupt-based servicing is possible in most MCUs. The specific details will depend on the manufacturer's design.

Like other serial formats, SPI transfers using the logic levels of the interface is limited to only a few inches. By adding appropriate drivers and receivers, SPI channels could be extended over longer distances, although it is mostly used for short intra-board distances. SPI has no error checking mechanism and is not defined as an standard.

Due to its simplicity, SPI has been adopted by numerous manufacturers of serial EEPROMs, real-time clock modules, data converters, LCDs, and other peripherals. Due to the absence of upper-level protocols and no standardization, each implementation can define its own particularities, and upper-level protocols are left to the application implementer.

## 9.4.2 The Inter-Integrated Circuit Bus: $I^2C$

The Inter-Integrated Circuit bus, or $I^2C$ is a synchronous serial protocol developed by Philips Semiconductor (now NXP Semiconductors) in the early 1980s to support board-level interconnection of IC modules and peripherals. The protocol uses two lines, SDA (Serial Data) and SCL (Serial Clock), (and ground) to establish a half-duplex, master/slave, multidrop bus capable of handling multiple masters and slaves. The serial clock line (SCL) synchronizes all bus transfers, while SDA carries the data being transferred.

$I^2C$ was designed to be an intra-system serial bus capable of accommodating all kind of peripherals found in embedded systems. These include MCUs, data converters (ADCs and DACs), display devices, memories, real-time clock calendars, GPIO modules, etc. A large number of IC peripheral manufacturers offer products compatible with $I^2C$.

Devices in an $I^2C$ bus are software addressable, with 7- or 10-bit address fields. Although the most common usage establishes a single master/multiple slave topology, the protocol allows for any device to be a master, as it incorporates collision detection and arbitration mechanisms necessary for a multi-master operation. Nominal maximum speeds can reach up to 5 Mbps, although in reality the limit is imposed by the total bus capacitance, with its maximum specified at 400 pF or 500 pF depending on the version. With input capacitances at 10 pF, plus that of cables, practical numbers call for a few dozen devices per bus. Bus extenders might be used to expand that number if the application requires doing so.

Being a bus for interconnecting ICs in a board, distances in $I^2C$ are expected to be short, in the order of a few inches. It might be possible to stretch its length to several feet, at the expense of reduced speed due to the effect of the increased capacitance.

The bidirectional multidrop capability of the SCL and SDA lines is achieved by driving them with open-collector or open-drain drivers. This calls for fitting bus lines pull-up resistors to complete their driver circuit. Figure 9.23 shows an $I^2C$ bus topology featuring an MCU as master and several other devices as slaves.

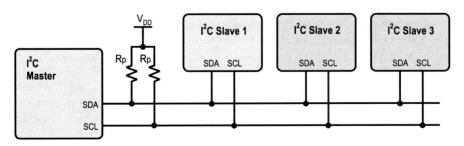

**Fig. 9.23** Topology of an $I^2C$ bus connection

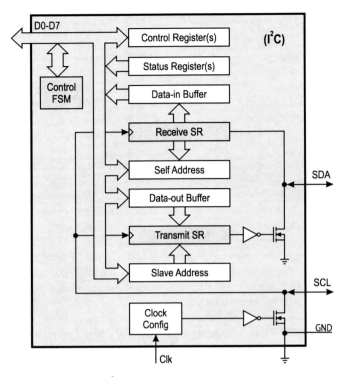

**Fig. 9.24**  Internal structure of an I²C interface

## I²C Interface Structure

Each device connecting to an I²C bus requires an interface able to produce the signalization specified by the protocol. An I²C interface uses two shift registers, one for receiving, one for transmitting, joined by a transceiver that allows for bidirectional operation in the SDA line. Figure 9.24 illustrates the internal structure of an I²C interface. Note how the SDA transceiver is implemented with a single NMOS and an inverter. A similar approach is used for SCL as this line must allow for receiving the clock signal if the device operates as a slave, or driving the clock signal when operated as master.

The transmitter section shift register can be loaded from either the slave address register with the address of the destination slave device, or from the data-out buffer with the character to be transmitted. The receiving shift register transfers incoming packets to either an address comparator to determine packet ownership, or to the data-in buffer for CPU access and when matched.

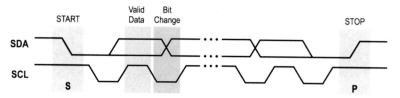

**Fig. 9.25** Timing diagram of $I^2C$ signals denoting the start, transfer, and stop conditions

## $I^2C$ Operation

The protocol in an $I^2C$ specifies three different conditions in the channel: *Start*, *transfer*, and *Stop*. A start condition (S) occurs when a master takes control of the bus to initiate a transfer. Before taking any action, the master listens to the SDA line to determine if the channel is idle. An idle status is indicated when both SCL and SDA remain high for at least one clock period. Upon channel availability, the master lowers the SDA line to set the start condition, making the channel busy. At this point all devices in the bus will be placed in listen mode for the incoming data. A start condition can also be sent in the middle of an active message. This action is called a *Restart* condition and can be used to change the transfer direction within a message without the current master actually releasing the bus.

A data transfer occurs when data bits are being sent over the channel. $I^2C$ supports eight-bit characters transfers. Character bits are transferred with the most significant bit (msb) first. A data bit is valid when the clock (SCL) is high, and the change from one bit to the next occurs when the clock is low. A transfer might contain one or multiple characters.

When a transfer is complete, the master issues a stop condition (P) by raising SDA while keeping SCL in high. Figure 9.25 illustrates the timing relations of SDA and SCL for all three conditions.

A message sent over an $I^2C$ bus begins right after a start condition with an address field sent by the master to select the destination slave device. The address field is ended by a read/write bit indicating whether the transfer would read or write the addressed slave. After this last bit the master releases the SDA line.

All slaves in the bus will receive the address field and compare it to their own address. Only the slave whose address matches that sent by the master will send an acknowledgment. The selected slave acknowledges the reception of its address by pulsing the SDA line.

The master issues a pulse in the SCL line to sample the SDA line and detect if the ACK condition is present. If the acknowledgment is received, the transmitter takes control of the SDA line and proceeds sending the next data byte, msb first. Note that in case of a write transfer, the master would be the transmitter and thus it would control both SDA and SCL. In a read transfer the slave would be the transmitter and thus it will control SDA while SCL remains controlled by the master.

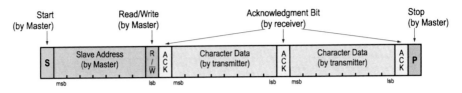

**Fig. 9.26**  Structure of an I²C message

If after sending the address field the master does not detect the SDA line going low to denote an acknowledgment, this would lead to a "Not Acknowledgment" condition (NACK). User software shall determine wether aborting the transmission (by sending a stop) or retrying the transfer.

The message following a successful address transfer might contain one or several bytes, each followed by an acknowledgment field. If after a byte transfer the receiver were not yet ready for the next character it would hold low SCL, serving as an indication to the master to wait. Next the transmitter sends each character in its message, each followed by an acknowledgement by the receiver. Note that after receiving the address field, the roles of transmitter and receiver will depend on the read/write bit.

There is no preestablished limit in the number of 8-bit characters that can be accommodated in a message. When the transmitter completes its message, the master issues a stop condition, setting the channel idle. If necessary, the master might issue a repeated start condition at the end of a message, maintaining control of the bus. Figure 9.26 illustrates an I²C message, denoting the order in which each field appears in the packet.

Besides the basic addressing and data transfer, I²C also provides clock synchronization mechanisms, clock stretching, and arbitration for contending masters.

## Clock Synchronization and Arbitration in the I²C Bus

Clock synchronization and arbitration are two mechanisms used in multi-master bus configurations to decide which master will take control of the bus in the event that more than one master begin transmitting on an idle channel. Arbitration relies on the synchronized clock operation of masters.

**Clock Synchronization:** Clock synchronization allows for synchronizing the clock signals of two or more masters that might be simultaneously operating at different speeds in the same bus. Consider two I²C master devices, namely *master 1* and *master 2*, attempting to simultaneously transfer information on the bus, as illustrated in Fig. 9.27. They both will attempt to drive the common clock line SCL. Let's call the master's clock line interfaces CLK1 and CLK2, respectively. Without loss of generality, lets assume CLK1 runs faster than CLK2.

When a high-to-low transition occurs in the SCL line, both devices will begin counting their low periods. As CLK1 runs faster than CLK2, *master 1* would disable

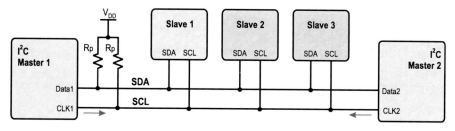

**Fig. 9.27** Multimaster I²C bus configuration

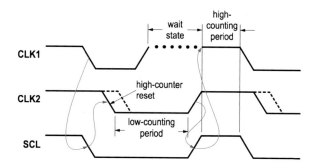

**Fig. 9.28** Multimaster signal timing for a clock synchronization process

its clock driver first trying to bring the SCL line high. However, as CLK2 is slower, *master 2* would still keep the SCL line low, as illustrated in Fig. 9.28. Recall that SCL is an open-collector or open-drain line that operates as a wired-AND.

When *master 1* senses that despite having released its clock drive, SCL is still low, it will enter into a *wait state* until detecting that SCL has actually gone high. When *master 2* finally completes it low period counting it would release its clock driver allowing SCL to finally go high. At this point both devices start counting their high periods.

In the high period, again, *master 1* will end first its high-period count, driving SCL low. At this point the slower master (*master 2*) would just reset its high-period counter and drive SCL low to begin counting its low period. This mechanism allows for synchronizing the clocks of all masters: the master with the slowest clock determines the length of the low period, while the fastest master sets the length of the high period. This synchronization mechanism is essential to enable an efficient I²C bus arbitration protocol.

**I²C Bus Arbitration:** An arbitration process takes place when multiple masters attempt to simultaneously seize control of the bus. The arbitration process determines which master will remain in control. The protocol is very simple: the fist master to place a logic high level conflicting with a low-level data placed by another master losses control of the bus.

To understand this rule it only takes to remember that SDA works as a wired-AND line. Thus, when two drivers place conflicting data, the SDA line will assume a low

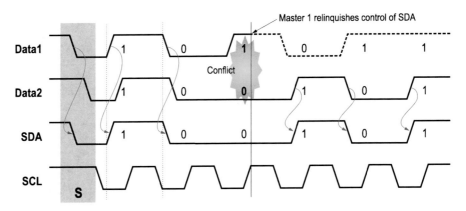

**Fig. 9.29** Timing of a bus arbitration process

level. Every time a master drives the SDA line placing a data bit on it, the master will also listen to the line to determine if it actually took the issued logic level. A master who places a logic-one level on SDA and listens a zero identifies that there is a conflict. As a consequence, the conflicting master relinquishes control of SDA line and enters into listening mode, like a slave, in case the other master were trying to address it. Otherwise it keeps waiting for a stop sequence to make a new attempt to send its data. Figure 9.29 illustrates a signal sequence exemplifying this type of transaction.

In this example Data1 and Data2 represent the data drivers in Master1 and Master2, respectively. At the beginning of the timing sequence both masters detect an idle channel, so both of them issue start conditions and begin driving the SDA line. As long as the placed data does no conflict, both masters continue to drive the SDA line. Recall that their SCL lines are synchronized. In the first conflict, the master placing a logic-high level, Master1 in this example, will not see its data on SDA and thus proceeds to relinquish control of the bus.

It deserves to mention that these features are transparent to the application developer when the MCU at hand has an I$^2$C peripheral module. The only instance when the programmer needs to consider such protocols is when the task at hand involves implementing an I$^2$C interface in software. This situation could arise when interfacing an MCU without I$^2$C hardware support to the bus, a not so common situation due to widespread availability of MCUs supporting the protocol. In such a case, a good source for detailed information is the official I$^2$C-bus specification published by Philips [74].

# 9.5 Serial Interfaces in MSP430 MCUs

MSP430 devices provide several serial on-chip communication resources that include UARTs, SPI, I$^2$C, and USB. These are made available through different modules that include:

**USARTs**   Universal Synchronous/Asynchronous Receiver/Transmitter modules, which provide basic UART and SPI connectivity.

**SCI**   Serial Communication Interface modules that provide fundamental SPI and I$^2$C capabilities.

**USCI**   Universal Serial Communication Interface, with wide support for UART, SPI, and I$^2$C modes. Series x5xx/x6xx feature enhanced USCIs as well.

**USB Modules**   Selected x5xx/x6xx devices feature Universal Serial Bus modules.

Not all resources are available in all families or all devices of a family. The datasheets of each specific device provides a comprehensive list of the serial capabilities they provide. Note that depending on the particular device, several types of serial adapters might coexist in the same MCU, including multiple instances of the same type. The descriptions in this section focus on the capabilities of each type of serial communication resource available for MSP430 devices.

## 9.5.1 The MSP430 USART

The Universal Synchronous/Asynchronous Receiver/Transmitter (USART) is one of the earliest serial peripheral interfaces supported by MSP430 devices. Featured in devices series x3xx, x4xx, x1xx, and x2xx, the USART is one of the most prevalent and classical serial interfaces in the MSP430.

The USART architecture in MSP430 devices has experimented little changes across generations. In all generations it supports both UART and SPI modes with the same peripheral module. This implies that a single on-chip module can be used as either one or the other. If both types of serial peripherals were required for a particular application, more than one module would be required. In addition to UART and SPI modes, the USART in generation x1xx devices also supported I$^2$C communication.

MSP430 series x5xx/x6xx devices discontinued supporting the classical USART, to stick only to the more recent Universal Serial Communication Interface (USCI) and its Enhanced version (eUSCI), discussed in Sect. 9.5.3.

### The MSP430 USART Module in UART Mode

In its UART mode, the MSP430 USART module behaves like a classical asynchronous interface with a few enhancements. Like UART configurations discussed earlier, in Sect. 9.3.4, this module supports 7- or 8-bit characters, full-duplex transfers, and

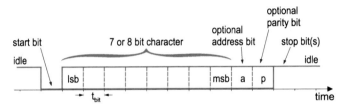

**Fig. 9.30** Structure of asynchronous packets in MSP430 UART

interrupt triggering capability for receive and transmit events. Particular enhancements include support for idle-line and address-bit protocols, and the inclusion of a baud rate generator capable of accommodating fractional rates with reduced channel error. The receiver and transmitter sections share the same baud rate generator module.

**UART Configuration and Operation:** The UART mode is selected by clearing the SYNC bit in the USART Control Register (UxCTL). This register also allows for configuring most UART parameters such as: parity enable/disable and type, number of stop bits, character length, and multiprocessor mode.

The USART is reset by setting the SWRST bit in the UxCTL. The SWRST bit is automatically set by a PUC, so by default the USART is disabled. The USART initialization and enabling must be completed while holding the module in its reset mode. If interrupt operation were desired, the corresponding interrupt enable bits must also be set, recalling that transmitter and receiver have independent enables.

The character format follows the general organization of an asynchronous packet, except that it might include an optional address bit after the most significant data bit. This address bit is inserted when the UART is configured in the multiprocessor mode, allowing for specifying wether the transmitted packet is an address or data field. Figure 9.30 shows the modified structure of an asynchronous packet.

**Multiprocessor Mode (MM):** When used in conventional point-to-point connections, the UART is configured in the idle-line format. This is the traditional form of an asynchronous point-to-point connection where each character in a message is individually framed and separated from the next by an idle channel period. The bit time is determined by the chosen baud rate.

The MSP430 UART also supports having three or more devices on the bus through either the idle-line or address-bit multiprocessor formats.

In the idle-line multiprocessor mode (MM = 0), characters within a message are also individually framed, but the idle time between characters is used to indicate the separation between characters and messages. A short idle time (less than ten-bit time) separates characters within a message and an idle-line interval of ten-bit times or longer separates messages. No address bit is inserted in this mode. Instead, the first character received after a valid idle-line interval is accepted as an address field.

When an address field is received, it is transferred to the data-in buffer and wakes-up the receiver. If interrupts were enabled, a receiver interrupt would be triggered.

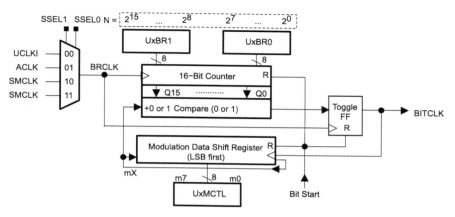

**Fig. 9.31**  Structure of the UART baud rate generator in MSP430 MCUs (*Courtesy of Texas Instruments, Inc.*)

Characters received after an address field do not generate interrupts nor get transferred to the input buffer. User software must take care of checking the received address field and compare it to any internally configured address, and if a match were detected, clear the receiver wake-up flag. This will cause the incoming characters to be transferred to the input buffer, making them available to the CPU.

Transmission in this mode only requires setting the transmitter awake control bit and place any character into the transmission buffer to send an idle frame. Afterwards, user software must just place the address information onto the transmission buffer, followed by the corresponding data characters.

In address-bit multi-processor mode (MM $= 1$) an address bit is inserted to each packet to denote wether the transmitted character is an address (a $= 1$) or data (a $= 0$) field, and the inter-character idle time has no meaning. If parity were enabled, the address bit wold be part of the parity computation. Character reception and transmission would happen similar to the cases described above for idle-time mode.

**Error Detection:** The UART can detect framing, parity, overrun, and break errors in the channel, and set the appropriate bits in the receiver control register (UxRCTL). A break error indicates the reception of ten or more low bits after missing a stop bit. Optionally, a receiver interrupt can be triggered by any of the detected errors. Recall that receiver interrupts are multi-sourced and therefore the ISR must check the UxRCTL register to determine the cause of the trigger.

**UART Baud Rate Generation:** The MSP430 UART has a baud rate generator capable of producing standard baud rates from non-standard clock frequencies. By combining a prescaler and a modulator, the generator is capable of supporting fractional divisors. This results in target baud rates with reduced error. Baud rates up to one third the source clock (BRCLK) are possible. Figure 9.31 shows the structure of the baud rate generation module.

To produce a desired baud rate, the source clock (BRCLK) is divided by a factor $N = $ BRCLK/baud rate. When a non-standard frequency is used, $N$ usually results in a non-integer number. To reduce the error that would otherwise be introduced if the fractional part of the quotient were discarded, the prescaler in the baud rate generator is loaded with the integer part of $N$, while the modulator is configured to perform an approximation of the bit time corresponding to the fractional part.

The fractional bit time approximation is achieved by choosing the length of each individual bit ($t_{bit(j)}$) to the nearest integer BRCLK count that produces the smallest cumulative packet time error. This computation is made by the programmer in an iterative procedure while determining the commands to configure the UART.

Equation (9.2) provides an expression for estimating the cumulative packet time error in the transmitter side up to bit $j$ ($\xi_{pkt}^t(j)$) obtained by setting or clearing each individual modulator bit $m_j$. The iterative evaluation of this formula tells wether bit $m_j$ needs to be set or clear, as we want to choose the condition yielding the minimum error. Input parameters include the desired baud rate (baud rate), the frequency fed to the baud rate generator (BRCLK), and the integer part of $N$ (UxBR). The modulation factor UxMCTL is built by taking the bit values that produce the minimum error at each position $j$.

$$\xi_{pkt}^t(j) = \frac{\text{baud rate}}{\text{BRCLK}} \left[ \text{UxBR} + \frac{1}{(j+1)} \sum_{i=0}^{j} m_i \right] - 1 \qquad (9.2)$$

The error of each individual bit time would be a measure of how much $t_{bit(j)}$ deviates from the nominal $t_{bit} = 1/$baud rate. Equation (9.3) provides an expression to quantify such an error.

$$\xi_{bit}^t(j) = \frac{\text{baud rate}}{\text{BRCLK}} \left( \text{UxBR} + m_j \right) - 1 \qquad (9.3)$$

Example 9.2 illustrates the process of computing the packet and bit errors an deciding the value of $m_j$ for $j = 0, 1, \ldots, 7$. Note that bit values for $j > 7$ simply reuse the already devised values of $m_j$ starting at $m_0$.

**Example 9.2** *Consider USART0 in UART mode in an MSP430F169, driven by a 32.768 KHz clock in ACLK. Determine the values of registers U0BR and U0MCTL to obtain a baud rate of 4,800 bps.*

**Solution:** *As the quotient $N = $ BRCLR/(baud rate) is non-integer ($N = 6.83$), we set U0BR $= 6$ and determine the bits in U0MCTL to approximate the 0.83 fractional part. Assuming the channel is configured for 8-bit characters, with parity even and one stop bit, each packet will consist of eleven bits. Table 9.2 shows the modulator bit values obtained by taking the bits that minimize the error computed with (9.2).*

*The third column indicates the exact expected value for bit transitions while columns $\xi_{pkt}^t(j)|_{m_j=0}$ and $\xi_{pkt}^t(j)|_{m_j=1}$ denote the errors when using $N$ or $N+1$ BRCLK cycles for the bit time. The absolute minimum of these two columns determines the value chosen for $m_j$. After bit eight, the values repeat from bit zero. The*

**Table 9.2** Timing values and relative errors in computing the modulation bits in the transmitter baud rate generator

| j | BIT | $t_{bit}(j)$ | $t_{bit}(n)$ | $\xi^t_{pkt}(j)\big|_{m_j=0}$ (%) | $t_{bit}(n+1)$ | $\xi^t_{pkt}(j)\big|_{m_j=1}$ (%) | $m_j$ |
|---|-----|---|---|---|---|---|---|
| 0 | Start | 208.3E-6 | 183.1E-6 | −12.11 | 213.6E-6 | 2.54 | 1 |
| 1 | D0 | 416.7E-6 | 396.7E-6 | −4.79 | 427.2E-6 | 2.54 | 1 |
| 2 | D1 | 625.0E-6 | 610.4E-6 | −2.34 | 640.9E-6 | 2.54 | 0 |
| 3 | D2 | 833.3E-6 | 793.5E-6 | −4.79 | 824.0E-6 | −1.12 | 1 |
| 4 | D3 | 1041.7E-6 | 1007.1E-6 | −3.32 | 1037.6E-6 | −0.39 | 1 |
| 5 | D4 | 1250.0E-6 | 1220.7E-6 | −2.34 | 1251.2E-6 | 0.10 | 1 |
| 6 | D5 | 1458.3E-6 | 1434.3E-6 | −1.65 | 1464.8E-6 | 0.45 | 1 |
| 7 | D6 | 1666.7E-6 | 1647.9E-6 | −1.12 | 1678.5E-6 | 0.71 | 1 |
| 8 | D7 | 1875.0E-6 | 1861.6E-6 | −0.72 | 1892.1E-6 | 0.91 | 1 |
| 9 | Parity | 2083.3E-6 | 2075.2E-6 | −0.39 | 2105.7E-6 | 1.07 | 1 |
| 10 | Stop | 2291.7E-6 | 2288.8E-6 | −0.12 | 2319.3E-6 | 1.21 | 0 |

*per bit error (not listed in the table) has only two possibilities, 2.54% when $m_j = 0$ and −12.11% when $m_j = 1$. The resulting register values are: U0BR = 06h and U0MCTL = FBh.*

The receiver side is also subject to errors. To identify the error sources let's review how the receiver works: the falling edge of a start bit wakes-up the UART receiver. The start bit is sampled at the middle of the bit time and thereafter the RxD line is sampled every $t_{bit}$. Note that both actions introduce error factors. The first is due to the mismatch between the occurrence of the start bit falling edge and the moment the UART actually accepts it. The second error factor is due the bit-to-bit timing variations introduced by the transmitter.

As in the case of the transmitter, a formula can be derived to estimate the cumulative bit error in a packet up to bit $j$, $\xi^r_{pkt}(j)$. The MSP430 User's Manual provides Eq. (9.4) [32].[4]

$$\xi^r_{pkt}(j) = \frac{\text{baud rate}}{\text{BRCLK}} \left( 2\left[ m_0 + \left\lfloor \frac{\text{UxBR}}{2} \right\rfloor \right] + \left[ j \cdot \text{UxBR} + \sum_{i=1}^{j} m_j \right] \right) - (j+1)$$

(9.4)

Iterative evaluation of (9.4), with both $m_j = 0$ and $m_j = 1$ allows identifying the values of $m_j$ yielding the smallest error. As in the case of the transmitter, these are the bits making up UxMCTL.

**Example 9.3** *For the case analyzed in Example 9.2, determine the value of U0MCTL in the receiver baud rate generator.*
**Solution:** *Iterative evaluation of Eq. (9.4) for $m_j = 0$ and $m_j = 1$ yields the values listed in Table 9.3, resulting in a value of U0MCTL = FEh.*

---

[4] This equation has not been verified.

**Table 9.3** Computing modulation bits in the receiver baud rate generator

| $j$ | BIT | $\xi^r_{pkt}(j)\big|_{m_j=0}$ (%) | $\xi^r_{pkt}(j)\big|_{m_j=1}$ (%) | $m_j$ | $\left|\xi^r_{pkt}(j)\right|$ (%) |
|---|---|---|---|---|---|
| 0 | start | −12.11 | 17.19 | 0 | 12.11 |
| 1 | D0 | −24.22 | −9.57 | 1 | 9.57 |
| 2 | D1 | −21.68 | −7.03 | 1 | 7.03 |
| 3 | D2 | −19.14 | −4.49 | 1 | 4.49 |
| 4 | D3 | −16.60 | −1.95 | 1 | 1.95 |
| 5 | D4 | −14.06 | 0.59 | 1 | 0.59 |
| 6 | D5 | −11.52 | 3.13 | 1 | 3.13 |
| 7 | D6 | −8.98 | 5.66 | 1 | 5.66 |
| 8 | D7 | −6.45 | 8.20 | 0 | 6.45 |
| 9 | parity | −18.55 | −3.91 | 1 | 3.91 |
| 10 | stop | −16.02 | −1.37 | 1 | 1.37 |

**Interrupt-based and Low-power Mode Operation:** The MSP430 USART has independent interrupt vectors for the transmit and receive sides. Whenever the transmit buffer can accept new characters for transmission, the UTXIFx is set. The interrupt will be triggered if both GIE and UTXIEx are set.

In the receiver side, whenever a new character is loaded into the receiver buffer (UxRXBUF), the URXIFGx is set. Its interrupt would be triggered if both GIE and URXIEx are enabled.

In addition to the receive and transmit events, error conditions can generate interrupts in the receiver side. Any detected errors: framing (FE), parity (PE), overrun (OE), or break (BRK) will set their corresponding error flags and also will set the RXERR flag. The URXIFGx would be set only if URXEIE is set, as this last setting allows the erroneous data to be transferred into the receive buffer. User software needs to check for any possible errors whenever the receiver interrupt is triggered. In the case of errors, error flags will remain set until UxRXBUF is read or the user's software clears them.

When operated in low-power mode with the DCO off, it is recommended to activate the Receive-Start Edge Detection mode. This allows turning-on the DCO to enable character reception out from the low-power mode. Enabling URXSE, URXIEx, and GIE would cause a start bit edge to trigger a receiver interrupt with URXIFGx clear. User software in the ISR must then cancel the low-power mode or turn on the chosen baud rate clock source to actually enable character reception.

The example below illustrates using USART0 in an MSP430F149 to implement a terminal echo. The code is written for TI CCE IDE.

**Example 9.4  (UART Echo at 115,200 bps on 8.0 MHz Xtal)** *Write a program to implement a terminal echo using an MSP430F149 running on an external 8 MHz clock. Use a low-power mode to save energy when no transfer is taking place. The port configuration is 8-bit per character, no parity, one stop bit (8N1).*

**Solution:** *This solution assumes the MCU has an 8.0 MHz HF crystal connected to XIN/XOUT pins. Pins P3.4 is used as TxD and P3.5 as RxD. A quick way to setup this system is on an 64-pin target board, installing the crystal, FET tool, and using an FTDI Chip TTL-232RG cable that allows for connecting the MSP430 UART pins to a USB port in the PC. A Hyperterminal (or similar) program can provide a quick interface on the PC. Below a brief code written by M. Buccini and G. Morton from Texas Instruments providing the MSP430 side of the solution [75].[5]*

```
;=============================================================================
;MSP-FET430P140 Demo - USART0, UART 115200 Echo ISR, HF XTAL ACLK
;By M. Buccini and G. Morton - 05/2005
;Copyright (c) Texas Instruments, Inc.
;-----------------------------------------------------------------------------
            .cdecls C,LIST,  "msp430x14x.h"
;-----------------------------------------------------------------------------
            .text                           ; Program Start
;-----------------------------------------------------------------------------
RESET       mov.w   #0A00h,SP               ; Initialize stack pointer
StopWDT     mov.w   #WDTPW+WDTHOLD,&WDTCTL   ; Stop WDT
SetupP3     bis.b   #030h,&P3SEL            ; P3.4,5 = USART0 TXD/RXD
SetupBC     bis.b   #XTS,&BCSCTL1           ; LFXT1 = HF XTAL
SetupOsc    bic.b   #OFIFG,&IFG1            ; Clear OSC fault flag
            mov.w   #0FFh,R15              ; R15 = Delay
SetupOsc1   dec.w   R15                    ; Addnl. delay to ensure start
            jnz     SetupOsc1              ;
            bit.b   #OFIFG,&IFG1           ; OSC fault flag set?
            jnz     SetupOsc               ; OSC Fault, clear flag again
            bis.b   #SELM_3,&BCSCTL2       ; MCLK = LFXT1
SetupUART0  bis.b   #UTXE0+URXE0,&ME1      ; Enable USART0 TXD/RXD
            bis.b   #CHAR,&UCTL0           ; 8-bit characters
            mov.b   #SSEL0,&UTCTL0         ; UCLK = ACLK
            mov.b   #045h,&UBR00           ; 8 MHz 115200
            mov.b   #000h,&UBR10           ; 8 MHz 115200
            mov.b   #000h,&UMCTL0          ; 8 MHz no modulation 115200
            bic.b   #SWRST,&UCTL0          ; **Initialize USART FSM **
                                           ;
Mainloop    bis.b   #CPUOFF+GIE,SR         ; Enter LPM0, intrpts enabled
            nop                            ; Needed only for debugger
                                           ;
;-----------------------------------------------------------------------------
USART0RX_ISR; Confirm TX buffer is ready, then Echo back RXed character
;-----------------------------------------------------------------------------
TX1         bit.b   #UTXIFG0,&IFG1         ; USART0 TX buffer ready?
            jz      TX1                    ; Jump is TX buffer not ready
            mov.b   &RXBUF0,&TXBUF0        ; TX -> RXed character
            reti                           ;
                                           ;
;-----------------------------------------------------------------------------
;           Interrupt Vectors
;-----------------------------------------------------------------------------
            .sect   ".reset"               ;
            .short  RESET                  ; POR, ext. Reset, Watchdog
            .sect   ".int09"               ;
            .short  USART0RX_ISR           ; USART0 receive
            .end
;=============================================================================
```

---

[5] See Appendix E.1 for terms of use.

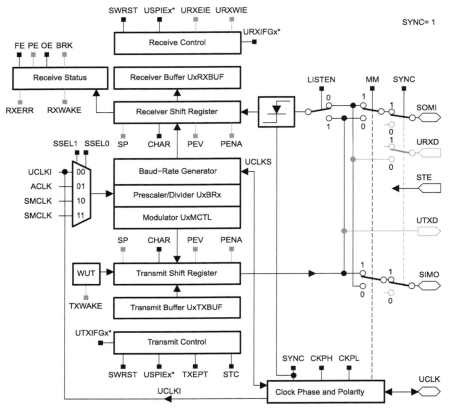

**Fig. 9.32** Block diagram of MSP430's USART in SPI mode (*Courtesy of Texas Instruments, Inc.*)

## The MSP430 USART Module in SPI Mode

The synchronous serial operation of the USART supports SPI mode transfers in three- or four-pin modes. The SPI mode features 7- or 8-bit characters, master or slave operation, 3- or 4-pin operation, separate transmit and receive shift registers, and configurable clock polarity.

The Interface lines are designated SIMO: Slave in, master out (same as SDO); SOMI: Slave out, master in (same as SDI); UCLK: Clock line (same as SCL); and STE: Slave transmit enable to allow for multi-master operation. Figure 9.32 shows the structure of the USART configured in the SPI mode.

Upon a reset, the SWRST bit is set, configuring the entire USART in reset mode. The SPI interface is enabled via the USPIEx bit in the Module Enable (MEx) register. MEx is configured with the USART in reset mode. Its programming is recommended to be performed after all other USART registers have been configured. The SPI

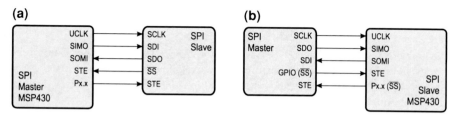

**Fig. 9.33** Connecting the MSP430 SPI in four-pin mode. **a** MSP430 as master. **b** MSP430 as slave

becomes usable after the USART is taken out from the reset mode, by clearing the SWRST flag.

An SPI transfer is initiated in the master when data is moved into the transmit data buffer (UxTXBUF). When the transmission shift register becomes empty, data loaded in UxTXBUF is transferred into the transmission shift register, flag UTXIFGx is set indicating the receiver can accept new data, and the actual transfer of the character in the shift register is initiated over the SIMO line. Recall that in an SPI, as data is shifted out from the master into the slave(s), slave data is shifted-in into the master RX register (see Fig. 9.21 on p. 499). The received data is transferred into the UxRXBUF and the URXIFGx is set, denoting the completion of the TX/RX operation. Recall that in SPI transmission and reception occur simultaneously, and thus a data reception is initiated in the same way.

When in four-pin master mode, the STE input prevents conflicts with other potential masters in the bus. When STE is high, SIMO and UCLK lines operate normally, while STE in low makes them work as inputs.

In slave mode, having STE in high prevents the RX/TX operation. A master MSP430 SPI drives the slave STE input via an I/O port, while a slave GPIO output connects to the master STE. The reciprocal connection is necessary when the MSP430 SPI is a slave. Figure 9.33 shows the MSP430 connections in each case.

Although the SPI module uses the USART baud rate generator, only the prescaler portion takes effect to establish the data rate. The master device provides UCLK using its baud rate generator. In slave mode, the baud rate generator goes unused. SSELxx control bits allow for selecting the clock source. The maximum transfer rate allowed for a given BRCLK frequency is BRCLK/2.

The clock polarity and phase can be configured through the CKPL and CKPH control bits in the USART transmit control register (UxTCTL).

The SPI module has a single interrupt vector triggered by flag UTXIFGx when both UTXIEx and GIE are enabled. UTXIFGx is automatically cleared when the UxTBUF is written, upon a PUC, or when SWRST = 1.

## 9.5.2 The MSP430 USI

The Universal Serial Interface (USI) is one of the simplest serial peripheral interfaces supported by MSP430 devices. Featured exclusively in series x2xx devices, the USI supports only synchronous serial communications in the SPI or $I^2C$ modes through a single module.

To operate in either mode, the USI is activated through the USI port enable control bits (USIPEx) in the USI control register USICTL0. These bits also define the mode and master/slave function of the module.

The USI clock module allows for choosing the phase and level of the clock signal. The USI clock can be fed from different sources that include SCLK, ACLK, SMCLK, SWCLK, TA0, TA1, or TA2, selected through USISSEL control bits in the USI clock control register (USICLKCTL). The clock division factor can be set to any power of two between 1 and 128.

### MSP430 USI in SPI Mode

In the SPI Mode, the USI closely resembles the basic interface described in Sect. 9.4.1 on p. 499, as it features a single shift register, a bit counter, and a clock divider to implement a module that can be used as either SPI master or slave. The SPI mode of the USI is selected by clearing bit USII2C in the USICTL1 register.

The shift register (USISR) directly serves as data IN/OUT buffer, and is configurable to support 8- or 16-bit words. The bit counter provides indication of transfer completion, setting the USI flag (USIIFG) upon TX/RX completion. Figure 9.34 shows a simplified diagram with the USI module configured as SPI.

As an SPI master, the USI module generates the clock signal for the slave(s), so in its configuration a suitable clock source must be chosen. The master mode is chosen by setting the master bit (USIMST) in the USI control register USICTL0.

To initiate a master mode transfer, the character to be transmitted must be written to USISR and the actual transmission or reception begins when the USICNTx bits in the USI bit counter register (USICNT) are written with the number of bits to be transferred.

In the slave mode transfers are initiated similarly to the master mode, except that no clock source is specified as the master provides the clock signal. Before the a slave can transfer data, its output must be enabled by setting the control bit USIOE. This allows using the USI SPI in a multi-slave configuration without causing output conflicts when not selected. Note that the GPIO enabled slave select signals become necessary for any multidrop configuration.

As the entire USI module provides a single interrupt vector, whenever a USI interrupt is triggered, status indicators must be checked by the user software to determine the type of even that caused the exception.

**Fig. 9.34** Simplified structure
of USI module in SPI mode

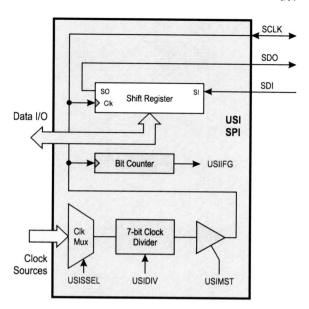

The example below illustrates the usage of the USI SPI to interface an external analog-to-digital converter (ADC) to an MSP430F20x2 in a simple voltage level indicator.

**Example 9.5 (USI SPI Interface of a TLC549 8-bit ADC)** *Provide a connection diagram for interfacing an external SPI serial ADC in 3-wire mode and an LED to an MSP430F2032. Write a C-language program solution to set the MSP430 as SPI master and turn the LED on when the analog input connected to the ADC falls below* $V_{DD}/4$.

**Solution:** *In this case, a suitable ADC is TI's TLC549, which provides a 3-wire simplex serial SPI interface [76]. As the ADC works as a read-only slave SPI peripheral, no SIMO line is needed nor is STE. The ADCs references VR+ and VR− are connected to* $V_{DD}$ *and ground, respectively. The analog input is assumed to change between* $V_{DD}$ *and GND. Figure 9.35 shows a connection diagram for this application.*

*The software solution is adapted from a TI's USI SPI Interface demo written by M. Buccini and L. Westlund [77].[6] The CPU clock is taken from the default DCO and the USI clock is chosen as SMCLK/4. The design is assumed to operated at* $V_{DD} = 3.3$ V.

---

[6] See Appendix E.1 for terms of use.

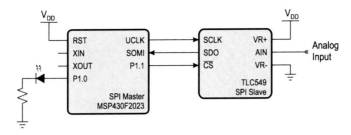

**Fig. 9.35** Connecting an SPI external ADC to an MSP430 SPI master

```
//=============================================================================
// Adapted from MSP430F20x2/3 Demo - USI SPI Interface to TLC549 8-bit ADC
// Original code by M. Buccini and L. Westlund - 10/2005
// IAR V3.40A/CCE V3.2.0 * Copyright (c) Texas Instruments, Inc.
//-----------------------------------------------------------------------------
#include <msp430.h>

int main(void)
{
  WDTCTL = WDTPW + WDTHOLD;              // Stop watchdog timer
  P1DIR |= 0x03;                        // P1.0 & P1.1 as outputs
  P1OUT = 0;                            // LED Off & ADC CS asserted

  USICTL0 |= USIPE7 + USIPE5 + USIMST + USIOE; // Port, SPI master
  USICTL1 |= USIIE;                     // Counter interrupt, flag left set
  USICKCTL = USIDIV_2 + USISSEL_2;      // /4 SMCLK
  USICTL0 &= ~USISWRST;                 // USI released for operation
  USICNT = 8;                           // init-load counter
  _BIS_SR(LPM0_bits + GIE);             // Enter LPM0 w/ interrupt
}
//-----------------------------------------------------------------------------
// USI interrupt service routine
#pragma vector=USI_VECTOR
__interrupt void universal_serial_interTface(void)
{
  P1OUT |= 0x02;                        // Disable TLC549
  if (USISRL < 0x3F)
    P1OUT |= 0x01;
  else
    P1OUT &= ~0x01;
  P1OUT &= ~0x02;                       // Enable TLC549
  USICNT = 8;                           // re-load counter
}
//=============================================================================
```

## MSP430 USI in I$^2$C Mode

In the I$^2$C Mode, the USI supports synchronous serial communication conformal to the Inter-Integrated Circuit protocol described in Sect. 9.4.2 on p. 501. The hardware configuration of the module expands that of SPI mode by including open-drain drivers for the bidirectional SDA and SCL lines, detectors for the start and stop

**Fig. 9.36** Simplified structure
of USI module in I²C mode

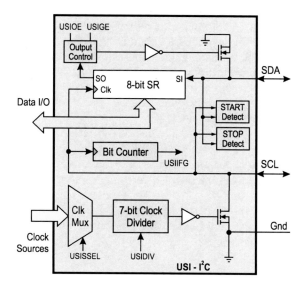

conditions plus the output control logic needed to handle synchronization and arbitration inherent to the I²C. Figure 9.36 shows a simplified diagram of the USI module in I²C mode.

The I²C mode of the USI is selected by setting bit USII2C in the USICTL1 register, while fixing the clock polarity and phase with USICKPL = 1, and USICKPH = 0. Moreover, character length must be fixed to eight bits and the SCL and SDA ports need to be enabled.

To set the module in master mode, the USIMST bit is set, and the module generates the transmission clock while USIIFG is clear. For slave mode operation, USIMST must be cleared. While as slave, the SCL line is set low by driving USIIFG and USISTTFG low while USICNTx is high. Flag USISTTIFG must be software cleared after setup to enable the reception of address fields at the beginning of each packet.

Transmissions require writing the output data into the shift register (USISRL) with the output enabled (USIOE = 1). Setting USICNTx to 08h (character length) will actually begin the transmission. When all eight characters are sent, USIIFG is set, and SCL is stopped. Reception of acknowledgement bits require software clearing USIOE.

I²C reception requires disabling the output driver and writing 08h to USICNTx to set the required character length. When a character arrives USIIFG will be set, triggering an interrupt, if USIIE and GIE are enabled. An acknowledgment (or NACK) is sent by setting (clearing) the MSB of the USISRL, enabling the output and writing 01h to the USICNTx register. Once the ACK is sent the USIIFG will be set, allowing to prepare for the reception of the next character.

Start and stop conditions are sent by clearing or setting the MSB of the shift register and using USIGE and USIOE to drive the corresponding transitions on the SDA line by making the output latch transparent while SCL is high.

To prevent holding low the SCL line when the modules is in slave mode and has detected that it has not been addressed by the master, setting the USISCLREL bit in the USI Bit Counter Register (USICNT). This allows releasing SCL without requiring to clear the USIIFG.

The USI $I^2C$ module can also be used in multi-master configurations, as it is capable of detecting a loss of arbitration in the case of contending masters. In an arbitration process, if the module looses arbitration by sending a one when the other contending master sent a zero, the USI arbitration lost flag (USIAL) in the USI Control Register 1 (USICTL1) will be set and the USIOE bit cleared, effectively relinquishing control of the bus. To detect such an event, user software must check the USIAL and USIIFG flags and configure the USI to slave receiver. The USIAL flag must be cleared by software.

### 9.5.3  The MSP430 USCI

The Universal Serial Communications Interface (USCI) is the most complete serial peripheral interface supported by MSP430 devices. Using a single hardware module, a USCI is able of supporting different serial communication formats including asynchronous and synchronous modes. USCI module configurations have been featured in MSP430 device families in series x4xx, x2xx, and x5xx/x6xx. Designators USCI_A, USCI_B are used to identify modules with different capabilities. Identical modules of the same kind coexisting in a particular MCU are labeled with a number following the designator (Ex. USCI_A0).

USCI_A modules can be seen as enhanced versions of the USART module described in Sect. 9.5.1 on p. 507, as both support UART and SPI communication modes. USCI_A enhancements include the ability to perform pulse shaping for IrDA[7] and automatic baud rate detection for supporting Local Interconnect Network[8] (LIN) protocols.

USCI_B modules support only synchronous serial communication modes in the form of SPI and $I^2C$, similar to the capabilities of the USI module described in Sect. 9.5.2 on p. 516.

Due to the similarities of USCI_A and USCI_B with the previous USART and USI modules, the discussion in the next sections will mainly focus in the enhancements brought up in the new modules, while referring to the earlier serial communication modules for the shared basic functionality aspects. For specific MSP430 devices, developers are strongly encouraged to refer to the family user's guide and device

---

[7] IrDA = Protocol for wireless infrared communication established by the Infrared Data Association.

[8] LIN is a serial communication format designed for localized vehicle networks.

data sheets for configuration details. Configuration and usage details change from one family to another and available resources change from one family member to another.

### MSP430 USCI in UART Mode

The UART mode of the USCI, supported by USCI_A modules, provides all the capabilities listed for the USART in UART mode, with added features to encode and decode IrDA bit streams in the UC0RX and UC0TX lines. Moreover, the USCI UART has autobaud detection capabilities, a feature that allows using it for LIN communications. Figure 9.37 shows a block diagram of USCI_A in UART mode.

**Configuration:** Before attempting to configure the USCI_A UART, the device needs to be placed in reset mode by setting the UCSWRST flag. The USCI is placed in its asynchronous mode by clearing UCSYNC bit in the UCAxCTL0 control register. The overall sequence to initialize the USCI can be outlined as follows:

Step 1:  Set UCSWRST to place the USCI in reset mode. UCSWRST is by default set by a PUC.

Step 2:  Initialize all USCI registers, including UCAxCTL1, while holding UCSWRST = 1.

Step 3:  Configure the USCI ports.

Step 4:  Clear UCSWRST via software to enable the USCI.

Step 5:  Enable interrupts via UCAxRXIE and/or UCAxTXIE.

The USCI asynchronous character format is consistent with that of the USART UART, illustrated in Fig. 9.30 on p. 508.

In its asynchronous mode the USCI can support point-to-point and multiprocessor transfers.

For common point-to-point connections, the USCI UART is configured in the idle-line mode, with no multiprocessor capability enabled. This is achieved by setting bits UCMODEx = 00 in control register UCAxCTL0.

**Idle-Line and Address-Bit Multiprocessor Modes:** When three or more devices need to be connected, the USCI UART can be configured in either idle-line or address-bit multiprocessor modes. Although these modes were explained in Sect. 9.5.1 on p. 507, they will be re-addressed here in terms of the USCI registers.

Bits UCMODEx in the UCAxCTL0 control register determine the mode of operation of the USCI.

For the USCI UART, the detection of an idle-line period is indicated with UCIDLE bit in the USCI_A status register (UCAxSTST). When in idle-line multiprocessor mode, the first received character after an idle period is an address. Recall that in this mode the address bit in the frame is not transmitted.

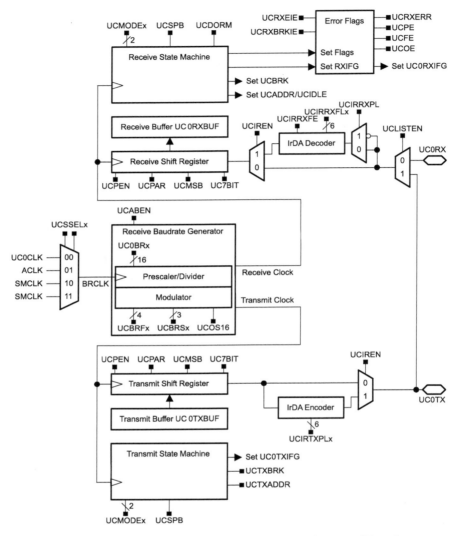

**Fig. 9.37** Block diagram of USCI_A configured in UART mode (*Courtesy of Texas Instruments, Inc.*)

Upon an address reception, the incoming value is placed into the receive buffer and bit UCAxRXIFG set and control bit UCDORM is set, placing the USCI in dormant mode. In this mode, only address fields are received. Incoming data characters, although assembled, are not transferred into the receiving buffer and no reception interrupt is triggered. User software must validate the received address and, when matched, clear UCDORM to make possible the reception of incoming data. Once

UCDORM is clear, data characters will be transferred to the in-buffer and the reception flag set.

Three different actions can be initiated from the transmitter: sending an address field, a data field, or an idle frame. Transmission of an address field is preceded by an idle period. This is achieved by setting the UCTXADDR flag and then loading the address character into the UCAxTXBUF. UCAxTXIFG must be set for this action, indicating the transmitter is ready. The UART will then generate an idle line frame of eleven bits followed by the address character. UCTXADDR will be automatically reset. Afterwards, data characters to be sent can be transferred to the transmission buffer (upon transmitter ready flag set) ensuring the time between characters does not exceed the idle-line time to prevent their misinterpretation as addresses.

In address-bit multiprocessor mode each transmitted character includes the address bit to denote wether it corresponds to an address or data field. When a character with an address bit set is received, the UCADDR bit is set and the character placed in the reception buffer. UCAxRXIFG and UCDORM bits are set. User software must validated the address and clear UCDORM to enable reception of the incoming data. UCDORM set will allow receiving only address characters.

A transmission event must send first an address character. This is achieved by setting the UCTXADDR bit and placing the address character in the transmission buffer when the transmitter is ready. The transmission of the address character automatically clears UCTXADDR, so subsequent transmitted characters will be sent as data. In address-bit multiprocessor mode the time between characters has no significance.

**Break Characters:** In either point-to-point or any of the multiprocessor modes a break character might be received, triggering (if enabled) a break exception. A break is just a frame with all bits in zero. Transmission of a break is achieved by setting the UCTXBRK bit, and writing 00h to the transmission buffer. The transmission of the break character clears UCTXBRK.

**Auto Baud Function:** Setting UCMODEx to 11 in UART mode will enable auto baud detection. In this mode control registers UCAxBR0, UCAxBR1, and UCAxMCTL are set based on the measurement of a synchronization character (055h) inserted in front of a data frame after a break character. Control bit UCABDEN must be set to enable automatic measurement of the frame length. If the receiver were unable to synchronize, UCSTOE would be set.

Bit UCDORM controls data reception in autobaud mode. When set, only the break and synch fields will be accepted. User software must clear UCDORM to enable reception of the remaining characters in the stream. Auto baud events will set the UCAxRXIFG. If USCI RX interrupt and GIE are enabled, the triggered ISR needs to check the status flags to determine the channel status. Autobaud is recommended only for half duplex operation, as the USCI cannot transmit while receiving a break-synch sequence.

The transmission of a break-synch sequence is achieved by setting UCTXBRK with UMODEx = 11, then writing 055h to UCAxTXBUF when UCAxTXIFG = 1. UCTXBRK is automatically reset when the synch character is transmitted.

**IrDA Pulse Shaping:** This feature is enabled by setting UCIREN in the IrDA Transmit Control Register (UCAxIRTCTL). This causes pulsing the encoder to send a pulse through the transmit line for every zero in the transmitted data. UCIRTXPLx bits specify the pulse width. The decoder detects high pulses when UCIRRXPL $= 0$. Otherwise it detects low pulses.

**Channel Operation:** The operation of the UART channel proceeds according to the standard rules for asynchronous communication, discussed earlier in this chapter. Transmission and reception require the corresponding section of the UART to indicate their readiness. If enabled, these conditions can trigger their corresponding interrupts. If any error (framing, parity, overrun, or break) were detected, the corresponding error flags, UCRXERR, and the receiver flag will be set, and if enabled, would cause a reception interrupt. User software must check the error flags to determine if the reception was erroneous, what type of error caused it (if any), and the consequent corrective actions (if any) after an error is detected.

**Baud Rate Generation:** The USCI_A baud rate generator can operate in either a low-frequency or an oversampling modes, depending on whether UCOS16 bit in the USCI_A Modulation Control Register is clear or set.

In the low-frequency mode, the baud rate generator operates in a way similar to the USART baud rate generator, discussed in Sect. 9.5.1, with a few exceptions. Like in the USART case, it uses a prescaler and a modulator to produce standard baud rates from non standard frequency sources, where the maximum baud rate is one third the BRCLK. The module's power consumption is reduced by limiting the bandwidth of the baud rate generator clock BRCLK. The prescaling factor is set through the baud rate control registers UCAxBR0 and UCAxBR1, while the modulation factor, unlike the USART case, is now selected from a set of eight predefined values selected with control bits UCBRSx in the modulation control register UCAxMCTL. A simple formula now provides the value to choose:

$$UCBRSx = round[8 \cdot (N - \lfloor N \rfloor)], \tag{9.5}$$

where $N = f_{\mathrm{BRCLK}}/(\text{baud rate})$. The device family user's manual lists the available modulation values and the recommended settings for standard baud rate values using typical frequencies.

In the oversampling mode (UCOS16 $= 1$), the generator also uses the approach of combining a prescaler and modulator to produce the desired baud rate. However, an intermediate bit clock 16 times the desired baud rate is generated from BRCLK and an additional divider produces the desired BITCLK. This approach limits the maximum baud rate to 1/16 the frequency of BRCLK.

The prescaler value is derived from the integer part of the quotient $f_{BRCLK}/(\text{baud}$ rate), while the modulation factor is chosen from a list of 16 pre-determined values selected with UCBRFx bits in the Modulation Control Register UCAxMCTL.

The value of UCBRFx is determined as:

$$UCBRFx = \text{round}\left(N - 16 \cdot \left\lfloor \frac{N}{16} \right\rfloor\right) \tag{9.6}$$

As in the case of the USART baud rate generator, the actual bit times obtained in either the low-frequency or oversampled mode are not exact, and different frequencies will yield different levels of error for different target baud rates. The User's Manual of the selected target device provides tables with recommended values of prescaler and modulation factors for standard baud rates derived from several common frequencies obtained from the MSP430 clock generator and external crystals. These tables also include the estimated transmission and reception errors.

The following two examples below illustrate the usage of the USCI_A UART to implement the same application as in Example 9.4 on p. 512. In this both cases USCI_A0 is configured to work at 9,600 bps, 8N1 running from the DCO at 1 MHz. In the first case using the baud rate generator in low-frequency mode, while the second uses its oversampled mode.

**Example 9.6 (USCI_A0 Echo at 9,600 bps on DCO at 1 MHz)** *Write a program to implement a terminal echo using an MSP430G2452 running on the DCO at 1.0 MHz. Use a low-power mode to save energy when no transfer is taking place. The port configuration is 8N.*

***Solution:*** *This solution assumes the MCU has is running from the internal DCO at 1.0 MHz. Pins P3.4 and P3.5 are used as TxD and RxD, respectively. A quick way to setup this system on the MSP430 Launchpad is using an FTDI Chip TTL-232RG cable connecting the MSP430 UART pins to a USB port in the PC. A Hyperterminal program or similar could be used to provide a quick interface on the PC. Below a brief code written by B. Nisarga from Texas Instruments providing the MSP430 side of the solution [78].[9] Observe that this solution is using the low-frequency mode of the baud rate generator with $N = 1\,MHz/9,600 = 104.167\,Hz$ and $UCBRSx = 1$. The DCO is configured using the standard calibration factor for 1 MHz provided in the header file, with a failsafe provision in case of value erasure.*

```
;========================================================================
;MSP430x24x Demo - USCI_A0, 9600 UART Echo ISR, DCO SMCLK - IAR V3.42A
;By B. Nisarga - 09/2007 * Copyright (c) Texas Instruments, Inc.
;------------------------------------------------------------------------
#include <msp430.h>
;------------------------------------------------------------------------
            RSEG    CSTACK              ; Define stack segment
;------------------------------------------------------------------------
            RSEG    CODE        ·       ; Assemble to Flash memory
;------------------------------------------------------------------------
RESET       mov.w   #SFE(CSTACK),SP     ; Initialize stack pointer
StopWDT     mov.w   #WDTPW+WDTHOLD,&WDTCTL  ; Stop WDT
CheckCal    cmp.b   #0xFF,&CALBC1_1 MHZ    ; Check calibration constant
            jne     Load                ; if not erased, load.
Trap        jmp     Trap                ; else, do not load, trap CPU!
Load        clr.b   &DCOCTL             ; Select lowest DCOx and MODx
            mov.b   &CALBC1_1 MHZ,&BCSCTL1 ; Set DCO to 1 MHz
```

---

[9] See Appendix E.1 for terms of use.

```
                 mov.b     &CALDCO_1 MHZ,&DCOCTL     ;
SetupP3          bis.b     #030h,&P3SEL             ; Use P3.4/P3.5 for USCI_A0
SetupUSCI0       bis.b     #UCSSEL_2,&UCA0CTL1      ; SMCLK
                 mov.b     #104,&UCA0BR0            ; 1 MHz 9600
                 mov.b     #0,&UCA0BR1              ; 1 MHz 9600
                 mov.b     #UCBRS0,&UCA0MCTL        ; Modulation UCBRSx = 1
                 bic.b     #UCSWRST,&UCA0CTL1       ; **Init USCI state machine**
                 bis.b     #UCA0RXIE,&IE2           ; Enable USCI_A0 RX interrupt
                                                    ;
Mainloop         bis.b     #CPUOFF+GIE,SR           ; Enter LPM0, interrupts enabled
                 nop                                ; Needed only for debugger
                                                    ;
;-----------------------------------------------------------------------------
USCI0RX_ISR;  Echo back RXed character, confirm TX buffer is ready first
;-----------------------------------------------------------------------------
TX0              bit.b     #UCA0TXIFG,&IFG2         ; USCI_A0 TX buffer ready?
                 jz        TX0                      ; Jump if TX buffer not ready
                 mov.b     &UCA0RXBUF,&UCA0TXBUF    ; TX -> RXed character
                 reti                               ;
                                                    ;
;-----------------------------------------------------------------------------
                 COMMON    INTVEC                   ; Interrupt Vectors
;-----------------------------------------------------------------------------
                 ORG       USCIAB0RX_VECTOR         ; USCI0 Rx Vector
                 DW        USCI0RX_ISR              ;
                 ORG       RESET_VECTOR             ; RESET Vector
                 DW        RESET                    ;
                 END
;=============================================================================
```

**Example 9.7  (USCI_A0 Echo at 9,600 bps on DCO at 1 MHz—C)** *Repeat Example 9.6 providing a solution in C Language. Change the baud rate generator to operate in oversampled mode while keeping unchanged all other parameters.*

**Solution:** *A porting of the solution in Example 9.6 running on CCE V3.2.0 and IAR V3.42A and using the baud rate generator oversampled mode is provided by B. Nisarga [79].[10] By keeping the same clock source (DCO at 1 MHz), the 16x requirement between BRCLK and the target baud rate is satisfied. In this case, $N = 1 MHz/9,600 Hz = 104.17$ and the modulation factor from (9.6) yields $UCBRSx = round(104.17 - 16\lfloor 104.17/16 \rfloor) = 8$. Again, a calibrated DCO is used for MCLK = 1 MHz.*

```
//=========================================================================
// MSP430x24x Demo - USCI_A0, 9600 UART, SMCLK, LPM0, Echo w/over-sampling
// By B. Nisarga - 09/2007 * Copyright (c) Texas Instruments, Inc.
//-------------------------------------------------------------------------
#include <msp430.h>

int main(void)
{
  WDTCTL = WDTPW + WDTHOLD;              // Stop WDT
  if (CALBC1_1MHZ==0xFF)                 // If calibration constant erased
  {
    while(1);                            // do not load, trap CPU!!
```

---

[10] See Appendix E.1 for terms of use.

```
    }
    DCOCTL = 0;                              // Select lowest DCOx & MODx values
    BCSCTL1 = CALBC1_1MHZ;                   // Set DCO
    DCOCTL = CALDCO_1MHZ;
    P3SEL = 0x30;                            // P3.4,5 = USCI_A0 TXD/RXD
    UCA0CTL1 |= UCSSEL_2;                    // SMCLK
    UCA0BR0 = 6;                             // 1 MHz 9600
    UCA0BR1 = 0;                             // 1 MHz 9600
    UCA0MCTL = UCBRF3 + UCOS16;              // Modln UCBRSx=8, over sampling
    UCA0CTL1 &= ~UCSWRST;                    // **Initialize USCI FSM**
    IE2 |= UCA0RXIE;                         // Enable USCI_A0 RX interrupt

    __bis_SR_register(LPM0_bits + GIE);      // Enter LPM0, interrupts enabled
}

// Echo back RXed character, confirm TX buffer is ready first
#pragma vector=USCIAB0RX_VECTOR
__interrupt void USCI0RX_ISR(void)
{
    while (!(IFG2&UCA0TXIFG));               // USCI_A0 TX buffer ready?
    UCA0TXBUF = UCA0RXBUF;                   // TX -> RXed character
}
//========================================================================
```

## MSP430 USCI in SPI Mode

The USCI SPI mode is functionally equivalent to that of the USART SPI explained in Sect. 9.5.1 on p. 501. Both modules offer three- and four-pin SPI operation with separate receive and transmit shift registers. As the SPI baud rate generator does not use the modulator section (disabled in the USART SPI), the USCI SPI mode omits it from its structure. Other than that, both modules are equivalent, and understanding the functionality of one of them allows to operate the other. The reader is referred to the description of the USART SPI for functional information. For programming details, register and control/status bit names, and available units, the reader is referred to the device-specific data sheets and user manuals.

## MSP430 USCI in I²C Mode

The synchronous $I^2C$ serial communication modules sported in USCI_Bx modules, are designed to fully comply with Phillips Semiconductor $I^2C$ specification V2.1. By using separate shift registers for the transmit and receive sections, the USCI_B $I^2C$ architecture, illustrated in Fig. 9.38, closely resembles the structure depicted in Fig. 9.24 on p. 502 for a standard interface for this communication modality.

It supports operation as either master or slave in single or multimaster buses. Moreover, this module, featured in MSP430 devices series x4xx, x2xx, and x5xx/x6xx, is designed for compatibility with the MCU low-power mode operation.

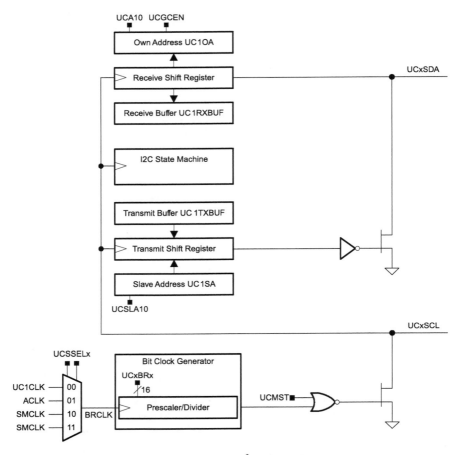

**Fig. 9.38** Block diagram of USCI_B configured in I$^2$C mode (*Courtesy of Texas Instruments, Inc.*)

Like any other USCI module, the I$^2$C is reset upon a PUC by setting control bit UCSWRST in the UCxCTL1 UCSI control register 1. It requires that all configuration be made with the module in reset mode to prevent undesirable behavior.

In adherence to Phillips standard, the module supports 7- and 10-bit address fields. For ten-bit addresses, the standard requires two consecutive 8-bit characters: the first containing the pattern $11110a_9a_8$ plus the R/$\overline{\text{W}}$ and acknowledged by the addressed slave, and the second containing address bits $a_7 \ldots a_0$, with no R/$\overline{\text{W}}$ field like a regular data character.

Unlike the simple operation of USI modules where many of the tasks related to using the bus rely on user software, USCI_B I$^2$C supports most aspects of the communication by hardware actions. Master transmit/receive, slave transmit/receive, clock synchronization, bus arbitration, address generation and clock stretching are all supported requiring minimal user software. SW actions fundamentally need to enable and configure the module in the desired mode, and afterwards essentially accessing the transmit and/or receive buffers upon activation of the receiver and transmitter ready flags. A few isolated instances require user software actions during transfers. These include a master initiating a transmission or reception transfer, a master aborting a transfer, a master issuing a restart condition, or a master/slave issuing a NACK condition when the acknowledgment of the last transmitted character is not received.

As the protocols needed to establish an $I^2C$ communication are described in detail in Sect. 9.4.2 on p. 501, we refer the reader to that section for understanding the protocol. The programming details specifying particular register names and status/control flags are dependent on the instance of USCI_Bx being used in a specific MSP430 device. The reader is referred to the corresponding family User's Manual for a detailed description of them.

A few observations deserve to be made about the module.

- Although each event occurring in an USCI module has it own indication flag and enable bits, the entire module has a single interrupt vector multiple-sourced by the transmitter, receiver, and status sections. This implies that USCI interrupt service routines need to check the corresponding flags to determine the actual cause of the interrupt.

- For convenience in identifying USCI $I^2C$ interrupts, all triggering flags are prioritized and combined in register UCBxIV, facilitating the implementation of a jump table based on the value of this register (see Sect. 7.3.3 on p. 318 for an example). Any R/W access to this register automatically clears the highest-pending interrupt flag.

- There are six fundamental events that could trigger an interrupt from the USCI module in $I^2C$ module. These include:

  1. **Transmitter Ready:** Indicated by UCTXIFG to denote that the transmission buffer (UCBxTXBUF) is ready to accept a new character. Automatically cleared when UCBxTXBUF is written.

  2. **Receiver Ready:** Signaled by UCRXIFG flag to indicate that a new character has been received and loaded into the reception buffer (UCBxRXBUF). Automatically cleared when UCBxRXBUF is read.

  3. **Arbitration Lost:** Indicates the device lost a bus arbitration process. Signaled with the UCALIFG flag.

  4. **No-acknowledgment (NACK):** Signaled with UCNACKIFG flag to denote an expected acknowledgment was not received. Cleared by the reception of a start condition.

5. **Start Condition:** Indicates the detection of a start condition and valid address by a slave device. This condition is exclusive for slave configured modules. Automatically cleared when a stop condition is detected.

6. **Stop Condition:** This status is also exclusive for slave devices and indicates the detection of a stop condition in the bus. Automatically cleared when a start condition is detected, regardless the address.

- When the module is used in conjunction with a DMA channel, the DMA transfer requests are triggered from the USCI via the UCTXIFG and UCRXIFG flags.

- To support low-power mode operation, the USCI clock source is automatically activated by transfer events. As only master configured modules can generate the SCL signal, this activation is not necessary in devices configured as slaves.

The examples below illustrate how to use USCI_B in its $I^2C$ mode to implement a master-slave bus connection between two MSP430x5xx devices. The first example will illustrate the master side, while the second the slave side.

**Example 9.8 (USCI_B0 I2C MSP430 Master-slave TX)** *Provide a program to transmit via an $I^2C$ bus master a multi-character message stored somewhere in memory.*

*Solution: The solution will use two MSP430F5438 connected one as master, the other as slave via their I2C interfaces (P3.1 = SDA and P3.2 = SCL). This solution provides a transmitter code demo written in C-language by D. Dang [80].[11] The code uses the internal DCO frequency through SMCLK to feed BRCLK. The actual data rate will be around 100 Kbps as BRCLK is further divided by 100. The MCU is placed in LPM0 after the data to be transmitted is placed in the transmission buffer. The transmission ISR illustrates how to process the multi-sourced USCI interrupt, in this case using a switch statement.*

```
//===========================================================================
// MSP430F543xA Demo - USCI_B0 I2C Master multi-byte TX to MSP430 Slave
// By D. Dang - 12/2009 * Copyright (c) Texas Instruments, Inc.
//---------------------------------------------------------------------------
#include <msp430.h>

unsigned char *PTxData;                  // Pointer to TX data
unsigned char TXByteCtr;

const unsigned char TxData[] =           // Table of data to transmit
{
  0x11,  0x22,  0x33,  0x44,  0x55
};

int main(void)
{
  WDTCTL = WDTPW + WDTHOLD;              // Stop WDT
  P3SEL |= 0x06;                        // Assign I2C pins to USCI_B0
  UCB0CTL1 |= UCSWRST;                  // Enable SW reset
```

---

[11] See Appendix E.1 for terms of use.

```
UCB0CTL0 = UCMST + UCMODE_3 + UCSYNC;       // I2C Master, synchronous mode
UCB0CTL1 = UCSSEL_2 + UCSWRST;              // Use SMCLK, keep SW reset
UCB0BR0 = 12;                               // fSCL = SMCLK/12 = ~100 kHz
UCB0BR1 = 0;
UCB0I2CSA = 0x48;                           // Slave Address is 048h
UCB0CTL1 &= ~UCSWRST;                       // Clear SW reset, and resume
UCB0IE |= UCTXIE;                           // Enable TX interrupt

while (1)
{
    __delay_cycles(50);                     // Delay  between transactions
    PTxData = (unsigned char *)TxData;      // TX array start address
                                            // Place breakpoint here to see
                                            // each transmit operation.
    TXByteCtr = sizeof TxData;              // Load TX byte counter

    UCB0CTL1 |= UCTR + UCTXSTT;             // I2C TX, start condition

    __bis_SR_register(LPM0_bits + GIE);     // Enter LPM0, enable interrupts
    __no_operation();                       // Remain in LPM0 until all data
                                            // is TX'd
    while (UCB0CTL1 & UCTXSTP);             // Ensure stop condition got sent
  }
}

//----------------------------------------------------------------------------
// The USCIAB0TX_ISR is structured such that it can be used to transmit any
// number of bytes by pre-loading TXByteCtr with the byte count.
// Also, TX Data points to the next byte to transmit.
//----------------------------------------------------------------------------
#pragma vector = USCI_B0_VECTOR
__interrupt void USCI_B0_ISR(void)
{
  switch(__even_in_range(UCB0IV,12))
  {
  case  0: break;                           // Vector  0: No interrupts
  case  2: break;                           // Vector  2: ALIFG
  case  4: break;                           // Vector  4: NACKIFG
  case  6: break;                           // Vector  6: STTIFG
  case  8: break;                           // Vector  8: STPIFG
  case 10: break;                           // Vector 10: RXIFG
  case 12:                                  // Vector 12: TXIFG
    if (TXByteCtr)                          // Check TX byte counter
    {
      UCB0TXBUF = *PTxData++;               // Load TX buffer
      TXByteCtr--;                          // Decrement TX byte counter
    }
    else
    {
      UCB0CTL1 |= UCTXSTP;                  // I2C stop condition
      UCB0IFG &= ~UCTXIFG;                  // Clear USCI_B0 TX int flag
      __bic_SR_register_on_exit(LPM0_bits); // Exit LPM0
    }
  default: break;
  }
}
//============================================================================
```

This second example illustrates how to use the MSP430 USCI in its I²C slave mode to receive the message string sent by the code in the previous example.

**Example 9.9 (USCI_B0 I2C MSP430 Master-slave TX)** *Provide a program to receive via an I²C bus slave a multi-character message and store it somewhere in memory.*

**Solution:** *The solution will use two MSP430F5438 connected one as master, the other as slave via their I2C interfaces (P3.1 = SDA and P3.2 = SCL). This solution provides the receiver code using a demo written in C-language by P. Thanigai and M. Morales [81].*[12] *The code uses the internal DCO frequency for the CPU, but as the I²C interface is configured in slave mode, no clock generator is needed. The MCU is placed in LPM0. The receiver ready state of the USCI is used to wake-up the CPU and place the received data in memory.*

```
//==============================================================================
// MSP430F543xA Demo - USCI_B0 I2C Slave multi-byte RX from MSP430 Master
// By P. Thanigai and M. Morales - 06/2009
// Copyright (c) Texas Instruments, Inc.
//------------------------------------------------------------------------------
#include <msp430.h>

unsigned char *PRxData;                    // Pointer to RX data
unsigned char RXByteCtr;
volatile unsigned char RxBuffer [128];     // Allocate 128 byte of RAM

int main(void)
{
  WDTCTL = WDTPW + WDTHOLD;                 // Stop WDT

  P3SEL  |= 0x06;                          // Assign I2C pins to USCI_B0
  UCB0CTL1 |= UCSWRST;                      // Enable SW reset
  UCB0CTL0 = UCMODE_3 + UCSYNC;            // I2C Slave, synchronous mode
  UCB0I2COA = 0x48;                        // Own Address is 048h
  UCB0CTL1 &= ~UCSWRST;                    // Clear SW reset and resume
  UCB0IE |= UCSTPIE + UCSTTIE + UCRXIE;    // Enable STT, STP & RX interrupt

  while (1)
  {
    PRxData = (unsigned char *)RxBuffer;   // Start of RX buffer
    RXByteCtr = 0;                         // Clear RX byte count
    __bis_SR_register(LPM0_bits + GIE);    // Enter LPM0, enable interrupts
                                           // Remain in LPM0 until master
                                           // finishes TX

    __no_operation();                      // Set breakpoint *here* and read
  }                                        // read out the RxData buffer
}

//------------------------------------------------------------------------------
// The USCI_B0 data ISR RX vector is used to move received data from the I2C
// master to the MSP430 memory. It wakes the CPU from LPM0 to process
// received data in the main program upon (re)start or stop conditions.
//------------------------------------------------------------------------------
#pragma vector = USCI_B0_VECTOR
__interrupt void USCI_B0_ISR(void)
```

---

[12] See Appendix E.1 for terms of use.

```
{
  switch(__even_in_range(UCB0IV,12))
  {
  case  0: break;                            // Vector  0: No interrupts
  case  2: break;                            // Vector  2: ALIFG
  case  4: break;                            // Vector  4: NACKIFG
  case  6:                                   // Vector  6: STTIFG
    UCB0IFG &= ~UCSTTIFG;
    break;
  case  8:                                   // Vector  8: STPIFG
    UCB0IFG &= ~UCSTPIFG;
    if (RXByteCtr)                           // Check RX byte counter
      __bic_SR_register_on_exit(LPM0_bits);
    break;
  case 10:                                   // Vector 10: RXIFG
    *PRxData++ = UCB0RXBUF;                   // Get RX'd byte into buffer
    RXByteCtr++;
    break;
  case 12: break;                            // Vector 12: TXIFG
  default: break;
  }
}
//============================================================================
```

## 9.5.4 The MSP430 Enhanced USCI

The Enhanced Universal Serial Communication Interface (eUSCI) is a modified version of the USCI module discussed in the previous section. The eUSCI is featured in MSP430x5xx/x6xx devices, and at the time of this printing had been released in the form of eUSCI_A and eUSCI_B. Like their USCI predecessors, eUSCI_A supports UART and SPI communications, while eUSCI_B supports $^2$C and SPI modes.

### The eUSCI_A

The eUSCI_A in UART mode is functionally equivalent to the USCI_A UART, except in the form of generating the baud rate clock. Its support for fractional dividers, although still using a prescaler and modulator, now features an 8-bit UCBRSx code for selecting one of 256 modulation patterns when operated in the low-frequency mode. This offers an improved baud rate approximation, reducing the error due to deviation from the desired baud rate in both the transmitter and receiver sides. The device family User's Manual provides revised tables for determining the values for configuration registers given the desired baud rate and clock frequency [59]. The tables also provide the recalculated expected errors resulting from using this new scheme.

The oversampled baud rate generation mode, also supported in the eUSCI_A UART remains the same as in the standard USCI.

In SPI mode, eUSCI_A is functionally and structurally equivalent to its predecessors in the original USCI module. Addresses, register names, and flags change, so the reader is referred to the datasheets and family's manuals of the specific device used for the specific nomenclature of configuration registers and flags.

## The eUSCI_B

The eUSCI_B in $I^2C$ mode provides improvements over its predecessor by including extra features that facilitate its usage when operated via interrupts and coordinated with DMA channels. Specifically, eUSCI_B adds the following features:

- An 8-bit byte counter (UCBxBCNT) with interrupt capability and automatic STOP assertion. This counter allows to keep track of the number of characters that have been transferred over the $I^2C$ bus since the last start condition was detected. A threshold value can be programmed to generate an interrupt when reached or automatically generating stop condition in the bus (master mode only).

- Up to four hardware slave addresses, each having its own interrupt and DMA trigger. This addition helps accelerating transactions when a slave module is needs to be operated responding to multiple addresses. Although this kind of operation is possible via user software, the addition of hardware addresses and their corresponding DMA channels allows for using this feature without leaving a low-power mode.

- Mask register for slave address and address received interrupt. This new feature allows extending the number of addresses to which a slave module can respond. When an address has been masked, the address match only looks at the unmasked address bits, resulting the masked bits as don't care conditions.

- Clock-Low-Timeout interrupt to avoid bus stalls. This feature allows detecting if the SCL line is held low by a clock stretching operation longer than a predefined period. An optional interrupt can be triggered with this condition.

As a result of the enhancements introduced into the eUSCI, the module now supports sixteen interrupt conditions, ten more than its predecessor. Despite these additional sources, the eUSCI_B continues to feature a single interrupt vector, making necessary to apply multi-sourced ISR techniques in its programming. Despite the enhancements, the module functionality remains compatible in most aspects to the standard $I^2C$ description. This allows to use the description of its predecessor to understand its functionality and referring to the user's manual and data sheet of the specific device used for details regarding addresses, registers, and flags.

The operation of eUSCI_B in SPI mode, like that of eUSCI_A remains unchanged with respect to that of the original USCI.

## 9.6 Chapter Summary

Serial communications have become the most widely used data transfer modality in modern embedded applications. This chapter provided a comprehensive discussion of the fundamentals of serial communication, establishing the conceptual and practical foundations for designing and using serial channels in embedded applications.

From the elemental protocols developed for RS-232 to the intricacies of the universal serial bus, the sections presented systematically discussed asynchronous and synchronous protocols for serial channels.

Formats and interfaces that include UART, SPI, $I^2C$, and USB were discussed. Physical standards for short and long range channels, at low-, medium-, and high-speed were discussed in detail.

Lastly, an exploration of the serial modules embedded in all MSP430 series devices were analyzed and illustrative examples provided to guide the reader through an learning experience of serial subjects.

## 9.7 Problems

9.1 A router's console port is used to communicate with a computer using a serial communication port. Both the router and the computer must be set using the same parameters. If the communication is set for 115,200 bits per second, 8 data bits, and 1 stop bit, how long would a 1 MB file take to be transferred from the computer to the router?

9.2 What is the purpose of the start bit in an asynchronous serial communication? How does the receiver synchronize with the transmitter?

9.3 Two computers are communicating using a serial communication channel. However, garbage is seeing at the receiving end. What is the possible cause of this problem?

9.4 In an SPI connection the master and slave need to communicate. What would the slave have to do to send data to the master? What would the master need to do to send data to the slave? Which station controls the data transfer and generates the clock? Can there be several slaves selected at the same time?

9.5 Draw a time diagram showing how would the following data be transmitted over a serial connection: 01100111.

9.6 Parallel communication had many advantages over serial communication. However, it was serial communication that ended on top. Can you explain why?

9.7 Provide a connection diagram and software modules to establish a master-slave $I^2C$ connection between two MSP430 launchpads connected one as master and the other as slave using the USI module. Provide a code implantation including both, master and slave sides in assembly language.

9.8 Show a derivation of the formula to estimate the transmitter baud rate error in the USCI module of a series x5xx MSP430 when configured in oversampled mode. Compare it against the formula in the device user's manual. Comment about your result.

# Chapter 10
# The Analog Signal Chain

Embedded systems interact with the outer real world, processing information received from it and delivering back an output to modify the performance of a device or to provide information needed to make decisions. In this environment, most signals to be processed are analog. Temperature is not simply hot or cold, pressure high or low, light intensity bright or dark. They all fall within a wide range of possible values. Hence, it becomes necessary to interface this analog world with our digital embedded systems.

Signal processing may be digital, analog, or a combination of both worlds, i.e., mixed signal. In fact, the "MSP" part of the microcontroller name MSP430 emphasizes this focus, "Mixed Signal Processor". Hence, several of the peripherals embedded in different models belong to the analog portion of the signal processing, the analog signal chain.

## 10.1 Introduction to Analog Processing

The term *signal chain*, short for signal-processing chain, refers to a series of signal-conditioning electronic components connected in tandem, or cascade. That is, the output of one portion of the chain supplies input to the next one. In mixed-signal systems, the term *analog signal chain*, or simply *analog chain*, focuses on the components that process the analog signals prior to the digital processor, or that extract analog information from the digital processor's output. The complete process is illustrated in Fig. 10.1.

Inputs to the system may be temperature, pressure, humidity, speed, flow rate and so on. Most often, the first block in the system transforms the information into an electrical signal such as voltage or current. This operation is known as *sensing*, and is based on *sensors*. These are components that receive as stimulus the physical magnitude of interest, or a change in that magnitude, and deliver a value or a modulation in an electrical signal related to the input. This electrical magnitude may be a voltage, current, resistance, charge, or an electrical or magnetic field.

M. Jiménez et al., *Introduction to Embedded Systems*,
DOI: 10.1007/978-1-4614-3143-5_10,
© Springer Science+Business Media New York 2014

**Fig. 10.1** Block description
for mixed-signal processing
chain

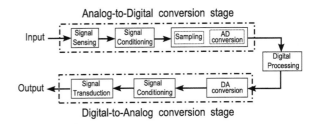

Often, the electrical signal provided by the sensing stage is not adequate for direct use in the chain, and should be processed with a *conditioning circuit*. For example, the sensing of pressure may produce a voltage in the millivolt range, insufficient for processing. Therefore, this signal should be amplified to bring it up to an appropriate volt range. Amplification is very common in conditioning. Other situations may require filtering, linearization, impedance conversion, and so on.

The conditioning stage output is fed into an *analog-to-digital converter* (ADC) stage whose output is then used as input to the digital microcontroller or processor. There are different ADCs and this component may include subsystems such as the sample-and-hold portion.

The ADC's output are digital words fed to the digital processor where they are utilized according to the user's program. This digital processor in turn will deliver $n$-bit words as output. Depending on the application, these words may be directly used (perhaps via a conditioning or interfacing circuit) or may need to be translated into an analog value, initiating a digital-to-analog conversion phase. The first step in this process is the *digital-to-analog converter* (DAC) which will deliver a voltage or current output whose value is proportional to the decimal equivalent of the digital input word. As in the previous phase, this analog output may need further conditioning such as filtering, smoothing and so on. The final step may be the transducer, which operates akin to the sensor, transforming an electrical signal energy into another type of energy. For example, the speaker translates current or voltage variations into acoustic information.

Important electronic components used in the analog-signal-chain are sensors, operational amplifiers, comparators, operational amplifiers, analog-to-digital and digital-to-analog converters, among others. In addition, supporting the operation of the main players we have reference voltage and reference current, LDO, sample-and-hold circuits, and so on. A brief introduction to some of the components in the analog chain devices is presented in the following sections.

## 10.2  Analog Signal Chain in MSP430

The MSP430 family of microcontrollers was projected as a mixed signal device. Therefore, several of the devices needed in the analog chain have been included in several models. Table 10.1 summarizes the available analog peripherals in MSP430

**Table 10.1**  Analog signal chain devices in MSP430

| Series | Comparator | OA | DAC | ADC | REF | LDO |
|--------|-----------|-----|------|-----|-----|-----|
| 'x3 | No | No | No | Yes | No | No |
| 'x1 | Yes | No | Yes | Yes | No | No |
| 'x2 | Yes | Yes | Yes[a] | Yes | No | No |
| 'x4 | Yes | Yes | Yes[a] | Yes | No | No |
| 'x5&'x6 | Yes | No | Yes[a] | Yes | Yes | Yes |

[a] SAR and Sigma-Delta converters

controllers for some models in the various families. The devices can be configured to receive external or internal inputs, and also to route their output to external terminals. The reader should consult the user guides and data sheets for full details on specific family members.

## 10.3  Sensors and Signal Conditioning

Many system applications require the measurement or detection of a physical or chemical or electrical quantity or condition. Detectors are often used as on/off switches to signal the presence of an environmental condition, even if the sensor function itself may be of analog type. Analog sensors are used to indicate the magnitude or change in this environmental condition by means of a value or change in an electrical property. Usual electrical magnitudes include voltage, current, resistance, capacitance, Hall-effect, and charge.

Usually, sensors are named after the application they are intended for, and not by the physical way in which they operate. Thus, we speak for example of proximity sensors. Also, many sensors are in fact integrated circuits, or MEMS systems that may or may not include conditioning circuits in the chip. Moreover, sometimes the term *transductor* is used as synonymous to sensor. But many authors and applied engineers use the term to indicate conversion from electrical energy to other type of energy, as for example in the case of speakers that translate current or voltage variations into acoustic waves. In this way, transductors are used in the output stage of the system. To finish this short discussions about terminology, engineers in other fields, like mechanical engineers, use "sensor" to speak of elements delivering a mechanical type energy.

### 10.3.1  Active and Passive Sensors

Sensors can be identified as active or passive. Active sensors generate a signal, normally current or voltage, in response to the environmental stimulation they are used for. Examples of such sensors are thermocouples and piezoelectric accelerometers. Most often the magnitude of the generated signal is quite small, requiring an amplification stage.

Passive sensors, on the other hand, produce changes in some passive quantity, like a resistance, or a capacitance, or an inductance. Some examples of passive sensors are the thermo-resistors or thermistors, capacitive humidity sensors, and others. Passive sensors require power sources to generate information. Often, the actual reading or processing is by means of voltage. For resistance type sensors, resistance sub-circuits like voltage/current dividers, Wheatstone bridge circuits or current sources are used, as illustrated in Fig. 10.2.

Capacitance is typically measured indirectly. For example, by using relaxation oscillator and measuring the charging time required to reach a threshold voltage, or by measuring the oscillator's frequency. A capacitive voltage divider using a fixed-frequency AC voltage signal is another measurement technique frequently used; more accurate instruments may use a capacitance bridge configuration.

## 10.3.2 Figures of Merit for Sensors

Sensors are not perfect. We need therefore ways to determine the fitness and validity for the intended applications. Some figures of merit associated to sensors are the following:

**Sensitivity:** This is mathematically defined by the derivative of output with respect to input, that is, the curve's slope. It represents the minimum input of physical parameter that will create a detectable standardized output change.

  • *Sensitivity error*:Departure from the ideal slope.

**Range:** stated by the maximum and minimum values that can be measured.

  • *Dynamic range* is the difference between the maximum and minimum measurable values. This range is also called *full scale*.

**Precision:**  This concept is associated to the degree of measurement reproducibility. While an ideal sensor would output exactly the same value every time its input is subjected to the same magnitude of the measured parameter, real sensors

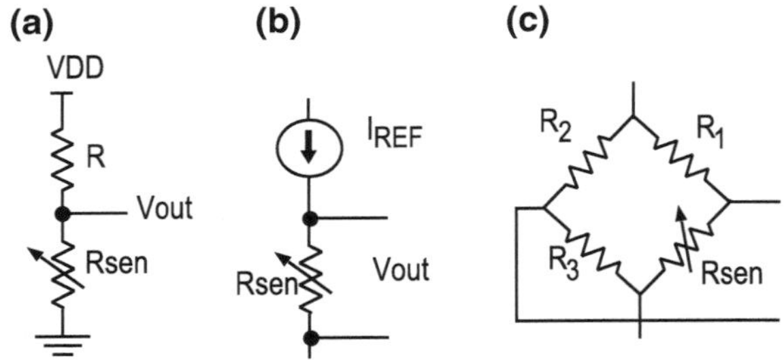

**Fig. 10.2** Converting resistance to voltage information: **a** voltage divider; **b** with reference current; **c** with resistive bridge

outputs are distributed around a mean value. Precision is usually described in statistical terms.

**Accuracy:**  The maximum difference that exhibited between the actual value and the value at the output of the sensor. True values are previously measured using a primary or good secondary standard. Accuracy can be expressed either as a percentage of full scale or in absolute terms.

**Resolution:**  The smallest detectable incremental change of input parameter that can be detected in the output signal. It can be expressed either as a proportion of the reading (or the full-scale reading) or in absolute terms.

**Offset:**  This parameter is defined as the output that will exist when it should be zero. In other words, it is a particular value of the accuracy error.

**Linearity:**  The linearity, or more exactly *non-linearity*, refers to the extent to which the actual measured curve departs from an ideal linear curve determined by a least squares fit, line and the actual measured or calibration line. Linearity is often specified in terms of percentage of nonlinearity, as defined by (10.1).

- *Dynamic Linearity:*

**Hysteresis:**  The property that tells how well the sensor or transducer follows changes of the input parameter regardless in which direction this occurs. Ideally the sensor should not have hysteresis.

**Response Time:**  This is the time required for the output to change from its previous state to a final settled value within a tolerance band, $\pm\varepsilon$, of the correct new value. This becomes very important for rapid changes in the input parameter values.

Notice that precision and accuracy are not synonymous. A sensor or instrument can be very precise, yielding consistently similar inaccurate readings (Fig. 10.3a). Conversely, it can be acceptably accurate, with imprecise readings (Fig. 10.3b). Of course, it is desirable to have both precise and accurate devices (Fig. 10.3c).

An ideal sensor has a linear response to changes of the input magnitude. In fact, in many instances the conditioning circuit consist precisely in producing a linear curve. Most often, however, the actual response deviates from a linear behavior (Fig. 10.4), which is really a least squares fit. The maximum deviation determines the non-linearity factor with respect to full-scale as

$$Nonlinearity(\%) = \frac{\Delta_{in(max)}}{IN_{FS}} \times 100\,\% \qquad (10.1)$$

**Fig. 10.3** Precision and accuracy: **a** "accurate" but imprecise readings; **b** precise but inaccurate readings; **c** precise and accurate readings

(a)

(b)

(c)

● True value
○ Readings/output values

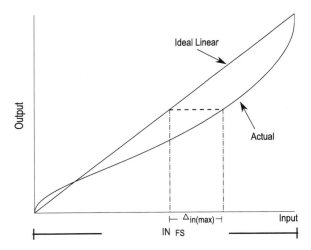

**Fig. 10.4** Definition of linearity in terms of input

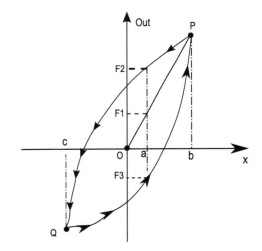

**Fig. 10.5** Hysteresis curve

where $\Delta_{in(max)}$ is the maximum input difference required to obtain an ideal output equal to the actual output, and $IN_{FS}$ is the full-scale input range.

The hysteresis information tells us how the sensor responds to direction of changes in the input values. Figure 10.5 illustrates this phenomenon. Starting from point O, when the input value x reaches the value "a", the output value will be F1. As long as the input value does not increase up to value "b", the output will be the same as before. However, the situation changes as soon as the input parameter gets to the value "b", and thus the curve to point P. When x decreases, it will follow the upper path, so for x = a the output will now be F2. when x goes down to x = c, reaching point Q, the path will now change for increasing values, and the output for x = a will now be F3. When the path reaches point P, the path will be again the upper one. Hence, we see that for input values between c and b, the output will depend on the path.

**Example 10.1**  *One of the most important magnitudes to measure is temperature. Even applications intended to measure other variables such as pressure, flow, or position, require temperature monitoring to ensure accuracy. The four most common temperature sensors are thermocouples, RTD (resistance-temperature-detectors), thermistors, and IC (integrated-circuit) sensors.*[1]

*A thermocouple is built with two wires of dissimilar metals or alloys, with a weld bead bonding them on one end. When there is a temperature difference between the bead and the open end of the thermocouple wires, or this end is heated, a voltage difference, called Seebeck voltage, appears between the open end of the two wires. This voltage depends on temperature difference between wire ends. The major strengths of thermocouples are the wide temperature range of application and the durability. Linearity depends on the type, and in some applications the function must be approximated by at least a fourth order polynomial or a look-up table be used. Their sensitivity is in the range of tens of microvolts per degree Celsius with an accuracy of ±0.5 °C. Practical use of thermocouples requires an isothermal block to compensate for parasitic thermocouples caused by the cooper connections.*

*RTDs provide excellent accuracy in a temperature-sensing environment, and are a good choice for accurate temperature measurement. Their temperature range is narrower than that of thermocouples but wider than those of thermistors. RTD are made of metals, which have a positive temperature coefficient; that is, resistance increases with temperature. Platinum RTDs offer the most stable version and offer better linearity, with an accuracy within ±4.3 °C over its temperature range, and a sensitivity of 0.00385 Ω/Ω/°C. For better accuracy, one may use a corrective equation or a look up table.*

*Thermistors are by far the mostly widely used sensors in temperature applications. They range in price from 10 cents to approximately $25, and find use in applications from automotive monitoring and exhaust-emissions control to ice detection, skin sensors, blood and urine analyzers, home appliances, mobile phones, and many more. Although thermistors are either NTC (negative temperature coefficient) or PTC (positive temperature coefficient), the former group is better for analog applications requiring precision temperature measurement, while the latter best suit switching applications. For computing purposes, one advantage of most NTC thermistors is that the curve can be approximated by the Steinhart-Hart equation*

$$\frac{1}{T} = A + B \ln R + C(\ln R)^3 \tag{10.2}$$

*where T is the temperature in Kelvin, R is the thermistor resistance and A, B and C are curve-fitting coefficients found with three points from the R-T curve within a 100 °C range. The equation attains a ±0.02 °C curve fit. A variation of this equation*

---

[1] Agilent's application note 290 at http://cp.literature.agilent.com/litweb/pdf/5965-7822E.pdf offers a very good introduction and tutorial on the use of these devices.

*is an approximation given by*

$$\ln R = a_o + a_1 \frac{1}{T} + a_3 \frac{1}{T^3} \tag{10.3}$$

*which stems from the solution of the previous one. Usually the coefficients for either form are provided by the manufacturer.*

*The above equations for the thermistor or else a look up table can be used when using this sensor. Also, a rather good linearization technique can be achieved with an appropriate resistor in parallel.*

*The fourth group of temperature sensors are the Integrated-Circuit Temperature sensors. Their advantages include user-friendly output formats and easy installation during printed-circuit-board assembly. They have however slow response. They come in a variety of combinations of input and output. In analog form, the common output form are in current or voltage, with sensitivities of the order of 1 μA/K and 10 mV/K. Since they are already integrated, they can be combined with more circuitry and offer a digital output already. In addition, these sensors offer very good output linearity.*

## 10.4 Operational Amplifiers

*Operational amplifiers* (OA) are one of the IC backbones for signal conditioning. The designer may select the adequate model based on the application needs and the figures of merit for the OA, such as the gain-bandwidth product, offset voltage and current, input and output impedances, S/N ratio, slew-rate, and CMRR, among others. Most OAs are compensated internally, but further compensation may be applied if necessary.

The voltage OA, or simply OA, whose symbol and transfer function are shown in Fig. 10.6, is basically a differential amplifier with a *non-inverting input* $V_+$ and an *inverting input* $V_-$. It is often biased using two DC power supplies VCC and VSS, usually VSS $= -$VCC. However, single power supply versions (i.e., VSS $= 0$ V) are also common and it is also possible to reduce the two source to one source version using appropriate circuiting. The power supply values, including ground for one supply cases, are called *rails*.

The output versus differential input voltage transfer curve is shown in Fig. 10.6b. As seen in the curve, for differential inputs large enough (in a positive or negative sense), the output voltage reaches a saturation value, which is always in the region $VCC \geq Vout \geq VSS$. Since the gain is very high, usually greater than 50,000, saturation is reached after several microvolts in difference. When the saturation values are equal to the rail values, the OA is said to be rail-to-rail OA.

Mathematically, in the linear region for a differential input $\varepsilon+ \geq (V_+ - V_-) \geq \varepsilon-$ the transfer curve can be expressed as $Ao(V_+ - V_-)$, where the differential gain $Ao$, called *open-loop gain*. This gain is very high, 80 dB or more, so the $\varepsilon$ values are in the microvolt range. For this reason, numerical calculations use the *ideal OA model*,

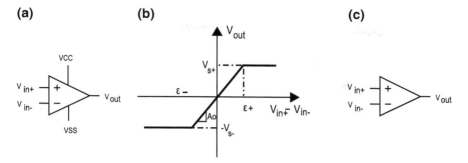

**Fig. 10.6** Operational amplifiers: **a** symbol showing DC supplies, **b** transfer curve, **c** simplified common symbol

also called *infinite gain OA model* because it assumes $Ao = \infty$, defined as

$$V_{out} = \begin{cases} V_{s+} & \text{for } (V_+ > V_-) \geq \varepsilon+ \\ (V_+ - = V_-) & \text{for } V_{s+} \geq Vout \geq V_{s-} \text{ (linear region)} \\ V_{s-} & \text{for } (V_+ < V_-) \leq \varepsilon- \end{cases} \qquad (10.4)$$

and

$$i_+ = i_- = 0 \qquad (10.5)$$

Negative feedback is required for stability in analog applications in the linear region. Working with positive feedback or in open-loop condition keeps the output saturated. This condition is applied when using the OA as comparator.

### 10.4.1 Basic OA Linear Configurations

Table 10.2 summarizes some important linear configurations used in analog applications, together with their *closed-loop gains* or formulas.

Resistance values in OA configurations cannot be very high because of associated stray capacitance problems. But they should be large enough to keep normal operating conditions on the OA without causing an excessive load to the signal. All configurations are valid for static inputs or slow signals. For fast changes, the inclusion of reactive elements may be required to deal with parasitics, as illustrated in the following example.

**Example 10.2** *Figure 10.7 shows some applications of basic configurations for sensor interface. Inset (a) presents a connection for a resistive type sensor, like a thermistor. In this configuration,* Radj *is a potentiometer used for calibration with respect to a particular value. In thermistors, the resistance at 25 °C is a common reference.*

**Table 10.2** Linear configurations using OA's

| Configuration | Circuit | Formula |
|---|---|---|
| Inverting amplifier | | $\frac{V_{out}}{V_{in}} = -\frac{R_f}{R_1}$ |
| Non inverting amplifier | | $\frac{V_{out}}{V_{in}} = 1 + \frac{R_f}{R_1}$ |
| Buffer[a] | | $\frac{V_{out}}{V_{in}} = 1$ |
| Inverting adder | | $V_{out} = -\left(\frac{R_f}{R_1}V_1 + \frac{R_f}{R_2}V_2 + \frac{R_f}{R_3}V_3\right)$ |
| Differential amplifier[b] (w/offset) | | $V_{out} = \frac{R_2}{R_1}(V_2 - V_1) + V_{sh}$ <br> if $\frac{R_2}{R_1} = \frac{R_4}{R_3}$ |
| Inverting integrator | | $V_{out} = -\frac{1}{RC}\int V_{in}dt$ |
| Current to voltage converter | | $V_{out} = R\,I_{in}$ |

[a] Also known as voltage follower or unity gain
[b] Also known as difference amplifier; subtractor when $R_2 R_3 \neq R_4 R_1$

*The second example in (b) shows the case for a current type sensor, using the current-to-voltage converter. Ideally, there is no current in the resistance Ro since this is in parallel with the inputs of the OA. The voltage type sensor, as in the thermocouple case, can be interfaced with a voltage amplifier as shown in (c).*

*An example of a current type sensor is the photodiode, which can be used in a current-to-voltage configuration. The voltage at the diode is practically zero, a condition which is known as photovoltaic mode. There are variants to this circuit, especially when using single supply OA's. When there are fast changes in input, the reactive parasitics of the resistor and the photodiode introduce frequency dependent components which degrade the performance considerably. Capacitor C is used to compensate for these problems by introducing a pole in the system.*

Voltage $V_{sh}$ in the difference, or differential, amplifier in Table 10.2 provides an offset or a shift to the output. The most general version presented in textbooks and in many applications has this terminal connected to ground. The offset is particularly useful for those single supply OA configurations to insure that the output falls within the applicable limits.

The formula for the differential depends on matching the ratios $R_2/R_1 = R_4/R_3$. For non-matching ratios, these amplifiers are also known as subtractor circuit with different weights for each voltage. Using nodal equations and the ideal OA model, it can be verified that

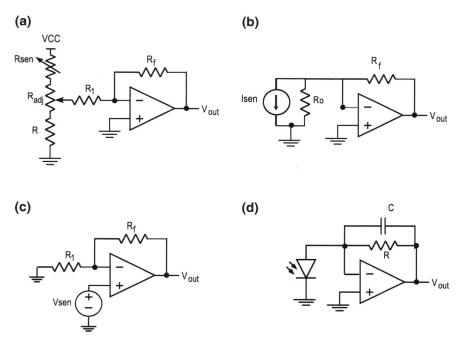

**Fig. 10.7** Sensor conditioning examples with basic OA amplifiers: **a** for resistive type sensor, **b** a current type sensors; **c** voltage type sensor; **d** a photodiode (current type) with capacitive compensation for fast changes

**Fig. 10.8** Two OA differential floating output amplifier

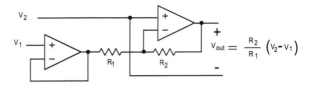

**Fig. 10.9** Instrumentation amplifier (with output shifting $V_{sh}$)

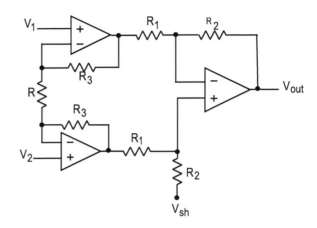

$$\frac{\frac{R_2}{R1}+1}{\frac{R_4}{R3}+1} = \frac{\frac{R_2}{R1}V_1 + V_{out}}{\frac{R_4}{R3}V_2 + V_{sh}} \qquad (10.6)$$

This is the general case. If the resistance ratios are matched, we arrive at the formula given in the table.

Another differential amplifier built with two operational amplifiers and only two resistors is shown in Fig. 10.8. The output, however, is a floating one. This configuration is sometimes applied in embedded systems.

The loading effect on inputs in the difference amplifier is solved with the so called *instrumentation amplifier*. A simple version consists in the use of buffers at the inputs. A more general solution is presented in Fig. 10.9. This is also based on the previously introduced differential amplifier and offers additional gain control using resistance R. Instrumentation amplifiers are available in IC form from different vendors, with only the R resistance and shifting reference, when a terminal is available, being provided by user. In some cases, $V_{sh}$ is grounded internally. For this instrumentation amplifier the output is given by

$$V_{out} = \left[1 + \frac{2\,R_3}{R}\right] \frac{R_2}{R_1} (V_2 - V_1) + V_{sh} \qquad (10.7)$$

**Example 10.3** *Differential and instrumentation amplifiers become very important in applications where for some reason or other we cannot use grounded signal* information. *The examples in Fig. 10.10 illustrate some configurations. Notice the use of the bridge configuration, where one or more of the resistances may vary with sensed magnitudes, depending on the applications.*

*Although (a) and (b) look similar, the first case the sensor is a resistance, while in the second one it is the current source. In the former case, the current is kept constant using a current reference circuit. Any change in the amplifier input voltage is due to a variation in the resistance. On the other side, the circuit in (b) has a constant resistance, and changes in input voltage to the amplifier are due to the current response of the sensor to variations in the magnitude being monitored.*

*The bridge circuit in (c) can be driven either by a voltage supply or a reference current.*

Important applications of OA fall in voltage-to-current conversions. One of the most popular circuits in this group is the *Howland current source*, of which two versions are shown in Fig. 10.11a, b.

It can be shown that for these circuits we have, respectively,

$$I_{out} = \frac{1}{R_3}(V2 - V1) \text{ if } \frac{R_2}{R_1} = \frac{R_4}{R_3} \tag{10.8}$$

for (a), and

$$I_{out} = \frac{R_2}{R_1 R_5}(V2 - V1) = \frac{1}{R_3}\left(1 + \frac{R_4}{R_5}\right)(V2 - V1) \text{ if } \frac{R_2}{R_1} = \frac{R_4 + R_5}{R_3} \tag{10.9}$$

for (b). Comparing the equations, it is seen that the improved version allows greater currents. The validity of these equations are limited by the output current and voltage limitations of the OA.

There are other voltage-to-current converters in the literature based on the original Howland source. Trimming is common in these circuits to achieve the matching conditions. OA limitations may force further compensation schemes; the reader can

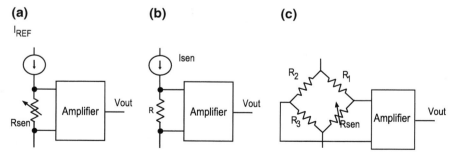

**Fig. 10.10** Sensor conditioning examples with differential or instrumentation amplifiers: **a** for resistive type sensor, **b** for current type sensors; **c** with resistive sensors in bridge

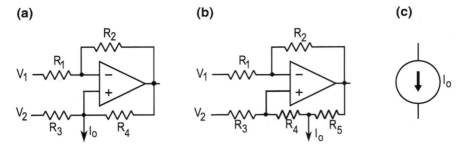

**Fig. 10.11** Howland current source: **a** basic configuration, **b** modified improved version, **c** equivalent representation

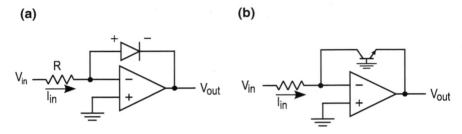

**Fig. 10.12**   Basic logarithm amplifiers: **a** diode based, **b** NPN bipolar based

find in literature several modifications to take into account limitations. Any input can be grounded, so positive or negative values for the current $I_o$ are possible.

The OA also has non linear applications. One of them is using the device as a comparator, either by not using feedback, in which case saturation is reached easily because of the high OA gain, or by forcing a positive feedback loop into the OA.[2] This use is sometimes practical although it is better to utilize comparator circuits, which are based on the same OA transistor construction but are designed to always be in saturation.

Other non linear applications use the OA in the linear region, but include non linear elements in the circuit. One example is seen in the logarithmic amplifiers shown in Fig. 10.12. These amplifiers require the current $I_{in}$ to be positive. Exchanging the resistance and the diode, or resistance and transistor, one gets exponential amplifiers.

The output relation for the logarithmic amplifiers are as follows. For the diode circuit,

$$V_o = -V_T \ln \left( \frac{I_{in}}{I_S} \right) = -V_T \ln \left( \frac{V_{in}}{R I_S} \right) \tag{10.10}$$

---

[2] In most applications, positive feedback is attained by connecting the OA output to the non-inverting input. However, the concept should be considered in the overall circuit, and some applications may use the local positive feedback connection while still providing a real negative feedback.

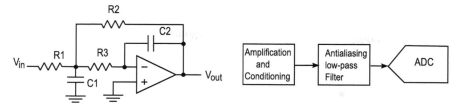

**Fig. 10.13** A low-pass filter example and insertion for antialiasing

where $I_S$ is the saturation diode current for (a), and the emitter-base junction saturation current in (b). $V_T$ is the thermal voltage. The first circuit serves also as basis for a commercial IC logarithm amplifier from Texas Instruments, the Log114,[3] which has many applications, including interfacing with photo diodes and other sensors to ADC's.

## 10.4.2 Antialiasing Filter

OA based active filters, more specifically, antialiasing filters, are very important in the analog signal chain. Figure 10.13 illustrates an example of a low-pass filter and its insertion in the analog chain. Bandpass filters are also sometimes used. The importance will become clearer in Sect. 10.7.

The subject on OA and their applications in the analog signal chain is quite extensive. For a more in depth treatment of operational amplifiers in different applications and how to deal with imperfections, the reader may consult the references at the end of the book.

## 10.4.3 Operational Amplifiers in MSP430

Several members of the MSP430'x2 and 'x4 series have two or three internal OA as peripherals, denoted as OA0, OA1 and OA2. All OA's are single supply, rail-to-rail output. Through software the user can control their speed (slew rate), for optimized settling time versus power consumption, configure single or multiple OA circuits, including differential amplifiers, and also program PGA (Programmable Gain Amplifier) with a feedback resistor ladder. In addition, the MSP430'x4xx series allow to configure for rail-to-rail input.

Operational configurations are achieved by software via readable/writable control registers 0 and 1, OAxCTL0 and OAxCTL0. There are slight variations in the control

---

[3] See document SBOS301A at http://www.ti.com/lit/ds/symlink/log114.pdf for details and application examples.

**(a)**                          **(b)**                                    **(c)**

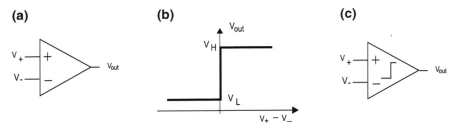

**Fig. 10.14** Comparator: **a** symbol, **b** transfer curve, **c** alternate symbol

registers of the two series, as well as among different members of the 'x4xx series. Moreover, some members of this series also have switch control registers for the OA's.

In this text we explain briefly the control registers for the 'x2xx series, applicable to some members of the 'x4xx series as well. We recommend the reader to consult the user guides and data sheets for more information.

## 10.5  Comparators

A comparator is basically a high gain differential amplifier, like the OA. In fact, the OA can be used as comparator, but it is preferable to utilize ad-hoc circuits if possible. The symbol(s) and output curves are shown in Fig. 10.14. For many engineers, this is the most basic element in the signal chain.

Ideally, the comparator output voltage goes to a high value $V_H$ whenever its non-inverting input is greater than the inverting, and to a low value $V_L$ when it is lower. Input equality is undefined, as explained later. In mathematical terms,

$$V_{out} = \begin{cases} V_{SH} & \text{for } (V_+ - V_-) > 0 \\ V_{SL} & \text{for } (V_+ - V_-) < 0 \end{cases} \qquad (10.11)$$

The values of $V_{sH}$ and $V_{sL}$ may be equal to the rail values or to other values by design. They are specified by the manufacturer.

Most comparators have either an open collector or open drain output, or else a push-pull output configuration. The former cases allows for connection to a subsystem with a different voltage supply. In this case, the output cannot be left floating; usually a pullup resistor is used.

A common application of comparators is that of threshold detector. Two simple examples are shown in Fig. 10.15. Circuit (a) is a simple threshold detector, which goes to VH whenever the input signal exceeds the threshold reference VREF. This operation is illustrated by Fig. 10.16. The circuit in Fig. 10.15b is a window comparator. Here, both comparator outputs are high only when $V_{REFL} < V_{in} V_{REFH}$.

**Fig. 10.15 a** A basic threshold detector; **b** window comparator

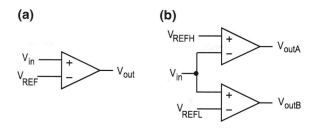

**Fig. 10.16** Threshold detection with comparator

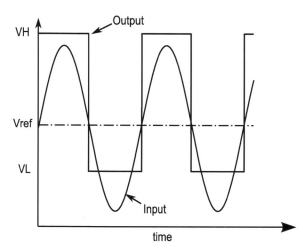

Comparator A goes low if the input signal exceeds the upper bound of this window, and comparator B goes low when it goes below the lower bound.

When the comparator operates with relatively low absolute differences between the two inputs, there are stability problems and the output oscillates. This also happens when processing slowly varying signals with even small amounts of superimposed noise. Noisy signals can occur in any application, and especially in industrial environments. To improve operation under these circumstances, *hysteresis* can be utilized. Also, it is possible to filter the output so that oscillation is limited to a narrower range of inputs.

In hysteresis comparator circuits, the transition between states occurs at different thresholds values $V_{IH}$ and $V_{IL}$ as illustrated in transfer function of Fig. 10.17. A comparator with this characteristic is said to be a *hysteresis comparator*, and it can be either non-inverting or inverting.

If the output in the non-inverting comparator is low, VL, it will stay there until $V_{in+} - V_{in-}$ increases and reaches the high threshold value VTH, when the output will change state to the high value VH. At this moment, the threshold for changing again to the low state VL is now VTL < VTH. The performance for the inverting hysteresis comparator is explained in similar terms. The *hysteresis value* is defined as

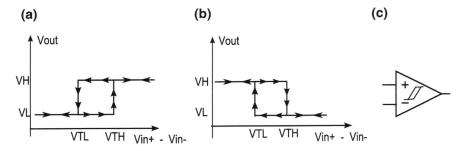

**Fig. 10.17**   Output for hysteresis comparator: **a** non inverting, **b** inverting, **c** symbol for a hysteresis comparator

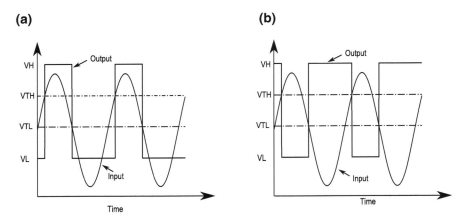

**Fig. 10.18**   Threshold detection with hysteresis: **a** non-inverting case, **b** inverting case

**Fig. 10.19**   Examples of hysteresis comparators: **a** non-inverting, **b** inverting

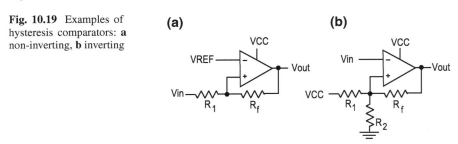

VTH-VTL. Figure 10.18 shows the changes in threshold detection when hysteresis is included.

Two comparator circuits with hysteresis are shown in Fig. 10.19. These positive feedback configurations are also applicable to OA with similar results. The non inverting circuit in this figure has the drawback of presenting a finite input impedance to the signal voltage, requiring an OA buffer if a high-impedance were required.

Let us now work some calculations, first for the non-inverting hysteresis comparator. From the nodal equations we find

$$V_{in+} = \frac{R_f V_{in} + R_1 V_{out}}{R_1 + R_f} \tag{10.12}$$

If the input voltage is large enough, the output will be $V_H$, and this state will remain while $V_{in+} > V_{REF}$. Using (10.12), we arrive at

$$VTL = \left(1 + \frac{R_1}{R_f}\right) V_{REF} - \frac{R_1}{R_f} VH \tag{10.13}$$

Similarly, when the input is low, the output is VL and will remain while $V_{in+} < V_{REF}$. Hence,

$$VTH = \left(1 + \frac{R_1}{R_f}\right) V_{REF} - \frac{R_1}{R_f} VL \tag{10.14}$$

Therefore, the hysteresis value for the non inverting hysteresis comparator becomes

$$Hyst_{\text{non inverting}} = VTH - VTL = \frac{R_1}{R_f}(VH - VL) \tag{10.15}$$

For the inverting case, we work similarly starting with the equation

$$V_{in+} = \frac{R_f R_2 VCC + R_1 R_2 V_{out}}{R_1 R_2 + R_1 R_f + R_2 R_f} \tag{10.16}$$

Notice however that now the high and low threshold values correspond to outputs as VH and VL, respectively. Therefore,

$$VTH = \frac{R_f R_2 VCC + R_1 R_2 VH}{R_1 R_2 + R_1 R_f + R_2 R_f} \tag{10.17}$$

$$VTL = \frac{R_f R_2 VCC + R_1 R_2 VL}{R_1 R_2 + R_1 R_f + R_2 R_f} \tag{10.18}$$

$$Hyst_{\text{inverting}} = \frac{R_1 R_2}{R_1 R_2 + R_1 R_f + R_2 R_f}(VH - VL) \tag{10.19}$$

**Example 10.4** *If $VH = VCC$ and $VL = 0\,V$, determine VTH, VTL and the Hysteresis value for the inverting and non inverting hysteresis comparators if all resistors are equal.*

*Solution:  Substituting values in the above equations we have: (a) For the non inverting case:*

$$VTL = 2V_{REF} - VCC \quad VTH = 2V_{REF} \quad and \ Hyst = VCC$$

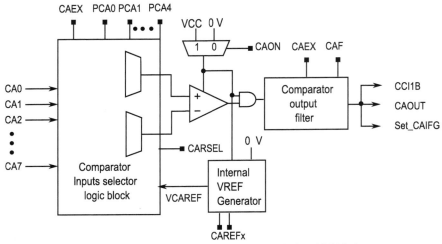

( Note:  Comparator A  provides only CA0 and CA1 -- PCA0 and PCA1 --)

**Fig. 10.20**   Block description of the comparators A and A+ in MSP430

*(b) For the inverting case*

$$VTL = VCC/3 \quad VTH = 2\,VCC/3 \quad and \; Hyst = VCC/3$$

Comparators are very important to the point that almost any ADC involves at least one device in its structure. Digital and mixed circuits usually contain comparators as well. Applications and considerations when dealing with comparators are discussed in literature.[4]

## 10.5.1 Internal Comparators in MSP430 (A and A+)

Several MSP430 models provide internal configurable comparators with interrupt capability. There are three models available: Comparators A, available in models of the MSP430x1xx family and some small models, Comparator A+ and, in recent families, Comparator B. We discuss briefly the first two models. The functional diagram for these comparators are shown in Fig. 10.20.

There are three registers associated to the MSP430 comparator as shown in Fig. 10.21. Control registers CACTL1 and CACTL2 allow to configure the operation. Port Disable Register CAPD bits are used to disable the buffer associated to

---

[4] The reader may consult http://www.analog.com/library/analogDialogue/archives/34-07/comparators/comparators.pdf.

**Fig. 10.21** Comparators
A and A+ control regis-
ters CACTL1, CACTL2 and
CAPD (Comparator Port Disable)

### CACTL0

| bit7 | bit6 | bit5 | bit4 | bit3 | bit2 | bit1 | bit 0 |
|------|------|------|------|------|------|------|-------|
| CAEX | CARSEL | CAREFx | | CAON | CAIES | CAIE | CAIFG |

### CACTL1

| bit7 | bit6 | bit5 | bit4 | bit3 | bit2 | bit1 | bit 0 |
|------|------|------|------|------|------|------|-------|
| CASHORT | P2CA4 | P2CA3 | P2CA2 | P2CA1 | P2CA0 | CAF | CAOUT |

Comparator A+ only  ← ⋮

### CAPD

| bit7 | bit6 | bit5 | bit4 | bit3 | bit2 | bit1 | bit 0 |
|------|------|------|------|------|------|------|-------|
| CAPD7 | CAPD6 | CAPD5 | CAPD4 | CAPD3 | CAPD2 | CAPD1 | CAPD0 |

the external pin input used by the comparator. This is convenient when the inputs are analog, since it reduces power consumption. The registers are cleared at reset.

The registers are readable and writable, with the exception of bit0 in CACTL2. This bit, CAOUT, reflects the comparator output and is shared with an I/O pin. The output also feeds the interrupt subsystem as SET_CAIFG signal to trigger an interrupt request, as well as signal **CCI1B** for capture input of Timer A.

The configurable features of comparator A and A+ are the following:

1. The system can be turned on or off, with the CAON bit.

   - When off, there is no power consumption from this subsystem.
   - Inputs should not float when working.

2. Comparator inputs may be connected to external inputs CA0, CA1, ..., or CA7, or to an internal reference voltage VCAREF.

   - Controllable by PCAx bits to connect to output I/O pin CAx.
   - CARSEL controls which the input to which VCAREF is applied.

3. The internal reference voltage VCAREF, when selected, can be set to 0.5VCC, 0.25VCC or diode voltage. The combination of bits CAREFx controls this feature. (CAREF0 puts VCAREF disconnected)

4. When working with analog inputs, the digital I/O pin x input buffer may be disabled with CAPDx.

5. The comparator output can be filtered with CAF to reduce oscillations.

6. In comparator A+, both inputs can be shorted with CASHORT and the comparator used for a sample and hold operation.

More detailed information about the MSP430 comparators can be found in the user guides and data sheets. The Comp_B module, not considered here, supports precision slope AD conversions, supply voltage supervision, and monitoring of external analog signals.

**Example 10.5**  *Consider using the comparator A+ in an MSP430g2553 for a simple threshold detection (see Fig. 10.15) with the analog signal connected to the +input and the reference voltage to the −input, using 0.5VCC as reference voltage. This value is available internally, so we can connect* **VCAREF** *to the negative input of comparator.*

*Solution:* From the data sheet, we see that CA1 is shared with the I/O pin P1.1; the comparator is automatically used when the comparator input is selected. From the user guide we find that to connect CA1 to the +input we need PCA4 = 1, PCA0 = 0, CAEX = 0. To connect the internal voltage reference to the −input, CAEX = 0, CARSEL = 1. The 0.5VCC reference is set with CAREF2 (=10). We disable the buffer for reduced power consumption. Thus, we configure with

```
SET_CA   mov.b #CAON+CARSEL+CAREF2,&CCTL1 ; system ON, Vref=0.5VCC to - input
         mov.b #P2CA4,&CCTL2           ; P1.1/CA1 to + input
         mov.b #CAPD1,&CAPD            ; Input buffer disabled
```

*A C-language version of the same code fragment:*

```
CCTL1 = CAON+CARSEL+CAREF2   // system ON, Vref=0.5VCC to - input
CCTL2 = P2CA4                // P1.1/CA1 to + input
CAPD  = CAPD1                //  Input buffer disabled
```

## 10.6  Analog-to-Digital and Digital-to-Analog: An Overview

The digitalization of a continuous analog signal consists in assigning digital codes to discrete samples of an application segment. The concept follows the principles established in Chap. 2, Sect. 2.10.

A properly continuous conditioned signal within limits determined by the available hardware, is first sampled at specific time instants. Each sampled value falls within a certain subinterval to which an $n$-bit word is associated. This sequence of words represents the function in the given interval. A faithful representation depends on various conditions that include the sampling rate and the number of bits $n$, used to encode the discrete signal values as well as other parameters, we must comply with.

The quantized values are transferred to a digital system, whatever this is, where they are processed by the program or hardware system designed by the user. In many applications, the objective of the process could be to simply reproduce the function of interest in another target.

**Example 10.6**  *Consider Fig. 10.22a illustrating the output of the sensing stage in the millivolts range, intended to be used for an AD converter requiring a 0–5 V input range. Hence, the signal is amplified 98.33 times to obtain an appropriate reproduction in this range, obtaining the function shown in (b).*

*Let us now encode this function with 4-bits. First, we sample the input signal at regular intervals. For our example, let us take 21 samples. The result is illustrated with black dots in Fig. 10.23. The x-axis values are now the integers corresponding to the order of the sample in the sequence of interest.*

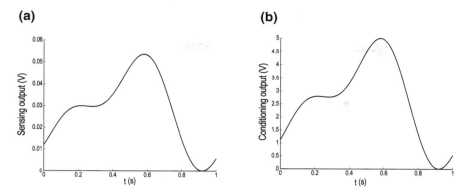

**Fig. 10.22** From analog to the digital process: **a** a sensing output, **b** after conditioning

**Fig. 10.23** Sampling and encoding with four bits the signal in Fig. 10.22b

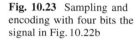

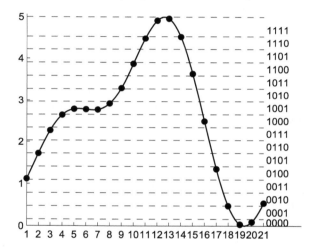

*With $n = 4\,bits$, we have $2^4 = 16$, 4-bit words. The voltage range (y-axis) is therefore subdivided in sixteen uniform subintervals, each of which is associated to a 4-bit code, as shown in the figure. Any sample falling within a subinterval is encoded with that code. For reasons to be explained later, the lowest and highest subintervals are not of the same size as the other ones.*

*To each sampled value we associate now the corresponding code. For example, sample 1 falls in the subinterval tagged with 0100. In this way, the function is then represented in digital form by the sequence of words {0100, 0101, 0111, 1000, 1001, 1001, 1001, 1001, 1011, 1100, 1110, 1111, 1111, 1110, 1100, 1000, 0100, 0001, 0000, 0000, 0010}.*

The process illustrated by the above example is called *quantization*. The inverse process is carried out by the digital to analog phase. The Digital-to-Analog converter takes an *n*-bit word and delivers an output value $KV_{ref}$, where K is proportional to the decimal equivalent of the digital word. This is done for each word in a sequence. An analog continuous signal will be constructed from the set. The accuracy will depend on the number of bits used in sampling, the number of samples, and the building method. The result may need to be further conditioned with filters, amplification/attenuation, or other operations. It may also be converted into a non-electrical magnitude using a transducer.

**Example 10.7** *Let us reconstruct the function from the previous 4-bit sequence, which in decimal terms is { 4,5,7,8, …, 0, 2}, using a 5 V reference voltage and a proportional constant $1/2^n$, where n is the number of bits. For example, the first point in reconstruction is $4 \times 5/16 = 1.25$ V.*

*Figure 10.24 shows the reconstructed (circles) and sampled points (x). Although the figure looks like the function, there are differences between the two sets. The difference between the true value and the reconstructed one is called quantization error. As an example, the true value at $t = 0$ is 1.1196 while the value obtained is 1.25.*

*This figure is discrete and we should fill the space between consecutive points with the intermediate values, that is, interpolate. Unprocessed results usually keep a constant value between reconstructions steps, yielding a curve like the one in Fig. 10.25a. One "smoothing" filter could provide straight line between consecutive points, as shown in (b), obtaining a piece-wise linear curve. There are of course other processing steps possible.*

*Although imperfect in this example, the process of going from a stepwise function to a continuous curve is called smoothing. As one might expect, a better result is obtained with more bits.*

**Fig. 10.24** Output values from DAC

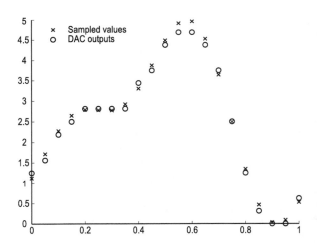

**(a)** **(b)**

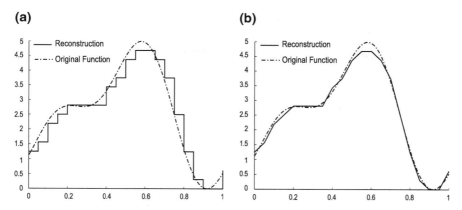

**Fig. 10.25** Reconstructing the analog function of Fig. 10.22b using 21 samples and 4-bit words: **a** step-wise form; **b** piece-wise linear form

## 10.7 Sampling and Quantization Principles

As illustrated with the previous examples, *sampling* starts the conversion of an analog signal into a digital representation. The assignment of digital codes is called *quantization*, in which the analog function is represented by a set of discrete values, each identified by an *n*-bit word. Let us take a closer look at the sampling and quantization procedures.

### 10.7.1 Sampling and the Nyquist Principle

The time interval between two consecutive samples is the *sampling period* or *sampling rate* $T_s$. Its inverse is the *sampling frequency* $f_s = 1/T_s$, measured in Hertz or cycles per second. It is also common to use "*samples per second* (sps)" as units for this frequency. In Example 10.7, $T_s = 0.05$ s and $f_s = 1/.05$ Hz $= 20$ Hz $= 20$ sps.

The first problem we should address is to take enough samples so as to be able to reconstruct back the function with enough accuracy. The answer to this problem was formulated by Henry Nyquist, from IBM Bell Laboratory, in the 1940s:

**Nyquist principle:** If $f_h$ is the highest frequency component of a signal in its bandwidth of interest, then the sampling frequency $f_s$ must satisfy

$$f_s \geq 2 \times f_h \tag{10.20}$$

The frequency $f_N = 2 \times f_h$ is called the *Nyquist frequency* and its inverse is the *Nyquist rate*.

The term "Nyquist rate" is in fact used by engineers both for the frequency and the associated period, the meaning being interpreted by context. Let us illustrate Nyquist principle with an example.

**Example 10.8** *Consider a single sine wave function with frequency $f_h = 3\,kHz$ to be sampled. The Nyquist frequency is $f_N = 6\,kHz$, with a Nyquist rate of $T_N = 167\,\mu s$. Nyquist principle states that for the function to be recovered adequately, the sampling rate must be less than or equal to $T_N$.*

*We can see in Fig. 10.26a that for a rate of 1.3 times the Nyquist rate, the original function is not recovered at all. For the other two cases the reconstruction is better.*

### 10.7.2 Sampling and Aliasing

*Aliasing* is the term used to describe the fact that when a signal of frequency $f$ is sampled with a frequency $f_s$, the same samples will apply to any other signal of frequency $k\,f_s \pm f$, where $k$ is an integer.

Figure 10.27 illustrates the aliasing phenomenon for a pure sine wave of frequency $f = 250\,Hz$ and a sampling frequency $f_s = 800\,Hz$, with $k = 1$.

The problem of aliasing arises whenever the bandwidth of interest also contains components which we do not want to sample. The maximum frequency that might be sampled should comply with $f \leq \frac{1}{2}f_s$. For example, though audio frequencies are in the range 0–20 kHz, digital audio broadcast systems (DAB) take samples at 32 kHz, and thus they can only reproduce up to 16 kHz faithfully.

To avoid the inconvenience of reproducing unwanted signals, we use low pass and bandpass *antialiasing filters*.

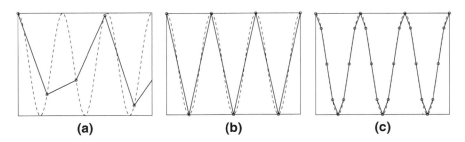

|     (a)     |     (b)     |     (c)     |

**Fig. 10.26** Reconstructing from samples: **a** at 1.3 times $T_N$ ($fs = fN/1.3$); **b** at Nyquist rate, and **c** four times $T_N$ ($f_s = 4f_N$)

**Fig. 10.27** Illustrating aliasing: **a** signal of frequency $f$ being sampled at frequency $f_s$; **b** original signal and another with frequency $f_s - f$; **c** original signal and another with frequency $f_s + f$

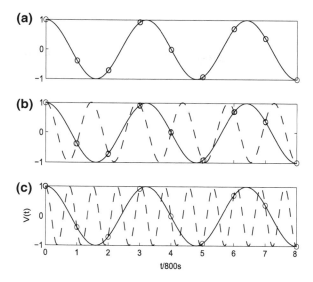

### 10.7.3 Quantization

In addition to sampling frequency, the user must decide the size of the words to encode the data. With quantization, we represent a continuous interval by a set of $2^n$ discrete values, called *quantization levels* each one associated to an $n$-bit word. These levels are representatives for a set of $2^n$ subintervals, so that any analog value within a subinterval is encoded by the same word as the quantization level. This introduces an error in the process, called *quantization error* defined by

$$\text{Quantization error} = \text{actual value} - \text{quantization level} \qquad (10.21)$$

The fundamental hardware devices that are used in the quantization process are the *Analog-to-Digital Converter* (ADC) and the *Digital-to-Analog Converter* (DAC). The first one associates the $n$-bit words to the analog value, the second one produces the analog value of the quantization level associated to a given $n$-bit word. The symbols for the ADC and DAC are shown in Fig. 10.28. Notice that the difference in the symbols is basically the direction of the information flow.

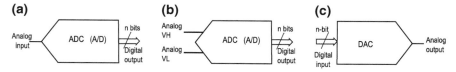

**Fig. 10.28** Basic quantization: **a** ADC symbol; **b** differential input ADC; **c** DAC symbol

Data converters work with voltage (or current) inputs in a range $Vmin < Vin < Vmax$. This range is not always equal to the one delivered by the sensing stage, so pre-conditioning may be necessary. It is the responsibility of the designer to insure a proper relationship. For the hardware devices, it is common to have $Vmin = 0$ or $Vmin = -Vmax$. The *full scale range* $V_{FS}$, or simply $FS$, is defined by

$$VFS = Vmax - Vmin \qquad (10.22)$$

Usually, the quantization levels are equally distributed in the range of interest. In this case, the difference between the levels is given by

$$\Delta = \frac{Vmax - Vmin}{2^n} = \frac{FS}{2^n} = LSB \qquad (10.23)$$

$\Delta$ is called *resolution* or *precision* of the quantization. It is also called the *Least Significant Bit resolution*, or simply $1\,LSB$ resolution. Notice that the resolution depends both on the full-scale FS and $n$. Thus, with FS $= 5\,V$ and $n = 8$, $\Delta = 19.5\,mV$, while FS $= 3\,V$ yields $\Delta = 11.72\,mV$ for the same $n$.

Because of the hardware characteristics, engineers usually prefer to talk of a resolution of $n$ bits, instead of $\Delta$. ADC's and DAC's have a fixed $n$ value, but may accept different FS magnitudes.

From (10.23), the quantization levels are the $2^n$ values $Vmin + k\Delta, k = 0, 1, \ldots, 2^n - 1$. That is,

$$Vmin, \quad Vmin + \Delta, \quad Vmin + 2\,\Delta, \ldots, \quad Vmin + \left(2^n - 1\right)\Delta \qquad (10.24)$$

Notice that $Vmax$ is not a quantization level, and no $n$-bit word is associated to it. Each quantization level is representative for a subinterval in $[Vmin, Vmax]$. The subdivision of this range can be done in different ways, two of which are illustrated in Fig. 10.29 for three bits, where subintervals are associated to 000, 001, ..., 111.

Inset (a) is a quantization with all subintervals width equal to $1\,LSB$. Here, the first transition from subinterval 000 to 001 takes place $1\,LSB$ after Vmin, and then each transition happens at $1\,LSB$ intervals. Now, any value $x_j \le x_{in} < x_{j+1}$ is encoded by the word for $x_j$, as illustrated in this figure for $x_{in}$ to which we associate 010. In this example $x_{in}$ is much closer to $x_3$. In fact, this subdivision has the disadvantage that the quantization error can be up to $1\,LSB$. To illustrate, consider an 8 V full scale, $Vmin = 0\,V$, and a resolution of 4 bits, $\Delta = 0.5\,V$. Then 0.49 V will be represented by 0000, i.e., 0 V.

Generally, a quantization error of $1\,LSB = 1\,\Delta$, is not acceptable. One solution is to modify the assignment of subintervals so that the maximum error is within $\pm\frac{1}{2}\,LSB$. This is done by choosing the subintervals with the quantization levels as midpoints, as illustrated by Fig. 10.29b. In this case, the length for the first subinterval is $\frac{1}{2}\,LSB$ while the highest one is $1\frac{1}{2}\,LSB$ long. All the others are $1\,LSB$ long. Therefore, with the exception of the last subinterval, the quantization error for

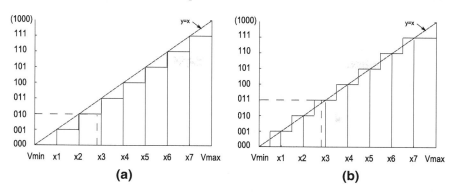

**Fig. 10.29** Basic quantization: **a** all regular intervals; **b** half-LSB compensated (Levels: Vmin, x1, ...x7)

any other value is within $\pm\frac{1}{2}$ *LSB*. This is the best solution, and thus this subdivision is the most common. Unless otherwise stated, this is the one we work with.

Figures 10.29a and b illustrate *straight binary codes*, where the lowest input voltage is assigned the all zeros word and then counting proceeds sequentially, using a full-scale input. Strictly speaking, this type of encoding is usually limited to unipolar voltages only, i.e., all voltages of the same sign. When $Vmin \neq 0$, it is also called *offset* or *biased binary code*.

In this situation, the decimal equivalent $N_Q$ for the word assigned encoding given analog value $x$ in the Half-LSB compensated subdivision would be

$$
N_Q = \begin{cases} \left[2^n\dfrac{x-Vmin}{V_{FS}}\right] & \text{if } x < \text{Vmax and } [\bullet] < 2^n \\[2mm] 2^n - 1 & \text{if } [\bullet] \geq 2^n \end{cases} \tag{10.25}
$$

where $[A]$ yields a rounding to the integer closest to the argument value $A$.

**Example 10.9** *Assume that the interval* $[-10\,V,\ 10\,V]$ *is quantized with a resolution of 4 bits and regularly spaced quantization levels. (a) What is the LSB resolution; (b) List the quantization levels; (c) How do you encode* $-1.28\,V$ *in an all-regular subinterval quantization and what will the quantization error be? (c) Encode* $-1.28$ *with half-LSB compensated quantization and find the quantization error. (d) Find the LSB resolution with 12 bits, and the encoding and quantization error for* $-1.28$.

**Solution:** *Since* $n = 4$, *there would be* $2^4 = 16$ *subintervals.*

*(a) The LSB resolution is found with (10.23):*

$$
\Delta = \frac{10 - (-10)}{2^4}\,V = 1.25\,V
$$

*(b) The quantization levels are given as* $Vmin + k\,\Delta$ *for k 0, 1, ...,* $2^n - 1$ *as:*

−10 V,  −8.75 V,  −7.50 V,  −6.25 V,  −5 V,  −3.75 V,  −2.50 V,  −1.25 V,
0 V,      1.25 V,    2.50 V,    3.75 V,    5.0 V, 6.25 V,    7.50 V,    8.75 V

*(c) Using (10.25), $[16 \times (-1.28 - (-10))/20] = [6.976] = 7$, so the encoding yields 0111. The quantization value is $-10 + 7 \times \Delta = -1.25$. Therefore, the quantization error is $-1.28$ V $- (-1.25$ V$) = -0.03$ V.*

*(d) For 12 bits, the LSB resolution is $20/2^{12} = 4.8828$ mV. The LSB-compensated subdivision yields $N_Q = 1,786$ for an encoding 0x6FA. The associated quantization value is $-1.2793$ for a quantization error of $-0.7$ mV.*

In straight binary codes, 1000...00 corresponds to the mid value. Very often, the minimum and maximum values are symmetrical around 0, with Vmin $= -$Vmax $= -$VR, as it was the case for the above examples. In these cases another encoding is generally preferred. Namely, a two's complement type, where the middle term 0000...00 is assigned to 0 V, and the distribution correspond to a two's complement representation in such a way that negative values are assigned a negative number, as illustrated by Fig. 10.30. This encoding is called *two's complement quantization* type, convenient for symmetrical bipolar ranges. The quantization levels in the x-axis have been labeled as shown to better illustrate the encoding scheme.[5]

It must be emphasized that the quantization levels used for an analog input is the same for both the offset and two's complement encoding cases, what is different is the encoding word. In twos complement quantization, the decimal signed integer $N_Q$ value associated to an analog $x$ is

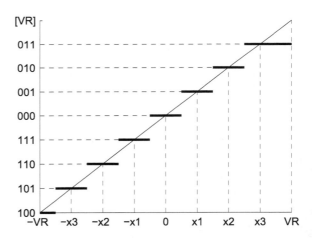

**Fig. 10.30** Two's complement quantization

---

[5] When Vmin $= -$Vmax $= -V_R$, we have $\Delta = 2V_R/2^n = V_R/2^{n-1}$. Since the quantization levels are $-V_R + m\Delta$ for $m = 0, 1, \ldots, (2^n - 1)$, the value 0 will always appear as a quantization level.

$$N_Q = \begin{cases} \left[2^{n-1}\frac{x}{V_R}\right] & \text{if } x < +V_R \text{ and } [\bullet] < 2^{n-1} \\[2ex] 2^{n-1} - 1 & \text{if } x = V_R \text{ or } [\bullet] = 2^{n-1} \end{cases} \tag{10.26}$$

Conversely, given the decimal equivalent $N_Q$ of an encoded word, the corresponding quantization level is

$$Level_{N_Q} = N_Q \Delta \tag{10.27}$$

**Example 10.10**  *The interval in Example 10.9 is symmetrical and was encoded using an offset straight binary quantization. Now let us consider a two's complement code for the same range of values. (a) Verify that the quantization levels found using (10.27) are the same as before. (b) Encode $-1.28$ V for a 4-bit and a 12-bit resolution and find the quantization errors. How do they compare with the previous encoding?*

**Solution:**
(a) *The values to consider for $N_Q$ are $-8, -7, -6, \ldots, 0, 1, \ldots, 7$ since we are talking of signed numbers encoded with four bits. On the other hand, $\Delta = 1.25$, as before. The reader can verify that the set of quantization values is the same by multiplying $N_Q\Delta$ for the different values of $N_Q$.*
(b) *Using $n = 4$ in (10.26) we have*

$$N_Q = \left[2^3 \times \frac{-1.28}{10}\right] = [-1.024] = -1$$

*which gives a quantization level of $-1 \times 1.25$ V $= -1.25$ V, just as before. The corresponding code is 1111.*
*For twelve bits,*

$$N_Q = \left[2^{11} \times \frac{-1.28}{10}\right] = -262$$

*yielding a quantization level $-262 \times 4.8828$ mV $= -1.2793$ V, the same one as before. The corresponding code is 0xEFA.*
*Neither the quantization levels nor the quantization errors changed with the different encoding schemes.*

Notice that the accuracy improves with more bits. However, there are physical limitations to the number of bits used, due both to hardware constraints and other factors such as the noise levels which can affect measurements.

## 10.7.4 *Quantization and Noise*

When the signal producing the $x$ value to be quantized is static, the quantization error gives a good information of how well the translation process is realized. However, when the signal is varying with time, especially at high frequencies, the $x$ value is more susceptible to be affected by noise. In this case, a better reference to consider *signal-to-noise ratio* (SNR). In general, SNR is defined in dB as

$$\text{SNR} = 20 \log \left( \frac{\text{RMS value of signal}}{\text{RMS value of noise}} \right) \text{ dB} \qquad (10.28)$$

The higher SNR the better. Now, even in an ideal quantization process, there is a finite SNR that will depend on the LSB resolution and the magnitude of the signal.

For perfectly quantized signals, the reference is a sinusoidal of magnitude $V_m$, so the interval to quantize is $[-V_m, V_m]$. Using $n$ bits in two's complement quantization, we infer that $V_m = 2^{n-1}(LSB)$ and the RMS value of the signal is then

$$\text{RMS of signal} = \frac{V_m}{\sqrt{2}} = \frac{2^{n-1}\text{LSB}}{\sqrt{2}} \qquad (10.29)$$

On the other hand, for quantized signals it has been shown that the RMS noise is $\text{LSB}/\sqrt{12} = \text{LSB}/(2\sqrt{3})$. Substituting this result and (10.29) into (10.28) we have

$$\text{SNR} = 20 \log \left( \frac{2^{n-1}LSB/\sqrt{2}}{LSB/2\sqrt{3}} \right) = 20 \log \sqrt{\frac{3}{2}} + n \times \log 2 \qquad (10.30)$$

This translates into

$$\text{SNR} = 1.76 + 6.02n \text{ dB} \qquad (10.31)$$

This result applies to an ideal quantization process, independent of the hardware itself, and as such is the best value we can achieve. For this reason, this is a reference for design, since non idealities from hardware will degrade this figure. These factors are considered next.

### 10.7.5 Quantization and Hardware Imperfections

The above theoretical discussion related to sampling and quantization process not only introduces the theoretical concepts, but it applies to an ideal signal-chain composed by ideal hardware components. Let us now look into the errors resulting from the use of real devices. The usual errors, and figures of merit, to look at are summarized next:

**Offset error:**  Any deviation from the point where the input voltage cause a transition from zero count. The offset error can usually be factored or calibrated out. Offset error may be expressed in percent of full scale voltage, Volts or in LSB. It is also called Zero Scale error.

  • *Bottom offset* is the input required to cause the transition to the first count.

**Full scale offset error:**   Is the error in the actual full-scale output transition point from the ideal value. It is also expressed in percent of full scale voltage, Volts or in LSB.

**Gain error:**   It is the deviation from the ideal slope of the transfer function. Although the ideal slope is usually 1, it may perfectly be otherwise, a gain G.

**Differential Non Linearity (DNL):**   This feature is also called Differential Linearity Error (DLE). This is the difference between the ideal and the actual *input code width*, the range of input values that produces the same digital output code. It is usually expressed in terms of the widest (positive) DNL and narrowest (negative) DNL. DNL is expressed in terms of LSB.

**Integral Non Linearity (INL):**   This feature is also called Integral Linearity Error (ILE). This term describes the departure from an ideal linear transfer curve. INL does not include quantization errors, offset error, or gain error. Expressed in terms of LSB.

**Signal-to-Noise Ratio (SNR):**   In addition to the maximum quantization SNR described by (10.31), the actual figure of merit includes the noise generated by hardware, the input circuit noise, and the jitter. These factors degrade the figure of the ideal case (10.31).

**Total Harmonic Distortion (THD):**   This one measures the effect of harmonics due to non-linearities. If the input signal were assumed to have a fundamental component at frequency $f1$, quantization would produce harmonics at frequencies $f2, f3, \ldots$ THD would then be measured using the RMS values as defined by (10.32) below. This definition excludes DC components.

**Signal-to-Noise and Distortion (SINAD):**   This is a combination of SNR and THD, as defined in (10.33). It usually tracks better SNR, and as such is used to define the next figure of merit.

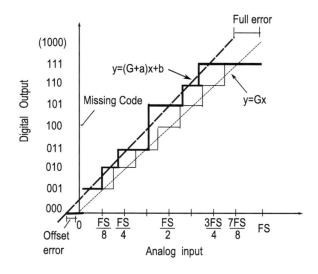

**Fig. 10.31** Ideal versus actual quantization: gain error $a$, missing code 100

**Fig. 10.32** Basic sample-
and-hold circuit configuration

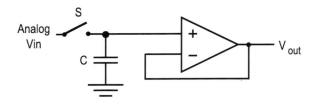

**Effective Number of Bits (ENOB):**   Using the association between SINAD and
SNR, this figure of merit states the number of bits of an ideal converter yielding
a similar quantization. ENOB is defined by Eq. (10.34).

$$\text{THD} = \sqrt{\frac{V_{f2}^2 + V_{f3}^2 + V_{f4}^2 + \cdots}{V_{f1}^2}} \tag{10.32}$$

$$\text{SINAD} = -10\log\left[10^{-\text{SNR}/10} + 10^{\text{THD}/10}\right] \tag{10.33}$$

$$\text{ENOB} = \frac{SINAD - 1.76}{6.02} \tag{10.34}$$

In the SINAD equation, notice that if the THD is sufficiently small, SINAD $\approx$
SNR. This is common with modern technology. If SINAD = SNR, then ENOB = $n$,
the number of bits.

Other definitions are illustrated in Fig. 10.31.

### 10.7.6 Sample & Hold Circuits

A *sample-and-hold circuit* (SH circuit) samples the analog value to be processed
by the ADC for conversion into an $n$-bit word. The basic principle is shown in
Fig. 10.32. Some versions add a buffer before the switch to prevent loading effects.
Ideally, when the switch S is closed at instant $t = t_k$, the input value $V_{in}(t_k)$ is stored
at the capacitor C. This is the sampling phase. The switch S then opens to enter to the
hold phase so the capacitor holds the stored value, which is readable at the output of
the right buffer. The output buffer prevents the capacitor from discharging through
the input resistance of the load. The switch S is a MOSFET switch, controlled by a
clock with a sampling frequency $f_s$. Thus, samples are taken at $1/f_s$ intervals.

The above ideal description does not take into account the switch resistances, for
example. There are many limitations that degrade the performance of the circuit. For
example, the capacitor has leakage currents that discharge it, as well as actual currents
going into the buffer, which are small, but not zero. Timing problems, imperfections
in the switches and so on result in an unavoidable degradation of the signal, but still

good enough for all practical purposes. Therefore high quality capacitors and fast switches are key elements in this operation.

One of the timing problems associated with sampling is *jittering*, which affects the periodicity of sampling and therefore introduces an instantaneous signal error. This error is proportional to the slew rate of the desired signal and the absolute value of the clock error, that is,

$$\Delta V = \left| \frac{dV}{dt} \right| \Delta t \tag{10.35}$$

For example, using $N$ bits to sample a sine signal of frequency $f$, $(2^n \, LSB)\sin(2\pi f \, t)$ with an error less than $1\,LSB$, we must maintain the time deviations $\Delta t$ such that it satisfies

$$\Delta V = \left| 2^N \, LSB \, 2 \, f \, \pi \, f \cos(2\pi f \, t) \right|_{t=0} \Delta t < 1 \, LSB \tag{10.36}$$

The instant $t = 0$ applies because this is where the slope is maximum. From this expression, then

$$\Delta t < \frac{1}{2^{N+1}\pi \, f} \tag{10.37}$$

To illustrate, sampling with 10 bits an audio frequency in the upper range of, say $f = 20\,\text{kHz}$, needs to keep an error of time sampling less than $7.8\,\text{ns}$ to be adequate. Remember that the sampling frequency must be greater than or equal to $40\,\text{KHz}$, with a sampling period of $25\,\mu\text{s}$, and therefore the absolute deviation with respect to sampling instants must be kept within $0.03\,\%$ to insure fidelity. In CD's, sampling is done with 15 bits, requiring an error less than $0.25\,\text{ns}$.

As seen with the above figures, jittering is particularly critical in high-frequency signal conversion, or where the clock signal is prone to interference. For this reason, it is more common to work this problem with hardware instead of software. In the MSP430, for example, converters work together with internal timers. Delta-sigma ADC's suffer less with this problem because they are slower.

Practical commercial SH circuits (amplifiers) are available in IC form which are designed to minimize jittering. National Semiconductor's—now TI—LF398-n is a good example, where the holding capacitor is provided by the user. The Maxim's DS1843 consists of a fully differential sampling capacitor, switches, and a differential output buffer which can be configured for single-ended operations. For applications requiring multiple channels, Maxim's versatile MAX5167 offers 32 buffered sample/hold circuits with internal hold capacitors.

Nowadays, most ADC integrated circuits have the SH circuit already incorporated. This is also the case in MSP430 microcontrollers.

## 10.8 Digital-to-Analog Converters

As mentioned before, an $n$-bit *digital-to-analog converter* (DAC) receives as input $n$ digital bit signals and delivers an analog value. This one is proportional to a decimal equivalent $N_D$ of the binary word and a reference voltage $V_{REF}$, which is taken as the full-scale value $V_{FS}$. The input-output relation is given as:

$$V_{out} = K \frac{N_D}{2^n} V_{REF} \tag{10.38}$$

where $N_D$ is the decimal equivalent to the input $n$-bit word. It is generally between 0 and $2^n - 1$, but signed two's complement equivalent is possible too. DACs working with two's complement coding inputs, are usually referred to as *bipolar DACs*. Notice the similarity with the definition of resolution in Eq. (10.23).

From (10.38) we can infer that a reference voltage signal is required. Modern DACs allow the user to select between an external reference or another one already built in the device. This reference is separate from the power supplies $VDD$ and $VSS$ needed. Terminals for these connections are often incorporated into the DAC symbol.

The DAC's output can be a voltage or a current signal. In the first case, the gain $K$ depends on the hardware circuit used. For a current output it is possible to set this gain with Ohm's law using a resistor connected to the output port; current-to-voltage converters are another alternative. The resistor values are limited by the acceptable output voltage and the current capacity of the DAC.

The $n$ bits at the input may be received all at once, as it is the case for parallel DACs, or else the bits are received one by one, for serial DACs. Both serial and parallel commercial DACs may be one or several channels type, meaning by this that inputs can be received from several sources and conversion outputs delivered at different terminals.

There are different types and architectures for DACs.[6] We mention one architecture which is very popular in DAC circuits, the R-2R ladder type.

### 10.8.1 R-2R Ladder

Figure 10.33a shows the basic configuration for an R-2R ladder. The figure shows a ladder for 4-bit words, but any number of bits can be considered by extending the R-2R ladder. A characteristic of this configuration is that the equivalent resistance seen to the left of any node $n_j$ is 2R.

---

[6] Reference [1] provides a good overview of this topic.

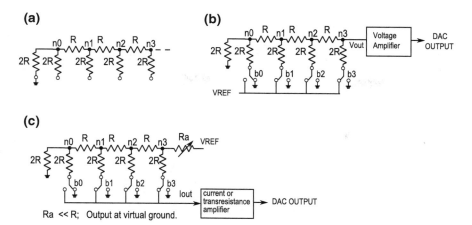

**Fig. 10.33**   R-2R ladder configuration

This ladder is so practical that commercial ICs with just the ladder (with some variations to add flexibility) are available from several vendors.[7]

Figure 10.33b shows how the circuit is used in voltage output DACs. The switches are operated by the input, with switch $b_j$ connected to ground when bit $b_j = 0$ and to $V_{REF}$ when the bit is 1. The case 1110 is illustrated in the figure. The Thevenin's equivalent for the circuit seen from the output gives a voltage

$$Vth = \frac{VREF}{2^n} \left( 2^{n-1}b_{n-1} + 2^{n-2}b_{n-2} + \ldots + 2\,b_1 + b_0 \right) \qquad (10.39)$$

On the other hand, Fig. 10.33c shows how the circuit is used in current output DACs. Here the output is taken at a virtual ground node, which can be obtained through a current-to-voltage amplifier or a current amplifier. The input resistance seen by $VREF$ is R + Ra. Assuming $Ra \ll R$, the input current will be $Iin \approx VREF/R$. The output current will then be given by an equation similar to (10.39).

More suitable for integrated CMOS technology, popular DAC configurations are based on capacitor arrays, the charge distribution DAC, like the one shown in Fig. 10.34.

In this configuration, switches $b_0$, $b_1$, ... are controlled by the word bits, Vout is taken from a unity gain amplifier to prevent loading to the capacitors. The equivalent capacitance when all are in parallel is $2^nC$. Conversion takes place in two phases. In the first one, all capacitors are discharged by closing S and connecting all switches to ground. In the second phase, S is open and the input digital word connects the switches to ground ( for $b_j = 0$) or to VREF (for $b_j = 1$) to produce the analog

---

[7] For example, the QS009 and QS014 from TTE Electronics, BCN 31, and 68 Ladder from BI Technologies—later acquired by TT, or the SWR2R series from Semiconwell, ranging from 4 to 20 bits.

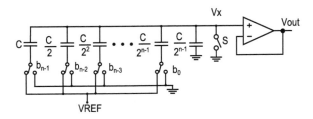

**Fig. 10.34** Charge distribution DAC

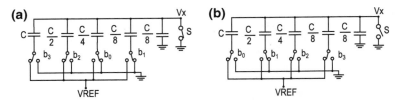

**Fig. 10.35** Charge distribution DAC example. **a** Discharging phase; **b** Conversion phase

equivalent Vx. Observe that the sum of all capacitors is

$$C + \frac{1}{2}C + \frac{1}{2^2}C + \ldots + \frac{1}{2^{n-1}}C + \frac{1}{2^{n-1}}C = 2\,C \qquad (10.40)$$

When all switches are connected to ground for the discharging phase, the total charge becomes $Q_T = 0\,C$. In the second phase, the conversion phase, the addition of the individual charges must not change, that is

$$\sum_{j=0}^{n-1} \left(b_{n-1-j}\,VREF - V_x\right)\frac{C}{2^j} + (0 - V_x)\frac{C}{2^{n-1}} = Q_T = 0$$

After some Algebra and using (10.40) we arrive at

$$V_{out} = V_x = \frac{VREF}{2^n}\left(b_{n-1}\,2^{n-1} + b_{n-2}\,2^{n-2} + \ldots + b_1\,2^1 + b_0\right) \qquad (10.41)$$

This configuration has the disadvantage that the ratio between the largest and the smallest capacitor is $2^n - 1$, which is unacceptable and gives rise to many problems. Several good solutions to this problem have been found, but discussion of them is beyond the scope of this book.

**Example 10.11** *Let us consider the charge distribution DAC with four bits, and convert 1100. The two phases are illustrated in Fig. 10.35a, b. The buffer is not included in the figure.*
   *In the second phase, we can write*

$$(VREF - Vx)C + (VREF - Vx)\frac{C}{2} + (0 - Vx)\frac{C}{4} + (0 - Vx)\frac{C}{8} + (0 - Vx)\frac{C}{8} = 0$$

*That is,*

$$Vx = \frac{3}{4}VREF \Rightarrow Vx = \frac{12}{16}VREF$$

12 *is the unsigned decimal for* 1100 *and Vx has the form* (10.41).

## 10.8.2 Smoothing Filters

When the application is dynamic, the input words to the DAC will change periodically. This will result in a stepwise output like the one shown in Fig. 10.25a. To modify this appearance, the DACs output is fed into a *smoothing filter*, also called reconstruction filter, so that the output become like that of Fig. 10.25b.

## 10.8.3 Using MSP430 DACs

With the exception of the 'x3xx series, several members of the other series have one or two DACs incorporated. Series 'x1xx,'x2xx and 'x4xx have model DAC12, while 'x5xx and 'x6xx have model DAC12_A. The block structure of the DAC12 is shown in Fig. 10.36. The DACs are called DAC12_0 and DAC12_1, when two are present. The block structure is similar for both DACs, except that for DAC12_1 the control signal DAC12GRP is don't care. Models with only one DAC do not include the DAC12_1 block nor the Group Load Logic Unit.

The DAC12 output is given as

$$DAC\_Output = K \times VREF\frac{DAC12\_xDAT}{2^n} \qquad (10.42)$$

where the dynamic range K, the voltage reference VREF, and the resolution $n$ can all be defined by the user. Although register DAC12_xDAT is 16 bits wide, only 12 or 8 are meaningful depending on the value of $n$. This register may be buffered through register DAC12_xLatch or directly applied to DAC and also receive data from the Analog to Digital converter ADC12.

Configuration is achieved through control registers DAC12_xCTLC shown in Fig. 10.37. Series 'x5/6xx DAC12_A has in addition a control register DAC12_xCTLC1, calibration control register DAC12_xCALCTL, a calibration data register DAC12_1CALDAT, and an interrupt vector generator register DAC12IV. These registers are all readable and writable, and are cleared on reset with the exception of DAC12_xCALCTL which is set to 9601h.

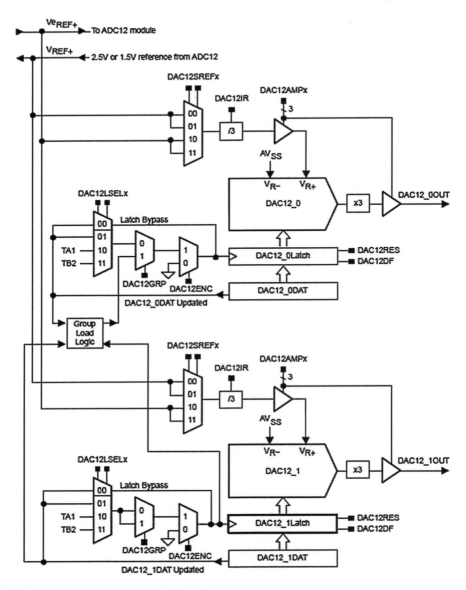

**Fig. 10.36** MSP430 dual DAC configuration

MSP430 DACs have interrupt capability. Yet, the interrupt flags DAC12GIF shares the interrupt vector address with th DMA, so the flag is not automatically reset. The flag is set when a conversion is achieved. Table 10.3 shows features of the different families.

**(a)**

| Bit15 | Bit 14 | Bit 13 | Bit12 | Bit11 | Bit10 | Bit9 | Bit8 |
|---|---|---|---|---|---|---|---|
| DAC12OPS | DAC12SREFx | | DAC12RES | DAC12LSELx | | DAC12CALON | DAC12IR |

| Bit7 | Bit6 | Bit5 | Bit4 | Bit3 | Bit2 | Bit1 | Bit0 |
|---|---|---|---|---|---|---|---|
| DAC12AMPx | | | DAC12DF | DAC12IE | DAC12IFG | DAC12ENC | DAC12GRP |

**(b)**

| Bit15 | Bit 14 | Bit 13 | Bit12 | Bit11 | Bit10 | Bit9 | Bit8 |
|---|---|---|---|---|---|---|---|
| 0 | 0 | 0 | 0 | DAC12DATA | | | |

| Bit7 | Bit6 | Bit5 | Bit4 | Bit3 | Bit2 | Bit1 | Bit0 |
|---|---|---|---|---|---|---|---|
| DAC12DATA | | | | | | | |

**Fig. 10.37** MSP430 DAC12 control register (DAC12_xCTL)—x = 0 or 1

**Table 10.3** Digital to analog converter DAC12, DAC12_A in MSP430

| Feature | 'x1xx | 'x2xx | 'x4xx | 'x5/6xx |
|---|---|---|---|---|
| 8- or 12-bit resolution | Yes | Yes | Yes | Yes |
| Straight binary or 2's complement data format | Yes | Yes | Yes | Yes |
| Data right/left justification | No | No | No | Yes |
| Adjustable dynamic range K | Yes | Yes | Yes | Yes |
| Values: | (1 or 3) | (1 or 3) | (1 or 3) | (1, 2 or 3) |
| Internal or external VREF selection | Yes | Yes | Yes | Yes |
| Internal Output availability* | No | Yes | Yes | Yes |
| Offset calibration capability | Yes | Yes | Yes | Yes |
| Data calibration capability | No | No | No | Yes |
| Interrupt vector generator | No | No | No | Yes |
| Programmable settling time versus power | Yes | Yes | Yes | Yes |
| Multiple DAC's synchronized update | Yes | Yes | Yes | Yes |

**Example 10.12** *Let us use an MSP430FG4617 to generate* $\approx 1$ V *from DAC12_0, using as reference voltage an external 3.3 V, with no extended dynamic range. From (10.25), we find that the corresponding encoding* $N_Q$ *is* 1241 = 4673 h.

*From the use'r guide ([slau056j]) we obtain the following information for the register bits:*

DAC12OPS = 0 *for output at P6.6, which comes by default at reset;*
DAC12SREFx bits = 10 or 11 *for the external reference;*
DAC12RE = 0 *for 12-bit resolution, which comes by default at reset;*
DAC12OPS = 0 *for output at P6.6, which comes by default at reset;*
DAC12IR = 1 *for one time reference voltage resolution.*

*Finally, let us work with medium speed and current for both input and output buffers, DAC12AMPx bits = 101. We are now ready to go:*

```
DAC120_setup    mov #DAC12REF_2+DAC12IR+DAC12AMP_5,&DAC12_0CTL
                                  ; Vref=3.3 external
```

```
DAC12_out         mov #1241,&DAC12_0DAT          ; 1 V at output
```

*or in C*

```
DAC12_0CTL = DAC12REF_2+DAC12IR+DAC12AMP_5
                              // DAC12_0 set up  Vref=3.3 external
DAC12_0DAT = 1241   // 1 V at output
```

*The data register is bypassing the latch one, so the value of* DAC12ENC *is a don't care. Many programmers prefer to set this bit.*


## 10.9  Analog-to-Digital Converters

An *Analog-to-digital Converter* (ADC, A/D), performs the opposite operation from the DAC, that is, it receives an analog input signal, and delivers an $n$-bit word. Most commercial ADCs range from 6-bit to 24-bit words, available from different vendors. Moreover, parallel and serial output ADCs are also in the market, with the latter ones being increasingly more popular.

There are different types of converters. Flash ADCs are the fastest ones, but very power hungry. The basic principle consists in identifying the subinterval of [0 V, VREF] in which the input value falls. To do that, it uses $2^n - 1$ comparators, whose outputs are input to a logic block that generates the $n$-bit word. An 8-bit ADC requires 255 comparators, for example. This structure has the great advantage of speed, but is limited to low number of bits, due to its size and power consumption. To deal with these disadvantages, several modifications and combinations have been proposed in the literature, such as the two-step ADC, pipeline ADCs, and others. Nevertheless, their discussion is out of the scope of this book.

There are several other architectures available. We discuss briefly three of them: The slope, the successive approximation, and the sigma-delta types. These architectures are also used as embedded peripherals in many microcontroller families.


## 10.9.1  Slope ADC

The simplest form of an ADC is the *slope ADC*, also called *integrating ADC*. Two versions are illustrated in Fig. 10.38, the single slope and the dual slope ADCs.

The logic units in these ADCs contain counters, output registers, the necessary logic to generate the signals to control the switches, synchronization, and so on. The Vin value is obtained from time measurements. In the single slope case, it measures the time $t_u$ that it takes the integrators output to reach the Vin value. On the other hand, the controller of the dual slope ADC connects the integrators output to $-$Vin during a fixed time $t_{uf}$ and then switches to VREF. We measure the time $t_d$ that the integrator needs to get back to 0 V. These principles are illustrated in Fig. 10.39.

In the single slope ADC, the moment in which the integrator's output $V_S$ reaches Vin is detected with a threshold detector. The comparator's change of state tells the digital unit to capture the word from a counter, and to generate a reset signal for the integrator. In this case, the time $t_u$ is found from

$$V_{in} = \frac{\text{VREF}}{RC} t_u \qquad (10.43)$$

This requires precision passive values and a good clock of stable frequency $f_{clk}$.

In the dual slope ADC, after $t_{uf}$ seconds the integrator's output is

$$V_S = \frac{\text{Vin}}{RC} t_{uf} \qquad (10.44)$$

The switch connects then to VREF and the measured time $t_d$ for the integrator's output to become $0\,V$ is related to the equation

$$-\frac{\text{VREF}}{RC} t_d + V_S = 0 \qquad (10.45)$$

Using this information we have

$$\text{Vin} = \frac{t_d}{t_{uf}} \text{VREF} \qquad (10.46)$$

The main disadvantage of slope ADCs is their speed. However, when speed is not a great concern, they have the advantage of simplicity, and can be realized with the comparators embedded in the microcontroller.

## 10.9.2 Successive Approximation ADC

A popular architecture, embedded as a peripheral in many microcontrollers, is the *successive approximation ADC*. The $n$ bits are determined from MSB to LSB in $n$

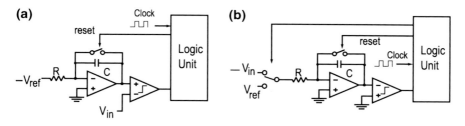

**Fig. 10.38** Basic diagrams of (**a**) single slope ADC, and (**b**) dual slope ADC

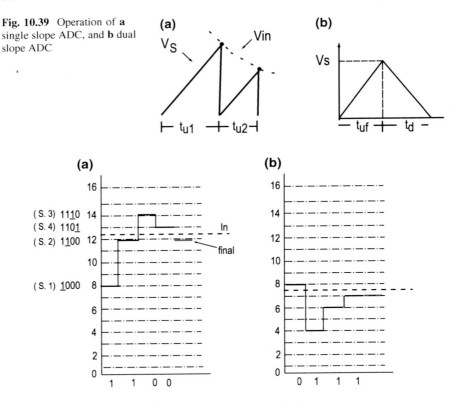

**Fig. 10.39** Operation of **a** single slope ADC, and **b** dual slope ADC

**Fig. 10.40** Two 4-bit examples illustrating successive approximation

steps by comparing with mid levels of successive region subintervals. This principle is illustrated with the following example.

**Example 10.13** *We illustrate the successive approximation for 4-bit words $b_3 b_2 b_1 b_0$ using Fig. 10.40. The dotted lines show the lower bounds of the subintervals in a half LSB compensated quantization. We discuss case (a) and leave (b) to the reader.*

*We first compare with the mid level 1000 of the full scale region and by comparison in which half the input is located, defining then the most significant bit. In our example, $b_3 = 1$ since the input is above this level. We proceed then to compare with the mid level of the already determined region, that is, with 1100. Again, the input is above the level, so $b_2 = 1$.*

*Hence, at this step we know $\frac{1}{4} FS$ region where our result is, which will be of the form 11xx. We go on to the third step selecting the mid-point of this region 1110 and compare again. Since the input is now below this level, $b_1 = 0$. In the fourth and final step, we find that the input is below the level 1101, so $b_0 = 0$. Hence the conversion yields 1100.*

**Fig. 10.41** Basic operational block diagram of a successive approximation ADC

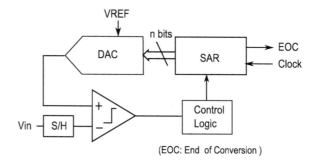

(EOC: End of Conversion)

Figure 10.41 shows the basic structure of a successive approximation ADC. The SAR block is the successive approximation register which keeps the intermediate values as well as the final one. This register feeds a DCA whose output is the one we use to compare with the input level. The process takes time, so there is a flag to indicate when the conversion has finished. The system is regulated by a clock, which may be independent of the system clock.

The main advantage of the SAR ADC is that the circuit complexity and power dissipation are less than those found in most other types of ADCs. One of the drawbacks is that eventually the comparator must do a comparison within 1 LSB of precision, and precautions must be taken to deal with noise.

Modern SAR structures use the charge distribution principles of a DAC to simplify the architecture. The charge distribution SAR ADC shown in Fig. 10.42 has the advantage that the capacitor array serves both as S/H and DAC stage. This configuration is now common in most cases in which the microcontroller has a SAR ADC peripheral. The switches are controlled by the SAR register. The total capacitance is 2C.

Conversion starts by sampling, with switch S open and all others connected to Vin. Then the hold phase opens S and all other switches connected to ground. The bits are found one by one starting with the MSB $b_{n-1}$.

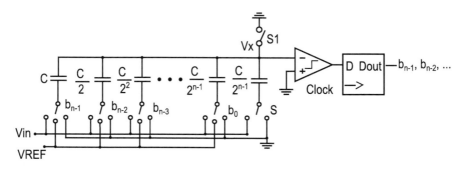

**Fig. 10.42** Basic operational block diagram of a successive approximation ADC with charge distribution

As we determine the output bits, the voltage Vx is changing, until it decreases to a value within $\pm\frac{1}{2}$ LSB, when the least significant bit is determined. Remark that once bit $b_j$ has been determined, the switch is connected to VREF if $b_j = 0$ and to ground if $b_j = 1$. The Vx value becomes, step by step,

$$Vx = -Vin \Rightarrow Vx = -Vin + \frac{VREF}{2} \Rightarrow Vx = -Vin + \frac{b_{n-1}VREF}{2} + \frac{VREF}{2^2}$$
$$\Rightarrow Vx = -Vin + \frac{b_{n-1}VREF}{2} + \frac{b_{n-2}VREF}{2^2} + \frac{VREF}{2^3} \cdots$$

until it reaches its final value of

$$Vx = -Vin + \frac{VREF}{2^n}\left(b_{n-1}2^{n-1} + b_{n-2}2^{n-2} + \ldots b_0\right)$$

The circuit in Fig. 10.42 is one among different configurations based on the same ADC topology and principles. Variants of it include having two reference voltages to deal with bipolar inputs and two's complement representations, input on the top instead of bottom, etc. Their discussion does not fall within the scope of this book.

### 10.9.3  Oversampled Converter: Sigma-Delta

The ADC architectures mentioned before belong to a class of converter types known as Nyquist rate converters, which work with a sampling frequency close to the Nyquist rate, that is $f_s \approx f_N$. Moreover, they work with quantization levels which are associated in a one to one relationship with the $n$-bit words. These converters are generally well suited for moderate resolutions in a high input bandwidth environment.

Nyquist rate converters have, however, two main disadvantages. One is the need of anti aliasing filters. The other is the level of quantization noise, which imposes severe constraints on the resolution and on the voltage references required, which may go to the microvolt range for large values of $n$. These disadvantages are addressed with *oversampling AD converters* for which $f_s > f_N$. These converters perform an inverse trade-off. That is, they allow high resolution but input bandwidths are not as high. The most popular oversampling converter is the *Sigma-Delta ADC* ($\Sigma\Delta$ ADC), sometimes also called *Delta-Sigma ADC*.[8]

The theoretical bases for this architecture go back to the 1940s, when the delta modulation concept was developed for PCM communication. The sigma-delta concept appeared in the 1950s, and the development of VLSI technology provided the definite impulse for this architecture. Because of its several advantages, it is one of the preferred methods for analog-to-digital conversion. Among them, the relaxation—or even no need—of requirements on anti aliasing filters, better management of noise, and high resolution. While 14 bits is in general a forced upper

---

[8] In fact, delta-sigma was first used, until 1970 when ATT engineers started using sigma-delta.

**Fig. 10.43** Block diagram of
a first order sigma delta ADC

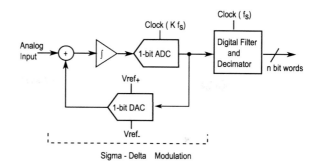

Sigma - Delta   Modulation

bound for SAR ADCs, $\Sigma\Delta$ ADCs can achieve resolutions well beyond 20 bits. On the downside, the speed is in general lower than those architectures, due to the fact that they emphasize changes and not absolute values. Therefore, signals with fast changes are not easily handled.

The common architectures are the first order and second order structures of Figs. 10.43 and 10.44. Theoretically, the 1-bit ADC/DAC may be of $m$ bits instead. However, for all applications and practical implementations it has been found, and theoretically proved, that 1 bit is very good.

The 1-bit ADC may be as simple as a comparator, and the 1-bit DAC an inverter, or a comparator again. The $\Sigma\Delta$ modulation portion works at a frequency much higher that the Nyquist frequency. The *Oversample Ratio* (OSR) is defined as

$$\text{OSR} = \frac{Kf_s}{f_N}$$

Frequency $f_s$, which controls the digital filter is close to the Nyquist frequency.

The bit stream delivered by the sigma-delta modulation subsystem can be regarded either as a digital or an analog signal. This bit stream is a one-bit serial signal with a

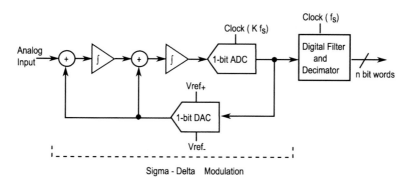

Sigma - Delta   Modulation

**Fig. 10.44** Block diagram of a second order sigma delta ADC

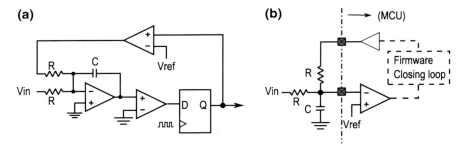

**Fig. 10.45** Two examples for delta modulation block: with OA integrator. (**a**) and comparator (**b**)

very high bit rate. Its major property is that its average level represents the average input signal level, which is extracted by the filter part. In fact, this filter is a low-pass one, and it does not need to be really digital. However, working it this way has the advantage that it can be implemented by software.

One interesting advantage of sigma-delta converters is that, since the input value is determined as an average value, it is not necessary to use extremely precise element values, like in the slope ADCs. Moreover, circuit approximations are valid, as illustrated in the following example.

**Example 10.14** *Figure 10.45 shows two simple realization principles of a $\Sigma\Delta$ converter. The first one uses three OpAmps, two of them as comparators. One comparator is used as a single bit ADC, the other as a DAC. In the second one, the sum and integration is done only with passive components.*[9]

*In the first configuration, an MSP430 with three OAs can be used. The external elements would be the resistors and the capacitor. Ditto for the right configuration, in which case the loop may be closed with software code.*

*These examples provide an orientation of how we can realize a sigma delta converter for microcontrollers which do not have such peripheral.*

### 10.9.4 Closing Remarks

We discussed briefly three types of ADCs, and mentioned the flash converter ADC too. Another interesting architecture that we are leaving out is the pipeline ADC. More architectures are available in the literature, but commercially these five are the most popular in IC form, while the first three are preferred for built in ADCs in microcontrollers, and also relatively easy to build within an MCU with only comparator included. Let us close our visit to this topic with some concluding remarks.

---

[9] Even for moderate clock frequencies, the circuit behaves like a summing integrator when the product CR is large enough.

**Flash ADC**   It is the fastest of all configurations. The choice if you need very high speed and power is not of concern. It is limited however to low resolutions, 8 bits at most. Conversion time is independent of resolution.

**Pipeline ADC**   Not discussed here, it is a good alternative for high speed applications, in the order of several Msps (Mega samples/s) to approximately 100 Msps. It offers better resolutions than the flash architecture with less power consumption.

**Successive Approximation ADC**   It is good in the medium range of resolution, like the pipeline, but with low power consumption. The speed, however, is limited to around 5 Msps, and diminishes as resolution goes higher. It needs anti aliasing.

**Sigma delta ADC**   It offers the highest resolution, from 16 to 26 bits. But it is slower than the successive approximation type. An advantage is that it does not require precision components and anti aliasing is not necessary or is very loose.

**Slope ADC**   It is the slowest type. It is good for monitoring DC signals, and it consumes little power. It also exhibits very good noise performance. Depending on the realization, though, it may require high precision elements.

## 10.9.5  Using the MSP430 ADCs

ADCs built-in in MSP430 microcontrollers are all SAR or Sigma Delta type converters. Some models are mentioned as having slope ADC incorporated. In fact, they have comparators and I/O ports with good performance by design for these applications. Low voltage series, on the other hand, have a set of analog peripherals which allow configuring 8-bit SAR ADC converters, but do not have the converter as a built in peripheral.

The actual ADCs that are available are the 10-bit SAR ADC converter, ADC10, 12-bit SAR ADC converter, ADC12, and the 16-bit and 24-bit sigma-delta converters SD16, SD16_A, SD24, and SD24_A.

### MSP430 ADC10 (ADC12) Features

ADC10 and ADC12 differ in very few aspects: resolution and management of results. Hence, we limit our introduction to the ADC10 case. A simplified block structure of the ADC10 system is shown in Fig. 10.46. We have omitted several intermediate blocks and the control signals, introduced below.

The ADC10 system has eight registers associated to it:

- Two 16-bit read-and-write control registers: ADC10CTL0 and ADC10CTL1. Some specific bits are read-only. They are reset on POR.
- Two 8-bit read-and-write input enable registers: ADC10AE0 and ADC10AE1. They are reset on POR.
- One 16-bit read-only register, ADC10MEM, to receive conversion results.

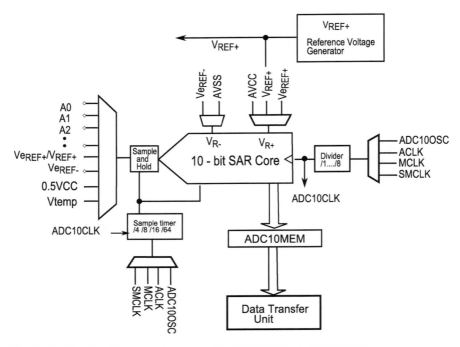

**Fig. 10.46** Simplified block configuration of the MSP430 10-bit ADC10 SAR converter

- Two read-and-write data transfer control registers: ADC10DTC0 and ADC10 DTC1. Some specific bits are read-only. They are reset on POR.
- On 16-bit read-and-write data transfer start address register: ADC10SA, initialized with 0200h on POR. Bit 0 is hardwired to 0.

With reference to Fig. 10.46, we have the following:

**Core**     The core of the system is a switched capacitor SAR converter. The SH unit in the figure is only functional, since it is practically part of the core (see Fig. 10.42). It may be turned on or off with the ADC10ON bit—Bit 4 in ADC10CTL0–. A read-only flag ADC10BUSY—bit 0 in ADC10CTL1—is set to indicate that sampling and conversion are in progress. The result is written to register ADC10MEM in a format selected with ADC10DF—bit 9 in ADC10CTL1:

- The default right justified straight binary in ADC10MEM. Bits b15 to b10 are 0. Result 0000h corresponds to the bottom of input range VR−. Result goes from 0000h to 03FFh.
- Left justified two's complement format, with bits b5 to b0 equal to 0. Result 0000h corresponds to the midpoint of input range. Results read from 8000h to 7FC0h.

The full scale is VR+ to VR−. When ADC10MEM is updated, it sets the interrupt flag ADC10IF (bit 2 in ADC10CTL0) which will generate an interrupt request if ADC10IE (bit 3) is set.

**Core References**    VR+ may be chosen from two external references, AVCC or a special $V_{eREF+}$, or else from an internally generated reference $V_{REF+}$. The external reference can be buffered or unbuffered. VR− may be chosen from two external references, the default AVSS or a special $V_{eREF−}$. Selection of reference is done with bits 15 to 13 from ADC10CTL0, SREFx.

**Internal Reference**    $V_{REF+}$ is generated internally when REFON (bit 5 in ADC10 CTL0) is set. It is by default 1.5 V, unless bit 6 REF2_5 is set, making it 2.5 V. It is very stable. This voltage is usually available at a pin if REFOUT is set, and can be buffered. The buffer is controlled with ADC10SR.

**Core Clock**    The clock for the SAR core is a signal ADC10CLK, which is generated from a divider. This divider receives an input from one of four sources: The MSP430 clocks ACLK, MCLK, or SMCLK, or else a built-in oscillator ADC10OSC. The selection is done with bits 4-3, ADC10SSELx, and division with bits ADC10DIVx, bits 7-5 all found in control register 1.

**Conversion**    ADC10 must be enabled with bit ENC (bit 1 in control register 0) for a conversion to take place. Conversion may be starte by software setting bit ADC10SC, or else by any of time A outputs, Timer_A.OUT0, Timer_A.OUT1 or Timer_A.OUT2. The trigger source may be selected with bits SHSx in ADC10CTL1. Once triggered, the sample and hold time is controlled with bits ADC10SHTx of ADC10CTL0, to be 4, 8, 16, or 64 times ADC10CLK. ADC10SC is automatically reset when conversion is terminated.

**Inputs**    The analog inputs to the ADC unit may be external inputs A0, A1, …, the number depending on the model. The external references and internal reference may serve as inputs. Internally generated values 0.5 VCC or from temperature sensor may also be inputs to the converter. The selection is done with bits 15-12, INCHx, from control register 1. In the same register, bits 2-1, CONSEQx, allow to control for single or multiple conversions from different channels. Inputs A1, A2, …are shared with I/O pins. Hence, the ADC10AE0 and ADC10AE1 registers are used for enabling the connections for these pins to the converter.

**Result and transfer**    The ADC10MEM is readable. However, we usually need to transfer the results to another place in memory. This is done with the ADC10 data transfer controllers. One convenient feature is that data transfer is done without the CPU intervention (it may even be off) directly interacting with DMA. These registers function as follows:

- ADC10DTC0: selects one or two blocks in destination, or continuous transfers.
- ADC10DTC1: provides the number of transfers per block. Transfer is disabled when the value is 0.
- ADC10SA is the start address for destination. To enable transfers, the user must write to this register.

Let us work some simple code examples to set up the ADC10 and work conversions.

**Example 10.15** *Let us set up ADC10 to work with input* A0 *(pin shared with* P2.0*). We use the converter oscillator* ADC10OSC *(default) for generation of signal* ADC10CLK, *the Sample and hold time will be eight times this signal. Straight binary conversion is performed. The CPU is in low power mode while the ADC10 is working. The following piece of code works for our purpose:*

```
; ==================================================================
; ADC10SHT_1 for 8x ADC10CLKs and ADC10IE to enable interrupt

setup_ADC   mov.w   #ADC10SHT_1+ADC10ON+ADC10IE,&ADC10CTL0 Input_A0
bis.b   #01h,&ADC10AEO          ; select A0 for input Setup_P2
bis.b   #01h,&P2SEL             ; pin P2.0 to ADC10 ; Mainloop
bis.w   #ENC+ADC10SC,&ADC10CTL0 ; Start sampling
                                              ; and conversion
            bis.w   #CPUOFF+GIE,SR            ; LPM0
            jmp     Mainloop                 ; Again


; ------------------------------------------------------------------
; - - - - - - - - - - - - - -
ADC10_ISR
            bic.w   #CPUOFF,0(SP)             ; CPU Active on return

            - -   - -                         ; ISR instructions
            reti                              ; return from interrupt

; ------------------------------------------------------------------
; - - - Interrupt vector - -

            ORG     ADC10_VECTOR
            DW      ADC10_ISR
; ==================================================================
```

*A C-code is the following:*

```
//================================================================
void main(void)
{
  // 8 times ADC10CLKs, ADC10ON, interrupt enabled
  ADC10CTL0 = ADC10SHT_1 + ADC10ON + ADC10IE;
  ADC10AE0 |= 0x01;                 // P2.0 ADC A0 option select
  for (;;)
  {
    ADC10CTL0 |= ENC + ADC10SC;     // Sampling & conv. start
    __bis_SR_register(CPUOFF + GIE); // LPM0,
  }
}


//----------------------------------------------------------------
// ADC10 interrupt service routine
#pragma vector=ADC10_VECTOR
__interrupt void ADC10_ISR(void)
{
 __bic_SR_register_on_exit(CPUOFF);  // Active on Return

  - - - - - - - - -                  // ISR instructions here
}
//================================================================
```

*These code examples are limited to setting of the ADC10.*

## MSP430 SD16_A Sigma-Delta Converter

The MSP430 has three variations for its sigma-delta ADC. The original one SD16 contains three independent channels, each one being a complete ADC with a single input. They can operate simultaneously or with specified delay in between, and are linked with logic.

The SD16_A has a single core but several inputs which may be sequentially but not simultaneously converted. It also has enhancements with respect to the previous model. A third module is an extension of this model containing several single input SD16_A cores linked with logic.

Figure 10.47 shows a block diagram of the SD16_A module, showing the control bits for different blocks. The converter has three registers associated to the core: SD16_A control register SD16CTL, SD16_A interrupt vector register SD16IV, and the SD16_A analog enable, SD16AE. Each channel x has additional registers for control, conversion and input control, SD16CCTLx, SD16MEMx, and SD16INTCTLx, where x stands for the channel number. Series 'x4xx models have also a preload register SD16PREx.

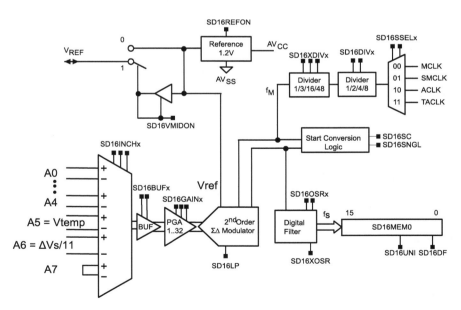

**Fig. 10.47** Simplified block configuration of the MSP430 Sigma-Delta converter SD16_A (*Adapted from MSP430 user guide, Texas Instruments Inc.*)

The reference voltage $V_{ref}$ for the converter can be externally supplied or generated internally with a 1.2 V reference voltage generator when REFON is set. This voltage is also buffered to be externally available.

The clock, on the other hand, is selected from one of the MSP430 signals MCLK, SMCLK, ACLK, and TACLK and divided first by 1, 2, 4, or 8 with the SD16DIVx bits, and then by 1, 3, 16, or 48 with the SD16XDIVx bits. This will generate the modulator frequency signal $f_M$. The outputs from the filter are generated at a frequency

$$f_S = f_M / OSR$$

The oversampling ratio OSR is set by the SD16OSRx bits as 32, 64, 128 and 256. SD16XOSR bit can give increased ratios of 512 or 1024. Unlike the ADC10 and ADC12, the core is automatically shut down when not in used.

The system has eight differential inputs selectable with the SD16INCHx bits. Input A5 is internally generated by a voltage divider yielding $\Delta Vs = (VCC - VSS)/11$, a useful feature to monitor state of battery. Input A6 is generated by the temperature sensor, and A7 has the inputs short circuited, which is useful for calibration.

The inputs go through a high impedance differential buffer, not available in all models, whose current capacity is configured with SD16BUFx. The buffer is convenient for low impedance inputs. This block is followed by a fully differential programmable gain amplifier configured with SD16GAINx bits. Gains are from 1 to 32 in powers of 2. The differential input voltage should be between $\pm V_{FSD}$, where

$$V_{FSD} = \frac{0.5 V_{ref}}{PGAGain} \tag{10.47}$$

conversion starts when the SD16SC bit is set. It may work a single conversion or the default continuous conversions depending on SD16SNGL bit. To fully realize a single conversion, the core has to perform a number of conversions which is specified by the SD16INTDLYx bits. On the other hand, continuous conversions do not stop until SD16SC is cleared by software, in which case conversions stop without completing the current one.

The interrupt flag SD16IF is set when a new result is available at register SD16MEM0. The flag is automatically reset when the result is read. Multiple channels version have one register and one interrupt flag for each channel. An overflow interrupt flag SD16OVIFG is set when a new result overruns a previous value which had not been read. These flags generate maskable interrupt requests when SD16IE is set.

The output format can be configured using the SD16DF and SD16UNI bits in anyone of the following: (a) bipolar two's complement, (b) bipolar offset binary, and (c) unipolar binary. These output formats are illustrated in Fig. 10.48. Remember that the upper bound of the full scale is not encoded. This is shown with parenthesis.

The following two listings, courtesy of Texas Instruments, Inc., show examples using this ADC converter. The first one is in assembly language and the second one

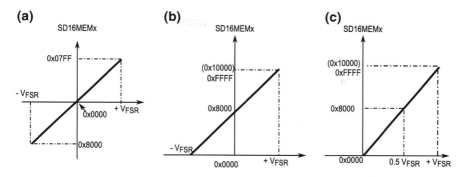

**Fig. 10.48** MSP430 Sigma-Delta converter SD16_A output formats: **a** bipolar two's complement, **b** bipolar offset binary, and **c** unipolar binary (*Courtesy of Texas Instruments, Inc.*)

in C language. In the latter case, the documentation is omitted for brevity since it contains the same information.

**Example 10.16** *This example program uses the SD16 module to perform a single A-to-D conversion on Channel 2. A interrupt from the SD16 is configured to occur upon end-of-conversion.*

*To test the program, apply a known voltage at CH2 (A2.0+, A2.0−) and insert a breakpoint at the line indicated in the code comments. The converted value can be examined in R12 with the debugger.*

*The assumed settings include ACLK = LFXT1 = 32768 Hz and MCLK = SMCLK = default DCO = 32 × ACLK = 1048576 Hz. An external clock crystal is required for ACLK and a 100 nF capacitor between Vref and AVss is recommended when using $V_{ref} = 1.2$ V.*

```
;=========================================================================
; Demo - SD16, Single Conversion on Single Channel Using ISR
;
; By H. Grewal * Texas Instruments, Inc. *  February 2005
; Built with IAR Embedded Workbench Version: 3.21A
;=========================================================================
#include    <msp430x42x.h>
;- - - - - - CPU Registers Used - - - - - - - - - - - - - - - - - -
#define      RESULTS R12       ; R12 - Store conversion result from SD16MEM2
                               ; R15 - Temporary working register


;-------------------------------------------------------------------------
            ORG    08000h                    ; Program Start
;-------------------------------------------------------------------------
RESET       mov.w  #600h,SP                  ; Initialize SP
StopWDT     mov.w  #WDTPW+WDTHOLD,&WDTCTL    ; Stop watchdog
SetupFLL    bis.b  #XCAP14PF,&FLL_CTL0       ; Configure load caps
            mov.w  #10000,R15                ;
Xtal_Wait   dec.w  R15                       ; Delay for 32 kHz crystal to
            jnz    Xtal_Wait        ; stabilize
SetupSD16   mov.w  #SD16REFON+SD16SSEL0,&SD16CTL
                                             ; 1.2V ref, SMCLK
```

```
                bis.w   #SD16SNGL+SD16IE,&SD16CCTL2
                                            ; Single conv, enable interrupt
                bis.b   #SD16INTDLY0,&SD16INCTL2
                                            ; 3rd sample generates interrupt
                mov.w   #03600h,R15         ; Delay  for 1.2V ref startup
L$1             dec.w   R15                 ;
                jnz     L$1                 ;
                eint                        ; Enable general interrupts
Mainloop        bis.w   #SD16SC,&SD16CCTL2  ; Start conversion
                bis.w   #CPUOFF,SR          ; Enter LPM0 (disable CPU),
                                            ; wait for conversion to complete
                nop                         ; Required for debug only
                jmp     Mainloop            ; SET BREAKPOINT HERE


;-----------------------------------------------------------------------
SD16_ISR     ; SD16 Interrupt Service Routine
;-----------------------------------------------------------------------
                add.w   &SD16IV,PC          ; Add offset to PC
                reti                        ; Vector 0: No interrupt
                jmp     SD_OV               ; Vector 2: Overflow
                jmp     SD_CH0              ; Vector 4: CH0 IFG
                jmp     SD_CH1              ; Vector 6: CH1 IFG
                                            ; Vector 8: CH2 IFG


;-----------------------------------------------------------------------
;- - - - - -SD16 Channel 2 Interrupt Handler;- - - - - - - - -
SD_CH2          mov.w   &SD16MEM2,RESULTS   ; Save CH2 conversion
                bic.w   #CPUOFF,0(SP)       ; Return active
SD_CH2_END   reti                           ; Return from interrupt


;-----------------------------------------------------------------------
;- - - - - -SD16 Memory Overflow Interrupt Handler - - - - - - SD_OV
reti                                        ; Return from interrupt


;-----------------------------------------------------------------------
;- - - - - -SD16 Channel 0 Interrupt Handler- - - - - - - - - -
SD_CH0          reti                        ; Return from interrupt


;-----------------------------------------------------------------------
;- - - - - -SD16 Channel 1 Interrupt Handler- - - - - - - - - -
SD_CH1          reti                        ; Return from interrupt


;-----------------------------------------------------------------------
                COMMON  INTVEC              ; Interrupt Vectors
;-----------------------------------------------------------------------
                ORG     RESET_VECTOR        ; RESET Vector
                DW      RESET               ;
                ORG     SD16_VECTOR         ; SD16 Vector
                DW      SD16_ISR            ;
                END
;=======================================================================
```

**Example 10.17** *A C-language version of the program in the previous example.*

```
//=======================================================================
// Demo - SD16, Single Conversion on Single Channel Using ISR
//-----------------------------------------------------------------------
// By H. Grewal * Texas Instruments, Inc. *  February 2005
// Built with IAR Embedded Workbench Version: 3.21A
```

```
//========================================================================
#include   <msp430x42x.h>
//------------------------------------------------------------------------
unsigned int result;

void main(void)
{
  volatile unsigned int i;           // Use volatile to prevent removal
                                     // by compiler optimization

  WDTCTL = WDTPW + WDTHOLD;          // Stop WDT
  FLL_CTL0 |= XCAP14PF;             // Configure load caps
  for (i = 0; i < 10000; i++);       // Delay for 32 kHz crystal to
                                     // stabilize

  SD16CTL = SD16REFON+SD16SSEL0;    // 1.2V ref, SMCLK
  SD16CCTL2 |= SD16SNGL+SD16IE ;    // Single conv, enable interrupt
  SD16INCTL2 |= SD16INTDLY0;        // Interrupt on 3rd sample
  for (i = 0; i < 0x3600; i++);      // Delay for 1.2V ref startup

  _EINT();                           // Enable general interrupts

  while (1)
  {
    SD16CCTL2 |= SD16SC;            // SET BREAKPOINT HERE
                                     // Set bit to start conversion
    _BIS_SR(LPM0_bits);              // Enter LPM0
  }
}

//------------------------------------------------------------------------
#pragma vector=SD16_VECTOR __interrupt void SD16ISR(void) {
  switch (SD16IV)
  {
  case 2:                            // SD16MEM Overflow
    break;
  case 4:                            // SD16MEM0 IFG
    break;
  case 6:                            // SD16MEM1 IFG
    break;
  case 8:                            // SD16MEM2 IFG
    result = SD16MEM2;              // Save CH2 results (clears IFG)
    break;
  }

  _BIC_SR_IRQ(LPM0_bits);            // Exit LPM0
}
//========================================================================
```

## 10.10  Chapter Summary

This chapter described the basic components of the analog signal chain. That is,
the components that process the analog signal prior to the digital processor, or that
extract analog information from the digital processor's output.

**Fig. 10.49** A subtractor
circuit

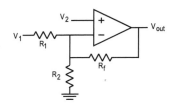

The chain starts with a sensor, which transforms a non-electrical signal into one that can be processed by electronic means. An overview of sensors was covered in this chapter.

The sensor's output may need conditioning so that it can satisfy the requirements of the electronic components that process the signal. This conditioning may include amplification, filtering, or other pre-processing. This stage uses operational amplifiers, another of the signal chain components. Different circuits were discussed for this stage.

The conditioned signal is then sampled and converted into a digital word using an Analog-to-Digital converter (ADC). The theoretical principles behind this transformation, as well as the basic architectures were reviewed.

The conversion from the digital output to an analog value was similarly considered. The Digital-to-Analog converter (DAC) was reviewed.

Finally, a tour on the embedded analog signal chain peripherals in the MSP430 devices was introduced, so the reader can gain familiarity with these important components of the MSP430 architecture.

## 10.11 Problems

10.1  In general, human speech generates frequencies from 50 Hz upward, to about 10 kHz. However, most energy is concentrated in the 300 Hz–3.4 kHz band, which is the one in which speech intelligibility is the best. On this basis, the telephone system and many other speech recognition systems sample the speech at a rate of 8 kHz. Justify the use of this frequency.

10.2  Starting with Eq. 10.6, with $V_r = 0$, find the output voltage expression for the subtractor when $R_4/R_3 \neq R_2/R_1$.

10.3  The circuit in Fig. 10.49 is another example of a subtractor circuit. It is often used to provide amplification to a sensor signal $V_1$ and an offset (*scaling*) to shift toward a desired origin or to keep the output voltage in a linear region, especially in cases of one supply OAs. The circuit can also be considered as a non-inverting amplifier with shifting. Find the expression for the output voltage.

10.4  With a resolution of $N$ bits and using as quantization levels those in Eq. (10.24), what is the quantization error for Vmax?

10.5 Show that in an $N$-bit quantization for symmetrical intervals, if $N_{QB}$ is the representation in biased encoding and $N_{QT}$ in two's complement encoding, then $N_{QB} - N_{QT} = 2^{N-1}$.

10.6 The Texas Instruments' TLV5604 is a 10-bit four channel serial digital-to-analog converter (visit http://www.ti.com/lit/ds/symlink/tlv5604.pdf for full reference). Sketch an appropriate interface connection with your MSP430 microcontroller and design a flowchart to deliver the digital information to the DAC.

10.7 The AD9760 (data sheet at http://www.analog.com/static/importedfiles/data_sheets/AD9760.pdf) is a 10-bit parallel DAC available from Analog Devices. This converter has a current output. Sketch an appropriate interface connection with your MSP430 microcontroller and design a flowchart to deliver the digital information to the DAC.

# References

1. Bryant, J., & Kester, W. (2011). *Ch. 3 DATA CONVERTER ARCHITECTURES: Sec. 3.1 DAC ARCHITECTURES*. Analog Devices: Inc

# Appendix A
# Software Planning Using Flowcharts

## A.1 Introduction

In this section we present an overview of flowcharts. Flowcharts are one of the oldest and most popular tools used to express an algorithm, regardless of whether it is a software or a hardware algorithm. Probably the reason flowcharts are so popular is because they give a general picture of the algorithm, as opposed to the sometimes obscure picture provided by other tools. Although, as mentioned before, flowcharts are used to express an algorithm that could later be implemented in either software or hardware, this appendix was written with the idea of a software implementation in mind.

Flowcharts are not intended to include every detail of the implementation of the algorithm because that is left to the actual coding of the instructions in the particular programming language of choice. Thus, flowcharts are expected to give a general idea of the algorithm and it will help to keep this in mind when developing algorithms using flowcharts. On the other hand, if you already know the programming language in which the algorithm will be implemented, then you could take advantage of it and introduce some details particular to that programming language, but bear in mind that these details might make it harder for the algorithm to be later implemented using some other programming language.

## A.2 Symbols

We will be introducing five (5) basic symbols used to develop flowcharts. The names of these symbols are:

1. Start/Finish
2. Assignment or process
3. Subprocess

M. Jiménez et al., *Introduction to Embedded Systems*,
DOI: 10.1007/978-1-4614-3143-5,
© Springer Science+Business Media New York 2014

**(a)**                                  **(b)**

**Fig. A.1**  Start and Finish Symbols

**Fig. A.2**  Assignment or
Process Symbol

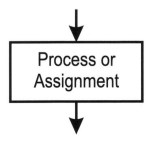

4. Connector
5. Decision

The Start/Finish symbol consists of an oval shape as illustrated in Fig. A.1. As its name implies, the Start/Finish symbol is used to show an entry point into the algorithm or an exit point out of the algorithm. Although there is nothing that hinders from doing otherwise, it is a good programming practice to specify a single point of entry and also a single point of exit and we strongly recommend it. Otherwise a rather difficult to follow algorithm will result with an also difficult to maintain implementation, whether it is a software or a hardware implementation. When it is being used as an entry point it is labeled Start. Conversely, when it is used as an exit point it is labeled Finish, Exit, or End. If the symbol is used as an entry point into a subprocess or subroutine, then the name of the subprocess is written after the corresponding label, e.g. Start Sub1, Finish Sub1. A start/Finish symbol have only one arrow, either leaving a Start or entering a Finish, as shown in Fig. A.1.

Our next symbol is the Assigment or Process. This is drawn as a rectangle and inside of the rectangle we indicate the corresponding arithmetic or logic assignments including any expressions. We could include as many assignment statements inside the rectangle as we want, but if it begins to look to crowded then it is a good practice to break it into several smaller rectangles which are easier on the eye. The assignment or process symbol is shown in Fig. A.2. Note that process symbols can have only one input arrows and one exit arrow.

The Subprocess symbol is shown in Fig. A.3. It looks a lot like the assignment symbol since it is also a rectangle, but it includes an additional vertical line on both the left and right side. This symbol is used to indicate whenever a subroutine or function call is invoked. After the subprocess returns, the algorithm continues with the next symbol in sequence in the flowchart. It is a good practice to use subprocesses in the algorithm development because it makes the algorithm less tedious and consequently

**Fig. A.3** Subroutine or Subprocess Symbol

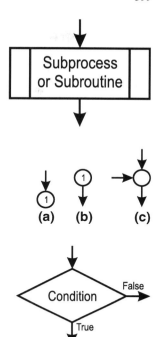

**Fig. A.4** Connector Symbol

**Fig. A.5** Decision Symbol

more appealing, especially if the subprocess is called upon several times. On the other hand, we should bear in mind that calling a subprocess and returning from it adds execution time to our algorithm. Like process symbols, subprocesses have only one input and one exit arrow.

The Connector symbol shown in Fig. A.4 is used whenever we run out of space in the page where we are developing our flowchart and need to continue on another part of the page or on a completely different page. Figure A.4a illustrates a connector leaving a page, while Fig. A.4b corresponds to the entry into a new page. Connectors are also used to denote junction points where multiple flows merge. This is the format illustrated in Fig. A.4c. Page connector symbols are used in pairs, both with the same identifier number or letter inside to indicate where the flow continues.

Our final but not less important symbol is the Decision symbol, see Fig. A.5. Its shape resembles a diamond and, as its name implies, it is used to alter the flow of the algorithm based on the value of some variable after a particular decision is made. The ability to make decisions allow for powerful algorithms as opposed to purely sequential ones. Decision blocks are the only ones that feature two outputs, one corresponding to each of the two possible decision outcomes: true or false, yes or no, left or right, etc.

Developing a flowchart for an algorithm involves these symbols and connecting them as needed with arrows indicating the flow direction.

**Fig. A.6** Flowchart for Toggling LED

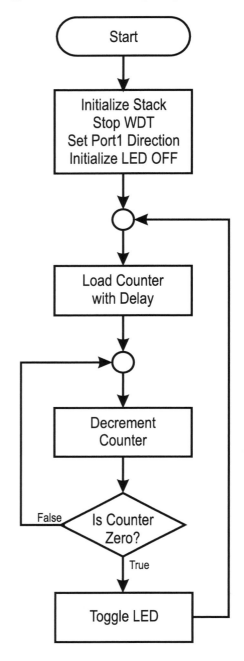

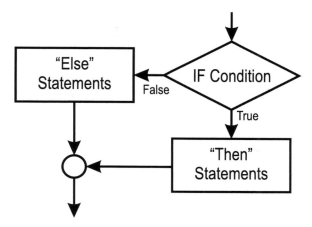

**Fig. A.7** If-Else or If-Then-Else Structure

## A.3 Putting it all Together

We will now develop an algorithm for toggling the red LED connected to pin P1.3 in the MSP430G2 Launchpad. The algorithm is illustrated in Fig. A.6 and a possible implementation using the MSP430 assembly language is shown in Listing A.1.

**Listing A.1** MSP430 Assembly Language Instructions for Figure A.6

```
1  RESET    mov     #0280h,SP              ; Initialize stackpointer
2  StopWDT  mov     #WDTPW+WDTHOLD, &WDTCTL  ; Stop WDT
3  SetupP1  bis.b   #1,&P1DIR             ; P1.0 as output
4  Main     xor.b   #1,&P1OUT             ; Toggle P1.0
5  Wait     mov     #50000,R15            ; Delay to R15
6  L1       dec     R15       ; Decrement R15
7           jnz     L1        ; Delay over?
8           jmp     Main      ; Again
```

Observe that the algorithm in Fig. A.6 does not have an exit symbol. This characteristic is common in the main loop of many embedded programs: they run forever, as long as the system is powered. Exceptions include the case when the system is placed in a low-power mode. Chapter 7 illustrates several flowchart instances like this. Now that we have introduced the basic flowchart symbols and have shown a simple example we are only left with showing a few examples of how to combine the basic symbols to obtain common algorithmic structures.

An If-Else structure, is also known as an If-Then-Else can be obtained by combining the symbols for the Decision and Process as shown in Fig. A.7. This is a powerful construct and is used widely as part of the design of algorithms.

We could combine the same symbols and come up with one of the most used iterative structures, the while construct, shown in Fig. A.8 in its two variants: do-while and while-do.

We can of course combine symbols with structures as shown in Fig. A.9. What needs to be kept in mind is the fact that flowcharts are a tool that should be used to

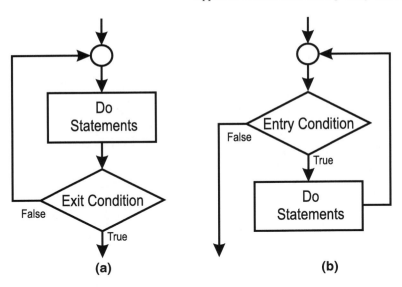

**Fig. A.8** While structures: **a**. "Do-while", **b**. "While-do"

aid in the design of algorithms to aid during the coding phase. As such we should strive to not make things more complex than they need to be, which includes both the flowchart and coding. The structure of the flowchart is such that it aims to have a single entry point and a single termination point. This includes not just the main program, but also any subroutine invoked by it or by any other subroutine. Avoid the so called spaghetti code which is both unreadable and unmaintainable. Structured techniques are the key for good algorithm development and programming practices. The power of these techniques are not to be underestimated.

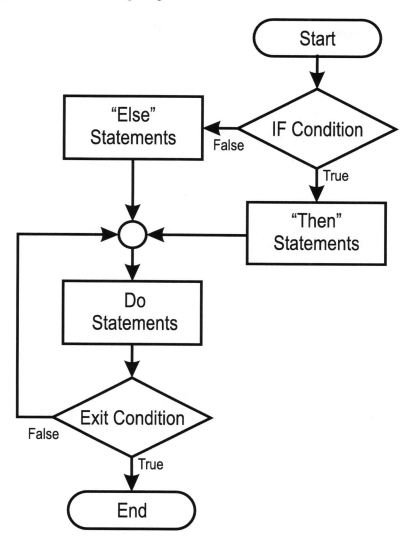

**Fig. A.9** Combined Structure

# Appendix B
# MSP430 Instruction Set

## B.1 Preliminary Notes

This appendix offers the complete instruction list for the MSP430 CPU and CPUX models. The contents of this appendix is courtesy of Texas Instruments Inc., and practically a reproduction of selected pages and chapters from MSP430x4xx Family (2007) and the MSP430x5xx Family (2008) user guides. Some additional paragraphs or sentences might be added for further clarifications, but otherwise the reproduction is verbatim. The appendix can be used as a quick reference for explanation and syntax of instructions, both for the normal and extended type.

Section B.2 describes the machine language instructions for the original MSP430 CPU. These may consist of one, two or three words, depending on the addressing modes and the type of instruction. Then the following section has the list of all the MSP430 instructions. Extended instructions are considered together with the CPUX assembly instructions in Appendix D.

The MSP430 instructions are the original 27 core instructions of the MSP430 CPU. For models with up to 64K memory range, these are the only ones available. For other models, these instructions may be used throughout the 1-MB memory range unless their 16-bit capability is exceeded. The MSP430X instructions are used when the addressing of the operands or the data length exceeds the 16-bit capability of the MSP430 instructions. There are three possibilities when choosing between an MSP430 and MSP430X instruction:

1. To use only the MSP430 instructions (except CALLA and the RETA instructions.) This can be done if a few, simple rules are met:

    (a) Placement of all constants, variables, arrays, tables, and data in the lower 64 KB. This allows the use of MSP430 instructions with 16-bit addressing for all data accesses. No pointers with 20-bit addresses are needed.
    (b) Placement of subroutine constants immediately after the subroutine code. This allows the use of the symbolic addressing mode with its 16-bit index to reach addresses within the range of PC +32 KB.

M. Jiménez et al., *Introduction to Embedded Systems*,
DOI: 10.1007/978-1-4614-3143-5,
© Springer Science+Business Media New York 2014

2. To use only MSP430X instructions: If the MSP430 instructions cannot be used. Two disadvantages of this method are the reduced speed due to the additional CPU cycles and the increased program space due to the necessary extension word for any double operand instruction.
3. Use the best fitting instruction where needed.

## B.2 MSP430 Machine Instructions for the CPU

The complete MSP430 machine language set consists of 27 core instructions, each one associated to a unique op-code decoded by the CPU. The emulated instructions are instructions replaced automatically by the assembler with an equivalent core instruction, but have no opcode themselves. There is no code or performance penalty for using emulated instruction.

There are four MSP430 machine instruction formats:

- Dual–operand
- Single–operand
- Jump
- The `reti` instruction: 1300h

The `reti` group has only one instruction, `reti`. With very few exceptions, all single–operand and dual–operand instructions can be byte or word instructions by using .B or .W extensions. Byte instructions are used to access byte data, byte peripherals or the least significant byte of registers. Word instructions are used to access word data or word peripherals. If no extension is used, the instruction is by default a word instruction.

An instruction may consist of one, two or three words. The leading word is the instruction word, and the only one conveying information of the type of instruction and operands. The other words are either the immediate value of an immediate addressing word, or are values associated to addresses in memory. The bit fields for the instruction word are shown in Fig. B.1. The source (src) is defined by the As and S-reg fields while the destination (dst) by the Ad and D-reg fields. The fields are described as follows:

**As** Group of bits defining the addressing mode used for the source (src)
**S-reg** The working register used for the source (src), or as defined by As
**Ad** Bit defining the addressing mode used for the destination (dst)
**D-reg** The working register used for the destination (dst), or as defined by Ad.
**B/W** Byte or word operation:
0: word operation
1: byte operation

Notice that destination addresses are valid anywhere in the memory map. However, for an instruction that modifies the contents of the destination, the user must ensure the destination address is writable. For example, a masked-ROM location

**(a)**

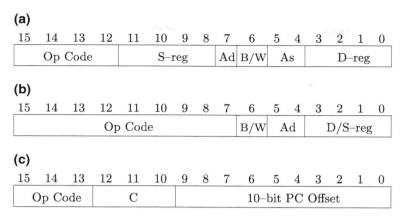

**(b)**

**(c)**

**Fig. B.1** Word Structure for Machine Instruction words (a) double-operand instruction format (b) single-operand instruction format (c) Jump instruction format

**Table B.1** Opcodes for dual-operand instructions

| Hex Code (Bits 15-12) | Instruction | Hex Code (Bits 15-12) | Instruction |
|---|---|---|---|
| 4 | mov.w, mov.b | A | dadd.w, dadd.b |
| 5 | add.w, add.b | B | bit.w, bit.b |
| 6 | addc.w,addc.b | C | bic.w, bic.b |
| 7 | subc.w, subc.b | D | bis.w, bis.b |
| 8 | sub.w, sub.b | E | xor.w, xor.b |
| 9 | cmp.w, cmp.b | F | and.w, and.b |

would be a valid destination address, but since the contents are not modifiable, the results of the instruction would be lost.

## *B.2.1 Operational Codes*

From Fig. B.1 we see that the length of opcodes depend on the type of instruction. Dual-operand instructions, involving a destination and a source, show the most significant nibble as the opcode. The MSP430 CPU has 12 instructions falling in this category, all of which may be either word or byte type. These instructions together with the corresponding opcode are shown in Table B.1.

The most significant nibble equal to 1 is given to single operand instructions. The reti = 1300h instruction, which has no operands, also has this property. The opcode for a single operand instruction consists of the nine most significant bits, and some of these instructions are word type only. The instructions, together with their opcode, and the W/B bit, are shown in Table B.2.

**Table B.2** OpCodes for single operand instructions

| Bits 15-12 | Bits 11-7 | B/W bit | Instruction |
|---|---|---|---|
| 0001 | 00000 | 0,1 | rrc, rrc.b |
| 0001 | 00001 | 0 | swpb* |
| 0001 | 00010 | 0,1 | rra, rra.b |
| 0001 | 00011 | 0 | sxt* |
| 0001 | 00100 | 0,1 | push, push.b |
| 0001 | 00101 | 0 | call* |

* Word instruction only

**Table B.3** OpCodes for jump instructions

| Bits 15-13 | Bits 12-10 | Instruction |
|---|---|---|
| 001 | 000 | jnz/jneq |
| 001 | 001 | jz/jeq |
| 001 | 010 | jnc/jlo |
| 001 | 011 | jc/jhs |
| 001 | 100 | jn |
| 001 | 101 | jge |
| 001 | 110 | jl |
| 001 | 111 | jmp |

Finally, the jump instructions have six bits in their opcodes. For all cases, the three most significant bits are 001. This leaves three additional bits to define the jump, for a total of eight jump instructions. In the hex form, the most significant nibble corresponds to 2 or 3, depending on the C bits. The opcodes and jumps are shown in Table B.3.

## B.2.2 Operand Field in Jump Instructions

Jump instructions work in what is called an offset-relative mode. The execution of the jump consists in adding two times the signed offset value defined by the ten least significant bits of the instruction word. That is

$$PC \leftarrow PC + 2 \times \text{(ten-bit-PC-offset)}$$

Here, the PC contents is that of the instruction address following the jump instruction. This is because execution phase occurs after the decode phase, when the PC is already updated to the new address. The ten-bit offset has a range from $-2^9 = -512$ to $2^9 - 1 = 511$ memory locations. Therefore, jumps can be done between $-1024$ and $+1022$ memory locations. For larger jumps, the emulated branch instruction should be used. The execution is illustrated in the following example.

**Fig. B.2** Illustration for Jumps. **a jc : 2C03, b jmp : 3FF7**

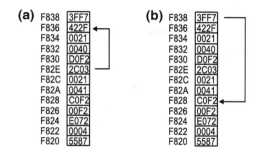

**Example B.1** *Figure B.2 illustrates what happens when two jump instructions are executed. Case (a) corresponds to the machine language instruction 2C03h, while case (b) to 3FF7h.*

*In the first case, let us separate the opcode and the operand as 2C03 : **001011–0000000011**. The opcode corresponds to jc/jhs mnemonics. The ten bit operand is +3. Hence, when executed, if the carry flag is set, this instruction will add +6 to the PC contents. After decoding, PC is already pointing to the next location, in this case F830, so after execution it ends up with F836.*

*The instruction 3FF7 : **001111–1111110111** has the opcode corresponding to the unconditional jump and the operand stands for −9, which means it goes back 18 memory locations. The PC contents is F83A at that moment, so we have F83A + FFEE = 1F828 which yields memory location F828 discarding the carry.*

## B.2.3 Operand Definition from Instruction Word Fields

The two-bit As field, together with the 4-bit S-Reg field, define the source, with the combinations shown in Table B.4 for both one- and two-operands instructions. The destination operand is defined by the Ad bit and the D-reg group in dual-operand instructions as shown in Table B.5.

In the register fields, for both source and destination, registers R2 and R3 play a special role. When R2 appears in the source register field, and the As bits indicate register mode, then R2 refers to the Status Register itself. Similarly for the D-register field if Ad = 0. For other addressing modes, R2 is playing the role of constant generator, as explained later. R3 always plays the role of a constant generator, as explained in the next section.

## B.2.4 Constant Generators

Registers R2 and R3 play a role of constant generators for particular cases of immediate mode values. Also, in absolute mode, &ADDR is equivalent to ADDR(R2),

**Table B.4**  Source definition

| As1–As0 (Bits 5–4) | S-reg (Bits 11–8; 3–0)[1] | Comment |
|---|---|---|
| 00 | Reg. Number | Register mode |
| 01 | Reg. Number | Indexed Mode X(Rn) |
| 01 | 0 | Symbolic Mode ADDR. X value stored in the word following the instruction word, where X = PC - ADDR |
| 01 | 2 | Absolute mode &ADDR. SR takes value 0, and works as ADDR(SR)[2] ADDR follows instruction word |
| 10 | Reg. Number | Indirect Register mode @Rn |
| 11 | Reg. Number | Indirect Autoincrement @Rn+ |
| 11 | 0 | Immediate mode[3] #N N follows the instruction word |

(1): Bits 11-8 in two operands, 3-0 in one operand instructions
(2): ADDR(SR) is invalid syntax in an instruction
(3): Technically, this is equivalent to @PC+

**Table B.5**  Destination definition

| Ad (Bit 7) | D-reg (3-0) | Comment |
|---|---|---|
| 0 | Reg. Number | Register mode |
| 1 | Reg. Number | Indexed Mode X(Rn) |
| 1 | 0 | Symbolic Mode ADDR. X value stored in the word following the instruction word, where X = PC - ADDR |
| 1 | 2 | Absolute mode &ADDR ADDR is last word |

where R2 takes the value 0. Constant generators are combined with As values in the way shown in Table B.6, corresponding to the immediate value in the respective line. The use of a constant generator saves a word in the machine instruction.

For example, `mov #1,R7` translates into **4317** ( 0010 **0011 0001** 0111 ), and `mov.b #4,R7` into **4267** ( 0010 **0011 0110** 0111 ).

The constant generators provide faster instructions, no special instructions are required, there is no additional code word for the six constants. The assembler uses the constant generator automatically if one of the six constants is used as an immediate source operand. When register R2 appears explicitly as as an operand, it refers to the status register itself.

R3 as a source is equivalent to #0. R3 is valid as destination, but no value is stored. In fact, `nop` is emulated by `mov R3, R3`.

**Table B.6** Immediate values and constant generators

| Register (Bits 11-8) | As bits (Bits 5-4) | Equivalent operand |
|---|---|---|
| R2 | 00 | SR in register mode* |
| R2 | 01 | (0)** |
| R2 | 10 | #4 |
| R2 | 11 | #8 |
| R3 | 00 | #0 |
| R3 | 01 | #1 |
| R3 | 10 | #2 |
| R3 | 11 | #-1 (=#0xFFFF) |

\* R2 refers to the register SR itself, used in Register Mode.
\*\* For absolute mode, as in &ADDR = ADDR(0)

## B.3 Instruction Cycles

Remember that an instruction cycle covers three basic phases: fetch-decode-execute. This instruction cycle is also called a CPU cycle. By hardware, there is an internal clock in the CPU such that the number of cycles involved in one CPU cycle corresponds in time to one MCLK cycle. Hence, knowing the frequency of operation, and the number of CPU cycles required to execute one instruction, we can keep track of the timing required to execute loops, subroutines, and other tasks.

The number of CPU clock cycles required for an instruction depends on the instruction format and the addressing modes used - not the instruction itself. The number of clock cycles refers to the MCLK.

### B.3.1 Interrupt and Reset Cycles

The CPU cycles involved in the processes of interrupts and requests are summarized in Table B.7.

**Table B.7** Interrupt and reset cycles

| Action | No. of Cycles |
|---|---|
| Return from interrupt (RETI) | 5 |
| Interrupt accepted | 6 |
| WDT reset | 4 |
| Reset (RST/NMI) | 4 |

**Table B.8**  Interrupt and reset cycles

| Addressing mode | No. of Cycles | | |
|---|---|---|---|
| | RRA, RRC SWPB, SXT | PUSH | CALL |
| Register Rn | 1 | 3 | 4 |
| Indirect @Rn | 3 | 4 | 4 |
| Autoincrement Rn+ | 3 | 5 | 5 |
| Immediate #N | N/A[1] | 4 | 5 |
| Indexed N(Rn) | 4 | 5 | 5 |
| Symbolic N | 4 | 5 | 5 |
| Absolute &N | 4 | 5 | 5 |

(1) Do not use immediate mode in these instructions

## B.3.2  Jump Instruction Cycles

All jump instructions require one code word, and take two CPU cycles to execute, regardless of whether the jump is taken or not.

## B.3.3  Single Operand Instruction Cycles

## B.3.4  Double Operand Instruction Cycles

In this case, the special situation in which the Program Counter (PC) register is used in destination must be considered apart. Thus, mov R5, R6 and mov R5, PC, both with register addressing mode in the destination operand, have different cycle behavior.

**Table B.9** Interrupt and reset cycles

| Source addr. mode | Destination addr. mode | No. of cycles |
|---|---|---|
| Register Rn: | | |
| | Register Rn | 1 |
| | Register PC | 2 |
| | Indexed N(Rn) | 4 |
| | Symbolic N | 4 |
| | Absolute &N | 4 |
| Indirect @Rn: | | |
| | Register Rn | 2 |
| | Register PC | 2 |
| | Indexed N(Rn) | 5 |
| | Symbolic N | 5 |
| | Absolute &N | 5 |
| Autoincrement @Rn+ and Immediate #N: | | |
| | Register Rn | 2 |
| | Register PC | 3 |
| | Indexed N(Rn) | 5 |
| | Symbolic N | 5 |
| | Absolute &N | 5 |
| Indexed N(Rn), Symbolic N, and Absolute &N: | | |
| | Register Rn | 3 |
| | Register PC | 3 |
| | Indexed N(Rn) | 6 |
| | Symbolic N | 6 |
| | Absolute &N | 6 |

# Appendix C
# Code Composer Introduction Tutorial

## Materials

- CCS.
- MSP-EXP430G2 LaunchPad development board.

## C.1 Introduction

Code Composer Studio (CCS) is TI's own compiler and debugging environment for the MSP430. Based on the Eclipse platform, CCS allows to leverage hundreds of plug-ins to accelerate code development. For all the upcoming experiments CCS will be used as the compiler, assembler, linker and code debugger for the MSP430 LaunchPad. The MSP-EXP430G2 LaunchPad is a complete MSP430 development tool. It includes all the hardware and software components to evaluate the MSP430 and develop a complete project in a convenient board. The module consists of two parts: a USB communications interface and chip programmer/debugger interface and an MSP430G2211/31 target board. The USB interface provides power to operate the ultra-low-power MSP430 so no external power supply is required to operate the module, making it portable and easy to use. It also generates the signal to program and debug code in the MSP430 memory. The MSP430 target board provides a fully functional MSP430 microcontroller with over 20 user accessible pins and several LED indicators.

M. Jiménez et al., *Introduction to Embedded Systems*,
DOI: 10.1007/978-1-4614-3143-5,
© Springer Science+Business Media New York 2014

## C.2 Procedure

CCS Tutorial (These procedure assumes the CCS Integrated Development Environment (IDE) is already installed. If the CCS IDE is not installed, then go to the installation instructions at the end.) At the time of printing, the current CCS version is 4.2.1.00004.

1. On the workstation go to 'Start > All Programs > Texas Instruments > Code Composer Studio'. Additional letters and numbers will follow after the word Studio, depending on the installed version.
2. If this is the first time using CCS, it will ask a location (workspace) to store your project files. You may choose the path shown. Alternatively, you may choose to use any other location as your workspace. Regardless of your choice, please make sure you are able to recognize this directory in the future. We will assume the workplace is at installed path Workspace.
3. To create a new project: select 'File > New > CCS Project'.
4. Name your project "EXP1". Use the default location (in the current workspace). Continue with default settings by clicking 'Next' until the 'Project Settings' page appears. If you get the 'Project Type' page before getting to the 'Project Settings' page, make sure that you choose MSP430 in the pull down menu.
5. Select "MSP430G2231" under 'Device Variant' and click 'Next'.
6. Make sure that you choose "Empty Assembly-only Project" and then click 'Finish'. If you have clicked 'Finish' instead of 'Next' in the previous step, you would have chosen the option that expects you to use C in your project and this would have caused problems later on.
7. Look inside the 'C/C++ Project' tab to make sure that EXP1 is the active project. If it is not, right click on the name of your project and choose 'Set as Active Project'. Now click 'File > New—Source File'. Name your source file "FlashLed.asm" under 'Souce File:'
8. Type the code shown below in Listing C.1 as is, except for the line numbers, into the source window and then save the work done by clicking 'File > Save'.

**Listing C.1**  Your first assembly language program.

```
1                          .cdecls C , LIST ,   "msp430g2231.h"
2    ; - - - - - - - - - - - - - - - - - - - - - - - - -
3                          .text  ; Progam Start
4    ; - - - - - - - - - - - - - - - - - - - - - - - - -
5    RESET      mov     # 0280h , SP  ; Initialize stackpointer
6    StopWDT    mov     # WDTPW+WDTHOLD , & WDTCTL  ; Stop WDT
7    SetupP1    bis.b   # 1,& P1DIR   ; P1.0 as output
8    Main       xor.b   # 1,& P1OUT   ; Toggle P1.0
9    Wait       mov     # 50000 , R15 ; Delay to R15
10   L1         dec     R15           ; Decrement R15
11              jnz     L1            ; Delay over?
12              jmp     Main          ; Again
13   ; - - - - - - - - - - - - - - - - - - - - - - - - -
14   ;                    Interrupt Vectors
15   ; - - - - - - - - - - - - - - - - - - - - - - - - -
```

```
16              .sect      ".reset"     ; MSP430 RESET Vector
17              .short     RESET   ;
18              .end
```

9. Connect the LaunchPad to the computer via the USB cable provided with it. If this is the first time the LaunchPad is connected, you will notice that Windows will install the appropriate driver. This driver is supplied as part of CCS. The rest of this experiment relies on the assumption that the driver is properly installed.

10. You can now click 'Target > Debug Active Project'. The Progress Information page is displayed while the code downloads. Once the download is completed the debug perspective opens automatically.

11. If there are no errors, you can now run your project by clicking Target—Run or by hitting the F8 key. If there are any errors, then they must be syntax errors as opposed to logic errors or programming errors. After typing ";" on any line, the assembler ignores the rest of the line, not the rest of the program. This means that it is not important if you have made a typographical error after a ";". Review each line to make sure that the code entered is as shown. The above code has been properly tested and it was run to make sure that the steps shown here will produce the same results for you.

12. The red LED should be toggling on and off in the MSP-EXP430G2 LaunchPad. If it does not, then you should 'debug' your program to find the errors.

## C.3 Exercises

1. Replace the #50000 with #100000. Save your changes by clicking 'File > Save'. You can now click 'Target > Debug Active Project'. The 'Progress Information' page is displayed. Now hit the F8 key to run your program. What change did you observe on the red LED toggling frequency?

Note that we made the changes on the FlashLed.asm file in the computer while the program was running on the LaunchPad. We could have also terminated the program by clicking 'Target > Terminate All' or Ctrl+Alt+T. We could have also halted the program by clicking 'Target > Halt' or Shift+F8. After terminating or halting the program, we can also make the desired changes and then click 'Target > Debug Active Project' and hit F8 with similar results. If you run into problems we advice you to exit CCS and then continue with your work.

2. Repeat the above exercise replacing the #50000 or #100000 with #2500. While your project is running, click 'View > Registers'. A tab should appear with the label "Registers (1)". Double click to expand it. Now click "Core Registers". Do the same with "Special_Function". You will probably see "Unable to read" under the column "Value". Halt your project by clicking 'Target > Halt' or pressing Shift+F8. You should now see some hexadecimal numbers under the "Value" column. Go ahead and change the value for register R5 by clicking the value for

R5. Now enter decimal number 256 and hit the Enter key. You should see that the new value is 0x0100, which is decimal 256 written in hexadecimal notation. Now change it to 0xabcd. You should see that it changed the contents to 0xABCD. Hit F8 to run the project. We just showed you a way to examine and change the contents of the MSP430 registers. If you need to examine or change the contents of other parts of the chip, you just need to follow a similar procedure. At this time we want to show you another feature included with CCS. While your program is running, click 'View > Disassembly'. You should see that a new window opens in CCS with the name Disassembly. At this time the window should be empty. Now halt your project and you should see something like this in the Disassembly Tab:

```
1                RESET , .text , _text , $.. / FlashLed . asm : 25 : 44$:
2   0xF800:  4031 0280        MOV.W      #0x0280 , SP
3                StopWDT :
4   0xF804:  40B2 5A80 0120   MOV.W      #0x5a80 , & Watchdog_ ...
5                SetupP1 :
6   0xF80A:  D3D2 0022        BIS.B      #1 , & Port_1_2_P1DIR
7   0xF80E:  D3C2 0021        BIS.B      #0 , & Port_1_2_P1OUT
8                Main :
9   0xF812:  E3D2 0021        XOR.B      #1 , & Port_1_2_P1OUT
10               Wait :
11  0xF816:  403F 5000        MOV.W      #0x5000 ,R15
12               L1 :
13  0xF81A:  831F     DEC.W    R15
14  0xF81C:  23FE     JNE      (L1)
15  0xF81E:  E0F2 0041 0021   XOR.B      #0x0041 , & Port_1_2_P1OUT
16  0xF824:  403F 5000        MOV.W      #0x5000 , R15
17               L2 :
18  0xF828:  831F     DEC.W    R15
19  0xF82A:  23FE     JNE (L2)
20  0xF82C:  E0F2 0081 0021   XOR.B      #0x0081 , & Port_1_2_P1OUT
21  0xF832:  3FEF     JMP      (Main)
```

You probably recognize the instructions in capital letters since you entered them when you were writing the code. There are, however, several things that were added to your program. For example, the second line, which is actually the first instruction, is displayed:

```
1   0xF800:  4031 0280    MOV.W   #0x0280,SP
```

At the beginning of the above line you see 0xF800: 4031 0280. The string of characters 0xF800 is indicating the address of the location in memory whose content is 4031 0280. This two character strings are the machine language version of the assembly language instruction MOV.W #0x0280,SP. The first of these two strings, 4031, indicates that this is a MOV instruction working on two bytes or 16 bits at a time, while the second string, #0x0280, specifies the source operand for the MOV instruction.

If you only leave the program memory addresses and their content we would see this

```
 1   0xF800    RESET  ,  . text ,  _text ,  $../FlashLed . asm : 25 : 43$
 2   0xF800    4031  0280
 3   0xF804    40B2  5A80  0120
 4   0xF80A    D3D2  0022
 5   0xF80E    D3C2  0021
 6   0xF812    E3D2  0021
 7   0xF816    403F  5000
 8   0xF81A    831F
 9   0xF81C    23FE
10   0xF81E    E0F2  0041  0021
11   0xF824    403F  5000
12   0xF828    831F
13   0xF82A    23FE
14   0xF82C    E0F2  0081  0021
15   0xF832    3FEF
```

Now click 'View > Memory' and, when the memory window opens, type 0xF800 in the textbox where it says 'Enter location here'. You should see a display similar to the above. At this time you should not be concerned with the actual meaning of the assembly language instruction or its machine language version. Our intention at this time was to show you that with CCS you can see the program memory and its actual content. This will come in handy in the future.

3. We would like now to toggle the red LED and the green LED at the same time. The only changes we need to make are the following: replace the #1 with #0x41 in the two lines where the #1 appears, i.e. lines 7 and 8 in Listing C.1. Now click 'Target > Debug Active Project' and then hit F8 (or 'Target > Run') to run your project.

4. For this part of the exercise we will assume that the red LED is toggling on and off. Locate connector J5. Now grab jumper P1.0 with your fingers and pull it out of the connector. The red LED should be off because no power is delivered to it although the program is still running. Replace the jumper and the red LED should once again begin to toggle on and off. If the green LED is toggling on and off, you could do the same only this time you need to work with jumper P1.6.

5. Another exercise is to turn each LED in turn. One way to accomplish this is as follows. Insert

```
mov.b    #0,&P1OUT    ; both LEDs off
```

after line 7 and before line 8. The above line should now be line 8.
Now insert the following instructions

```
      xor.b    #01000001b,&P1OUT; P1.0 off, P1.6 on
      mov.w    #050000,R15
L2    dec.w    R15
      jnz L2
      xor.b    #01000001b,&P1OUT ; P1.0 on, P1.6 off
```

just before the new line 13, i.e. just before

```
      jmp Main
```

Your code should now look like Listing C.2:

**Listing C.2**  Modified code for turning each LED in turn.

```
1               .cdecls C,LIST, "msp430g2231.h"
2  ;------------------------------------------------------------
3               .text    ; Program Start
4  ;------------------------------------------------------------
5  RESET    mov.w    #0280h , SP ; Initialize stackpointer
6  StopWDT  mov.w    #WDTPW+WDTHOLD , & WDTCTL ; Stop WDT
7  SetupP1  bis.b    #0x41,&P1DIR ;     P1.0, P1.6 as output
8           mov.b    #0 , & P1OUT ;   both LEDs off
9  Main     xor.b    #1 , & P1OUT ;    Toggle P1.0
10 Wait     mov.w    #050000 , R15 ; Delay to R15
11 L1       dec.w    R15          ; Decrement R15
12          jnz      L1           ; Delay over?
13          xor.b    #0x41 , & P1OUT ; P1.0 off , P1.6 on
14          mov.w    #050000 , R15
15 L2       dec.w    R15
16          jnz      L2
17          xor.b    #1 , & P1OUT ; P1.0 on, P1.6 off
18          jmp      Main ; Again
19 ;------------------------------------------------------------
20 ;                    Interrupt Vectors
21 ;------------------------------------------------------------
22          .sect    ".reset" ; MSP430 RESET Vector
23          .short   RESET ;
24          .end
```

# C.4 Installing CCS

1. Obtain the CCS setup application. You should be able to obtain the CCS application at http://www.ti.com. If it is in a compressed format, then you should extract the files to a destination of your choice.
2. Double click the CCS setup application. The install wizard displays progress information.
3. Follow the wizard prompts during the installation.
4. After reading the license agreement, accept it if you still want to install CCS and then click 'Next'.
5. If this is the first time CCS will be installed on your computer, choose the default path settings for the installation location. Click 'Next'. If you have already installed CCS on your computer, make sure you choose a different path for this installation.
6. When prompted for the product configuration, choose the 'MSP430-only Core tools' and then click 'Next'.
7. The setup wizard will display the components it is about to install of the install. Click 'Next'.
8. Now you will be shown a summary of the changes the setup wizard is about to make. Click 'Next' if you agree with the information.

9. The wizard displays progress information, wait until the installation is completed.
10. Click 'Finish' when prompted.

The installation instructions shown above are general in nature and assume you are using the Windows 7 operating system (OS). Depending on your OS and the CCS version, the setup wizard may behave differently. You should make sure that your computer and OS meet the requirements for installing CCS. You should also make sure that your OS is aware of your computer hardware such as the chipset. For example, CCS may not work properly on a computer built for Windows Vista or Windows 7 if the computer is running an older version of Windows like XP, because Windows XP may not have the appropriate driver for the machine. You should also be aware that, on some operating systems, like Windows 7, you may have to make the installation from the Administrator account.

# Appendix D
# CPUX: An Overview of the Extended MSP430X CPU

## D.1 Introduction

The MSP430X (CPUx) is a 20-bit version of Texas Instruments MSP430 microcontroller. It was introduced with the second generation of MSP430 chips, i.e. with the MSP430x4xx family, and it has been present with all subsequent versions, except for the MSP430x1xx.

CPUx's 20-bit address bus, see Fig. D.1, allows to reach the 1-MB address space without paging. Remember that a traditional MSP430 is a 16-bit processor both at its address and data buses and can reach a 64-KB address space.

CPUx is backward compatible with the MSP430 CPU and it can address bytes, 16-bit words, and 20-bit words. It maintains its orthogonal RISC architecture allowing any CPU register to be used as an operand. Several instructions have been extended for 20 bit operation. However, using a prefix any instruction can be extended to 20 bit. CPUx has fewer interrupt overhead cycles and fewer instruction cycles in some cases than the MSP430 CPU.

As shown in Table D.1, there are several instructions that were extended to take advantage of the increased address space, namely: MOVA, CALLA, ADDA, SUBA, CMPA. ADDA, SUBA, and CMPA are restricted to the immediate and register addressing modes only, i.e. you cannot use the rest of the addressing modes with these three (3) instructions.

On the other hand, as summarized in Table D.2 there are several instructions for performing multi-bit shifts (1, 2, 3, or 4 bits): RRCM, RRAM, RLAM, RRUM. You can also push or pop several registers with the instructions PUSHM and POPM.

By using an additional word of op-code called an extension word, all addresses, indexes, and immediate values are extended to 20 bit. For example, BISX sets bits in the destination word that are set in the source word, BISX.B sets bits in the destination byte that are set in the source byte, and BISX.A sets bits in the destination address-work that are set in the source address-word.

You can do similar operations with ADDX, ADDCX, ANDX, BICX, BISX, BITX, CMPX, DADDX, MOVX, POPM, PUSHM, PUSHX, RLAM, RRAM,

M. Jiménez et al., *Introduction to Embedded Systems*,
DOI: 10.1007/978-1-4614-3143-5,
© Springer Science+Business Media New York 2014

**Fig. D.1** MSP430X (CPUX) Block Diagram. *Courtesy of Texas Instruments, Inc.*

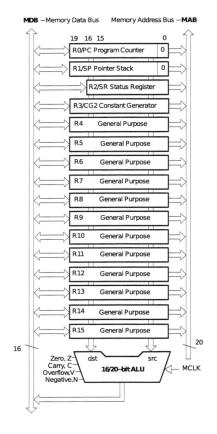

**Table D.1** MSP430X new instructions

|   | Instruction | Restriction |
|---|-------------|-------------|
| 1 | MOVA  |  |
| 2 | CALLA |  |
| 3 | ADDA  | Only immediate and register addressing modes. |
| 4 | SUBA  | Only immediate and register addressing modes. |
| 5 | CMPA  | Only immediate and register addressing modes. |

RRAX, RRCM, RRCX, RRUM, RRUX, SUBX, SUBCX, SWPBX, SXTX, and XORX.

Table D.3 shows the instructions that have been extended in the MSP430X and the original MSP430 instructions.

**Table D.2** MSP430X Multibit instructions

| | Instruction | Description |
|---|---|---|
| 1 | RRCM(.W/.A) | Rotate right n bits through carry. |
| 2 | RRAM(.W/.A) | Rotate right n bits arithmetically. |
| 3 | RLAM(.W/.A) | Rotate left n bits arithmetically. |
| 4 | RRUM(.W,.A) | Rotate right n bits unsigned. |
| 5 | PUSHM(.W/.A) | Push n 16 or 20 bit registers onto the stack. |
| 6 | POPM(.W/.A) | Pop n 16 or 20 bit register from the stack. |

**Table D.3** MSP430X Extended and original instructions

| | Extended instruction | Original CPU instruction | Extended instruction | Original CPU instruction |
|---|---|---|---|---|
| 1 | ADDX(.B/.W/.A) | ADD | PUSHX(.B/.W/.A) | PUSH |
| 2 | ADDCX(.B/.W/.A) | ADDC | RRAX(.B/.W/.A) | RRA |
| 3 | ANDX(.B/.W/.A) | AND | RRCX(.B/.W/.A) | RRC |
| 4 | BICX(.B/.W/.A) | BIC | SUBX(.B/.W/.A) | SUB |
| 5 | BISX(.B/.W/.A) | BIS | SUBCX(.B/.W/.A) | SUBC |
| 6 | BITX(.B/.W/.A) | BIT | SWPBX(.W/.A) | SWPB |
| 7 | CMPX(.B/.W/.A) | CMP | SXTX(.W/.A)[1] | SXT |
| 8 | DADDX(.B/.W/.A) | DADD | TSTX(.B/.W/.A) | TST |
| 9 | MOVX(.B/.W/.A) | MOV | XORX(.B/.W/.A) | XOR |
| 10 | POPX(.B/.W/.A) | POP | | |

1 The operation of the extended instruction is not completely similar to that of the original MSP430 one

## D.2 Differences Between the MSP430X and MSP430

1. Address bus

   (a) MSP430 has a 16 bit address bus.
   (b) MSP430X has a 20 bit address bus.

2. Address space

   (a) MSP430 can access up to $2^{20}$ locations.
   (b) MSP430X can access up to $2^{16}$ locations.

3. Memory accesses

   (a) MSP430 memory accesses are 8 and 16 bits.
   (b) MSP430X memory accesses are 8, 16, and 20 bits.

4. Size of registers

   (a) MSP430 registers are 16 bit long.
   (b) MSP430X registers are 20 bit long, except for the SR register which is 16 bit long.

5. Register Accesses

   (a) MSP430 register accesses are 8 and 16 bits.
   (b) MSP430X register accesses are 8, 16, and 20 bits.

6. R/W RAM

   (a) MSP430 starts at location 0x0200.
   (b) MSP430X starting location varies.

7. Interrupt Vector Table

   (a) MSP430: 0xFFFF - 0xFFC0.
   (b) MSP430X: 0xFFFF - 0xFF80.

8. Constants generated with constant generators

   (a) MSP430: (0) for absolute mode, 0x0000, 0x0001, 0x0002, 0x0004, 0x0008, and -1 as 0xFFFF.
   (b) MSP430X: (0) for absolute mode, 0x00000, 0x00001, 0x00002, 0x00004, 0x00008, and -1 as 0xFF, 0xFFFF, and 0xFFFFF.

9. Interrupt latency

   (a) MSP430: 6 cycles.
   (b) MSP430X: 5 cycles.

10. Interrupt return

   (a) MSP430: 5 cycles.
   (b) MSP430X: 3 cycles.

11. Four (4) bit extension word in opcode

   (a) MSP430: Does not apply.
   (b) MSP430X: Present.

12. Can extend all addresses, indexes, and immediate values to 20 bit

   (a) MSP430: Does not apply.
   (b) MSP430X: True.

13. Multibit (1, 2, 3, or 4 bits) shift instructions

   (a) MSP430: Does not apply.
   (b) MSP430X: True.

14. Instructions for pushing or popping several (1 to 16) registers

   (a) MSP430: Does not apply.
   (b) MSP430X: Present.

15. Can perform direct access and branching throughout entire memory range without paging

(a) MSP430: Does not apply.

(b) MSP430X: True.

16. Instructions length

(a) MSP430: 1, 2, or 3 16-bit words.

(b) MSP430X: 1, 2, 3, or 4 16-bit words.

17. BR and CALL reset upper four PC bits to 0

(a) MSP430: Does not apply.

(b) MSP430X: True.

18. Appends bits 19 - 16 to stored SR on stack during interrupt

(a) MSP430: Does not apply.

(b) MSP430X: True.

19. Byte-write to register clears (except SXT instruction)

(a) MSP430: bits 15 - 8.

(b) MSP430X: bits 19 - 8.

20. Word-write to register clears (except SXT instruction)

(a) MSP430: Does not apply.

(b) MSP430X: bits 19 - 16.

21. Register to Address-word (.A) clears

(a) MSP430: Does not apply.

(b) MSP430X: bits 15 - 4 on second 16 bit word.

22. Address-word to Register

(a) MSP430: Does not apply.

(b) MSP430X: bits 15 - 4 on second 16 bit word not used.

23. Instructions limited to immediate and register addressing modes

(a) MSP430: Does not apply.

(b) MSP430X: ADDA, SUBA, and CMPA

24. Effect of X instructions

(a) MSP430: Does not apply.

(b) MSP430X:
    i. Reduced speed due to additional CPU cycles.
    ii. Increased space due to extension word for double operand instructions.

# Appendix E
# Copyright Notices

This book contains code examples and figures that were either developed as original material for this book or obtained from external open sources available elsewhere. Materials obtained from such sources are identified next to their instances and their source listed in the Bibliography section of this book. This appendix reproduces the copyright notices of the original distribution of those materials obtained from external sources. The authors of this book have included such examples and figures as illustrative instances of the subjects discussed in each case. Although the authors have made every effort to verify their correctness, these materials are provided "as is". Any express or implied warranties, including, but not limited to, the implied warranties of fitness for any particular purpose are disclaimed. Under no circumstance or event shall the authors or the copyright owners be liable for any direct, indirect, incidental, exemplary, or consequential damages arising from the use of these materials.

## E.1 Copyright Notice for Code Examples from Texas Instruments

M. Jiménez et al., *Introduction to Embedded Systems*,
DOI: 10.1007/978-1-4614-3143-5,
© Springer Science+Business Media New York 2014

# References

1. Sale, T. (2000). Lorentz Ciphers and the Colossus. *Technical report, Bletchley park museum*, Retrieved June, 2008, from http://www.codesandciphers.org.uk/lorenz.
2. Pultorak, J. (2004). Block I Apollo Guidance Computer (AGC): How to build one in your basement. *Technical report, NASA office of logic design*, Retrieved June, 2008, from http://www.klabs.org/history/build_agc/.
3. Daniels, R. G. (1996). A participant's perspective. *IEEE Micro, 16*(6), 21–31.
4. Holt, R. M. (1998). Architecture of a microprocessor. Retrieved June, 2008, from http://www.microcomputerhistory.com/f14paper.htm.
5. Penumuchu, C. V. (2007). *Simple real-time operating system: A kernel inside view for a beginner*. Victoria, VC, Canada: Trafford Publishing, Inc.
6. Koopman, P. (1996). Embedded system design issues (the rest of the story). In *Proceedings of the International Conference on Computer Design (ICCD 96)*, (pp. 310–317), October 1996.
7. Scheffer, L. (2006). *EDA for IC system design, verification, and testing*. Boca raton, FL: CRC Press.
8. Madisetti, V. K. (1996). Rapid digital system prototyping: current practice, future challenges. *IEEE Design and Test of Computers, 13*(3):12–22.
9. Peckol, J. K. (2008). *Embedded systems: A contemporary design tool*. Hoboken, NJ: John Wiley & Sons, Inc.
10. Roth, K., & McKenney, K. (2007). Energy Consumption by Consumer Electronics in U.S. Residences. Technical report, Consumer Electronics Association/TIAX LLC.
11. Malik, S., Tiwari, V., & Wolfe, A. (2001). Power estimation in embedded systems: A hardware/software codesign approach. *Readings in hardware/software codesign* (pp. 249–258). Norwell, MA: Kluwer Academic Publishers.
12. Kmetovicz, R. E. (1992). *New product development: Design and analysis*. New York, NY: John Wiley & Sons, Inc.
13. Lofgren, J. D. (2004). A generic test and maintenance node for embedded system test. In *Proceedings of the International Test Conference 2004*, (pp. 143–153). October 2004.
14. Lindvall, M., Komi-Sirviö, S., Costa, P., & Seaman, C. (2003). A State of the Art Report: Embedded Software Maintenance. Technical Report 0704–0188, Data and Analysis Center for Software, 775 Daedlian Dr., Rome NY, January 2003.
15. Mano, M. M., & Ciletti, M. D. (2006). *Digital design* (4th ed.). New York, NY: Pearson College Div.
16. Daniels, W. L. (2002). *Fundamentals of embeded software: where C and assembly meet*. Upper Saddle River, NJ: Prentice Hall Inc.
17. Parhami, B. (2000). *Computer arithmetic: Algorithms and hardware design*. New York, NY: Oxford University Press.

M. Jiménez et al., *Introduction to Embedded Systems*,
DOI: 10.1007/978-1-4614-3143-5,
© Springer Science+Business Media New York 2014

18. IEEE Computer Society. IEEE ANSI/IEEE Standard 754–1985, Standard for Binary Floating Point Arithmetic. IEEE Society, 1985.
19. Fletcher, W. I. (1980). *An engineering approach to digital design*. Englewood Cliffs, N.J.: Prentice Hall, Inc.
20. Stallings, W. (2009). *Computer organization and architecture*. Upper Saddle River, NJ: Prentice Hall Inc.
21. Korpela, J. K. (2006). *Unicode explained*. Sebastopol, CA: O'Reilly Media, Inc.
22. Davies, J. H. (2008). *MSP430 Microcontroller basics*. Burlington, MA: Elsevier, Inc. Newnes.
23. Inc. Texas Instruments. (2012). Msp430x2xx family user's guide, slau144i. Electronic, Texas Instruments Inc, Post Office Box 655303 Dallas, Texas 75265, January 2012.
24. Inc. Texas Instruments. (2011). Msp430x5xx/msp430x6xx family user's guide, slau208i. Technical report, Texas Instruments, Inc., Post Office Box 655303 Dallas, Texas 75265, September 2011.
25. Inc. Texas Instruments. (2006). Msp430x1xx family user's guide, slau049f. Technical report, Texas Instruments, Inc., Post Office Box 655303 Dallas, Texas 75265, 2006.
26. Inc. Texas Instruments. (2002). Msp430x3xx family user's guide, slau012a. http://www.ti.com/lit/ug/slau012a/slau012a.pdf.
27. Inc. Texas Instruments. (2010). Msp430x4xx family user's guide, slau056j. Technical report, Texas Instruments, Inc., Post Office Box 655303 Dallas, Texas 75265, June 2010.
28. Kelley, A. l., & Pohl, I. (1990). *A book on C: Programming in C*. Boston: Addison-Wesley.
29. Msp430 optimizing c/c++ compiler v 3.0 user's guide, march 2008, 2008.
30. Inc. IAR Systems. (2006). Msp430 iar c/c++ compiler reference guide for texas instruments' msp430 microcontroller family. Technical report.
31. Kubes, D. (2012). The 5-minute guide to c pointers. Retrieved August, 2012, from http://denniskubes.com/2012/08/16/the-5-minute-guide-to-c-pointers/.
32. Inc. Texas Instruments. (2012). Msp430x2xx family user's guide, slau144i. Retrieved December, 2004, from http://www.ti.com/lit/ug/slau144i/slau144i.pdf. (Revised January 2012).
33. Dannenberg, A. (2006). msp430x22x4_ta_08. Retrieved April, 2006, from, http://www.ti.com/lit/zip/slac123.
34. Iar c/c++ compiler reference guide for texas instruments+ msp430 microcontroller family, march 2010, 2010.
35. Buccini, M., & Westlund, L. (2005). Msp430f20xx demo - timer_a, toggle p1.0, ccr0 cont. mode isr, dco smclk. Retrieved October, 2005, from http://www.ti.com/lit/zip/slac080i.
36. Westlund, L. (2006). Msp430f20x2 demo - adc10, sample a1, avcc ref, set p1.0 if > 0.5*avcc. Retrieved May, 2006, from http://www.ti.com/lit/zip/slac080i.
37. Buccini, M. & Westlund, L. (2005). Msp430f20xx demo - usicnt used as a one-shot timer function, dco smclk. Retrieved September, 2005, from http://www.ti.com/lit/zip/slac080i.
38. Inc. Energizer Holdings. Energizer 522 alkaline 9v battery, form no. ebc - 1108k. http://data.energizer.com/PDFs/522.pdf.
39. Kundert, K. (2012). Power supply noise reduction. Retrieved January, 2004, from http://www.designers-guide.org. (Last retrieved November 2012).
40. National Semiconductor (Now Texas Instruments). (2012). Lp3470: Timy power-on reset circuit. Retrieved September, 2009, from http://www.ti.com/lit/gpn/lp3470. (Last retrieved December 2012).
41. Dannenberg, A. (2005). Msp430f22x4 toggle p1.0 demo. Retrieved April, 2005, from http://www.ti.com/lit/zip/slac123.
42. Atmel Corporation. (2012). Doc9108: Microcontroller in a harsh environment. Retrieved October, 2007, from http://www.atmel.com/Images/doc9108.pdf. (Last retrieved December 2012).
43. Kleidermacher, D. (2005). Minimizing interrupt response time. *Information Quarterly*, *4*(1), 52–54.
44. Keith, Q. (2006). Msp430 software coding techniques. Technical report, Texas Instruments, 2006. Revised version, August 2006.
45. Inc. Texas Instruments. (2012). Msp430 low cost pinosc capacitive touch overview. Retrieved September, 2011, from http://processors.wiki.ti.com/index.php/MSP430_Low_Cost_PinOsc_Capacitive_Touch_Keypad. (Last retrieved December 2012).

46. Bierl, L. (2002). Interfacing the 3-v msp430 to 5-v circuits (slaa148). Retrieved October, 2002, from http://www.ti.com/litv/pdf/slaa148.

47. Inc. Maxim Integrated. (2012). Logic-level translation (an3007). Retrieved July, 2004, from http://pdfserv.maximintegrated.com/en/an/AN3007.pdf. (Last retrieved November 2012).

48. Inc. Texas Instruments. (2011). Msp430f22x2/f22x4 mixed signal microcontroller. Retrieved July, 2011, from http://www.ti.com/docs/prod/folders/print/msp430f2274.html.

49. Fornaciari, W., Gubian, P., Sciuto, D., & Silvano, C. (June 1988). Power estimation of embedded systems: A hardware/software codesign approach. *IEEE Transactions on Very Large Scale Integration (VLSI) Systems, 6*(2), 266–275.

50. Seshasayee, N. (2012). Understanding thermal dissipation and design of a heatsink (slva462). Retrieved May, 2011, from http://www.ti.com/litv/pdf/slva462. (Last retrieved November 2012).

51. Bachman, P., & Haiduk, R. (2011). The effect of forced air cooling on heat sink thermal ratings. Retrieved July, 2011, from http://www.crydom.com/en/Tech/HS_WP_FA.pdf.

52. Ganssle, J. G. (2008). A guide to debouncing. Retrieved June, 2008, from http://www.ganssle.com/debouncing.htm The Ganssle Group PO Box 38346 Baltimore, MD 21231.

53. Slater, M. (1989). *Microprocessor-based design: A comprehensive guide to effective hardware design.* Englewood Cliffs, NJ: Prentice Hall, Inc.

54. Kingbright USA. Super-bright red three digit numeric display. On the World Wide Web. March 2011.

55. Fairchild Semiconductor Co. 2n3904 / mmbt3904 / pzt3904 npn general purpose amplifier. On the World Wide Web, October 2011.

56. National Semiconductor (Now Texas Instruments). Mm5452/mm5453 liquid crystal display drivers. Retrieved December, 2008, from http://www.ti.com/product/mm5452.

57. Ltd. Hitachi. Dot matrix liquid crystal display controller/driver. Retrieved September, 1999, from http://www.semiconductor.hitachi.com/.

58. Inc. Texas Instruments. (2010). Msp430x4xx family user's guide. Retrieved January, 2010, from http://www.ti.com/litv/pdf/slau056j.

59. Inc. Texas Instruments. (2012). Msp430x5xx/msp430x6xx family user's guide. Retrieved August, 2012, from http://www.ti.com/litv/pdf/slau208k.

60. Luxeon Star LEDs. Sr-05rebel 10mm square led assembly. Retrieved September, 2012, from http://www.luxeonstar.com//v/vspfiles/downloadables/sr-05.pdf.

61. Inc. Texas Instruments. (2012). Msp430f11x2, msp430f12x2 mixed signal microcontroller (rev. d). Retrieved August, 2004, from http://www.ti.com/product/msp430f1222. (Last downloaded November 2012).

62. ON Semiconductor. Npn small-signal darlington transistor data sheet. Retrieved January, 2012, from http://www.onsemi.com/pub/Collateral/BSP52T1-D.PDF.

63. Inc. Texas Instruments. (1997). Uln2803a darlington transistor array (slrs049c). Retrieved February, 1997, from http://www.ti.com/lit/ds/symlink/uln2803a.pdf.

64. Hughes, A. (2005). *Electric motors and drives: Fundamentals, types and applications* (3rd edn.). New York: Newones.

65. Condit, R.,& Jones, D. W. (2012). An907: Stepping motors fundamentals. Retrieved February, 2004, from http://ww1.microchip.com/downloads/en/AppNotes/00907a.pdf. (Last retrieved December 2012).

66. Inc. Motorola. (2012). Mc 3479 stepper motor driver. Retrieved July, 1996, from http://www.futurlec.com/Motorola/MC3479P.shtml. (last retrieved December 2012).

67. Inc. Texas Instruments. (2010). Drv8811 stepper motor controller ic. Retiieved May, 2010, from http://www.ti.com/lit/ds/symlink/drv8811.pdf.

68. Soltero, M., Zhang, J., Cockril, C., Zhan, Z., Kinnaird, C., & Kugelstadt, T. (2012). Rs-422 and rs-485 standards overview and system configurations (slla070d). Retrieved May, 2010, from http://focus.ti.com/lit/an/slla070d/slla070d.pdf. (Last retrieved December 2012).

69. Texas Instruments' Application Report. Interface circuits for tia/eia-485 (rs-485) (slla036d). Retrieved August, 2008, from http://www.ti.com/lit/an/slla036d/slla036d.pdf. (Last retrieved Decemeber 2012).

70. Texas Instruments Technical Staff. Comparing bus solutions (slla067b). Retrieved October, 2009, from http://www.ti.com/lit/an/slla067b/slla067b.pdf. (Last retrieved December 2012).

71. Universal serial bus specification, revision 2.0. Retrieved April, 2000, from http://www.usb.org. (Last retrieved December 2012).

72. Universal serial bus 3.0 specification, revision 1.0. Retrieved June, 2011, from http://www.usb.org. (Last retrieved December 2012).

73. Kollman, R. & Betten, J. (2012). Powering electronics from the usb port. Retrieved April, 2002, from http://www.ti.com/sc/analogapps. (Last retrieved December 2012).

74. NXP Semiconductors. I2c-bus specification and user manual - rev. 5 (um10204). Retrieved October, 2012, from http://www.nxp.com/documents/user_manual/UM10204.pdf. (Last retrieved January 2013).

75. Buccini, M., & Morton, G. (2005). Msp-fet430p140 demo - usart0, uart 115200 echo isr, hf xtal aclk. Retrieved May, 2005, from http://www.ti.com/lit/zip/slac015.

76. Texas Instruments Technical Staff. Tlc549c 8-bit analog-to-digital converters with serial control - slas067c. Retrieved September, 1996, from http://www.ti.com/lit/ds/symlink/tlc549.pdf.

77. Buccini, M. & Westlund, L. Msp430f20x2/3 demo - usi spi interface to tlc549 8-bit adc. Retrieved October, 2005, from http://www.ti.com/lit/zip/slac149f.

78. Nisarga, B. (2007). Msp430x24x demo - usci_a0, 9600 uart echo isr, dco smclk. Retrieved September, 2007, from http://www.ti.com/lit/zip/slac149f.

79. Nisarga, B. (2007). Msp430x24x demo - usci_a0, 9600 uart, smclk, lpm0, echo with over-sampling. Retrieved September, 2007, from http://www.ti.com/lit/zip/slac149f.

80. Dang, D. (2009). Msp430f543xa demo - usci_b0 i2c master tx multiple bytes to msp430 slave. Retrieved December, 2009, from http://www.ti.com/lit/zip/slac357d. Built with CCE Version: 3.2.2 and IAR Embedded Workbench Version: 4.11B.

81. Thanigai, P., & Morales, M. (2009). Msp430f543xa demo - usci_b0 i2c slave rx multiple bytes from msp430 master. Retrieved June, 2009, from http://www.ti.com/lit/zip/slac357d. Built with CCE Version: 3.2.2 and IAR Embedded Workbench Version: 4.11B.

82. Baker, B. (2011). *Designing with temperature sensors, part one: Sensor types*. Austin: Texas Instruments (EDN).

83. Baker, B. (2011). *Designing with temperature sensors, part two: Thermistors*. Austin: Texas Instruments (EDN).

84. Baker, B. (2011). *Designing with temperature sensors, part three: Rtds*. Austin: Texas Instruments (EDN).

85. Baker, B. (2011). *Designing with temperature sensors, part four: Thermocuples*. Austin: Texas Instruments (EDN).

86. Baker, B. (2012). *Designing with temperature sensors, part five: Ic-temperature sensors*. Austin: Texas Instruments (EDN).

87. Inc. Agilent Technologies. Application note 290: Practical temperature measurements. Technical report, Agilent Technologies, Inc., January 2012.

88. Bonnie, B. (2005). *A Baker's dozen: Real analog solutions for digital designers*. Boston, MA: Newnes.

89. Mancini, R. (Ed.). (2003). *Op Amps for everyone*. Boston, MA: Newnes.

90. Moghimi, R. (2000). Curing comparator instability with hysteresis. *Analog/AnalogDialogue*.

91. Inc. Maxim Integrated. Ds1843:fast sample-and-hold circuit. Technical report, Maxim Semiconductor Inc, Maxim Integrated Products, 120 San Gabriel Drive, Sunnyvale, CA 94086, 2012.

92. Inc. Maxim Integrated. Max5167: 32-channel sample/hold amplifier with output clamping diodes. Technical report, Maxim Integrated Products, Inc., Maxim Integrated Products, 120 San Gabriel Drive, Sunnyvale, CA, 2000.

93. Vega, C. A. (2005). A switched opamp comparator to improve the conversion rate of low-power low-voltage successive approximation adcs. Master's thesis, U. of Puerto Rico at Mayaguez, 2005.

94. Hauser, M. W. (1991). Principles of oversamplinga/d conversion. *Journal of Radio Engineering Society, 39*(1/2), 3–26.

# Author Biography

Manuel Jiménez-Cedeño  Professor Jiménez has over twenty-five years of experience teaching and developing embedded systems applications using microprocessors. After earning his Ph.D. degree in Electrical Engineering from Michigan State University in 1999, he joined the University of Puerto Rico at Mayaguez where he currently holds a position as professor in the Electrical and Computer Engineering Department. His current research and teaching interests include embedded systems design, rapid systems prototyping, and automated characterization of electronic devices and systems. His work has produced over one hundred refereed articles published in conferences and journals around the world. Professor Jiménez is a professional member of the IEEE and ASEE.

Rogelio Palomera-García  Obtained his degree as Docteur es Sciences from the Swiss Federal Institute of Technology at Lausanne, Switzerland in 1979. After six years in a research and graduate teaching position at the Research and Higher Education Center at Ensenada (CICESE) in Mexico, he moved to the University of Puerto Rico at Mayaguez where he has been teaching since 1985. His work and teaching interests include analog, digital, and embedded electronics among other topics. He is a member of the IEEE and the Japan Institute of Information, Electronic and Communication Engineers (IECE).

Isidoro Couvertier-Reyes  Obtained his degree of Doctor of Philosophy in Electrical Engineering from the Louisiana State University-Baton Rouge in 1996. From that moment he has been teaching at the University of Puerto Rico at Mayaguez. His work and teaching interests include Engineering Education, Computer Networking, Microprocessors, Computer Arithmetic, Computer Architecture, and Digital Logic. His MS degree is from the University of Wisconsin-Madison and his Baccalaureate from the University of Puerto Rico at Mayaguez. He is a senior member of the IEEE.

M. Jiménez et al., *Introduction to Embedded Systems*,
DOI: 10.1007/978-1-4614-3143-5,
© Springer Science+Business Media New York 2014

# Index

M. Jiménez et al., *Introduction to Embedded Systems*,
DOI: 10.1007/978-1-4614-3143-5,
© Springer Science+Business Media New York 2014

Printed by Publishers' Graphics LLC